Electronic Structure, Dynamics, and Quantum Structural Properties of Condensed Matter

NATO ASI Series

Advanced Science Institutes Series

A series presenting the results of activities sponsored by the NATO Science Committee, which aims at the dissemination of advanced scientific and technological knowledge, with a view to strengthening links between scientific communities.

The series is published by an international board of publishers in conjunction with the NATO Scientific Affairs Division

A	Life Sciences	Plenum Publishing Corporation
B	Physics	New York and London
C	Mathematical and Physical Sciences	D. Reidel Publishing Company Dordrecht, Boston, and Lancaster
D	Behavioral and Social Sciences	Martinus Nijhoff Publishers
E	Engineering and Materials Sciences	The Hague, Boston, and Lancaster
F	Computer and Systems Sciences	Springer-Verlag
G	Ecological Sciences	Berlin, Heidelberg, New York, and Tokyo

Recent Volumes in this Series

Volume 115—Progress in Gauge Field Theory
edited by G. 't Hooft, A. Jaffe, H. Lehmann, P. K. Mitter, I. M. Singer, and R. Stora

Volume 116—Nonequilibrium Cooperative Phenomena in Physics and Related Fields
edited by Manuel G. Velarde

Volume 117—Moment Formation in Solids
edited by W. J. L. Buyers

Volume 118—Regular and Chaotic Motions in Dynamic Systems
edited by G. Velo and A. S. Wightman

Volume 119—Analytical Laser Spectroscopy
edited by S. Martellucci and A. N. Chester

Volume 120—Chaotic Behavior in Quantum Systems: Theory and Applications
edited by Giulio Casati

Volume 121—Electronic Structure, Dynamics, and Quantum Structural Properties of Condensed Matter
edited by Jozef T. Devreese and Piet Van Camp

Series B: Physics

Electronic Structure, Dynamics, and Quantum Structural Properties of Condensed Matter

Edited by

Jozef T. Devreese

and

Piet Van Camp

University of Antwerp
Antwerp, Belgium

Plenum Press
New York and London
Published in cooperation with NATO Scientific Affairs Division

Proceedings of the Antwerp Advanced Study Institute on
Electronic Structure, Dynamics, and Quantum Structural Properties
of Condensed Matter,
held July 16-27, 1984,
at the Conference Center, Priorij Corsendonk, Belgium

Library of Congress Cataloging in Publication Data

Antwerp Advanced Study Institute on Electronic Structure, Dynamics, and
Quantum Structural Properties of Condensed Matter (1984)
   Electronic structure, dynamics, and quantum structural properties of con-
densed matter.

   (NATO ASI series. Series B, Physics; v. 121)
   "Proceedings of the Antwerp Advanced Study Institute on Electronic Struc-
ture, Dynamics and Quantum Structural Properties of Condensed Matter, held
July 16-27, 1984, at the Conference Center, Priorij Corsendonk, Belgium"
—T.p. verso.
   "Published in cooperation with NATO Scientific Affairs Division"—T.p.
verso.
   Includes bibliographies and index.
   1. Electronic structure—Congresses. 2. Condensed matter—Congresses. 3.
Molecular dynamics—Congresses. I. Devreese, J. T. (Jozef T.) II. Van Camp, P.
E. (Piet E.) III. Title. IV. Series.
QC176.8.E4A57   1984   530.4'1   84-26483
ISBN 0-306-41912-2

PREFACE

The 1984 Advanced Study Institute on "Electronic Structure, Dynamics and Quantum Structural Properties of Condensed Matter" took place at the Corsendonk Conference Center, close to the City of Antwerpen, from July 16 till 27, 1984.

This NATO Advanced Study Institute was motivated by the research in my Institute, where, in 1971, a project was started on "ab-initio" phonon calculations in Silicon.

It is my pleasure to thank several instances and people who made this ASI possible. First of all, the sponsor of the Institute, the NATO Scientific Committee. Next, the co-sponsors: Agfa-Gevaert, Bell Telephone Mfg. Co. N.V., C & A, Esso Belgium, CDC Belgium, Janssens Pharmaceutica, Kredietbank and the Scientific Office of the U.S. Army.

Special thanks are due to Dr. P. Van Camp and Drs. H. Nachtegaele, who, over several months, prepared the practical aspects of the ASI with the secretarial help of Mrs. R.-M. Vandekerkhof.

I also like to thank Mrs. M. Cuyvers who prepared and organized the subject and material index and Mrs. H. Evans for typing-assistance.

I express particular gratitude to Mrs. F. Nedée, who, like in 1981 and 1982, has put the magnificent Corsendonk Conference Center at our disposal and to Mr. D. Van Der Brempt, Director of the Corsendonk Conference Center, for the efficient way in which he and his staff took care of the practical organization at the Conference Center.

J.T. Devreese
Professor of Theoretical Physics

October 30, 1984

CONTENTS

Introductory lecture:
    Computation of Electronic Structure: Its Role in
    the Development of Solid State Physics 1
  V. Heine

I.   AB-INITIO METHODS

I.1. Dielectric Response and Many-Body Theory

Basic Concepts in Dielectric Response and Pseudopotentials . . 9
  J.T. Devreese and F. Brosens
      1. The one-electron approximation 9
        1.1. Introduction 9
        1.2. The Hartree-Fock approximation in jellium . 14
        1.3. The polarizability matrix 17
        1.4. The dielectric matrices 19
        1.5. Approximations for exchange and correlation 23
        1.6. Adiabatic electronic energy 27
        1.7. Harmonic phonons in a periodic lattice . . . 29
      2. Dynamical exchange effects in the dielectric
        function of jellium 36
        2.1. General considerations 36
        2.2. Dynamical exchange decoupling 38
        2.3. Variational approximation 40
        2.4. Comparison with various approximations . . . 42
        2.5. Concluding remarks 44
      3. The pseudopotential concept 45
        3.1. Basic assumptions 45
        3.2. The pseudo Schrödinger equation 47
        3.3. Some pseudopotential approaches 50
      4. Pseudopotential perturbation theory 56
        4.1. Pseudo wave functions and band energies . . 56
        4.2. Ionic pseudopotential contribution and
            structure factor 58
        4.3. The electron density 61

4.4. The total pseudopotential 66
4.5. The Fermi surface 72
5. Total energy and pseudopotential perturbation
   theory 76
   5.1. The energy-wavenumber characteristic 76
   5.2. The vibration spectrum 83
   5.3. The dielectric matrices 85
6. Concluding remarks 97
Appendix A. Electrostatic energy of ions in a
            uniform background 98
Appendix B. Dynamical matrix by Ewald-Fuchs method . 101

Exchange-Correlation Potential for the Quasi-Particle
   Bloch States of a Semiconductor 113
W. Hanke, N. Meskini and H. Weiler
   1. Introduction 114
   2. The Green's functions, single- and two-
      particle excitations, and their interrelation
      with density-functional theory 116
   3. Quasi-particle states in a semiconductor 130
   4. An analytical energy- and local-density-
      dependent model for self-energy 140
   5. Summary 154
   6. References 155

Ab-Initio Calculation of the Phonon Frequencies in Covalent
   Semiconductors using the Dielectric Screening Method . . 157
J.T. Devreese, P.E. Van Camp and V.E. Van Doren
   1. Introduction 157
   2. Theory 160
   3. Discussion 163
      3.1. Computational procedure 163
         a. Electron-ion potential 163
         b. Electronic band structure 164
         c. Total energy 167
         d. Convergence of the sum over the conduct-
            ion bands 167
         e. Convergence of the phonon frequencies
            in terms of the dimension of ε 168
         f. Convergence in terms of the dimension
            of H 169
      3.2. Results 169
   4. Conclusions 171
   5. References 172

I.2. Total Energy Methods

Current Ideas and Methods for Calculation of Ground
    State Properties of Solids 175
    R.M. Martin
        1. Introduction 175
        2. General theory 179
            2.1. First derivatives of the energy:
                 generalized forces 180
            2.2. The stress theorem 182
            2.3. High order derivatives 183
            2.4. Beyond the variational principle 184
        3. Solutions to the many-body problem 185
        4. Density functionals 186
            4.1. LDF functional 189
            4.2. Potential V_I 189
            4.3. Solutions of the Schrödinger equation . . . 190
            4.4. Sum over all filled states 193
            4.5. Self-consistency 194
            4.6. Stress 196
        5. Results 197
            5.1. Hydrogen 197
            5.2. Simple metals 199
            5.3. Rare gases 202
            5.4. Transition metals 202
            5.5. Semiconductors 204
        6. References 220

Recent Results in Semiconductor Dynamics by Ab-Initio
    'Direct' Approach 227
    K. Kunc
        1. Introduction 227
        2. Total energy calculations 230
        3. Static equilibrium 232
        4. Frozen phonons 236
            4.1. The concept 236
            4.2. Zone center TO(Γ) mode 237
            4.3. Summary of Ref. 13 239
            4.4. Composition of phonon energies in Ge 241
            4.5. Literature guide 242
        5. Forces and plane force constants 242
            5.1. Forces and frozen phonons 242
            5.2. Force constants 244
            5.3. Ab-initio phonon force constants 246
            5.4. Anharmonicity 248
            5.5. Spatial extent of forces 253
            5.6. Internal strain 260
            5.7. Phonon dispersion in chains of atomic
                 plains 260

5.8. Forces in covalent crystals 264
5.9. Defects in solids 268
6. Electric fields in ab-initio treatment of
polar crystals 271
6.1. Effective charges 273
6.2. Macroscopic field 280
7. Planar forces in polar crystals 286
7.1. Longitudinal vibrations 288
7.2. Transverse forces 290
7.3. Phonon dispersion in GaAs 294
8. Latest developments 295
8.1. External macroscopic electric field 297
8.2. Inverse dielectric matrix 299
8.3. Discussion 303
9. Conclusion 304
Appendix: Pseudoatoms Ga and As 305
References 310

Stress: Concepts and Applications 313
    O. Nielsen and R.M. Martin
1. Introduction 313
2. Ab-initio pseudopotential calculations 315
3. Stress calculations 317
4. Method of calculation 319
5. Ab-initio calculations on Si, Ge and GaAs 323
6. Elastic properties of Si, Ge and GaAs 325
7. Conclusion 331
8. References 331

Pseudopotentials and Total Energy Calculations: Applica-
    tions to Crystal Stability, Vibrational Properties,
    Phase Transformations and Surface Structures 335
    S.G. Louie
1. Introduction 335
2. Theoretical techniques 337
A. Density functional formalism 337
B. Ab-initio pseudopotentials 338
C. Basis sets 341
D. Total energy expressions 342
3. Bulk properties 344
A. Static structural properties 344
B. Vibrational properties 348
1. Phonon dispersions 350
2. Anharmonic terms and phonon-phonon
interactions 353
C. Structural phase transformations 361
4. Surfaces and chemisorption systems 367
A. Surfaces of semiconductors and insulators . . 368
B. Surfaces of transition metals 377

5. Beyond local density approximation 388
6. Summary and conclusions 392
7. Acknowledgments 394
8. References 394

II. EXPERIMENTAL ANALYSIS

Surface Phonon Calculations in Metals and Comparison
  with Experimental Techniques 401
  V. Bortolani, A. Franchini and G. Santoro
        1. Introduction 402
        2. Bulk phonons 403
        3. Angular forces 406
        4. Surface phonons 418
        5. Atom-surface scattering cross section 428
        6. Atom-surface potential 430
        7. Calculation of the [111] surface 434
        8. References 448

Intervalley Electron-Phonon and Hole-Phonon Interactions
  in Semiconductors: Experiment and Theory 451
  F.H. Pollak and O.J. Glembocki
        1. Introduction 451
        2. Experimental results 452
          2.1. Absorption coefficient 456
          2.2. Stress effects 460
          2.3. Stress along [111] 461
          2.4. Stress along [001] 465
          2.5. Experimental results 466
          2.6. Evaluation of electron-phonon and hole-
               phonon scattering matrix elements 473
        3. Theoretical calculations 476
          3.1. The rigid-pseudoion model 477
          3.2. Application to diamond-type materials . . . 479
        4. Summary 488
        5. References 490

Nonlinear Electronic and Dynamical Response of Solids
  in the Ultrashort Time Domain 495
  W.E. Bron
        1. General introduction 495
        2. Optical phonon lifetimes 496
          2.1. Theoretical background 496
          2.2. Nonlinear response of solids 502
          2.3. Experimental results 506
        3. Third order nonlinear electronic susceptibility . 511
          3.1. Theoretical background 511
          3.2. Experimental results 512

    4. Conclusion 516
    5. Acknowledgment 517
    6. References 517

III. MOLECULAR DYNAMICS

Molecular-Dynamic Simulations of Many-Particle Systems:
    New Faces on Old Problems 521
  D.C. Wallace
    1. Introduction 521
    2. Description of equilibrium molecular dynamics . . 522
       2.1. The ordinary MD system 522
       2.2. Classical canonical ensemble 524
       2.3. The logic of calculating thermodynamics
            from MD 527
    3. Interpretation of equilibrium molecular
       dynamics 529
       3.1. The appearance of MD data 529
       3.2. Statistical analysis of MD data 529
       3.3. Generalized canonical ensemble 535
       3.4. Ensemble corrections for the MD ensemble . . 538
       3.5. Remaining finite-size effects 541
       3.6. Phase-space trajectory of an MD system . . . 543
    4. Molecular-dynamic calculations for metallic
       sodium 545
       4.1. Pseudopotential theory 545
       4.2. Crystalline sodium and anharmonicity 547
       4.3. Fluid sodium and melting 553
    5. More exotic applications of molecular dynamics . 558
    6. Acknowledgments 561
    7. References 561

AUTHOR INDEX 565

MATERIAL INDEX 579

SUBJECT INDEX 581

COMPUTATION OF ELECTRONIC STRUCTURE: ITS ROLE IN THE DEVELOPMENT

OF SOLID STATE PHYSICS

Volker Heine

Cavendish Laboratory
Madingley Road
Cambridge CB3 0HE, UK

How often have I read a paper about a piece of computational physics which finishes with the words "...and we obtain good agreement with experiment". If you know the answer from experiment, I want to cry, why are you wasting so much time calculating it???

Of course I realise every craftsman has to test his tools, but in opening this important NATO Advanced Studies Institute and Summer School I want to focus for a few minutes on the role and general methodology of our subject. I have two reasons. The first is to optimise our own work and maximise our contribution to research. Secondly I am concerned that the subject of computational physics has such a low status. Most physicists consider computation as an occupation for failed theorists and failed experimentalists. This attitude, which has all sorts of consequences for funding and jobs, will only change if we articulate clearly the role of computational physics. And with the capabilities of modern machines, I believe we are at the beginning of an enormous and exciting development so that there is a lot to say.

Let me start with the role of computational physics and in particular the computation of electronic structure of solids including all the properties which can readily be calculated from it. This Advanced Studies Institute focuses on what I believe is the most important advance in recent years, namely the calculation of total energy etc. through solution of the Schrödinger equation so that one has the forces on individual atoms whose equilibrium positions can therefore be found. Sometimes the forces and stresses are found directly and sometimes from differentiating the

energy. One of the most convincing tests of convergence is when
these two approaches agree. Thus the arrangement of atoms at a
surface or around a defect can be found. Dynamical processes can
be modelled so that transition or diffusion pathways can be
obtained with the relevant activation energies. A particular case
is the calculation of vibration frequencies which have been such an
important research tool in physics and chemistry for so long. Both
absolute calculation of energy etc. and perturbation methods have
been used. An analysis of the electronic wave functions gives
additional information about the bonding, while the Hellman-
Feynman theorem indicates the understanding to be derived
from analysis of charge densities. Unfortunately much of this
information is often thrown away and not fully mined. We know that
bonding in real solids cannot be represented by short-ranged
pairwise interatomic forces except in the simplest cases. Thus a
full solution of the Schrödinger equation is needed.

We conclude that computation of electronic structure can
follow solid state processes at the atomic level in a degree of
detail that the experimentalists would dearly love to observe
directly but usually cannot. At the same time it yields extra
information about bonding etc through the wave functions which can
contribute an understanding of the results obtained. I believe
there is plenty of scope for this in what I see still as a growth
area, namely materials science. There is an enormous diversity of
materials, and one is concerned with two types of question. The
first is to understand why one material is different from another,
e.g. why do niobium compounds occur rather frequently in a list of
high temperature superconductors. The second is to understand the
complex phenomena observed in materials, such as the invar effect
or the degradation of a semiconducting laser over time, or our two
old friends corrosion and catalysis. (I call them old friends
because I sometimes wonder where our funding would be if we didn't
have them to stand by us!)

I want to make the point that I believe there are many
fundamental discoveries and insights to be made in materials
science. The word 'fundamental' is often misused. I have heard
high energy particle physicists claim that their subject is
fundamental, but I do not find it fundamental to materials science,
geology, biology or hardly anything on this earth except in a
rather empty sense which philosophers call reductionism. A more
apt use of the word fundamental is to recognise that understanding
is organised on various levels each with its own laws and concepts
fundamental at that level. The second law of thermodynamics can be
derived from statistical mechanics but that does not affect its
role as a fundamental law within that important and coherent
collection of concepts and relationships which we call thermodynamics. The concept of the pseudopotential is fundamental
to much of the subject of the Advanced Studies Institute: the fact

2

that some people think the calculations can perhaps be done better
now directly in terms of augmented plane waves or muffin tin
orbitals does not destroy the pseudopotential as a way of thinking
about electronic structure and how to make new advances. Similarly
throughout materials science there are fundamental insights needed
at every level. For example we have superficial understanding of
the invar effect, but not yet deep enough to explain why certain
classes of alloys show it whereas most do not. There is still a
concept missing. I know some people take a rather opposite view,
that materials science is only dotting i's and crossing t's and
grinding out a whole lot of uninteresting numbers. I once
co-authored a paper on systematic trends in the sizes of atoms, and
the referee started his comments with the words "This isn't
physics". His attitude was that one has a complete understanding
of atomic size because one could take a Hartree-Fock program to
calculate any atom, with the size coming out correctly to be
whatever it came out and that was all there was to be said on the
matter. I think that is incorrect. As long as there are
systematic trends in the atomic sizes through the periodic table,
it is part of science to uncover the factors which explain them in
a systematic way. Similarly among high temperature superconductors
why are niobium compounds prominent and why complicated crystal
structures? Unfortunately some computational physics is done too
much in the spirit of that anonymous referee of producing numbers
for their own sake. In the computer one can vary and control the
conditions of the system to a degree rarely possible in a
laboratory experiment, and as I have already said, the wave
functions contain a lot of extra information about bonding etc.
Thus there is plenty of fundamental understanding required in
materials science, and computational physics is uniquely placed to
delve into it. Let me finish this point with another example. We
know that the mechanical strength of a material depends on
dislocations multiplying, moving and interacting in complicated
ways. Well, why in terms of fundamental interatomic bonding are
metals generally ductile and non-metals brittle? What is the
unifying concept that transcends different crystal structures and
the difference between transition metals and sp bonded metals?

Computational physics historically has grown more out of
theoretical physics groups than from the experimental side, at
least that is so in the field of electronic structure of solids.
Over most of my scientific life computation has tended to be
something tacked onto the end of a piece of theory to bolster its
correctness and help get it past the referee. I exaggerate a bit
in order to make the point that one has to see computation these
days as a third mode different both from theory and experiment. It
is a third investigative technique in its own right, often more
akin to experiment than to theory. For example computation has
been used, as I am sure we will hear in talks later in the Advanced
Studies Institute, to investigate the atomic reconstruction

occurring at a silicon surface. One puts the atoms into the
computer and observes what equilibrium configuration they take up,
in much the same spirit as an experimentalist would in his
laboratory. I therefore call it a computer experiment. Rutherford
once described the amassing of pointless measurements as mere
"stamp collecting" and computational physics can degenerate in the
same way. In order to reap understanding, a series of experiments
has to be carefully designed with some aim or concept in mind,
which is equally true in the computer and the laboratory. I have
often heard computer experiments criticised as unrealistic, but
that in general is an ignorant comment. You don't investigate the
initial stages of corrosion by tossing a hunk of steel out into the
weather. Part of the scientific method is to control the number of
variables so that one studies pure flat single crystals of iron in
an ultrahigh vacuum with a small amount of oxygen admitted. A
computer experiment carried this process to its logical conclusion,
but it is important to observe there must be no compromise in
applying the laws of nature to the computer model, such as quantum
mechanics, statistical mechanics etc. For example the Axial Next
Nearest Neighbour Ising (ANNNI) model is enjoying a current vogue
as a system showing incommensurate phases and an infinite range of
long period polytypes. It is not expected to correspond to
any single real material, but it shows a number of interesting
phenomena which are observed in real systems such as incommen-
surateness, an infinite number of phases which conform to certain patterns, a tendency to lock in below a certain temperature
etc. Computation can help establish what the behaviour of the
model is, which theory will hopefully then explain. Provided
statistical mechanics through the Monte Carlo method has been
correctly applied to the Ising spins, the ANNNI model is as real a
system as an actual crystal in the laboratory and its behaviour worthy of understanding. Of course the hope is that the
understanding so obtained can then be applied usefully to investigate more complex real materials. Thus in computational physics "achieving understanding" sometimes demands simplifying the
model as much as possible to exhibit the necessary essentials, and
at other times requires including every effect possible in order to
model accurately a required situation. One needs to be clear what
the objective is in each particular case.

    The ease of altering the system in a computation allows one
to exploit fully the analysis of trends. This is often where
"understanding" takes over from mere numbers. It should be a
routine safeguard in computation of electronic structure because
one cannot expect the absolute correctness of some individual band
gap or cohesive energy etc using the local density approximation to
the density functional method. Computationally one can investigate
unstable materials as well as those that exist in the laboratory:
for example in studying the stability of alloy phases, it can be
just as illuminating to understand why one combination is unstable

as why another compound is stable. In this way the nature of the
instability can be analysed. There may also be other liberties one
can take, such as treating the valence Z of an element as a
continuous variable Z.

I have tried to describe some of the broad principles running
through computational physics. They are fairly obvious on the
whole, but their intelligent application in each specific project
may not be trivial as I know from discussions with collaborators.
Computational physics of electronic structure has one vital
advantage: it stands squarely in the main stream of solid state
physics and not on a little ego-trip of its own. This has always
been so: where would solid state physics be without the band
structure calculations on aluminium, silicon or iron at a time when
only a few pieces of Fermi surface or symmetry points in k-space
were accessible to experiment. The examples I have mentioned and
many others in the literature show that it is just as relevant
today.

I. AB-INITIO METHODS
I.1 Dielectric Response and Many-Body Theory

BASIC CONCEPTS IN DIELECTRIC RESPONSE AND PSEUDOPOTENTIALS

J. T. Devreese* and F. Brosens**

Department of Physics
Universitaire Instelling Antwerpen
Universiteitsplein 1
B-2610 Antwerpen-Wilrijk

Chapter 1
THE ONE-ELECTRON APPROXIMATION

1.1 Introduction

A system of M electrons in a potential V_0 (which might be due to the ions and to external perturbations), is described by the hamiltonian:

$$H = \sum_j^M \frac{p_j^2}{2m} + \sum_j^M V^0(\vec{r}_j) + \frac{e^2}{2} \sum_j^M \sum_{i \neq j}^M \frac{1}{|\vec{r}_i - \vec{r}_j|} \tag{1.1}$$

Due to the Coulomb repulsion between the electrons, this hamiltonian is obviously **not** a sum of single particle hamiltonians H_j. In principle, this means that the total wave function depends on the positions of all the electrons, the coordinates of which are all correlated. For a system with many

---

*And: Institute of Applied Mathematics, R. U. C. A., Groenenborgerlaan 171, B-2020 Antwerpen,
and: University of Technology, Eindhoven (The Netherlands).

**Research Associate of the National Fund for Scientific Research, Belgium.

electrons, like a solid, it seems unrealistic to aim at the exact solution of the many-particle Schrödinger equation. A further complication is the Pauli principle, which imposes the antisymmetry of the total wave function with respect to the interchange of any pair of electrons. For practical purposes, the **one-electron approximation** is commonly used, which means that in some sense (to be specified later on), the electron-electron interaction for some particular electron at $\vec{r}_i$ is replaced by an averaged interaction of all other electrons at $\vec{r}_j$, giving rise to some potential $V(\vec{r}_i)$.

1.1.1 The Hartree approximation

If it were possible to write the hamiltonian of a system of interacting electrons as a sum of single-particle hamiltonians, it would be possible to write the many-electron wave function as a product of single-particle wave functions:

$$\Psi(\vec{r}_1, \vec{r}_2, \cdots, \vec{r}_M) = \prod_{j=1}^{M} \phi_j(\vec{r}_j) \tag{1.2}$$

and the electron probability density would then be given by

$$n(\vec{r}) = \sum_i |\phi_i(\vec{r}_i)|^2 \tag{1.3}$$

Indeed, one readily checks that the product wave function (Eq. 2) is an eigen state of a sum of hamiltonians $\sum_j H_j$, where H_j only acts on the electron characterized by j, provided each one-electron wave function $\phi_j(\vec{r}_j)$ is an eigen function of H_j.

But, since the hamiltonian (Eq. 1) is not a sum of single-particle hamiltonians, the true wave function can not be written in the product form of Eq. 2, which furthermore does not have the required antisymmetry property. Nevertheless, a product wave function of any form, and thus also of the form of Eq. 2, can be used in the variational principle of quantum mechanics:

$$E_g \leq E \equiv \frac{<\Psi|H|\Psi>}{<\Psi|\Psi>} \tag{1.4}$$

which for any state $|\Psi>$ gives an upper bound for the exact ground state energy E_g.

Imposing the normalization of the functions ϕ_j by the introduction of the Lagrange multipliers E_j^H, one then has to minimize the expression:

10

$$< \phi_1(\vec{r}_1)\cdots\phi_M(\vec{r}_M)|H|\phi_M(\vec{r}_M)\cdots\phi_1(\vec{r}_1) > - \sum_j E_j^H < \phi_j|\phi_j > \qquad (1.5)$$

with respect to variations in the function ϕ_j. This approach is known as the **Hartree approximation** to the ground state energy, giving the Hartree equation for the single particle wave functions with corresponding **Hartree single particle energies** E_j^H:

$$\left[T + V^0(\vec{r}) + V^{Coul}(\vec{r})\right]\phi_j(\vec{r}) = E_j^H \phi_j(\vec{r}) \qquad (1.6)$$

$$T = -\frac{\hbar^2 \Delta}{2m} \qquad V^{Coul}(\vec{r}) = \int d^3 r_1 \frac{e^2 n(\vec{r}_1)}{|\vec{r} - \vec{r}_1|} \qquad (1.7)$$

where V^{Coul} is the standard Coulomb interaction. The Hartree approximation for the total energy then gives:

$$E^H = \sum_i < \phi_i|\frac{p_i^2}{2m} + V^0(\vec{r}_i)|\phi_i > + \frac{1}{2}\sum_{i \neq j} < \phi_i\phi_j|\frac{e^2}{|\vec{r}_i - \vec{r}_j|}|\phi_j\phi_i > \qquad (1.8)$$

1.1.2 The Hartree-Fock approximation

In the Hartree approximation, discussed above, the trial wave function did not satisfy the antisymmetry, imposed by the Pauli exclusion principle. Therefore, it seems more appropriate to consider a trial wave function which is a Slater determinant of single particle wave functions. This approximation is known as the **Hartree-Fock approximation**. The resulting Hartree-Fock energy E^{HF} differs from the Hartree energy by an additional (negative) energy (called the **exchange energy**):

$$E^{HF} = E^H + E^{exc}$$
$$E^{exc} = -\frac{1}{2}\sum_{i \neq j} \int d^3 r_1 \int d^3 r_2 \phi_i(\vec{r}_1)^* \phi_j(\vec{r}_2)^* \frac{e^2}{|\vec{r}_1 - \vec{r}_2|}\phi_j(\vec{r}_1)\phi_i(\vec{r}_2) \qquad (1.9)$$

where only electrons of the same spin are to be included in the sum, since the exclusion principle applies for particles in the same state. The corresponding Hartree-Fock equations for the single particle wave functions differ from the Hartree equations by an "exchange potential":

$$[T + V^0(\vec{r}) + V^{Coul}(\vec{r}) + V_i^{exc}(\vec{r})]\phi_i(\vec{r}) = E_i^{HF}\phi_i(\vec{r})$$

$$V_i^{exc}(\vec{r}) = -\sum_j \int d^3 r_1 \frac{e^2}{|\vec{r}-\vec{r}_1|} \frac{\phi_i(\vec{r})^*\phi_j(\vec{r}_1)^*\phi_j(\vec{r})\phi_i(\vec{r}_1)}{\phi_i(\vec{r})^*\phi_i(\vec{r})} \qquad (1.10)$$

which explicitly depends on the state $|\phi_i>$ upon which its acts. Note that the energy E^{exc} is only formally the difference between the Hartree energy and the Hartree-Fock energy, because the single particle wave functions are different in both approximations. Note also that in the **homogeneous** electron gas, plane waves are solutions of both the Hartree and the Hartree-Fock equations.

The Hartree-Fock calculations have been carried out for the ground state energy of all atoms[1], but for solids these calculations[2] are very complicated. Moreover, they have shown rather unsuccessful in the determination of band gaps.

1.1.3 The density functional formalism

A quite different approach is based on the Hohenberg-Kohn[3] theorem, which states that the **ground state energy of a (non-degenerate) many particle system in an external applied potential is a unique functional** $E[n]$ **of the density** $n(\vec{r})$. Furthermore, they proved that this universal functional $E[n]$ has as its minimum value the correct ground state energy E_g for the correct density $n(\vec{r})$. (Note that it was not proved whether an arbitrary density distribution can be realized by some external potential).

The Kohn-Sham[4] theorem then allows to obtain a more tractable single-particle Schrödinger equation. Many-particle effects are included by an effective exchange and correlation potential, which is derived variationally.

According to the Hohenberg-Kohn theorem, the ground state energy in terms of the exact ground state wave function Ψ_g is given by:

$$E_g[n] = T_s[n] + \int d^3 r\, V^0(\vec{r}) n(\vec{r}) + \frac{e^2}{2} \int d^3 r_1 \int d^3 r_2 \frac{n(\vec{r}_1)n(\vec{r}_2)}{|\vec{r}_1-\vec{r}_2|} + E^{XC}[n]$$

$$(1.11)$$

in which $T_s[n]$ is the sum of the single-particle kinetic energies, and $E^{XC}[n]$ denotes the exchange and correlation contribution to the energy, which itself is a unique functional of the density, independent of the applied potential V^0. (The density will of course depend on this potential.)

Carrying out the variational calculation with respect to the density, which is expressed as a sum of "single particle densities":

$$n(\vec{r}) = \sum_{j}^{M} |\phi_j(\vec{r})|^2 \quad ; \quad \frac{\delta n(\vec{r})}{\delta \phi_j(\vec{r}_1)^*} = \delta(\vec{r} - \vec{r}_1)\phi_j(\vec{r}) \qquad (1.12)$$

one obtains the Kohn-Sham equations for the functions $\phi_j(\vec{r})$, which have to be solved self-consistently:

$$\left[T + V^0(\vec{r}) + V^{Coul}(\vec{r}) + V^{XC}(\vec{r})\right]\phi_j(\vec{r}) = E_j\phi_j(\vec{r}) \qquad (1.13)$$

$$V^{XC}(\vec{r}) = \frac{\delta E^{XC}[n]}{\delta n(\vec{r})} \qquad (1.14)$$

The resulting exchange and correlation potential is manifestly also a functional of the density. But the single particle equation (Eq. 13) is still formal, because the unique exchange and correlation functional E^{XC} is still (and maybe will always remain) unknown.

Consider then the sum of the eigen energies E_n:

$$\sum_n E_n = \sum_n < \phi_n|T + V^0|\phi_n >$$
$$+ e^2 \int d^3r_1 \int d^3r_2 \frac{n(\vec{r}_1)n(\vec{r}_2)}{|\vec{r}_1 - \vec{r}_2|} + \int d^3rn(\vec{r})V^{XC}(\vec{r}) \qquad (1.15)$$

Comparing this expression with the ground state energy (Eq. 11), one readily observes that the Kohn-Sham approach gives an upper bound E for the total energy, given by:

$$E = \sum_n E_n - \frac{e^2}{2} \int d^3r_1 \int d^3r_2 \frac{n(\vec{r}_1)n(\vec{r}_2)}{|\vec{r}_1 - \vec{r}_2|} + \int d^3rn(\vec{r})[\epsilon^{XC}(\vec{r}) - V^{XC}(\vec{r})]$$
$$(1.16)$$

where the total exchange and correlation energy is written in the often used form:

$$E^{XC} = \int d^3rn(\vec{r})\epsilon^{XC}(\vec{r}) \qquad (1.17)$$

Of course, ϵ^{XC} is also a functional of the density, and could be called an exchange and correlation energy density.

Given the fact that the exchange and correlation energy functional is not known, approximations have to be made, e.g. the local density approximation and/or the $n^{1/3}$ approximation or another suitable scheme. The meaning of the famous $n^{1/3}$ –usually called $\rho^{1/3}$– approximation is most easily understood in the framework of the homogeneous electron gas.

1.2 The Hartree-Fock approximation in jellium

In the homogeneous electron gas –or jellium– the electrons are assumed to be embedded in a uniform compensating positive background. It is well known that in this case the plane waves:

$$\phi_{\vec{k}}(\vec{r}) = \frac{1}{\sqrt{\Omega}} e^{i\vec{k}.\vec{r}} \tag{1.18}$$

are solutions of the Hartree-Fock equations. The wave numbers $\vec{k}$ have to satisfy the periodic boundary conditions, and Ω denotes the volume of the crystal.

These states fill up the Fermi sphere, the radius k_F of which is obtained by imposing that the number of states in the Fermi sphere equals the number of electrons, giving:

$$\frac{1}{n} = \frac{3\pi^2}{k_F^3} \tag{1.19}$$

The exchange potential which enters in the Hartree-Fock equation (Eq. 10), can be evaluated analytically in this simple case:

$$V_{\vec{k}}^{exc} = -\frac{e^2 k_F}{\pi} \left[1 + \frac{k_F^2 - k^2}{2kk_F} \log\left| \frac{k + k_F}{k - k_F} \right| \right] \tag{1.20}$$

Some typical values are:

$$
\begin{aligned}
V_0^{exc} &= -\frac{2}{\pi} e^2 k_F = -\frac{2}{\pi} e^2 (3\pi^2 n)^{1/3} \qquad &\text{at } k = 0 \\
V_{k_F}^{exc} &= -\frac{1}{\pi} e^2 k_F = -\frac{1}{\pi} e^2 (3\pi^2 n)^{1/3} \qquad &\text{at } k = k_F
\end{aligned}
\tag{1.21}
$$

The exchange potential (Eq. 20) is plotted in Fig. 1.

Averaging the exchange potential (Eq. 20) over all the occupied states $k \leq k_F$ (replacing the sum by an integral $\sum_{\vec{k}} \rightarrow (2\Omega/(2\pi)^3) \int d^3k$ because of the periodic boundary conditions) one finds

14

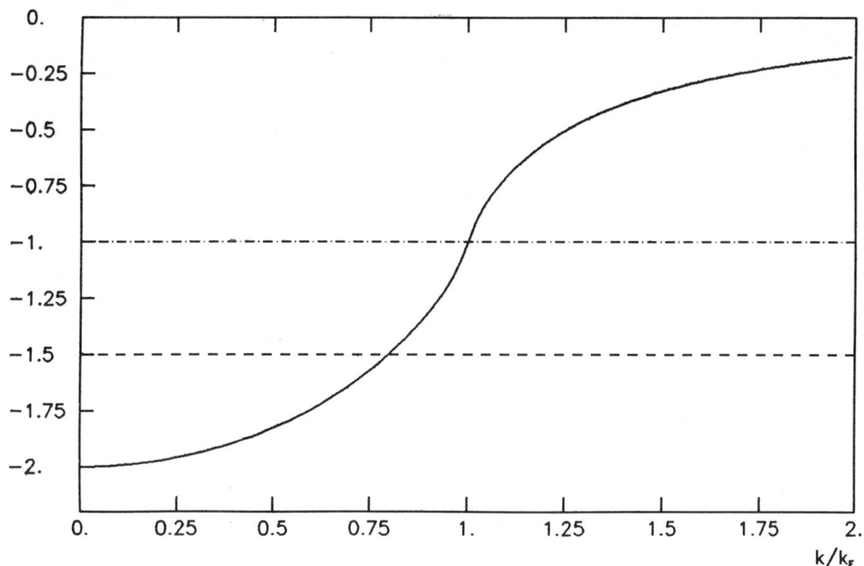

Figure 1. Exchange potential in the HF approximation (in units $e^2 k_F/\pi$).
Dashed line: Slater approximation;
Dash-dotted line: Kohn-Sham approximation.

$$\overline{V}^{exc} = \frac{1}{n}\frac{1}{(2\pi)^3}\int d^3k\, V_{\underline{k}}^{exc} = -\frac{3}{2\pi}e^2 k_F = -\frac{3}{2\pi}e^2(3\pi^2 n)^{1/3} \qquad (1.22)$$

which is the **Slater exchange potential**. The exchange energy per electron
is half this value:

$$\epsilon^X = \frac{1}{2}\overline{V}^{exc} = -\frac{3}{4\pi}e^2(3\pi^2 n)^{1/3} \qquad (1.23)$$

The electron gas density is often expressed in terms of the Wigner-Seitz
parameter r_s, which is the radius, expressed in units of the Bohr radius a_0,
of a sphere with the averaged volume occupied by one electron:

$$\frac{1}{n} = \frac{4\pi}{3}a_0^3 r_s^3 \qquad k_F = \left(\frac{9\pi}{4}\right)^{1/3}\frac{1}{r_s a_0} \qquad (1.24)$$

giving:

$$\epsilon^X = -\frac{3e^2}{4\pi a_0}\left(\frac{9\pi}{4}\right)^{1/3}\frac{1}{r_s} = -\frac{0.91633}{r_s}\text{Rydbergs}$$

This term can be combined with the averaged kinetic energy per particle:

15

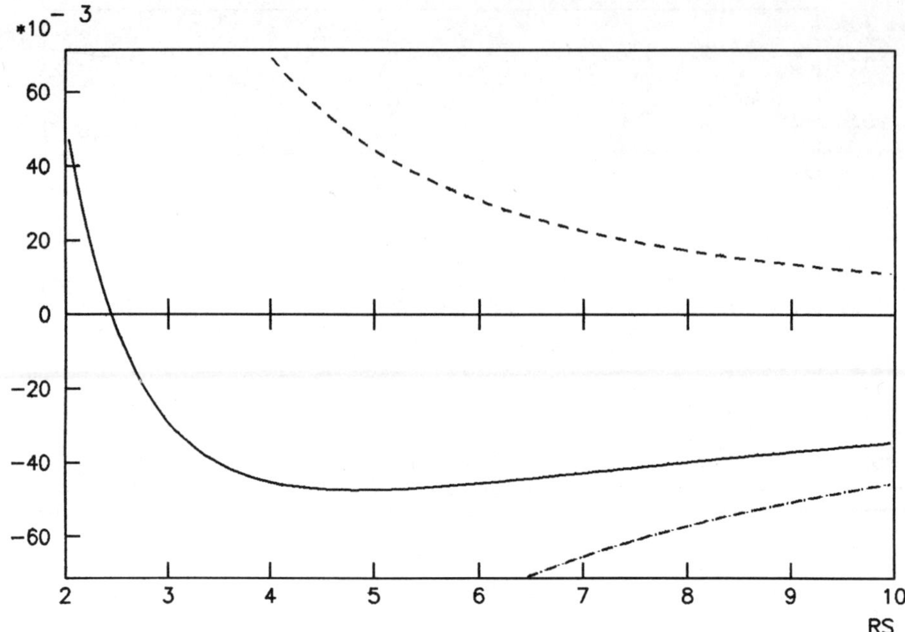

Figure 2. HF-energy per particle (atomic units).
Dashed and dash-dotted line: kinetic and exchange contribution.

$$\epsilon^{kin} = \frac{3}{5}\left(\frac{9\pi}{4}\right)^{2/3}\frac{1}{r_s^2}\frac{e^2}{2a_0} = \frac{2.2099}{r_s^2}\text{Rydbergs}$$

giving for the Hartree-Fock energy ϵ^{HF} per particle:

$$\epsilon^{HF} = \left(\frac{2.2099}{r_s^2} - \frac{0.91633}{r_s}\right)\frac{e^2}{2a_0} \qquad (1.25)$$

This HF-energy is plotted in Fig. 2 as a function of r_s.

Already some 50 years ago, attempts have been made to estimate corrections[5] upon the HF approximation. A concise review, including the diagrammatic expansion of the correlation energy[6] and the Wigner lattice[7] is e.g. given by Mahan[8] .

If one would like to introduce an effective potential to describe deviations of the electron gas from uniformity, one would intuitively expect that $V_{k_F}^{exc}$, rather than the Slater potential, gives the most accurate description, since density fluctuations in the first place are expected at the Fermi sphere. This is confirmed in the Kohn-Sham approximation. Indeed, the total exchange energy for the homogeneous electron gas is given by:

$$E^X = -\frac{3}{4\pi}e^2 \int d^3r (3\pi^2)^{1/3} n^{4/3} \tag{1.26}$$

and thus the Kohn-Sham exchange potential for a homogeneous electron gas becomes:

$$V^X = \frac{\delta E^X}{\delta n} = -\frac{1}{\pi}e^2 (3\pi^2 n)^{1/3} \tag{1.27}$$

which differs by a factor 2/3 from the Slater potential (Eq. 22).

In the **local density approximation**, which will be used by several lecturers at this ASI, the basic assumption is that the gradient of the electron density is small compared to the average density. It is assumed that the functional form of the exchange and correlation potential is not affected by the density fluctuations. The local density $n(\vec{r})$ is subsequently inserted in the exchange potential of the homogeneous electron gas, and also in some suitable expression for the correlation part.

1.3 The polarizability matrix[9]

Assuming that the potential, acting upon an electron in the crystal, is known, one can examine the response of the electrons if the external potential V^0 is changed by a small amount δV^0. This perturbation induces a polarization of the electrons. Consequently, the Coulomb potential and the exchange and correlation potential are then also modified. The electron thus interacts with a total potential $V + \delta V$, where:

$$\delta V = \delta V^0 + \delta V^{Coul} + \delta V^{XC} \tag{1.28}$$

From perturbation theory, it is well known that the first order correction to the wave function for an electron with unperturbed states $|\phi_m>$ is given by:

$$|\delta\phi_m> = \sum_{l \neq m} |\phi_l> \frac{<\phi_l|\delta V|\phi_m>}{E_m - E_l} \tag{1.29}$$

The induced density to first order is given by:

$$\delta n(\vec{r}) = \sum_{m\ occupied} [\phi_m{}^*(\vec{r})\delta\phi_m(\vec{r}) + c.c.] \tag{1.30}$$

In the following, a distinction has to be made between occupied and unoccupied states $|\phi_m>$ in the equilibrium configuration. Therefore we introduce an occupation factor:

$$p_m = 1 \qquad \text{if } \phi_m(\vec{r}) \text{ is occupied}$$
$$0 \qquad \text{otherwise}$$

$$(1.31)$$

Using this notation, and introducing the first order correction to the wave function, the induced electron density becomes:

$$\delta n(\vec{r}) = \sum_m p_m \sum_{l \neq m} \left[\phi_m^*(\vec{r})\phi_l(\vec{r}) \frac{< \phi_l|\delta V|\phi_m >}{E_m - E_l} + c.c. \right] \qquad (1.32)$$

By interchanging the indices l and m in the complex conjugated part, this expression is readily converted into:

$$\delta n(\vec{r}) = \sum_{l,m}{}' (p_m - p_l)\phi_m^*(\vec{r})\phi_l(\vec{r}) \frac{< \phi_l|\delta V|\phi_m >}{E_m - E_l} \qquad (1.33)$$

where $\sum'_{m,l}$ is a shorthand notation for $\sum_m \sum_{l \neq m}$. The polarizability $P(\vec{r}_1, \vec{r}_2)$ in real space is defined as:

$$\delta n(\vec{r}_1) = \int d^3 r_2 P(\vec{r}_1, \vec{r}_2)\delta V(\vec{r}_2) \qquad (1.34)$$

In the expression for the induced density, $< \phi_l|\delta V|\phi_m >$ is now explicitly written as an integral. The polarizability then takes the form:

$$P(\vec{r}_1, \vec{r}_2) = \sum_{l,m}{}' (p_m - p_l) \frac{\phi_m^*(\vec{r}_1)\phi_l(\vec{r}_1)\phi_l^*(\vec{r}_2)\phi_m(\vec{r}_2)}{E_m - E_l} \qquad (1.35)$$

Integration of Eq. 35 over $\vec{r}$ gives zero, because of the orthogonality of the wave functions. This means that the total induced density is zero, or equivalently that the number of electrons is conserved (in a finite system).

In solid state physics, it is usually prefered to work in the wave number representation, in which the polarizability is defined as the matrix which gives the induced density from the total perturbing potential:

$$\delta n_{\vec{q}_1} = \sum_{\vec{q}_2} \Pi(\vec{q}_1, \vec{q}_2)\delta V_{\vec{q}_2} \qquad (1.36)$$

where the Fourier transform is defined as:

18

$$\delta n(\vec{r}) = \sum_{\vec{q}} \delta n_{\vec{q}} e^{i\vec{q}\cdot\vec{r}} \qquad (1.37)$$

$$\delta n_{\vec{q}} = \frac{1}{\Omega} \int d^3r \, \delta n(\vec{r}) e^{-i\vec{q}\cdot\vec{r}} \qquad (1.38)$$

and similarly for the potentials. The matrix elements in Eq. 33 for the induced density can be written as:

$$< \phi_l|\delta V|\phi_m > = \sum_{\vec{q}_2} \left[\int d^3r \, \phi_l{}^*(\vec{r}) e^{i\vec{q}_2\cdot\vec{r}} \phi_m(\vec{r}) \right] \delta V_{\vec{q}_2} \qquad (1.39)$$

By taking the Fourier transforms of the induced density, one is led to the following well-known[10] form for the polarizability matrix:

$$\Pi(\vec{q}_1,\vec{q}_2) = \frac{1}{\Omega} {\sum_{l,m}}' \frac{p_m - p_l}{E_m - E_l} < \phi_m|e^{-i\vec{q}_1\cdot\vec{r}}|\phi_l > < \phi_l|e^{i\vec{q}_2\cdot\vec{r}}|\phi_m > \qquad (1.40)$$

The symmetry properties:

$$\Pi(\vec{q}_1,\vec{q}_2) = \Pi(-\vec{q}_2,-\vec{q}_1) = \Pi(\vec{q}_2,\vec{q}_1)^* \qquad (1.41)$$

follow immediately by interchanging l and m, and taking the complex conjugate. Note that the $\vec{q}_1 \to 0$ and the $\vec{q}_2 \to 0$ limits of the polarizability matrix are zero, which reflects the conservation of the number of electrons, as mentioned above in connection with the polarizability matrix in real space.

1.4 The dielectric matrices

1.4.1 Defining equations

The determination of the polarizability is only the first step in the calculation of the response of the electron system. The second step is the derivation of the change δV in the electron potential, due to the external perturbation δV^0. This interrelation is described by the dielectric matrix. But in this respect, an important distinction has to be made. The dielectric matrix, usually defined from the Maxwell equations (schematically $\vec{D} = \epsilon\vec{E}$), relates the **external** perturbing potential to the **total** perturbing potential (from both the perturbation and the induced density). For the Fourier components, this relation is:

$$\delta V^0_{\vec{q}_1} = \sum_{\vec{q}_2} \epsilon(\vec{q}_1, \vec{q}_2)\big[\delta V^0_{\vec{q}_2} + \delta V^{Coul}_{\vec{q}_2}\big] \qquad (1.42)$$

where the Coulomb contribution as before is due to the induced density, and is determined from the Poisson equation:

$$\delta V^{Coul}_{\vec{q}} = \frac{4\pi e^2}{q^2}\delta n_{\vec{q}} \qquad (1.43)$$

Note that this definition of the dielectric function immediately implies the symmetry relation:

$$\epsilon(-\vec{q}_1, -\vec{q}_2) = \epsilon(\vec{q}_1, \vec{q}_2)^* \qquad (1.44)$$

The quantity needed to calculate the total induced potential, as seen by a test charge, is the inverse dielectric matrix:

$$\delta V^0_{\vec{q}_1} + \delta V^{Coul}_{\vec{q}_1} = \sum_{\vec{q}_2} \epsilon^{-1}(\vec{q}_1, \vec{q}_2)\delta V^0_{\vec{q}_2} \qquad (1.45)$$

The inverse dielectric matrix by definition satisfies the relations:

$$\sum_{\vec{q}_2} \epsilon(\vec{q}_1, \vec{q}_2)\epsilon^{-1}(\vec{q}_2, \vec{q}_3) = \delta_{\vec{q}_1, \vec{q}_3} = \sum_{\vec{q}_2} \epsilon^{-1}(\vec{q}_1, \vec{q}_2)\epsilon(\vec{q}_2, \vec{q}_3) \qquad (1.46)$$

The **"test-charge–test-charge"** dielectric matrix allows to calculate the potential, due to a perturbation, as seen e.g. by the ions in the system. (The perturbation itself might be due to motions of the ions, since the ions are external in relation to the electron system.)

An electron in the system however, would feel a potential δV containing the exchange and correlation potential. We denote the corresponding **"electron–test-charge"** dielectric function by $\tilde{\epsilon}$:

$$\delta V^0_{\vec{q}_1} + \delta V^{Coul}_{\vec{q}_1} + \delta V^{XC}_{\vec{q}_1} = \sum_{\vec{q}_2} \tilde{\epsilon}^{-1}(\vec{q}_1, \vec{q}_2)\delta V^0_{\vec{q}_2} \qquad (1.47)$$

1.4.2 Induced exchange and correlation potential

To determine the dielectric matrices from the polarizability one first has to evaluate the Coulomb potential and the exchange and correlation

potential, due to the induced density. For the Coulomb part, the Poisson
equation provides the solution mentioned before, and thus:

$$\delta V_{\vec{q}_1}^{Coul} = \frac{4\pi e^2}{q_1^2} \sum_{\vec{q}_2} \Pi(\vec{q}_1, \vec{q}_2) \delta V_{\vec{q}_2} \tag{1.48}$$

The exchange and correlation part has to remain more formal in a
one-electron treatment, since this is a many-electron problem in nature.
Nevertheless, the formal treatment of the previous section can be worked out
further along the same lines. In a similar way as for the Coulomb interaction
(Eq. 48), one might introduce an exchange and correlation polarizability:

$$\delta V_{\vec{q}_1}^{XC} = \frac{4\pi e^2}{q_1^2} \sum_{\vec{q}_2} \Pi^{XC}(\vec{q}_1, \vec{q}_2) \delta V_{\vec{q}_2} \tag{1.49}$$

The evaluation of this polarizability is straightforward in principle. The
exchange and correlation potential, as discussed before, is a functional of
the density:

$$V^{XC}[n + \delta n; \vec{r}] = V^{XC}[n; \vec{r}] + \int d^3 r_1 \delta n(\vec{r}_1) \frac{\delta V^{XC}[n; \vec{r}]}{\delta n(\vec{r}_1)} \tag{1.50}$$

which gives for the change in the exchange and correlation potential, due to
a density fluctuation $\delta n(\vec{r}_1)$:

$$\delta V^{XC}(\vec{r}) = \int d^3 r_1 \delta n(\vec{r}_1) \frac{\delta V^{XC}(\vec{r})}{\delta n(\vec{r}_1)} \tag{1.51}$$

The Fourier transform of this equation is:

$$\delta V_{\vec{q}_1}^{XC} = \sum_{\vec{q}_2} \delta n_{\vec{q}_2} \frac{1}{\Omega} \int d^3 r_1 \int d^3 r_2 e^{-i\vec{q}_1 \cdot \vec{r}_1} e^{i\vec{q}_2 \cdot \vec{r}_2} \frac{\delta V^{XC}(\vec{r}_1)}{\delta n(\vec{r}_2)} \tag{1.52}$$

Since the density fluctuations are related to the potential (see Eq. 36) by the
polarizability matrix, some elementary algebraic manipulations yield:

$$\frac{4\pi e^2}{q_1^2} \Pi^{XC}(\vec{q}_1, \vec{q}_2) = \sum_{\vec{q}} \Pi(\vec{q}, \vec{q}_2) \frac{1}{\Omega} \int d^3 r_1 \int d^3 r\, e^{-i\vec{q}_1 \cdot \vec{r}_1} e^{i\vec{q} \cdot \vec{r}} \frac{\delta V^{XC}(\vec{r}_1)}{\delta n(\vec{r})}$$

$$\tag{1.53}$$

This expression can be written in matrix form:

$$\Pi^{XC}(\vec{q}_1, \vec{q}_2) = -\sum_{\vec{q}} G^{XC}(\vec{q}_1, \vec{q})\Pi(\vec{q}, \vec{q}_2) \qquad (1.54)$$

$$\frac{4\pi e^2}{q_1^2} G^{XC}(\vec{q}_1, \vec{q}) = -\frac{1}{\Omega} \int d^3 r_1 \int d^3 r\, e^{-i\vec{q}_1 \cdot \vec{r}_1} e^{i\vec{q}\cdot\vec{r}} \frac{\delta V^{XC}(\vec{r}_1)}{\delta n(\vec{r})} \qquad (1.55)$$

Because the exchange and correlation potential is already a functional derivative of the exchange and correlation energy, this expression involves the second functional derivative, which in general is unknown, just as E^{XC} is. Nevertheless, one can first proceed formally, assuming this exchange and correlation matrix is known.

1.4.3 Electron–test-charge dielectric matric

Given the induced density, due to the total pertubation δV, the sum of the Coulomb potential and the exchange and correlation potential becomes:

$$\delta V_{\vec{q}_1}^{Coul} + \delta V_{\vec{q}_1}^{XC} = \frac{4\pi e^2}{q_1^2} \sum_{\vec{q}_2} \left[\Pi(\vec{q}_1, \vec{q}_2) + \Pi^{XC}(\vec{q}_1, \vec{q}_2) \right]$$
$$\times \left[\delta V_{\vec{q}_2}^0 + \delta V_{\vec{q}_2}^{Coul} + \delta V_{\vec{q}_2}^{XC} \right] \qquad (1.56)$$

where the total potential δV in the right hand side was written out in its constituting components. This is equivalent to:

$$\sum_{\vec{q}_2} \left[\delta_{\vec{q}_1, \vec{q}_2} - \frac{4\pi e^2}{q_1^2} \left(\Pi(\vec{q}_1, \vec{q}_2) + \Pi^{XC}(\vec{q}_1, \vec{q}_2) \right) \right] \left[\delta V_{\vec{q}_2}^{Coul} + \delta V_{\vec{q}_2}^{XC} \right]$$
$$= \frac{4\pi e^2}{q_1^2} \sum_{\vec{q}_2} \left(\Pi(\vec{q}_1, \vec{q}_2) + \Pi^{XC}(\vec{q}_1, \vec{q}_2) \right) \delta V_{\vec{q}_2}^0 \qquad (1.57)$$

or:

$$\sum_{\vec{q}_2} \left[\delta_{\vec{q}_1, \vec{q}_2} - \frac{4\pi e^2}{q_1^2} \left(\Pi(\vec{q}_1, \vec{q}_2) + \Pi^{XC}(\vec{q}_1, \vec{q}_2) \right) \right]$$
$$\times \left[\delta V_{\vec{q}_2}^0 + \delta V_{\vec{q}_2}^{Coul} + \delta V_{\vec{q}_2}^{XC} \right] = \delta V_{\vec{q}_1}^0 \qquad (1.58)$$

Therefore, the **electron–test-charge** dielectric matrix is given by:

$$\tilde{\epsilon}(\vec{q}_1, \vec{q}_2) = \delta_{\vec{q}_1, \vec{q}_2} - \frac{4\pi e^2}{q_1^2} \left(\Pi(\vec{q}_1, \vec{q}_2) + \Pi^{XC}(\vec{q}_1, \vec{q}_2) \right) \qquad (1.59)$$

1.4.4 Test–charge–test–charge dielectric matrix

For the test–charge–test–charge dielectric matrix, only the contribution $\delta V_{\vec{q}}^{0} + \delta V_{\vec{q}}^{Coul}$ is to be related with $\delta V_{\vec{q}}^{0}$. Let us start this derivation from the inverse electron–test–charge dielectric matrix:

$$\delta V_{\vec{q}_1}^{0} + \delta V_{\vec{q}_1}^{Coul} + \delta V_{\vec{q}_1}^{XC} = \sum_{\vec{q}_2} \tilde{\epsilon}^{-1}(\vec{q}_1, \vec{q}_2)\delta V_{\vec{q}_2}^{0} \tag{1.60}$$

The exchange and correlation potential can be eliminated, using the defining equation (Eq. 49) for Π^{XC}, and the definition (Eq. 46) of $\tilde{\epsilon}^{-1}$:

$$\begin{aligned}
\delta V_{\vec{q}_1}^{XC} &= \frac{4\pi e^2}{q_1^2} \sum_{\vec{q}_2} \Pi^{XC}(\vec{q}_1, \vec{q}_2)\delta V_{\vec{q}_2} \\
&= \frac{4\pi e^2}{q_1^2} \sum_{\vec{q}_2, \vec{q}_3} \Pi^{XC}(\vec{q}_1, \vec{q}_2)\tilde{\epsilon}^{-1}(\vec{q}_2, \vec{q}_3)\delta V_{\vec{q}_3}^{0}
\end{aligned} \tag{1.61}$$

The test–charge–test–charge dielectric matrix is readily found if one inserts this expression for the exchange and correlation potential in the defining equation (Eq. 45) for the electron–test–charge dielectric matrix:

$$\epsilon^{-1}(\vec{q}_1, \vec{q}_2) = \tilde{\epsilon}^{-1}(\vec{q}_1, \vec{q}_2) - \frac{4\pi e^2}{q_1^2} \sum_{\vec{q}_3} \Pi^{XC}(\vec{q}_1, \vec{q}_3)\tilde{\epsilon}^{-1}(\vec{q}_3, \vec{q}_2) \tag{1.62}$$

From the expression (Eq. 59) found above for $\tilde{\epsilon}$, the matrix Π^{XC} can be written in terms of $\tilde{\epsilon}$ and Π, and one obtains for the **test–charge–test–charge** dielectric matrix:

$$\epsilon^{-1}(\vec{q}_1, \vec{q}_2) = \delta_{\vec{q}_1, \vec{q}_2} + \frac{4\pi e^2}{q_1^2} \sum_{\vec{q}} \Pi(\vec{q}_1, \vec{q})\tilde{\epsilon}^{-1}(\vec{q}, \vec{q}_2) \tag{1.63}$$

1.5 Approximations for exchange and correlation

Although in principle the evaluation of the exchange and correlation matrix $G^{XC}(\vec{q}_1, \vec{q}_2)$ would be possible if the functional for the exchange and correlation energy were known, in practice one has to make some approximations. In the local-density approximation, the exchange and correlation potential is assumed to be given by:

$$V_{LDA}^{XC}(\vec{r}) \approx \frac{\partial [n(\vec{r})\epsilon^{XC}(n(\vec{r}))]}{\partial n(\vec{r})} \tag{1.64}$$

which means that ϵ^{XC} is supposed to be a **function** of the density, instead of a functional. From Eq. 64, one obtains:

$$\frac{\delta V_{LDA}^{XC}(\vec{r}_1)}{\delta n(\vec{r}_2)} \approx \delta(\vec{r}_1 - \vec{r}_2)\frac{\partial^2 [n(\vec{r}_1)\epsilon^{XC}(n(\vec{r}_1))]}{\partial n^2(\vec{r}_1)} \tag{1.65}$$

The exchange and correlation polarizability in the local-density approximation therefore takes the form (see Eq. 54-55):

$$\Pi_{LDA}^{XC}(\vec{q}_1, \vec{q}_2) \approx - \sum_q G_{LDA}(\vec{q}_1, \vec{q} + \vec{q}_1)\Pi(\vec{q} + \vec{q}_1, \vec{q}_2) \tag{1.66}$$

where:

$$\frac{4\pi e^2}{q_1^2} G_{LDA}(\vec{q}_1, \vec{q} + \vec{q}_1) = -\frac{1}{\Omega} \int d^3r\, e^{i\vec{q}\cdot\vec{r}} \frac{\partial^2 [n(\vec{r})\epsilon^{XC}(n(\vec{r}))]}{\partial n^2(\vec{r})} \tag{1.67}$$

is independent of $\vec{q}_1$.

The meaning of this approximation can be elucidated by considering the uniform electron gas as the equilibrium system. In this case, the equilibrium density is independent of the position, and only the $\vec{q} \to 0$ term in Eq. 67 contributes. As a result, one obtains:

$$\Pi_{LDA}^{XC}(\vec{q}_1, \vec{q}_2)\,|_{unif.} = -G(q_1)_{LDA}\Pi(\vec{q}_1, \vec{q}_2) \tag{1.68}$$

$$G_{LDA}(q) = -\frac{q^2}{4\pi e^2}\frac{\partial^2 [n\epsilon^{XC}]}{\partial n^2} \tag{1.69}$$

Note that the exchange contribution is thus given by (see Eq. 23):

$$\frac{4\pi e^2}{q^2} G_{LDA}^{X}(q) = \frac{1}{3}\frac{e^2}{\pi}\frac{(3\pi^2)^{1/3}}{n^{2/3}}$$

which clearly illustrates the phenomenon that the exchange and correlation correction creates an exchange and correlation hole which lowers the Coulomb repulsion.

Instead of assuming the LDA-form (Eq. 69), one could introduce a function $G(q)$ which immitates this exchange and correlation hole, and which is to be determined later on from many-body theory:

$$\Pi^{XC}(\vec{q}_1, \vec{q}_2) = -G(q_1)\Pi(\vec{q}_1, \vec{q}_2) \tag{1.70}$$

This amounts to the approximation:

$$\delta V_{\vec{q}}^{XC} = -G(q)\delta V_{\vec{q}}^{Coul} \tag{1.71}$$

The determination of $G(q)$ from the dielectric theory, as discussed in the next chapter, is quite essential in order to satisfy some known sum rules. Indeed, from Eq. 69, $G_{LDA}(q)$ is obviously divergent at large wave vectors. Such a divergence can not exist, in view of e.g. a sum rule[11] which relates the pair correlation function $g(r)$ at the origin to the large wave vector limit of $G(q)$:

$$g(o) = 1 - \frac{3}{2}\lim_{q\to\infty}G(q) \tag{1.72}$$

On the other hand, the local density approximation is valid in the limit of small wave vectors (i.e. large wavelengths); this allows one to establish the $q \to 0$ limit of the exchange part of $G(q)$:

$$\lim_{\vec{q}\to 0}\frac{4\pi e^2}{q^2}G^X(q) = \frac{1}{3}\frac{e^2}{\pi}\frac{(3\pi^2)^{1/3}}{n^{2/3}} \tag{1.73}$$

Introducing the Fermi wave vector instead of the density, the limiting behaviour of $G^X(q)$ is:

$$G^X(q)\big|_{q\to 0} \to \frac{q^2}{4k_F^2} \tag{1.74}$$

for the homogeneous electron gas, including exchange only. In the next chapter, it will be shown that this relation constitutes an alternative formulation of the compressibility sum rule.

In the approximation, as introduced in Eq. 71, the electron–test-charge dielectric matrix becomes:

$$\tilde{\epsilon}(\vec{q}_1, \vec{q}_2) = \delta_{\vec{q}_1, \vec{q}_2} - \frac{4\pi e^2}{q_1^2}[1 - G(q_1)]\Pi(\vec{q}_1, \vec{q}_2) \tag{1.75}$$

In terms of the inverse of this matrix, the test-charge–test-charge dielectric matrix can then easily be calculated:

$$\epsilon^{-1}(\vec{q}_1, \vec{q}_2) - \delta_{\vec{q}_1, \vec{q}_2} = \frac{1}{1 - G(q_1)}\left[\tilde{\epsilon}^{-1}(\vec{q}_1, \vec{q}_2) - \delta_{\vec{q}_1, \vec{q}_2}\right] \tag{1.76}$$

Both dielectric matrices differ from each other due to the exchange and correlation effects, represented by the function $G(q)$, which is a diagonal approximation for $G^{XC}(\vec{q}_1, \vec{q}_2)$.

Note that the symmetry relations for the polarizability, imply the following symmetry relations for the inverse dielectric matrices:

$$\frac{q_2^2}{1 - G(q_2)}\tilde{\epsilon}^{-1}(-\vec{q}_2, -\vec{q}_1) = \frac{q_1^2}{1 - G(q_1)}\tilde{\epsilon}^{-1}(\vec{q}_1, \vec{q}_2) \qquad (1.77)$$

$$q_2^2 \epsilon^{-1}(-\vec{q}_2, -\vec{q}_1) = q_1^2 \epsilon^{-1}(\vec{q}_1, \vec{q}_2) \qquad (1.78)$$

In the homogeneous electron gas, the dielectric matrices are diagonal because of the **translation invariance** of the system. The dielectric functions then become:

$$\tilde{\epsilon}(q) = 1 + Q_0(q)[1 - G(q)]$$
$$\epsilon(q) = 1 + \frac{Q_0(q)}{1 - G(q)Q_0(q)} \qquad (1.79)$$

where $Q_0(\vec{q})$ is the Lindhard function, defined as:

$$Q_0(q) = -\frac{4\pi e^2}{q^2}\Pi(q) \qquad (1.80)$$

It should be emphasized that these derivations can be fully extended for time dependent perturbations. For that purpose it is sufficient to introduce a frequency ω, apart from the wave vectors, in the densities and potentials, in the polarizability matrices and dielectric matrices, and in the exchange and correlation function which then becomes $G(q, \omega)$. This latter point will be discussed in more detail in the next chapter. Note also that, strictly speaking, **causality** and the related **dispersion relations** only hold for the **inverse** dielectric matrices, since they describe the response. For the dielectric matrix itself, to the best of our knowledge no proof of the dispersion relations has been given.

1.6 Adiabatic electronic energy

In the first section of this chapter on the one-electron approximation, a variational expression (Eq. 16) was found for the ground state energy of the electron system in terms of the band energies, with "double counting" corrections:

$$E = \sum_m p_m E_m - \frac{1}{2} \int d^3 r_1 \int d^3 r_2 \frac{n(\vec{r}_1)n(\vec{r}_2)}{|\vec{r}_1 - \vec{r}_2|} + \int d^3 r\, n(\vec{r})\left[\epsilon^{XC}(\vec{r}) - V^{XC}(\vec{r})\right]$$

$$(1.81)$$

(We now explicitly wrote the occupation factor p_m in the summation.) If some external perturbation is applied, as discussed in the previous section, standard perturbation theory gives for the band energy correction to second order:

$$\delta E_m = \, <\phi_m|\delta V|\phi_m> + \sum_{l \neq m} \frac{|<\phi_l|\delta V|\phi_m>|^2}{E_m - E_l} \qquad (1.82)$$

In terms of the Fourier transforms of δV, this energy correction is:

$$\delta E_m = \sum_{\vec{q}} \delta V_{\vec{q}} <\phi_m|e^{i\vec{q}\cdot\vec{r}}|\phi_m>$$

$$+ \sum_{\vec{q}_1, \vec{q}_2} \delta V_{-\vec{q}_1} \delta V_{\vec{q}_2} \sum_{l \neq m} \frac{<\phi_l|e^{-i\vec{q}_1\cdot\vec{r}}|\phi_m><\phi_m|e^{i\vec{q}_2\cdot\vec{r}}|\phi_l>}{E_m - E_l} \qquad (1.83)$$

Summing this energy correction over the occupied states, one readily sees by interchanging l and m that p_m in the last contribution at the r.h.s can be replaced by $(p_m - p_l)/2$, and the summation over the states in this last contribution simply turns out to be half the polarizability matrix. The sum over the states in the first term at the r.h.s. is recognized to be the Fourier transform of the equilibrium density:

$$\sum_m p_m \delta E_m = \frac{\Omega}{2} \sum_{\vec{q}} \delta V_{\vec{q}} n_{-\vec{q}} + \frac{\Omega}{2} \sum_{\vec{q}_1, \vec{q}_2} \delta V_{-\vec{q}_1} \Pi(\vec{q}_1, \vec{q}_2) \delta V_{\vec{q}_2} \qquad (1.84)$$

which we want to evaluate to second order. For the last term, a first order calculation of the induced potentials is thus sufficient, but for the first term the induced potential would be required to second order. At first glance, this objective seems to go beyond the dielectric approach, as discussed in the previous section. But noting that the variational ground state energy is stationary with respect to the first variation in the wave functions, the first order corrections in the electron density in Eq. 84 do not contribute to the total energy correction in Eq. 81, provided that they are also omitted in the "double counting" terms. This consideration allows to avoid some elaborate algebra to point out which terms contribute to Eq. 81. The consequences are:

a. only keep the external perturbation in the first term of the band energy correction;

b. only keep the term $-\frac{e^2}{2} \int d^3r_1 \int d^3r_2 \delta n(\vec{r}_1)\delta n(\vec{r}_2)/|\vec{r}_1 - \vec{r}_2|$ in the correction for double counting in the Coulomb interaction;

c. only keep the term $-\frac{1}{2} \int d^3r_1 \int d^3r_2 \delta n(\vec{r}_1)\delta n(\vec{r}_2)\delta V^{XC}/\delta n(\vec{r}_2)$ in the exchange and correlation double counting correction.

Writing the last two contributions with their Fourier transforms, one is left with:

$$\delta E = \frac{\Omega}{2}\left[\sum_{\vec{q}} \delta V_{\vec{q}} n_{-\vec{q}} + \sum_{\vec{q}_1,\vec{q}_2} \delta V_{-\vec{q}_1} \Pi(\vec{q}_1,\vec{q}_2)\delta V_{\vec{q}_2} - \sum_{\vec{q}} \delta n_{-\vec{q}}[\delta V_{\vec{q}}^{Coul} + \delta V_{\vec{q}}^{XC}] \right]$$

$$(1.85)$$

But in the second term at the r.h.s., the sum over $\vec{q}_2$ simply gives the induced density, and thus only the interaction between the external perturbation and the induced density survives if this term is combined with the last one. One then ends up with the fairly simple result:

$$\delta E = \Omega \sum_{\vec{q}} \delta V_{\vec{q}}^0 n_{-\vec{q}} + \frac{\Omega}{2} \sum_{\vec{q}} \delta V_{\vec{q}}^0 \delta n_{-\vec{q}} \tag{1.86}$$

where $n_{\vec{q}}$ is the density in the *equilibrium* configuration. This energy correction is easily written in terms of the inverse dielectric function, using

$$\delta n_{\vec{q}} = \frac{q^2}{4\pi e^2}\delta V_{\vec{q}}^{Coul} = \frac{q^2}{4\pi e^2}\sum_{\vec{q}_2}[\epsilon^{-1}(\vec{q},\vec{q}_2) - \delta_{\vec{q},\vec{q}_2}]\delta V_{\vec{q}_2}^0 \tag{1.87}$$

In terms of the external perturbation and the dielectric properties of the system, the energy correction to second order then becomes:

$$\delta E = \Omega \sum_{\vec{q}} \delta V_{-\vec{q}}^0 n_{\vec{q}} + \frac{\Omega}{2} \sum_{\vec{q}_1,\vec{q}_2} \frac{q_1^2}{4\pi e^2}\delta V_{-\vec{q}_1}^0[\epsilon^{-1}(\vec{q}_1,\vec{q}_2) - \delta_{\vec{q}_1,\vec{q}_2}]\delta V_{\vec{q}_2}^0 \tag{1.88}$$

1.7 Harmonic phonons in a periodic lattice

In the previous section, the correction to the electronic energy, due to a perturbing potential, was evaluated to second order. If this perturbation results from ion displacements, there is an additional correction to the total energy of the lattice, arising from the interaction between the displaced ions. In the adiabatic approximation, the previous derivation of Eq. 88 for the electronic part of the energy remains valid. If the ions do not overlap, the ion-ion interaction energy for ions of valence Z at distance R is just the Coulomb interaction $Z^2 e^2/R$. Their interaction energy is then:

$$E^{i-i} = \frac{\Omega}{2} \sum_{\vec{q}} \frac{4\pi e^2}{q^2} |n_{\vec{q}}^{ion}|^2 \tag{1.89}$$

in which the Fourier transform of the ion density is given by:

$$n_{\vec{q}}^{ion} = \frac{Z}{\Omega_0} S(\vec{q}) \tag{1.90}$$

$$S(\vec{q}) = \frac{1}{N} \sum_j e^{-i\vec{q}\cdot\vec{R}_j} \tag{1.91}$$

where $S(\vec{q})$ is the geometrical structure factor. Denoting by $S^{(0)}(\vec{q})$ the structure factor in the equilibrium configuration, the correction to the energy from the interacting ions thus becomes:

$$\delta E^{i-i} = N \sum_{\vec{q}} \frac{2\pi Z^2 e^2}{\Omega_0 q^2} \left[|S(\vec{q})|^2 - |S^{(0)}(\vec{q})|^2 \right] \tag{1.92}$$

Since also the electron-ion potential V^0 is a sum of terms from each ion:

$$V^0(\vec{r}) = \sum_i v^0(\vec{r} - \vec{R}_i)$$

its Fourier transform is a product of the structure factor and a form factor:

$$V_{\vec{q}}^0 = \sum_{\vec{q}} S(\vec{q}) v_{\vec{q}}^0 \quad ; \quad v_{\vec{q}}^0 = \frac{1}{\Omega_0} \int d^3 r\, v^0(\vec{r}) e^{-i\vec{q}\cdot\vec{r}} \tag{1.93}$$

The correction to the total energy, including the ion-ion interaction and the electronic contribution, thus becomes:

$$\delta E^{tot} = N \sum_{\vec{q}} \frac{2\pi Z^2 e^2}{\Omega_0 q^2} \left[|S(\vec{q})|^2 - |S^{(0)}(\vec{q})|^2 \right]$$

$$+ \Omega \sum_{\vec{q}} v_{\vec{q}}^0 n_{-\vec{q}} \left[S(\vec{q}) - S^{(0)}(\vec{q}) \right]$$

$$+ \frac{\Omega}{2} \sum_{\vec{q}_1,\vec{q}_2} \frac{q_1^2}{4\pi e^2} v_{-\vec{q}_1}^0 v_{\vec{q}_2}^0 \left(\epsilon^{-1}(\vec{q}_1,\vec{q}_2) - \delta_{\vec{q}_1,\vec{q}_2} \right) \tag{1.94}$$

$$\times \left[S(-\vec{q}_1) - S^{(0)}(-\vec{q}_1) \right] \left[S(\vec{q}_2) - S^{(0)}(\vec{q}_2) \right]$$

which has to be evaluated to second order in the displacements from equilibrium, in order to determine the harmonic phonons.

Since the energy has to be stationary with respect to the displacements to first order, we first examine the first order correction. Denoting the ion displacements by $\vec{\eta}_j$:

$$\vec{R}_j = \vec{R}_j^0 + \vec{\eta}_j \tag{1.95}$$

the geometrical structure factor becomes:

$$S(\vec{q}) = S^{(0)}(\vec{q}) + S^{(1)}(\vec{q}) + S^{(2)}(\vec{q}) + \cdots \tag{1.96}$$

$$S^{(1)}(\vec{q}) = -\frac{i}{N} \sum_j \vec{q}.\vec{\eta}_j e^{-i\vec{q}.\vec{R}_j^0} \tag{1.97}$$

$$S^{(2)}(\vec{q}) = -\frac{1}{2N} \sum_j (\vec{q}.\vec{\eta}_j)^2 e^{-i\vec{q}.\vec{R}_j^0} \tag{1.98}$$

Collecting the terms to first order in the displacements, which have to be zero, one thus finds:

$$N \sum_{\vec{q}} \frac{2\pi Z^2 e^2}{\Omega_0 q^2} \left[S^{(0)}(-\vec{q})S^{(1)}(\vec{q}) + c.c \right] + \Omega \sum_{\vec{q}} v_{\vec{q}}^0 n_{-\vec{q}} S^{(1)}(\vec{q}) = 0 \tag{1.99}$$

The equilibrium structure factor only differs from zero at a reciprocal lattice vector, corresponding to the Bravais cell. The equilibrium electron density also has the lattice periodicity. Therefore only contributions arise from the reciprocal lattice vectors, (which we denote by $\vec{G}$). After explicitly inserting the expression for the structure factor to first order one obtains:

$$\sum_{\vec{G}} \sum_{j} (\vec{G}\cdot\vec{\eta}_j) e^{-i\vec{G}\cdot\vec{R}_j^0} \left[\frac{4\pi Z^2 e^2}{\Omega_0 G^2} N S^{(0)}(-\vec{G}) + \Omega v_{\vec{G}}^0 n_{-\vec{G}} \right] = 0 \qquad (1.100)$$

This simply expresses that the sum of the forces on the ions in equilibrium is zero. In fact, this condition can be expressed in a more stringent way, since the force on each individual ion should be zero:

$$\sum_{\vec{G}} \vec{G} e^{-i\vec{G}\cdot\vec{R}_j^0} \left[\frac{4\pi Z^2 e^2}{\Omega_0 G^2} S^{(0)}(-\vec{G}) + \Omega_0 v_{\vec{G}}^0 n_{-\vec{G}} \right] = 0 \qquad (1.101)$$

As a consequence, the only remaining terms in the energy (Eq. 49) to second order are:

$$\delta E^{tot} = N \sum_{\vec{q}} \frac{2\pi Z^2 e^2}{\Omega_0 q^2} \left[S^{(0)}(-\vec{q}) S^{(2)}(\vec{q}) + c.c. + |S^{(1)}(\vec{q})|^2 \right]$$

$$+ \Omega \sum_{\vec{q}} v_{\vec{q}}^0 n_{-\vec{q}} S^{(2)}(\vec{q}) \qquad (1.102)$$

$$+ \frac{\Omega}{2} \sum_{\vec{q}_1,\vec{q}_2} \frac{q_1^2}{4\pi e^2} v_{-\vec{q}_1}^0 v_{\vec{q}_2}^0 \left(\epsilon^{-1}(\vec{q}_1,\vec{q}_2) - \delta_{\vec{q}_1,\vec{q}_2} \right) S^{(1)}(-\vec{q}_1) S^{(1)}(\vec{q}_2)$$

Again we use the fact that the electron density and the structure factor in equilibrium only differ from zero if the wave vector is a reciprocal lattice vector, which is relevant for the terms in $S^{(2)}$. Furthermore, the wave vectors in the dielectric matrix only differ by a reciprocal lattice vector, which we now use explicitly:

$$\delta E^{tot} = N \sum_{\vec{G}} \left[\frac{4\pi Z^2 e^2}{\Omega_0 G^2} S^{(0)}(-\vec{G}) + \Omega_0 v_{\vec{G}}^0 n_{-\vec{G}} \right] S^{(2)}(\vec{G})$$

$$+ N \sum_{\vec{q}} \frac{2\pi Z^2 e^2}{\Omega_0 q^2} |S^{(1)}(\vec{q})|^2$$

$$+ \frac{\Omega}{2} \sum_{\vec{q},\vec{G}} \frac{q^2}{4\pi e^2} v_{-\vec{q}}^0 v_{\vec{q}+\vec{G}}^0 \left(\epsilon^{-1}(\vec{q},\vec{q}+\vec{G}) - \delta_{\vec{G},0} \right) S^{(1)}(-\vec{q}) S^{(1)}(\vec{q}+\vec{G})$$

$$(1.103)$$

This expression is a good starting point for a normal mode analysis. Consider a lattice with ν ions per primitive cel, in which they have the relative coordinates $\vec{p}_m$, with $m = 0, 1, \cdots, \nu-1$. Denoting the N/ν positions of the primitive cells by $\vec{X}_p$, the equilibrium structure factor is given by:

$$S^{(0)}(\vec{G}) = \frac{1}{N} \sum_{p,m} e^{-i\vec{G}\cdot(\vec{X}_p + \vec{\rho}_m)} = \frac{1}{\nu} \sum_m e^{-i\vec{G}\cdot\vec{\rho}_m} \tag{1.104}$$

The ion displacements are then characterized by the two indices p and m, referring respectively to the position of the Bravais cell and the relative dispacements in the cell. Expanding these displacements:

$$\vec{\eta}_{p,m} = \sum_{\vec{Q}} \vec{a}_m(\vec{Q}) e^{i\vec{Q}\cdot(\vec{X}_p + \rho_m)} \tag{1.105}$$

where $\vec{Q}$ *is limited to the first Brillouin zone*, one readily derives the complementary relation:

$$\vec{a}_m(\vec{Q}) = \frac{\nu}{N} \sum_p \vec{\eta}_{p,m} e^{-i\vec{Q}\cdot(\vec{X}_p + \vec{\rho}_m)} \tag{1.106}$$

This relation is obtained by multiplying the expansion (Eq. 105) for $\vec{\eta}_{p,m}$ with $e^{-i\vec{Q}_1\cdot\vec{X}_p}$, and summing over the Bravais cells. It then turns out that the only contribution arises if $\vec{Q}_1 - \vec{Q}$ is a reciprocal lattice vector. If $\vec{Q}$ and $\vec{Q}_1$ are restricted to the first Brillouin zone, this is only possible for $\vec{Q}_1 = \vec{Q}$.

The Lagrangian of the system:

$$L = \frac{M}{2} \sum_{p,m} \left(\frac{\partial \vec{\eta}_{p,m}}{\partial t}\right)^2 - \delta E^{tot} \tag{1.107}$$

then gives rise to the equation of motion:

$$M \frac{\partial^2 \vec{\eta}_{p,m}}{\partial t^2} + \frac{\partial(\delta E^{tot})}{\partial \vec{\eta}_{p,m}} = 0 \tag{1.108}$$

In looking for oscillations with frequency $\omega_{\vec{Q}}$ for the collective coordinates $\vec{a}_m(\vec{Q})$, one readily obtains the equation:

$$\omega_{\vec{Q}}^2 \vec{a}_m(\vec{Q}) = \frac{\nu}{NM} \sum_p e^{-i\vec{Q}\cdot(\vec{X}_p + \vec{\rho}_m)} \frac{\partial(\delta E^{tot})}{\partial \vec{\eta}_{p,m}} \tag{1.109}$$

The derivatives of the energy with respect to the displacements only involve the derivatives of the structure factors:

$$\frac{\partial S^{(2)}(\vec{G})}{\partial \vec{\eta}_{p,m}} = -\frac{1}{N}\vec{G}(\vec{G}.\vec{\eta}_{p,m})e^{-i\vec{G}.\vec{\rho}_m} \qquad (1.110)$$

$$\frac{\partial S^{(1)}(-\vec{q})S^{(1)}(\vec{q}+\vec{G})}{\partial \vec{\eta}_{p,m}}$$

$$= \frac{1}{N^2}\sum_{p_1,m_1} \vec{q}\big((\vec{q}+\vec{G}).\vec{\eta}_{p_1,m_1}\big)e^{i\vec{q}.(\vec{X}_p+\vec{\rho}_m)}e^{-i\vec{q}.(\vec{X}_{p_1}+\vec{\rho}_{m_1})}e^{-i\vec{G}.\vec{\rho}_{m_1}}$$

$$+ \frac{1}{N^2}\sum_{p_1,m_1} (\vec{q}+\vec{G})(\vec{q}.\vec{\eta}_{p_1,m_1})e^{-i\vec{G}.\vec{\rho}_m}e^{-i\vec{q}.(\vec{X}_p+\vec{\rho}_m)}e^{i\vec{q}.(\vec{X}_{p_1}+\vec{\rho}_{m_1})}$$

$$(1.111)$$

where factors of the form $e^{i\vec{G}.\vec{X}_p}$, which are 1, have been omitted. Using the expression (Eq. 106) for $\vec{a}_m(\vec{Q})$, one observes that:

$$\frac{\nu}{NM}\sum_{p} e^{-i\vec{Q}.(\vec{X}_p+\vec{\rho}_m)}\frac{\partial S^{(2)}(\vec{G})}{\partial \vec{\eta}_{p,m}} = -\frac{1}{NM}\vec{G}(\vec{G}.\vec{a}_m(\vec{Q}))e^{-i\vec{G}.\vec{\rho}_m} \qquad (1.112)$$

The Fourier transform of the derivatives of the first order structure factor is slightly more involved. The sum over the N/ν Bravais cell positions $\vec{X}_p$ only gives contributions if $\vec{Q}-\vec{q}$ is a reciprocal lattice vector in the first term. Similarly, in the second term $\vec{Q}+\vec{q}$ has to be a reciprocal lattice vector:

$$\frac{\nu}{NM}\sum_{p} e^{-i\vec{Q}.(\vec{X}_p+\vec{\rho}_m)}\frac{\partial S^{(1)}(-\vec{q})S^{(1)}(\vec{q}+\vec{G})}{\partial \vec{\eta}_{p,m}}$$

$$= \delta_{\vec{q},\vec{Q}-\vec{G}_1}\frac{1}{MN^2}\sum_{p_1,m_1}(\vec{Q}-\vec{G}_1)\big((\vec{Q}+\vec{G}-\vec{G}_1).\vec{\eta}_{p_1,m_1}\big)$$

$$\times e^{-i\vec{G}_1.\vec{\rho}_m}e^{-i\vec{Q}.(\vec{X}_{p_1}+\vec{\rho}_{m_1})}e^{i(\vec{G}_1-\vec{G}).\vec{\rho}_{m_1}}$$

$$+ \delta_{\vec{q},\vec{G}_1-\vec{Q}}\frac{1}{MN^2}\sum_{p_1,m_1}(\vec{G}+\vec{G}_1-\vec{Q})\big((\vec{G}_1-\vec{Q}).\vec{\eta}_{p_1,m_1}\big)$$

$$\times e^{-i(\vec{G}_1+\vec{G}).\vec{\rho}_m}e^{i\vec{G}_1.\vec{\rho}_{m_1}}e^{-i\vec{Q}.(\vec{X}_{p_1}+\vec{\rho}_{m_1})}$$

$$(1.113)$$

In the summations over the displacements $\vec{\eta}_{p_1,m_1}$, one now recognizes the collective coordinates $\vec{a}_{m_1}(\vec{Q})$:

$$\frac{\nu}{NM}\sum_p e^{-i\vec{Q}\cdot(\vec{X}_p+\vec{p}_m)}\frac{\partial S^{(1)}(-\vec{q})S^{(1)}(\vec{q}+\vec{G})}{\partial\eta_{p,m}}$$

$$=\delta_{\vec{q},\vec{Q}-\vec{G}_1}\frac{1}{MN\nu}\sum_{m_1}(\vec{G}_1-\vec{Q})\big((\vec{G}_1-\vec{G}-\vec{Q})\cdot\vec{a}_{m_1}(\vec{Q})\big)e^{-i\vec{G}_1\cdot\vec{p}_m}e^{i(\vec{G}_1-\vec{G})\cdot\vec{p}_{m_1}}$$

$$+\delta_{\vec{q},\vec{G}_1-\vec{Q}}\frac{1}{MN\nu}\sum_{m_1}(\vec{G}+\vec{G}_1-\vec{Q})\big((\vec{G}_1-\vec{Q})\cdot\vec{a}_{m_1}(\vec{Q})\big)e^{-i(\vec{G}_1+\vec{G})\cdot\vec{p}_m}e^{i\vec{G}_1\cdot\vec{p}_{m_1}}$$

$$\tag{1.114}$$

Note that the term involving the derivative of $|S^{(1)}(\vec{q})|^2$ is readily obtained from the expression above by putting $\vec{G}$ equal to zero.

Combining terms, we then end up with the matrix equation:

$$\omega_{\vec{Q}}^2 a_{m_1,\alpha}(\vec{Q})=\sum_{m_2,\beta}D_{m_1,\alpha;m_2,\beta}(\vec{Q})a_{m_2,\beta}(\vec{Q})\tag{1.115}$$

where α and β denote Cartesian components, and where the dynamical matrix is derived with straightforward algebra:

$$D_{m_1,\alpha;m_2,\beta}(\vec{Q})=$$

$$-\delta_{m_1,m_2}\frac{1}{M}\sum_{\vec{G}}\left[\frac{4\pi Z^2 e^2}{\Omega_0 G^2}S^{(0)}(-\vec{G})+\Omega_0 v_{\vec{G}}^0 n_{-\vec{G}}\right]G_\alpha G_\beta e^{-i\vec{G}\cdot\vec{p}_{m_1}}$$

$$+\frac{1}{\nu M}\sum_{\vec{G}_1}\frac{4\pi Z^2 e^2}{\Omega_0|\vec{Q}-\vec{G}_1|^2}(\vec{G}_1-\vec{Q})_\alpha(\vec{G}-\vec{Q})_\beta e^{-i\vec{G}_1\cdot(\vec{p}_{m_1}-\vec{p}_{m_2})}$$

$$+\frac{\Omega_0}{2\nu M}\sum_{\vec{G},\vec{G}_1}\frac{|\vec{Q}-\vec{G}_1|^2}{4\pi e^2}\Big[v_{-\vec{Q}+\vec{G}_1}^0 v_{\vec{Q}-\vec{G}_1+\vec{G}}^0\big(\epsilon^{-1}(\vec{Q}-\vec{G}_1,\vec{Q}-\vec{G}_1+\vec{G})-\delta_{\vec{G},0}\big)$$

$$\times(\vec{G}_1-\vec{Q})_\alpha(\vec{G}_1-\vec{G}-\vec{Q})_\beta e^{-i\vec{G}_1\cdot\vec{p}_{m_1}}e^{i(\vec{G}_1-\vec{G})\cdot\vec{p}_{m_2}}$$

$$+v_{\vec{Q}-\vec{G}_1}^0 v_{\vec{G}_1+\vec{G}-\vec{Q}}^0\big(\epsilon^{-1}(\vec{G}_1-\vec{Q},\vec{G}_1+\vec{G}-\vec{Q})-\delta_{\vec{G},0}\big)$$

$$\times(\vec{G}_1+\vec{G}-\vec{Q})_\alpha(\vec{G}_1-\vec{Q})_\beta e^{-i(\vec{G}_1+\vec{G})\cdot\vec{p}_{m_1}}e^{i\vec{G}_1\cdot\vec{p}_{m_2}}\Big]$$

$$\tag{1.116}$$

This expression is quite complicated, but can substantially be simplified by noting that the sum or the difference of two reciprocal lattice vectors is still a reciprocal lattice vector. The last term with the double summation can be made more transparant if $\vec{G}$ in its first contribution is replaced by $\vec{G}_1-\vec{G}_2$, and by $\vec{G}_2-\vec{G}_1$ in the last contribution, with subsequently an interchange of $\vec{G}_1$ and $\vec{G}_2$.

$$D_{m_1,\alpha;m_2,\beta}(\vec{Q}) =$$

$$-\,\delta_{m_1,m_2}\frac{1}{M}\sum_{\vec{G}}\left[\frac{4\pi Z^2 e^2}{\Omega_0 G^2}S^{(0)}(-\vec{G}) + \Omega_0 v^0_{\vec{G}}n_{-\vec{G}}\right]G_\alpha G_\beta e^{-i\vec{G}\cdot\vec{p}_{m_1}}$$

$$+\,\frac{1}{\nu M}\sum_{\vec{G}_1}\frac{4\pi Z^2 e^2}{\Omega_0|\vec{Q}-\vec{G}_1|^2}(\vec{G}_1-\vec{Q})_\alpha(\vec{G}_1-\vec{Q})_\beta e^{-i\vec{G}_1\cdot(\vec{p}_{m_1}-\vec{p}_{m_2})}$$

$$+\,\frac{\Omega_0}{2\nu M}\sum_{\vec{G}_1,\vec{G}_2}v^0_{\vec{G}_1-\vec{Q}}v^0_{\vec{Q}-\vec{G}_2}(\vec{G}_1-\vec{Q})_\alpha(\vec{G}_2-\vec{Q})_\beta e^{-i\vec{G}_1\cdot\vec{p}_{m_1}}e^{i\vec{G}_2\cdot\vec{p}_{m_2}} \qquad (1.117)$$

$$\times\left[\frac{|\vec{Q}-\vec{G}_1|^2}{4\pi e^2}\left(\epsilon^{-1}(\vec{Q}-\vec{G}_1,\vec{Q}-\vec{G}_2)-\delta_{\vec{G}_1,\vec{G}_2}\right)\right.$$

$$\left.+\frac{|\vec{Q}-\vec{G}_2|^2}{4\pi e^2}\left(\epsilon^{-1}(\vec{G}_2-\vec{Q},\vec{G}_1-\vec{Q})-\delta_{\vec{G}_1,\vec{G}_2}\right)\right]$$

From the symmetry relations for the dielectric matrix (Eq. **77**), both terms in the double sum are equal. Furthermore, in the first term of this expression one can omit the term in $\vec{G}=0$, since in this limit the electron contribution cancels against the ion contribution. Considering then the lattice terms and the electron terms separately, one obtains:

$$D_{m_1,\alpha;m_2,\beta}(\vec{Q}) = D^{ion}_{m_1,\alpha;m_2,\beta}(\vec{Q}) + D^{el}_{m_1,\alpha;m_2,\beta}(\vec{Q}) \qquad (1.118)$$

In the ion contribution we write the structure factor explicitly:

$$D^{ion}_{m_1,\alpha;m_2,\beta}(\vec{Q}) = \frac{1}{\nu M}\frac{4\pi Z^2 e^2}{\Omega_0}\left[\sum_{\vec{G}}\frac{(\vec{G}-\vec{Q})_\alpha(\vec{G}-\vec{Q})_\beta}{|\vec{Q}-\vec{G}|^2}e^{-i\vec{G}\cdot(\vec{p}_{m_1}-\vec{p}_{m_2})}\right.$$

$$\left.-\,\delta_{m_1,m_2}\sum_{\vec{G}\neq0}\frac{G_\alpha G_\beta}{G^2}\sum_{m_3}e^{-i\vec{G}\cdot(\vec{p}_{m_1}-\vec{p}_{m_3})}\right] \qquad (1.119)$$

which can be calculated very fast with the Ewald-Fuchs procedure, outlined in Appendix B. The electron contribution becomes:

$$D^{el}_{m_1,\alpha;m_2,\beta}(\vec{Q}) = \frac{\Omega_0}{\nu M}\sum_{\vec{G}_1,\vec{G}_2}v^0_{\vec{G}_1-\vec{Q}}v^0_{\vec{Q}-\vec{G}_2}\frac{|\vec{Q}-\vec{G}_1|^2}{4\pi e^2}(\vec{G}_1-\vec{Q})_\alpha(\vec{G}_2-\vec{Q})_\beta$$

$$\times\,e^{-i\vec{G}_1\cdot\vec{p}_{m_1}}e^{i\vec{G}_2\cdot\vec{p}_{m_2}}\left(\epsilon^{-1}(\vec{Q}-\vec{G}_1,\vec{Q}-\vec{G}_2)-\delta_{\vec{G}_1,\vec{G}_2}\right)$$

$$-\,\delta_{m_1,m_2}\frac{\Omega_0}{\nu M}\sum_{\vec{G}\neq0}\nu v^0_{\vec{G}}n_{-\vec{G}}G_\alpha G_\beta e^{-i\vec{G}\cdot\vec{p}_{m_1}}$$

$$(1.120)$$

This dielectric approach has been used by Van Camp, Van Doren and Devreese[12] to calculate the first ab initio phonon spectrum in Silicon.

Chapter 2

DYNAMICAL EXCHANGE EFFECTS IN THE DIELECTRIC FUNCTION OF JELLIUM

2.1 General considerations

The calculation of many properties of simple metals is based on the "jellium model". In this model, one merely studies the interaction among the electrons themselves, whereas the lattice of positive metal ions is replaced by a rigid uniform background. The large amount of research on this hypothetical system is not only due to its conceptual simplicity. The main reason for the persisting interest is the fact that in many metals, the conduction electrons approximately have a homogeneous distribution in space. Futhermore, as indicated in the previous chapter, many theories for inhomogeneous systems, use the jellium model as a starting point for further investigations. A basic quantity for the study of the jellium model is the frequency- and wave vector-dependent longitudinal dielectric function $\epsilon(q, \omega)$, which not only allows to study the dielectric response, but also to calculate several other properties, such as e.g. the ground state energy, the dynamic structure factor $S(q, \omega)$, the pair correlation function $g(r)$, ...

Since the pioneering work[1] of Lindhard, Bohm, Pines, Nozières, ..., the dielectric function in the so-called Random Phase Approximation (RPA), has been the standard reference basis for further improvements. In this approximation, one calculates the response of the electrons subjected to an external field, assuming that each electron moves in the Hartree field of all the other electrons. In the long-wavelength-limit, the RPA succeeds basically in explaining the dispersion of the energy of the collective excitations, and simultaneously to combine the ideas of screening, collective excitations and single-particle excitations. However, despite its merits, the RPA is less satisfactory whenever short-range correlations are important. For instance, energy loss measurements of fast electrons and X-ray scattering experiments revealed appreciable deviations in the structure factor from the RPA predictions[2]. This failure of the RPA in the short-range description of the electron-electron interactions is also reflected in the fact that the pair correlation function $g(r)$, as obtained from the RPA dielectric function, becomes negative near the origin[3] for metallic densities.

A large variety of approximations has been proposed to improve upon the RPA. In these approximations one tries to calculate the "local field correction" $G(q, \omega)$, occurring in the dielectric function which is usually written in the form:

$$\epsilon(q, \omega) = 1 + \frac{Q_0(q, \omega)}{1 - G(q, \omega)Q_0(q, \omega)} \tag{2.1}$$

In this expression, $Q_0(q, \omega)$ is the Lindhard function:

$$Q_0(q, \omega) = \frac{4\pi e^2}{q^2} \int d^3p \, \frac{N_{\vec{q}}(\vec{p})}{\omega^+ - \vec{p}.\vec{q}/m} \qquad (2.2)$$

where $\omega^+ = \omega + i\delta$ accounts for the adiabatic switching of external perturbations, and $N_{\vec{q}}(\vec{p})$ is a geometrical factor, related to the equilibrium distribution function $f^0(p)$ of the electron gas:

$$N_{\vec{q}}(\vec{p}) = \frac{1}{\hbar}\left[f^0(\vec{p} + \frac{\hbar\vec{q}}{2}) - f^0(\vec{p} - \frac{\hbar\vec{q}}{2}) \right] \qquad (2.3)$$

where:

$$f^0(p) = \begin{array}{ll} \dfrac{2}{(2\pi\hbar)^3} & \text{if } p \le p_F \\[2ex] 0 & \text{if } p > p_F \end{array} \qquad (2.4)$$

The Lindhard function $Q_0(q, \omega)$ and the RPA dielectric function are related via:

$$\epsilon_{RPA}(q, \omega) = 1 + Q_0(q, \omega) \qquad (2.5)$$

The function $G(q, \omega)$ is intended to describe the exchange and correlation potential on each electron, due to the other electrons. Originally, not much effort was done to include the frequency dependence in $G(q, \omega)$, and most studies concentrated on **static** approximations $G(q)$, which were nevertheless often used in the dynamic dielectric function[4].

The interest in the explicit frequency dependence of $G(q, \omega)$ was rather exceptional[5], and restricted to some limiting cases.

In the last decade however, the dynamics of the exchange and correlation hole became an important topic, in view of the more accurate measurements of the dynamical structure factor at large wave vector, and because sum rules and causality arguments revealed that a static treatment of $G(q, \omega)$ necessarily leads to theoretical inconsistencies[6,7,8].

A variety of different approaches has been used to derive expressions for $G(q, \omega)$. But not many explicit calculations of the full frequency- and wave vector-dependence have been performed. The present authors have been involved in one of the first explicit frequency dependent calculations[9] of $G(q, \omega)$ This approach was based on the equation of motion for the Wigner distribution function. In the present lecture, the same approach will be followed, because of its physical transparancy, and because it easily reveals the connection between various approximations. Only a detailed discussion of the Mori formalism[10] could not be achieved in terms of the Wigner distribution function. The Mori formalism is particularly useful in the study of the dynamics of a system with **known static properties**, since it guarantees by construction that the relevant sum rules in the high- and low-frequency limits are satisfied. The lack of detailed knowledge of the static correlations in the electron liquid however, complicates the interpretation of the physical significance of the applications[11] of the Mori formalism in this domain.

In the present lecture notes, only a general outline will be given of the formalism which we developed, and of the results obtained. A detailed discussion of the underlying physics, comparisons with other theories and with experiment, and mathematical details can be found elsewhere[12].

2.2 Dynamical exchange decoupling

Consider a gas of electrons, interacting with some scalar potential $\phi(\vec{r}, t)$. In the absence of magnetic and spin-spin interactions, the hamiltonian is given by:

$$
\begin{aligned}
H = &-\frac{\hbar^2}{2m} \sum_\sigma \int d^3r\, \psi_\sigma^\dagger(\vec{r}) \Delta \psi_\sigma(\vec{r}) \\
&+ \sum_\sigma \int d^3r\, \psi_\sigma^\dagger(\vec{r}) e\Phi(\vec{r}, t)\psi_\sigma(\vec{r}) \\
&+ \frac{1}{2} \sum_{\sigma\sigma'} \int d^3r \int d^3r'\, \psi_\sigma^\dagger(\vec{r})\psi_{\sigma'}^\dagger(\vec{r}') \frac{e^2}{|\vec{r} - \vec{r}'|} \psi_{\sigma'}(\vec{r}')\psi_\sigma(\vec{r})
\end{aligned}
\tag{2.6}
$$

where $\Phi(\vec{r}, t)$ is the sum of the interaction potential with the background of ions, and the external applied potential $\phi(\vec{r}, t)$. The annihilation and creation field operators $\psi_\sigma(\vec{r}, t)$ and $\psi_\sigma^\dagger(\vec{r}, t)$ for particles of spin σ obey the standard anti-commutation rules for fermion fields.

The Wigner distribution function $f_\sigma(\vec{p}, \vec{R}, t)$ is the quantum analogue of the classical Boltzmann distribution function for particles of spin σ with momentum $\vec{p}$ in position $\vec{R}$ at time t. It is defined as[13]:

$$
f_\sigma(\vec{p}, \vec{R}, t) = \frac{1}{(2\pi\hbar)^3} \int d^3r\, e^{-i\vec{p}\cdot\vec{r}/\hbar} < \psi_\sigma^\dagger(\vec{R} - \frac{\vec{r}}{2})\psi_\sigma(\vec{R} + \frac{\vec{r}}{2}) >_t
\tag{2.7}
$$

where the brackets $< \cdots >_t$ denote expectation values at time t. Similar to classical mechanics, the density and current density are obtained from:

$$n(\vec{R}, t) = \sum_\sigma \int d^3p\, f_\sigma(\vec{p}, \vec{R}, t)$$

$$\vec{j}(\vec{R}, t) = \sum_\sigma \int d^3p\, \frac{\vec{p}}{m} f_\sigma(\vec{p}, \vec{R}, t) \tag{2.8}$$

The equation of motion for $f_\sigma(\vec{p}, \vec{R}, t)$ is obtained by calculating the commutator of $\psi_\sigma^\dagger(\vec{x}')\psi_\sigma(\vec{x})$ with the hamiltonian. As a result, one obtains an expression containing the two-particle Wigner distribution function:

$$f^{(2)}_{\sigma_1,\sigma_2}(\vec{p}_1, \vec{p}_2; \vec{R}_1, \vec{R}_2; t) = \frac{1}{(2\pi\hbar)^6} \int d^3r_1 \int d^3r_2 e^{-i\vec{p}_1 \cdot \vec{r}_1/\hbar} e^{-i\vec{p}_2 \cdot \vec{r}_2/\hbar}$$

$$< \psi_{\sigma_1}^\dagger(\vec{R}_1 - \frac{\vec{r}_1}{2})\psi_{\sigma_2}^\dagger(\vec{R}_2 - \frac{\vec{r}_2}{2})\psi_{\sigma_2}(\vec{R}_2 + \frac{\vec{r}_2}{2})\psi_{\sigma_1}(\vec{R}_1 + \frac{\vec{r}_1}{2}) >_t \tag{2.9}$$

One can proceed by deriving the equation of motion for the two-particle distribution function and so on. Continuing this way, one finds the well-known BBGKY hierarchy, in which distribution functions of higher order enter successively. A detailed study of the two-particle Wigner distribution function has e.g. been made by Niklasson[14].

The main problem is to find good approximations to break this hierarchy, in order to find the Wigner distribution function and subsequently the induced electron density. We adopt the Hartree-Fock decoupling, written schematically as:

$$< \psi_1^\dagger\psi_2^\dagger\psi_3\psi_4 > \approx < \psi_1^\dagger\psi_4 >< \psi_2^\dagger\psi_3 > - < \psi_1^\dagger\psi_3 >< \psi_2^\dagger\psi_4 > \tag{2.10}$$

Considering then the Fourier transform of $f_\sigma(\vec{p}, \vec{R}, t) - f_\sigma^0(p)$ with respect to space and time, which we denote by $f_\sigma(\vec{p}, \vec{q}, \omega)$, one obtains the equation of motion:

$$f_\sigma(\vec{p}, \vec{q}, \omega) = \frac{-\frac{1}{2}N_{\vec{p}}(\vec{q})U_{\vec{q},\omega} + X_\sigma(\vec{p}, \vec{q}, \omega)}{\omega^+ - \vec{p}.\vec{q}/m} \tag{2.11}$$

where $U_{\vec{q},\omega}$ is the total Hartree potential:

$$U_{\vec{q},\omega} = e\Phi_{\vec{q},\omega} + \frac{4\pi e^2}{q^2} \sum_\sigma \int d^3p\, f_\sigma(\vec{p}, \vec{q}, \omega) \tag{2.12}$$

and where $X_\sigma(\vec{p}, \vec{q}, \omega)$ accounts for the exchange interaction in the electron dynamics:

$$X_\sigma(\vec{p}, \vec{q}, \omega) = \frac{1}{2} \int d^3p' \frac{4\pi e^2 \hbar^2}{|\vec{p} - \vec{p}'|^2} \left[N_{\vec{q}}(\vec{p}) f_\sigma(\vec{p}', \vec{q}, \omega) - N_{\vec{q}}(\vec{p}') f_\sigma(\vec{p}, \vec{q}, \omega) \right]$$

$$(2.13)$$

If one would be able to solve this integral equation for $f_\sigma(\vec{p}, \vec{q}, \omega)$, then the dielectric function with dynamical exchange effects would be obtained from the induced density:

$$n_{\vec{q}, \omega} = \sum_\sigma \int d^3p f_\sigma(\vec{p}, \vec{q}, \omega) \tag{2.14}$$

which is related to the external potential $e\phi_{\vec{q}, \omega}$ through the defining equation for the dielectric function:

$$e\phi_{\vec{q}, \omega} + \frac{4\pi e^2}{q^2} n_{\vec{q}, \omega} = \frac{e\phi_{\vec{q}, \omega}}{\epsilon(q, \omega)} \tag{2.15}$$

2.3 Variational approximation

The exchange term $X_\sigma(\vec{p}, \vec{q}, \omega)$ obviously forms the complicating factor to solve the equation of motion (Eq. 11-13) for $f_\sigma(\vec{p}, \vec{q}, \omega)$. Indeed, if one neglects this exchange term, one readily obtains the Lindhard solution:

$$f_\sigma^L(\vec{p}, \vec{q}, \omega) = -\frac{1}{2} e\phi_{\vec{q}, \omega} \frac{1}{Q_0(q, \omega)} \frac{N_{\vec{q}}(\vec{p})}{\omega^+ - \vec{p}.\vec{q}/m} \tag{2.16}$$

which corresponds to the RPA dielectric function.

Attempting to account for the dynamical exchange effects, neglected in the RPA, we applied a variational technique, which consists in deriving a functional $F[f_\sigma(\vec{p}, \vec{q}, \omega)]$ with the property that the equation of motion (Eq. 11-13) follows from the extremum condition:

$$\frac{\delta F}{\delta f_\sigma(\vec{p}, \vec{q}, \omega)} = 0 \tag{2.17}$$

Taking as a trial function:

$$f_\sigma^{trial}(\vec{p}, \vec{q}, \omega) = f_\sigma^L(\vec{p}, \vec{q}, \omega) \gamma_{\vec{q}, \omega} \tag{2.18}$$

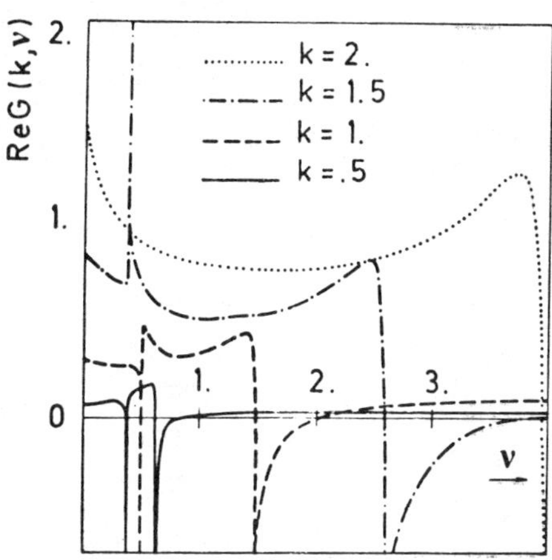

Figure 1. Frequency dependence of the real part of $G(q, \omega)$.
Units: $k = q/2k_F$, $\nu = \hbar\omega/2E_F$.

with $\gamma_{\vec{q},\omega}$ independent of the momentum $\vec{p}$, this extremum condition resulted in an algebraic equation for $\gamma_{\vec{q},\omega}$, which is readily solved [Ref. 9]. Given the variational expression for $\gamma_{\vec{q},\omega}$, an explicit expression for f^{trial} is obtained, from which the induced density and hence the dielectric function follows. The result, which we obtained for $G(q, \omega)$ is:

$$G(q,\omega) = \frac{4\pi e^2}{q^2} \frac{2\pi e^2 \hbar^2}{Q_0(q,\omega)^2} \int d^3p \int d^3p' \frac{N_{\vec{q}}(\vec{p})N_{\vec{q}}(\vec{p}')}{|\vec{p}-\vec{p}'|^2}$$
$$\times \frac{1}{\omega^+ - \vec{p}.\vec{q}/m}\left[\frac{1}{\omega^+ - \vec{p}'.\vec{q}/m} - \frac{1}{\omega^+ - \vec{p}.\vec{q}/m}\right]$$

(2.10)

To the best of our knowledge, this equation was the first expression for $G(q, \omega)$, which was evaluated with respect to both the wave vector and the frequency dependence [Ref. 9]. Some results are shown in Fig. 1, where the real part of $G(q, \omega)$ is plotted as a function of frequency (expressed in units proportional to the Fermi-frequency) for different values of the wave vector (in units of twice the Fermi wave vector).

Although Eq. 19 is a rather complicated sixfold integral, we could analytically reduce it to a tractable double integral. Details on the numerical and analytical techniques, figures and tables, and a more extensive discussion

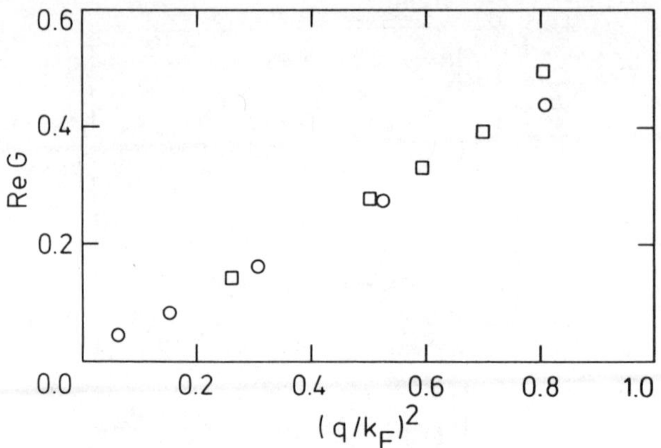

Figure 2. Fitted values of the real part of $G(q, \omega)$ for different wave vectors.
Circles: fitted data for Na $(r_s = 3.96)$.
Squares: fitted data for Al $(r_s = 2.07)$.

on the physical implications, can be found elsewhere [Ref. 12]. The main results are:

1) The **continuity equation**, and all other **sum rules** checked up to now are exactly satisfied;

2) $G(q, \omega)$ is a **universal** function of q/k_F and ω/E_F **for all densities**. This seems to be confirmed by experimental energy loss spectra in aluminum and sodium[15] as illustrated in Fig. 2.

3) The **plasmon energy at finite wave vector** is substantially **lowered** as compared to the RPA. As shown in Fig. 3, this appreciably improves the agreement with experimental data in aluminum.

2.4 Comparison with various approximations

The variational result (Eq. 19) was later obtained by Tripathy and Mandal[16] with a quite different method. They applied the decoupling (Eq. 10) in the equation of motion for the double-time-retarded commutator of the charge density fluctuation operators, and imposed conservation of frequency moments to all orders in the Hartree-Fock approximation for the static properties. Their method is an extension of an earlier approximation[17] by Toigo and Woodruff, who imposed only conservation of the first frequency

42

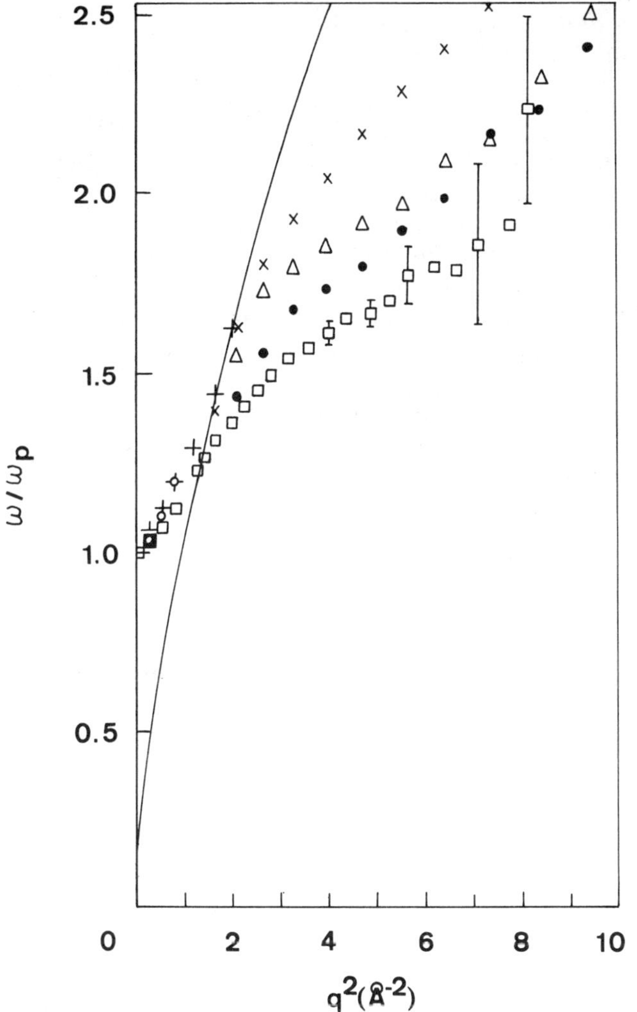

Figure 3. Plasmon dispersion in Al versus q^2.
Squares: experiment (Batson, Chen, Silcox, Phys. Rev. Lett. **37**, 937 (1976).
Triangles: from P. Vashishta and K. S. Singwi, Phys. Rev. **B6**, 875 (1972).
Crosses: maxima in the RPA dynamical structure factor.
Black dots: including dynamical exchange effects.

moment. Unfortunately, the explicit evaluations of the dielectric function by Tripathy and Mandal turned out to be seriously in error[18].

The integral equation (Eq. 11-13) has also been studied by an iterative approach to first order in the exchange effects [Ref. 5], starting from the Lindhard distribution (Eq. 16), and treating the exchange contribution X_σ

(Eq. 13) as a perturbation. However, no explicit evaluation was performed. The result obtained to first order is:

$$\epsilon_{iterative}(q, \omega) = 1 + Q_0(q, \omega)[1 + G(q, \omega)Q_0(q, \omega)] \qquad (2.20)$$

Comparing this iterative result with the dielectric function (Eq. 1), one immediately observes that the iterative dielectric function is the expansion of the variational dielectric function to first order in the exchange effects. It is interesting to note that the **static** limit of the iterative dielectric function has been evaluated by Geldart and Taylor[19], by considering the diagrams for the proper polarizability to first order in the electron-electron interaction. This diagrammatic expansion was evaluated **dynamically**[20] by Holas, Aravind and Singwi, again yielding the iterative dielectric function. The main problem of this approximation to first order is the fact that the imaginary part of the dielectric function becomes negative[21] for a finite range of wave vectors and frequencies.

A related approach should be mentioned, which was developed by Brener and Fry[22], and which consists in an iterative treatment of a set of self-consistent equations[23] for the self-energy, polarization and Green's function. Their dielectric function can be obtained from the iteration scheme mentioned above, if one starts from the Lindhard distribution function (Eq. 16) with the modification that the Hartree-Fock self-energy is included in the energy denominators. A disadvantage of this approach is that in each step of the iteration **the continuity equation is violated**. One therefore should pursue this iteration scheme untill full convergence is reached.

A remarkable point in our treatment of dynamical exchange effects, is the fact that $G(q, \omega)$ (Eq. 19) is logarithmically divergent at the boundaries of the particle-hole continuum. The physical significance of these singularities is not quite clear. Recently it has been shown[24] that a generalization of the trial distribution function (Eq. 18), including momentum corrections to order p^2, eliminates these logarithmic singularities, and leaves $G(q, \omega)$ essentially unaltered outside the singular regions. This might be an indication that the singularities are an artefact of the theory.

2.5 Concluding remarks

The variational solution of the decoupled equation of motion (Eq. 11-13) for the Wigner distribution function might serve as a starting point for further studies of exchange and correlation in the dielectric function. Its connection with several other approximations has been examined, showing that many of them are particular cases or additional approximations to this variational approach. The improvement upon the RPA from dynamical exchange effects, and the fact that all checked sum rules are satisfied, gives confidence in the variational and related approaches, even in the study of the equation of motion for the two-particle distribution function [Ref. 14].

Up to now, very few conclusions can be drawn concerning the dynamic correlations beyond the dynamical exchange decoupling. Even in the static limit, most recent studies make the approximation that $G(q, \omega)$ is frequency-independent, although dynamical sum rules are implicitly imposed[25].

Also the numerical methods up to now do not include the frequency dependence. For instance, the calculation of the dielectric function by Lantto et al.[26] using a Jastrow variational many-body wave function[27], explicitly assumes a static approximation $G(q)$ for $G(q, \omega)$. One of the consequences is that the large wave vector limit of this approximation for $G(q)$ is not correctly related to the pair correlation function at the origin. As soon as frequency dependent sum rules are used, one can hardly avoid the unpleasant complication of accepting the frequency dependence in $G(q, \omega)$, even in numerical treatments of the static limit.

Chapter 3

THE PSEUDOPOTENTIAL CONCEPT

3.1 Basic assumptions

Not only in solid state physics, but also in the physics of atoms and molecules, it is well known that the chemical bond between atoms is essentially determined by the valence electrons of the atoms, i.e. the electrons in the outer shells. The electrons in the fully occupied shells, i.e. the core electrons, essentially do not contribute to the chemical bond. The periodicity in the table of the chemical elements is based on the fact that the core electrons are (to a first approximation) chemically inactif.

This does not mean that the core electrons are unimportant in atomic, molecular and solid state physics. They determine the actual states and energies of the valence electrons. Thus, in a quantum mechanical description of the electronic structure of atoms, solids or molecules, these core electrons have to be included in the hamiltonian. Consider e.g. an atom of charge $Z_a e$, with M electrons. The hamiltonian then contains a sum of M kinetic terms, M attractive terms between the nucleus and the electrons, and $M(M-1)/2$ terms for the Coulomb repulsion of the electrons:

$$H = \sum_{j=1}^{M} \frac{p_j^2}{2m} - Z_a e^2 \sum_{j=1}^{M} \frac{1}{r_j} + \frac{e^2}{2} \sum_{j=1}^{M} \sum_{i \neq j}^{M} \frac{1}{|\vec{r}_i - \vec{r}_j|} \qquad (3.1)$$

(In this hamiltonian, no external fields are included. Spin-orbit coupling and relativistic corrections are neglected).

As is well known, the Schrödinger equation with only one electron in the Coulomb field of a nucleus can be solved. For He, with 2 electrons, the ground state energy can be calculated variationally with reasonable accuracy by relatively simple analytical techniques. E.g. taking as a trial wave function a product of two hydrogen-like wave functions in the Hartree approximation already gives the ionisation energy correct to within 6%. For Li, with 3 electrons, a variational treatment with analytical methods becomes already quite complicated. And in order to solve the Schrödinger equation for atoms with more electrons, one has to rely on numerical techniques. For atoms with many electrons (say 30 or more), also the numerical techniques become quite involved.

If one has the programs to solve the Schrödinger equation in atoms, one can think about a quantum mechanical description of simple chemical systems. In NaCl e.g., one already has to deal with 28 electrons, and only 8 out of them determine the chemical bond. In more complex systems, the total number of electrons becomes fairly large, and the number of valence electrons is substantially lower. This is a rather frustrating situation, if one realizes that the core electrons are essentially chemically inactif, and remain intimately bound to their nucleus.

Given the relative stability of the core electrons, irrespective of the environment of the atom, one of course tries to avoid the calculation of the core states over and over again, especially in calculations for complex molecules and solids. One of the important techniques to take advantage of this core stability is the **pseudopotential method**. In the last decennia, it has been applied to a variety of chemistry and solid state problems, e.g. the study of the electronic structure of atoms[1,2,3], molecules[4,5,6,7,8] and solids[9,10], the chemical bond at surfaces[11,12,13,14], structural stability and phase-transitions[15,16,17], electron-phonon coupling in superconductivity[18],...

The pseudopotential concept has strongly evolved together with its applications, and so many models have been developed, that it is not always easy to recover the basic ideas and assumptions.

A first approximation is the **one-electron** approximation. One assumes that the electrons in a system (atom, molecule or solid), can be described by a one-electron Schrödinger equation:

$$H|\Psi_n> = (T + V)|\Psi_n> = E_n|\Psi_n> \qquad T = -\frac{\hbar^2}{2m}\Delta \qquad (3.2)$$

where T is the operator for the kinetic energy, and V the effective potential acting upon the electron. This approximation has been discussed in a previous section. The potential of course contains the Coulomb attraction from the nuclei of all the atoms, and the interaction with all the other electrons. In order to determine this effective potential, one therefore needs the electron density.

46

A first problem which one encounters is thus the need to solve a set of Schrödinger equations *self-consistently*. The potential is required to calculate the wave functions and the electron density. The density in turn determines the effective potential.

A second problem is related to the *exclusion principle*: two electrons are not allowed to occupy the same state.

If we indicate the valence electrons by the indices n, and use the indices c for the core electrons, these core electrons also have to satisfy the one-electron Schrödinger equation:

$$H|\Psi_c> = (T + V)|\Psi_c> = E_c|\Psi_c> \qquad (3.3)$$

(Note that the potential in general will be different from the potential in a free ion.) All these states have to satisfy the ortogonality relations:

$$< \Psi_c|\Psi_n> = 0 \qquad < \Psi_c|\Psi_{c'}> = \delta_{c,c'} \qquad < \Psi_n|\Psi_{n'}> = \delta_{n,n'} \qquad (3.4)$$

A second important assumption is the **small-core** approximation, in which one assumes that the core electrons are so strongly bound to the nucleus, that their wave functions are not substantially influenced by the presence of other atoms in the neighborhood. As long as one is limiting the calculations to atoms without d- or f-electrons, this approximation turns out to be quite adequate. This means that the core states in a molecule or a crystal can be assumed to be the same as in the free atom. However, if one has to deal with d- or f-electrons, this assumption might become invalid, and it is necessary to recalculate them, considering them on the same level as the valence electrons. (Note that this small-core approximation does not mean that the core energies are unchanged. They are assumed to change by a constant, determined by the potential from the environment at the atomic position.)

3.2 The pseudo Schrödinger equation

Given the core states $|\Psi_c>$, the orthogonalisation condition for the valence electrons at first glance constitutes a complication. Indeed, this condition imposes strong limitations on the valence wave functions inside the core region. On the other hand, for the study of the chemical bond, the region of interest is outside this core region.

The basic idea of the pseudopotential method is then to take advantage of this orthogonality [19]. By adding a set of core states $|\Psi_c>$ with arbitrary coefficients $< \Psi_c|\phi_n>$ to the wave function $|\Psi_n>$, a new function $|\phi_n>$ is obtained, which can be made much smoother in the core region than the wave function, which is then given by:

$$|\Psi_n> = |\phi_n> - P|\phi_n> \tag{3.5}$$

where P is the projection operator on the core states. Note that $|\Psi_n>$ becomes an orthogonalized plane wave[20] (OPW) if $|\phi_n>$ is replaced by a plane wave.

$$P = \sum_c |\Psi_c> < \Psi_c| \tag{3.6}$$

The orthogonality condition $< \Psi_n|\Psi_c> = 0$ is now automatically satisfied. Inserting the wave function in the one-particle Schrödinger equation, one obtains an equation for the state $|\phi_n>$:

$$(T + W_n)|\phi_n> = E_n|\phi_n>$$
$$W_n = V - HP + E_nP = V - \sum_c |\Psi_c> < \Psi_c|(E_c - E_n) \tag{3.7}$$

This **pseudo Schrödinger equation**, with a **pseudo hamiltonian** $T+W_n$ and a **pseudo wave function** $|\phi_n>$ has the same energy eigenvalues as the original Schrödinger equation. The knowledge of the pseudo wave function allows to derive the original wave function if the core states are known. However, the pseudopotential operator W is an **integral** operator because of the projection operator:

$$P|\phi_n> = \sum_c |\Psi_c> \int d^3 r \Psi_c{}^*(\vec{r})\phi_n(\vec{r}) \tag{3.8}$$

A further complication is the fact that this operator contains the energy eigen values E_n, which one is trying to find.

Despite these complications, what one has gained is a rather smooth pseudo wave function, which has a large degree of arbitrariness. Indeed, if one considers the pseudo hamiltonian, acting on a core state:

$$(H-HP+E_nP)|\Psi_c> = H|\Psi_c> -H|\Psi_c> +E_n|\Psi_c> = E_n|\Psi_c> \tag{3.9}$$

one realizes that the *core states are eigenstates of the pseudo hamiltonian*, but *the eigen energies are those of the valence electrons*. This means that the addition of a linear combination of core states to the pseudo wave function has no influence on the valence electron energy. Moreover, such an addition does not influence the wave function $|\Psi_n>$. Indeed, consider a new pseudo wave function $|\phi_n'>$:

$$|\phi_n'> = |\phi_n> + \sum_{c'} b_{c'}|\Psi_{c'}> \tag{3.10}$$

for which one finds:

$$(1 - P)|\phi'_n> = |\Psi_n> = (1 - P)|\phi_n> \qquad (3.11)$$

This arbitrariness has e.g. been exploited by Harrison to construct a pseudo wave function which is as smooth as possible [Ref. 9].

Several refinements and alternative formulations of the pseudopotential concept have been worked out. E.g. the arbitrariness in the pseudo wave function, can be transferred into an **arbitrariness in the pseudopotential**, to which a **unique pseudo wave function** belongs. This leads to the Austin-Heine-Sham pseudopotential[21]. Indeed, any pseudopotential of the form:

$$W^{AHS} = V + \sum_c |\Psi_c><f_c| \qquad (3.12)$$

with an arbitrary $f_c(\vec{r})$, is a **valid pseudopotential**. This means e.g. that the energy dependent factors $E_n - E_c$ in Eq. (3.7) can be replaced by any function one prefers. Also for the pseudo hamiltonian $T + W^{APS}$, the eigenvalues are the valence energies. But for each function $f_c(\vec{r})$ the eigen states $|\phi_n>$ are uniquely defined.

To prove that the eigen values of $(T + W^{AHS})$ are the valence energies, consider an eigen state $|\phi>$ with energy E:

$$(T + W^{AHS})|\phi> = E|\phi>$$

Taking the inner product with a valence state $|\Psi_n>$ gives:

$$<\Psi_n|\phi> E_n = <\Psi_n|\phi> E$$

Therefore $E_n = E$ (in which case the assertion is proven), or $|\phi>$ is orthogonal to $|\Psi_n>$, if $E_n \neq E$. The eigenstate $|\phi>$ is thus orthogonal to all valence states with different energy.

The uniqueness of the eigen state $|\phi_n>$ of $T + W^{AHS}$ with energy E_n is obtained from the expansion:

$$|\phi_n> = |\Psi_n> + \sum_c |\Psi_c><\Psi_c|\phi_n>$$

Taking the inner product of the pseudo Schrödinger equation with some core state $|\Psi_c>$ gives:

$$E_c <\Psi_c|\phi_n> + <f_c|\phi_n> = E_n <\Psi_c|\phi_n>$$

For the special case $|f_c> = (E_n - E_c)|\Psi_c>$, one has to do with the earlier pseudopotential (3.7), with an arbitrariness in the pseudo wave function as discussed before. But in all other cases, one obtains a unique pseudo wave function. This is easily seen if one for instance introduces some complete set of eigen states $|\lambda>$:

$$(E_c - E_n) \sum_\lambda < \Psi_c|\lambda><\lambda|\phi_n> + \sum_\lambda < f_c|\lambda><\lambda|\phi_n> = 0$$

This is a inhomogeneous set of linear equations in the expansion coefficients $<\lambda|\phi_n>$, and the solutions are thus non-zero and unique, except for the special pseudopotential (3.7).

The main result of the pseudopotential method, in its different formulations, is the fact that *the spectrum of the pseudo hamiltonian $T+W$ consists of the valence energies only*, whereas the spectrum of the original hamiltonian consists of both the core energies and the valence energies. This advantage becomes clear if one for instance considers the variational principle of quantum mechanics. With the original hamiltonian one obtains for an arbitrary function $\Psi(\vec{r})$:

$$\frac{< \Psi|T + V|\Psi >}{< \Psi|\Psi >} \geq E_{c1} \tag{3.13}$$

where E_{c1} is the lowest core energy. Applying however the variational principle with the pseudo hamiltonian, one finds for an arbitrary function $\Phi(\vec{r})$:

$$\frac{< \Phi|T + V|\Phi >}{< \Phi|\Phi >} \geq E_{v1} \tag{3.14}$$

where E_{v1} is the lowest valence energy[*]. This is illustrated in Fig. 1.

This means that the pseudopotential $W_n = V - (H - E_n)P$ contains a strongly repulsive interaction. The replacement of the orthogonalisation on the core states by an interaction term in the hamiltonian, in fact leads to the introduction of a repulsive term. This repulsive interaction is limited to the core region, because it results from the projection operator on the core states. Outside the core region, the pseudopotential is equal to the original potential. This is illustrated in Fig. 2, where one of the early pseudopotential calculations[22] calculations for Si is shown.

3.3 Some pseudopotential approaches

It is precisely the repulsive term in the core, which is responsible for the fact that the pseudopotential is an integral operator. Since such operators

---

[*] Some confusion might arise from the term valence energy, which means the energy associated with the valence electrons of the atom; but in a solid the energies of these valence electrons constitute the conduction bands and/or the higher valence bands.

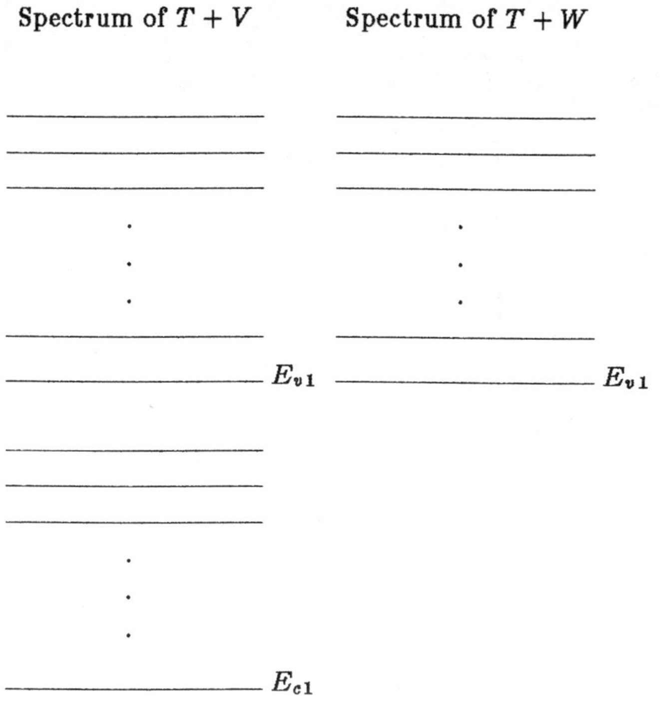

Figure 1. Schematic representation of the eigenvalues
of the hamiltonian and of the pseudo hamiltonian.

are quite difficult to handle in practice, several approximations have been proposed in the literature, to represent this repulsive contribution by a local operator, e.g. the **point-ion** model potential[23].

One of the simplest early approximations is the so-called **empty-core** approximation[24], in which one assumes that the ionic part of the pseudopotential is zero inside a core volume with radius r_c, and becomes the normal Coulomb potential $-Ze^2/r$ for $r > r_c$. The core radius r_c is a fitting parameter. The model can of course easily be extended, by giving it some more adjustable parameters, e.g. the depth of the pseudopotential in the core region can be given a finite value. These pseudopotentials, which rely on fitting procedures to experimental data, are commonly called **empirical model potentials**, and are extremely useful and accurate for many practical purposes. E.g. the empty core pseudopotential reproduces the phonon spectrum of Na to within a few percent[25] if the core radius is determined by a least square procedure, minimizing the difference between the theoretical and the observed spectrum, as is shown in Fig. 3.

However, in the following sections we will not discuss this model ap-

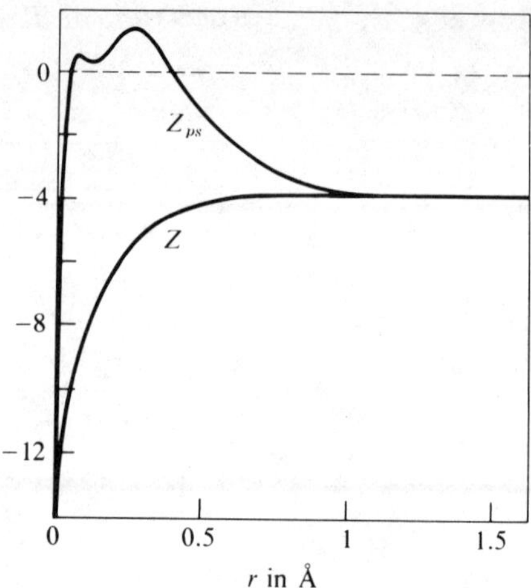

Figure 2. Potential and pseudopotential for the *s* states of a Si^{4+} ion [Ref. 22], expressed in units of e^2/r (i.e. $V(r) = Z(r)e^2/r$, and similarly for Z_{ps}).

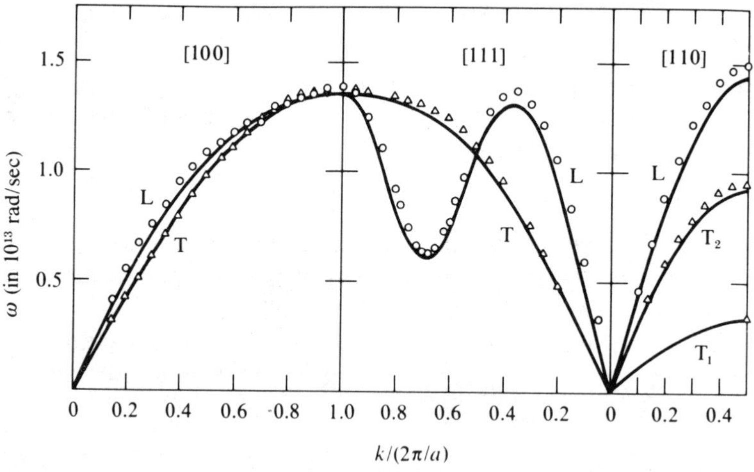

Figure 3. Vibration spectrum of K with the empty-core model for $r_c = 1.13$Å [Ref. 25].

proach any further, since the fitting procedures, although they increase the accuracy, tend to hide the effects and the relative importance of the various physical processes involved. We will rather discuss the line of the **ab-initio** approach, stressing the basic principles in their essential form. In order to see the general trends and backgrounds, (but at the expense of accuracy), complicating mathematical details and numerical problems will sometimes be omitted. (They can be found in the literature).

In the **ab-initio** pseudopotential theory, one essentially tries to take advantage of the arbitrariness in the pseudopotential or in the pseudo wave function, avoiding the introduction of adjustable parameters. For instance, Harrison developed an OPW-like scheme to make the pseudo wave function as smooth as possible [Ref. 9]. With this approach, detailed information is required on the core energies and core wave functions.

An alternative development in the theory rather concentrated on the Austin-Heine-Sham type of pseudopotential, to construct a **model** pseudo-potential, which is rather flat, but which gives the same **electron scattering** as the true potential[26]. This model potential has been parametrized, and tables of the parameters have been published[27] for many elements of the periodic table. According to Harrison[28], "these parameters have not really been superseded since" [Ref. 28,p. 343].

Recently, due to the development of computer technology and computer science, there is a growing tendency to determine the core contribution to the pseudopotential numerically, from the solutions of the Schrödinger equation in the atom[29]. One of the difficulties with this approach is the lack of transferability, not only to different chemical environments, but also among different research groups. This problem might at present have been overcome by the **norm-conserving** pseudopotentials[30], which have been tabulated for the entire periodic table[31] (even including relativistic corrections).

The meaning of the concept of norm-conservation can be understood from the Phillips-Kleinman type of approach, discussed so far. From the relation between the wave function and the pseudo wave function, one obviously obtains:

$$< \Psi_n | \Psi_n > = < \phi_n | \phi_n > - < \phi_n | P | \phi_n > \qquad (3.15)$$

Therefore, if the pseudo wave function is normalized, the true wave function is not. In the Phillips-Kleinman approach, the density is given by:

$$n(\vec{r}) = \sum_n \frac{|\Psi_n(\vec{r})|^2}{< \Psi_n | \Psi_n >} = \sum_n \frac{|\Psi_n(\vec{r})|^2}{1 - < \phi_n | P | \phi_n >} \qquad (3.16)$$

if the pseudo wave function is normalized. The difference between the true electron density $n(\vec{r})$ and the electron pseudo density $\sum_n |\phi_n(\vec{r})|^2$, is usually called the **orthogonalisation hole** (or depletion hole in the model potential language). This orthogonalisation hole is intimately related to the non-local character of the pseudopotential (i.e. the fact that it is an integral operator). Indeed, both the non-locality and the orthogonalisation hole are a consequence of the projection operator on the core states. This intuitive view has been confirmed by Shaw and Harrison[32], who proved that:

$$< \phi_n |P| \phi_n > = < \phi_n | \frac{\partial W}{\partial E_n} | \phi_n > \qquad (3.17)$$

Of course, this relation follows quite naturally from the general pseudopotential, given explicitly in Eq. (3.7). But the general proof for model potentials of the Austin-Heine-Sham type, is less trivial. (Note that this relation implies that in a local approximation to the pseudopotential, one should neglect the orthogonalisation hole together with the non-locality).

The concept of the orthogonalisation hole, appearing in the pseudopotentials of the Phillips-Kleinman type, thus results from the fact that the pseudo wave function overestimates the electron charge inside the core region. The **norm-conserving** Bachelet-Hamann-Schlüter (BHS) pseudopotentials differ from the Phillips-Kleinman pseudopotentials in at least two important aspects:

1)The pseudo wave function, **when normalized,** becomes identical to the true valence wave function beyond some core radius R_c.

2)The pseudopotential inside the core region correctly mimics the full potential inside the core region **at the eigenvalue energy**.

For details how this two-fold goal is achieved, we refer to the original paper.

In Fig. 4 we have plotted the BHS pseudopotential for Na, Mg, Al and Si as a function of distance for the angular momenta $l = 0, 1, 2$, as compared to the Heine-Animalu pseudopotential, for which the pseudopotential is a (l-dependent) constant inside the core region.

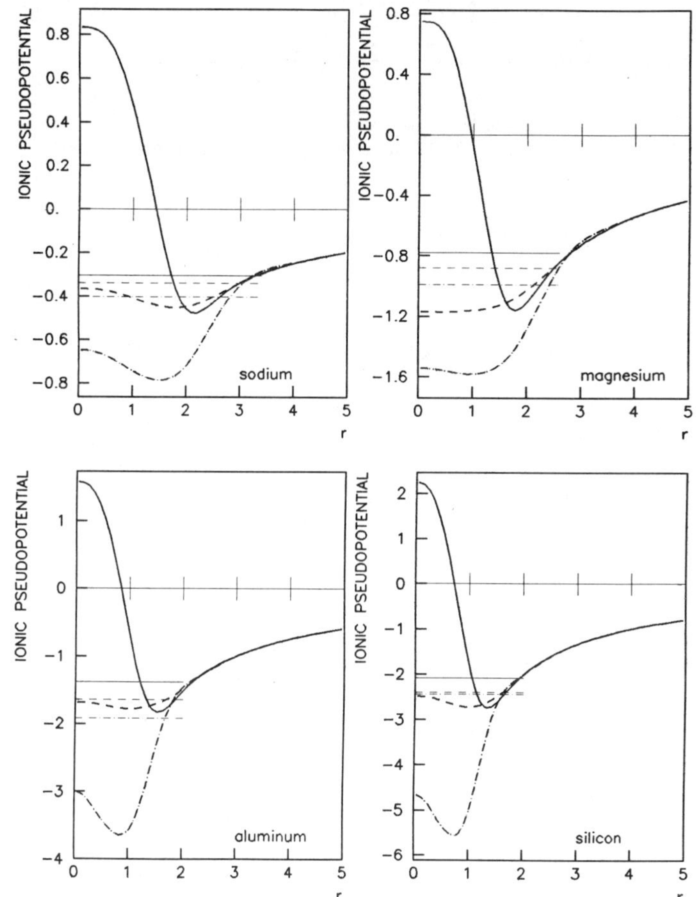

Figure 4. BHS ionic pseudopotential as a function of distance (atomic units).
Full line: l=0; dashed line: l=1; das-dotted line: l=2.
he tiny constant curves give the corresponding Hcinc-Animalu pseudopotentials.

55

Chapter 4

PSEUDOPOTENTIAL PERTURBATION THEORY

Whatever form for the pseudopotential one might have chosen, the general scheme for an actual calculation should proceed along the following lines:

1) assume some reasonable electron density $n(\vec{r})$;

2) determine the corresponding potential $V(\vec{r})$;

3) determine the pseudopotential W;

4) solve $(T + W)|\phi_n > = E_n|\phi_n >$;

5) evaluate $|\Psi_n > = (1 - P)|\phi_n >$;

6) calculate the corresponding density $n(\vec{r})$;

7) restart from step 2) until convergence is reached.

This self-consistent scheme is of course quite involved numerically, but for simple metals it turns out that the pseudopotential can be considered as a perturbation upon the kinetic energy operator. In this case, the self-consistency can be reached analytically up to second order in the perturbation. Therefore, we first follow this perturbative scheme, as outlined by Harrison[1] in order to get some feeling about the different processes involved.

4.1 Pseudo wave functions and band energies

One of the best known applications of pseudopotential theory, is in the theory of simple metals. The basis for its success lies in the large freedom in the form of the pseudo wave function. In simple metals, the electrons behave quasi free, and one might hope that the **pseudo** wave functions can be fairly well described by **plane waves**, upon which the pseudopotential acts as a perturbation. This possibility is a direct consequence of the smoothness of the pseudo wave function, in contrast to the true wave function, for which "plane waves do not work"[2]. Suppose thus that the unperturbed pseudo wave functions are given by

$$|\phi_{\vec{k}}^0 > = |\vec{k} > \qquad \phi_{\vec{k}}^0(\vec{r}) = \frac{1}{\sqrt{\Omega}} e^{i\vec{k}.\vec{r}} \qquad (4.1)$$

where Ω is the volume of the crystal, and the wave vectors $\vec{k}$ satisfy periodic boundary conditions. For the occupied states, the wave vectors are limited in magnitude by the Fermi wave vector $k \leq k_F$. The unperturbed energies, corresponding to these states, are:

$$E_{\vec{k}}^0 = \frac{\hbar^2 k^2}{2m} \tag{4.2}$$

The pseudo wave functions to first order in some perturbation W are then given by:

$$|\phi_{\vec{k}}\rangle = |\vec{k}\rangle + \sum_{\vec{q} \neq 0} |\vec{k} + \vec{q}\rangle \frac{\langle \vec{k} + \vec{q}|W|\vec{k}\rangle}{E_{\vec{k}}^0 - E_{\vec{k}+\vec{q}}^0} \tag{4.3}$$

and the energy to second order is given by:

$$E_{\vec{k}} = E_{\vec{k}}^0 + \langle \vec{k}|W|\vec{k}\rangle + \sum_{\vec{q} \neq 0} \frac{|\langle \vec{k} + \vec{q}|W|\vec{k}\rangle|^2}{E_{\vec{k}}^0 - E_{\vec{k}+\vec{q}}^0} \tag{4.4}$$

For the calculation of the energy to second order, the off-diagonal elements of the pseudopotential are thus only required to first order (which is a standard result of perturbation theory).

As an illustration of the usefulness of the pseudopotential concept in terms of perturbation theory, let us perform this first order calculation. The pseudopotential contains the electron potential V, and a repulsive term from the core states. The electron potential V, is split up in two contributions: one term due to the ions with their core electrons, and a second term due to the other valence electrons, the distribution of which remains to be calculated:

$$V(\vec{r}) = V^{ion}(\vec{r}) + V^{val}(\vec{r}) \tag{4.5}$$

It then follows:

$$\begin{aligned}
\langle \vec{k} + \vec{q}|W|\vec{k}\rangle &= \langle \vec{k} + \vec{q}|W^0|\vec{k}\rangle + \langle \vec{k} + \vec{q}|V^{val}|\vec{k}\rangle \\
\langle \vec{k} + \vec{q}|W^0|\vec{k}\rangle &= \langle \vec{k} + \vec{q}|V^{ion}|\vec{k}\rangle - \langle \vec{k} + \vec{q}|HP - E_{\vec{k}}P|\vec{k}\rangle
\end{aligned} \tag{4.6}$$

As mentioned earlier, the pseudopotential matrix elements are required only to first order for the calculation of the energy to second order:

$$\begin{aligned}
\langle \vec{k} + \vec{q}|W^0|\vec{k}\rangle = {}&\langle \vec{k} + \vec{q}|V^{ion}|\vec{k}\rangle \\
&- \sum_c \langle \vec{k} + \vec{q}|\Psi_c\rangle\langle \Psi_c|\vec{k}\rangle (E_c - E_{\vec{k}}^0 - \langle \vec{k}|W|\vec{k}\rangle)
\end{aligned} \tag{4.7}$$

It should be emphasized that this replacement of the band energy $E_{\vec{k}}$ by its first order expansion, breaks the exactness of the formulation from the original pseudopotential, and the arbitrariness in the pseudo wave function no longer exists. The pseudopotential is now of the Austin-Heine-Sham type (see Eq. 3.12), still giving the exact valence energies, but the pseudo wave function is now unique.

4.2 Ionic pseudopotential contribution and structure factor

The ionic part $< \vec{k} + \vec{q}|W^0|\vec{k} >$ of the pseudopotential depends on the position and the kind of atoms which constitute the solid. The potential v^{ion}, due to an atom in position $\vec{R}_j$ has to be known from some atomic calculation. The total ionic potential V^{ion} is then a sum of these potentials v^{ion}, centered at positions $\vec{R}_j$:

$$V^{ion}(\vec{r}) = \sum_j v^{ion}(\vec{r} - \vec{R}_j) \tag{4.8}$$

(For a lattice with atoms of different kind, a similar expression is readily obtained). The matrix elements $< \vec{k} + \vec{q}|V^{ion}|\vec{k} >$ then become:

$$< \vec{k} + \vec{q}|V^{ion}|\vec{k} > = S(\vec{q}) < \vec{k} + \vec{q}|v^{ion}|\vec{k} > \tag{4.9}$$

with:

$$< \vec{k} + \vec{q}|v^{ion}|\vec{k} > = \frac{1}{\Omega_0} \int d^3r e^{-i\vec{q}.\vec{r}} v^{ion}(\vec{r}) \tag{4.10}$$

and:

$$S(\vec{q}) = \frac{1}{N} \sum_j e^{-i\vec{q}.\vec{R}_j} \tag{4.11}$$

where $\Omega_0 = \Omega/N$ is the volume per atom, and N is the number of atoms in the crystal. The **geometrical structure factor** $S(\vec{q})$ is a very powerful tool in the study of solids, since it contains all the information about the lattice structure. It is well known from elementary text books [see e.g. Kittel, *Introduction to Solid State Physics*, $4^t h$ ed. (1971), p. 76], that for Bravais lattices this factor is unity if $\vec{q}$ is a reciprocal lattice vector, and zero otherwise. For more complicated lattice structures, a quite detailed analysis can e.g. be found in Ref. 1.

Apart from the ion potential, the ionic contribution to the pseudopotential also contains a term with the projection operator on the core states.

The core states $|\Psi_c>$ are centered around the ion positions, and therefore a core state can be labeled by its quantum numbers t and the position of the ion to which it belongs. One then obtains e.g.:

$$< \vec{k} + \vec{q}|\Psi_c > = \frac{1}{\sqrt{N}} e^{-i(\vec{k}+\vec{q}).\vec{R}_j} < \vec{k} + \vec{q}|t >$$

$$< \vec{k} + \vec{q}|t > = \frac{1}{\sqrt{\Omega_0}} \int d^3r \Psi_t(\vec{r}) e^{-i(\vec{k}+\vec{q}).\vec{r}} \tag{4.12}$$

and consequently:

$$\sum_c < \vec{k} + \vec{q}|\Psi_c > < \Psi_c|\vec{k} > (E_c - E_{\vec{k}}^0 - < \vec{k}|W|\vec{k} >) =$$

$$S(\vec{q}) \sum_t < \vec{k} + \vec{q}|t > < t|\vec{k} > (E_t - E_{\vec{k}}^0 - < \vec{k}|W|\vec{k} >) \tag{4.13}$$

where the index t in E_t refers to the quantum numbers of the core states, without referring to the position of the ion. Therefore, in the ionic contribution to the pseudopotential, the geometrical structure factor separates out, and one obtains a product of the geometrical structure factor, and a form factor $< \vec{k} + \vec{q}|w^0|\vec{k} >$ which is independent of the lattice structure:

$$< \vec{k} + \vec{q}|W^0|\vec{k} > = S(\vec{q}) < \vec{k} + \vec{q}|w^0|\vec{k} > \tag{4.14}$$

with:

$$< \vec{k} + \vec{q}|w^0|\vec{k} > = < \vec{k} + \vec{q}|v^{ion}|\vec{k} >$$

$$- \sum_t < \vec{k} + \vec{q}|\Psi_t > < \Psi_t|\vec{k} > (E_t - E_{\vec{k}}^0 - < \vec{k}|W|\vec{k} >) \tag{4.15}$$

As mentioned at the end of the previous section, the factors containing the energy can be replaced by an arbitrary expression of the Austin-Heine-Sham type. This fact can be further exploited, as is e.g. done in the Heine-Animalu pseudopotential, based on the Heine-Abarenkov derivation (see Ref. 26-27 of the previous chapter). Heine and Animalu determined the ionic contribution to the pseudopotential by imposing that it correctly describes the electron scattering from the atom. This potential is parametrized as $w^0(\vec{r}) = -\sum_l A_l(E)P_l$ inside the core and $w^0(\vec{r}) = -Ze^2/r$ outside the core. P_l is a projection operator extracting components of total angular momentum l, and the coefficients $A_l(E)$ have to be determined such that the correct scattering is obtained. At this stage, this pseudopotential is still a valid psudopotential. However, they then further assume $A_l(E) = A_2(E)$

for $l \geq 3$, and neglect the energy dependence in these coefficients:

$$w_{HA}^0 = -A_2 - (A_0 - A_2)P_0 - (A_1 - A_2)P_1 \quad \text{for } r \leq R_M$$
$$= -Ze^2/r \qquad \text{for } r > R_M \tag{4.16}$$

The coefficients A_0, A_1, A_2 and the core radius R_M have been tabulated [see Ref. 26-27 of the previous chapter] for a large variety of elements. This is very useful for practical applications, since one can avoid the cumbersome and time consuming task of repeating the core calculations.

The explicit expression for the matrix elements between plane waves becomes:

$$< \vec{k} + \vec{q}|w_{HA}^0|\vec{k} > = -\frac{4\pi Z e^2}{\Omega_0 q^2} \cos q R_M - \frac{4\pi A_2 R_M}{\Omega_0 q^2} q R_M j_1(q R_M)$$
$$-\frac{4\pi}{\Omega_0}\frac{(A_0 - A_2)R_M^2}{k^2 - |\vec{k} + \vec{q}|^2}\Big[k j_1(k R_M)j_0(|\vec{k} + \vec{q}|R_M)$$
$$-|\vec{k} + \vec{q}|j_0(k R_M)j_1(|\vec{k} + \vec{q}|R_M)\Big] \tag{4.17}$$
$$-\frac{4\pi}{\Omega_0}\frac{(A_1 - A_2)R_M^2}{k^2 - |\vec{k} + \vec{q}|^2}3\cos\theta\Big[|\vec{k} + \vec{q}|j_1(k R_M)j_0(|\vec{k} + \vec{q}|R_M)$$
$$-k j_0(k R_M)j_1(|\vec{k} + \vec{q}|R_M)\Big]$$

where $j_n(x)$ is a spherical Bessel function, and θ denotes the angle between $\vec{k}$ and $\vec{k} + \vec{q}$. Note that no divergence occurs for $\vec{k} = \vec{k} + \vec{q}$, since:

$$\lim_{k_1 \to k_2} \frac{k_1 j_1(k_1 r)j_0(k_2 r) - k_2 j_0(k_1 r)j_1(k_2 r)}{k_1^2 - k_2^2}$$
$$= \frac{k_1 r j_0^2(k_1 r) + k_1 r j_1^2(k_1 r) - j_0(k_1 r)j_1(k_1 r)}{2k_1} \tag{4.18}$$

$$\lim_{k_1 \to k_2} \frac{k_2 j_1(k_1 r)j_0(k_2 r) - k_1 j_0(k_1 r)j_1(k_2 r)}{k_1^2 - k_2^2}$$
$$= \frac{k_1 r j_0^2(k_1 r) + k_1 r j_1^2(k_1 r) - 3j_0(k_1 r)j_1(k_1 r)}{2k_1} \tag{4.19}$$

These Heine-Animalu pseudopotentials are also **model potentials**, in the sense that some parameters are fitted, but they are fitted to **atomic properties**. For applications in solids, this is a fundamental difference with the fitting procedures of model potentials with parameters to be determined from properties of the solid, since they still allow a first principle study of the typical solid state effects one wants to describe.

One important aspect, which is directly related to the fact that the ionic pseudopotential results from projecting out the core states, is its non-locality. Indeed, the matrix elements $< \vec{k}+\vec{q}|w^0_{HA}|\vec{k} >$ explicitly depend on both states $\vec{k}$ and $\vec{k}+\vec{q}$, and not only on the difference $\vec{q}$ between initial and final state. This effect is shown in a few plots of the ionic Heine-Abarenkov pseudopotential $< \vec{k}+\vec{q}|w^0_{HA}|\vec{k} >$ for Na, Mg, Al and Si, as a function of q for the cases $\vec{q}$ parallel and antiparallel to $\vec{k}$, with $|\vec{k}| = k_F$.

As discussed in the chapter on the pseudopotential concept, an extension of the Heine-Animalu pseudopotentials was provided by Bachelet, Hamann and Schlüter, who calculated **norm-conserving** ionic pseudopotentials from atomic calculations for all the elements, including relativistic effects, and fitted their results with a relatively small set of analytic functions, the parameters of which were tabulated. The resulting matrix elements $< \vec{k}+\vec{q}|w^0_{BHS}|\vec{k} >$ are plotted for the same materials as the Heine-Animalu pseudopotential matrix elements.

Both pseudopotentials turn out to be comparable for $q \leq k_F$, as is expected since the small wave vector limit of the pseudopotential corresponds to the Fourier transform of the pseudopotential at large distance, where it behaves as $-Ze^2/r$. In the intermediate wave vector region, the BHS ionic pseudopotential matrix elements are not only larger in magnitude than the HA pseudopotential matrix elements, but also show a more pronounced effect of the non-locality. With increasing wave vector, the ionic BHS-pseudopotential decays substantially faster than the HA-pseudopotential. The reason for this behaviour is the fact that the BHS-pseudopotential inside the core smoothly joins the pseudopotential outside the core. The HA-pseudopotential is discontinuous at the core radius. As a result, the Fourier transforms at large wave vector have a quite pronounced oscillatory behaviour (analogous to the "Gibbs overshoot").

4.3 The electron density

Consider now the contribution of the valence electrons to the pseudopotential:

$$< \vec{k}+\vec{q}|V^{val}|\vec{k} > = \frac{1}{\Omega} \int d^3r\, e^{-i(\vec{k}+\vec{q}).\vec{r}} V^{val}(\vec{r}) e^{i\vec{k}.\vec{r}} \qquad (4.20)$$

Obviously, this contribution is independent of $\vec{k}$ if the potential is local, and the matrix elements is just the Fourier transform of the valence potential.

The Coulomb contribution to this Fourier transform is $4\pi e^2 n_{\vec{q}}/q^2$, as obtained from the Poisson equation. If one introduces the effects of the

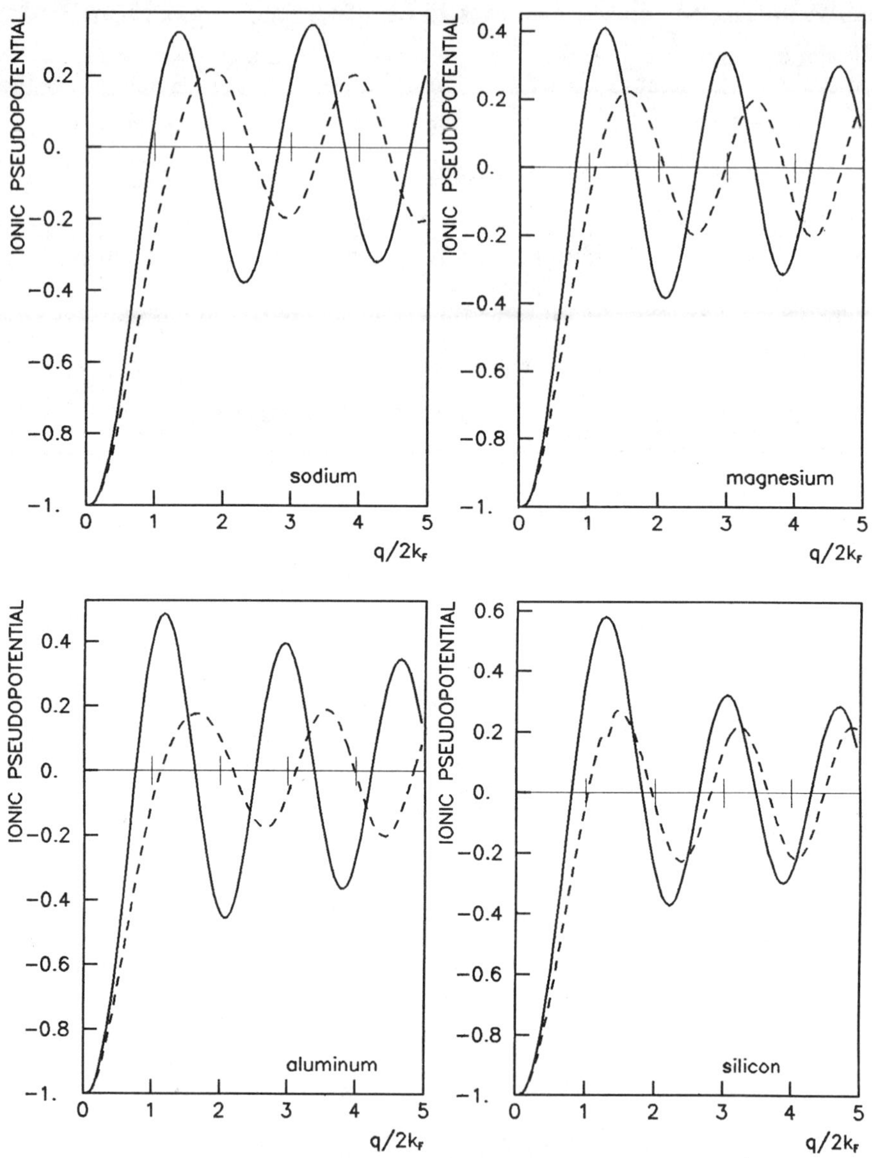

Figure 1. $< \vec{k} + \vec{q}|w_{HA}^{0}|\vec{k} >$ (in units $4\pi Z e^{2}/q^{2}\Omega_{0}$) for Na, Mg, Al and Si for $\vec{q}$ antiparallel to $\vec{k}$ (full curve) and for $\vec{q}$ parallel to $\vec{k}$ (dashed curve), as a function of q for $k = k_{F}$.

exchange and correlation hole by some function $G(q)$, one obtains:

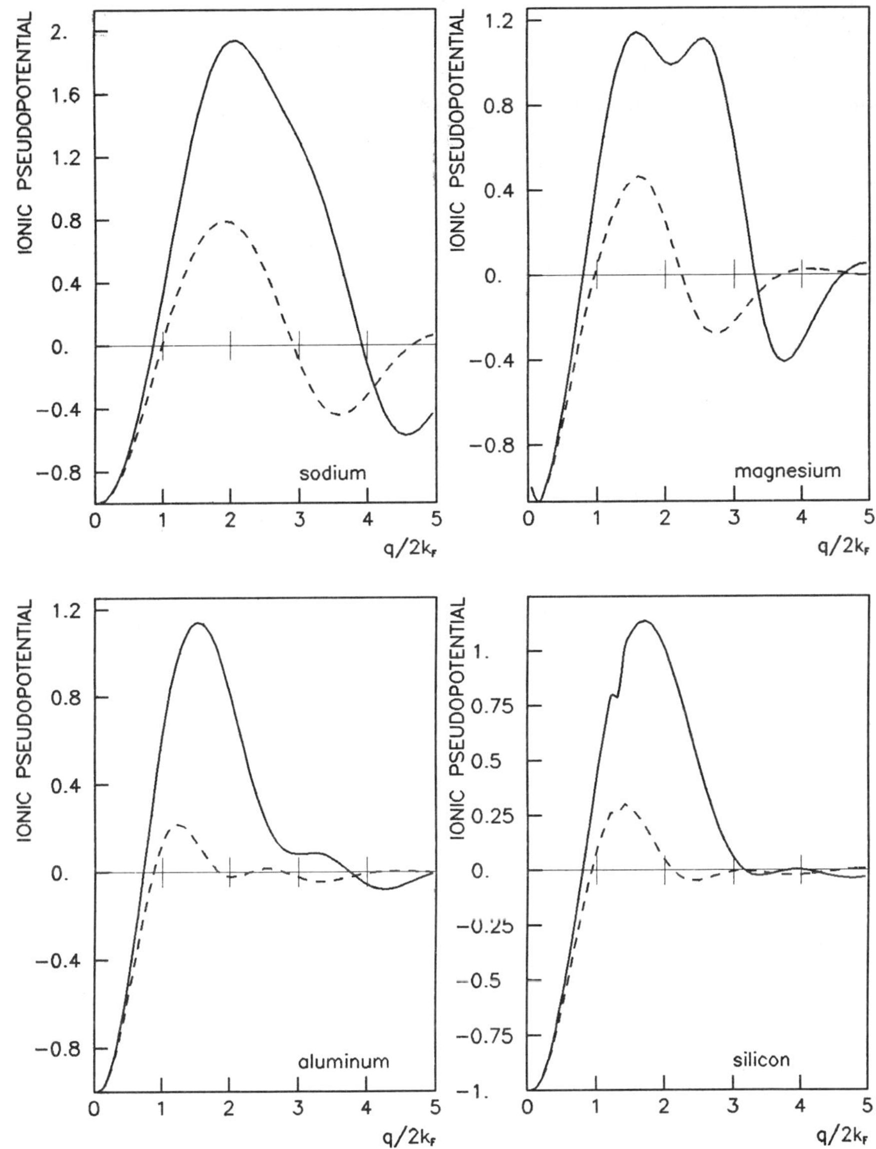

Figure 2. $< \vec{k} + \vec{q}|w^0_{BHS}|\vec{k} >$ (in units $4\pi Z e^2/q^2 \Omega_0$) for Na, Mg, Al and Si for $\vec{q}$ antiparallel to $\vec{k}$ (full curve) and for $\vec{q}$ parallel to $\vec{k}$ (dashed curve), as a function of q for $k = k_F$.

$$< \vec{k} + \vec{q}|V^{val}|\vec{k} > = V^{val}_{\vec{q}} = \frac{4\pi e^2}{q^2}[1 - G(q)]n_{\vec{q}} \qquad (4.21)$$

in which the induced density $n_{\vec{q}}$ has to be calculated. (Note that density here means the number of electrons per unit volume, not the charge density).

Of course, as discussed earlier, the determination of $G(q)$ remains a still unsolved many-body problem. In fact, it is quite questionable whether the exchange and correlation hole can be described in such a simple way by a local function. But in practice, one can proceed by using the best known approximations.

One should keep in mind that the calculation of the density has *not* to start from the pseudo wave function, but from the wave function:

$$\Psi_{\vec{k}}(\vec{r}) = \phi_{\vec{k}}(\vec{r}) - \sum_c \Psi_c(\vec{r}) < \Psi_c | \phi_{\vec{k}} > \qquad (4.22)$$

which should be normalized:

$$< \Psi_{\vec{k}} | \Psi_{\vec{k}} > = 1 - < \phi_{\vec{k}} | P | \phi_{\vec{k}} > \approx 1 - < \vec{k} | P | \vec{k} > \qquad (4.23)$$

The density is then given by:

$$n(\vec{r}) = \sum_{k \leq k_F} \frac{|\Psi_{\vec{k}}(\vec{r})|^2}{< \Psi_{\vec{k}} | \Psi_{\vec{k}} >} \qquad (4.24)$$

which can be worked out as:

$$n(\vec{r}) \approx \sum_{k \leq k_F} \frac{1}{1 - < \vec{k} | P | \vec{k} >} \left[|\phi_{\vec{k}}(\vec{r})|^2 - \frac{1}{\sqrt{\Omega}} \sum_c \left(\Psi_c(\vec{r}) e^{-i \vec{k} \cdot \vec{r}} < \Psi_c | \vec{k} > + c.c. \right) \right.$$

$$\left. + \sum_{c,c'} \Psi_c(\vec{r}) \Psi_{c'}{}^* < \vec{k} | \Psi_{c'} > < \Psi_c | \vec{k} > \right]$$

$$(4.25)$$

Since the zero order pseudo wave functions are plane waves, each one of these zero order terms contains a uniform contribution $1/\Omega$ to the density. This gives a uniform distribution outside the core region:

$$n^{uniform} = \frac{1}{\Omega} \sum_{k \leq k_F} \frac{1}{1 - < \vec{k} | P | \vec{k} >} \qquad (4.26)$$

Inside the core region a so-called **orthogonalisation hole** is formed, which can be evaluated from the core wave functions. In a first approximation, given the small radius of the core as compared to the volume of the primitive cel, this orthogonalisation hole can be considered as a point charge, increasing the valence Z of the ions to:

$$Z^* = Z\left(1 + \frac{1}{N}\sum_{k \leq k_F}\frac{<\vec{k}|P|\vec{k}>}{1 - <\vec{k}|P|\vec{k}>}\right) \qquad (4.27)$$

Instead of going through the (rather involved) numerical calculation of the density distribution inside the core, it is useful to follow Heine-Abarenkov, who considered a uniform depletion hole inside the core volume. They expressed the resulting effective valence as $Z^* \approx Z(1+\alpha)$, where α is the ratio between the core volume and the atomic volume in the crystal, and tabulated the value of $\alpha/2$ for a large range of elements. The values for Z^* from this simple relationship agree to within a few percent with the Z^* values from the full numerical calculation, starting from the core wave functions. E.g. for Li, Na, K and Al, the value of α is 0.10,0.10,0.13 and 0.05 resp., whereas the corresponding values from the full calculation are 0.07,0.07,0.14 and 0.08. In terms of Z^*, the density as given by Heine-Abarenkov to first order thus becomes:

$$n(\vec{r}) \approx \frac{Z^* - Z}{\Omega_0} + \sum_{k \leq k_F}|\phi_{\vec{k}}(\vec{r})|^2 + n^{ort}(\vec{r}) \qquad (4.28)$$

where the first term accounts for the increase in the uniform distribution, arising from the orthogonalisation hole at the ion sites. This orthogonalisation distribution is represented by n^{ort}. For a uniform distibution at the core sites within a sphere with radius R_c, its Fourier transform is:

$$n^{ort}_{\vec{q}} = -\frac{Z^* - Z}{\Omega_0}S(\vec{q})\frac{3j_1(qR_c)}{qR_c}$$

where:

$$R_c^3 = \frac{3}{4\pi}\frac{Z^* - Z}{Z}\Omega_0 \qquad (4.29)$$

which gives rise to an orthogonalisation potential:

$$V^{ort}_{\vec{q}} = S(\vec{q})v^{ort}_{\vec{q}}; \qquad v^{ort}_{\vec{q}} = -\frac{4\pi(Z^* - Z)e^2}{\Omega_0 q^2}[1 - G(q)]\frac{3j_1(qR_c)}{qR_c} \qquad (4.30)$$

It should be noted that, when using the BHS potential, discussed in the previous chapter, the pseudo wave function and the true wave function are simultaneously normalized. This means that the complications, due to the orthogonalisation hole, do not show up in this approach. This amounts simply in putting $Z^* = Z$.

The induced density can be obtained from:

$$|\phi_{\vec{k}}(\vec{r})|^2 \approx \frac{1}{\Omega} + \frac{1}{\Omega} \sum_{\vec{q} \neq 0} \left[e^{i\vec{q}\cdot\vec{r}} \frac{< \vec{k} + \vec{q}|W|\vec{k} >}{E_{\vec{k}}^0 - E_{\vec{k}+\vec{q}}^0} + c.c. \right] \qquad (4.31)$$

and the Fourier transform of the induced density for $\vec{q} \neq 0$ thus becomes:

$$
\begin{aligned}
n_{\vec{q}}^{ind} &= \frac{2}{\Omega} \sum_{k \leq k_F} \frac{< \vec{k} + \vec{q}|W|\vec{k} >}{E_{\vec{k}}^0 - E_{\vec{k}+\vec{q}}^0} \\
&= \frac{2}{\Omega} \sum_{k \leq k_F} \frac{< \vec{k} + \vec{q}|W^0|\vec{k} > + V_{\vec{q}}^{ort}}{E_{\vec{k}}^0 - E_{\vec{k}+\vec{q}}^0} \\
&\quad + n_{\vec{q}}^{ind}[1 - G(q)] \frac{2}{\Omega} \frac{4\pi e^2}{q^2} \sum_{k \leq k_F} \frac{1}{E_{\vec{k}}^0 - E_{\vec{k}+\vec{q}}^0}
\end{aligned}
\qquad (4.32)
$$

For free electrons, the summation in the last term can be done analytically, (using the standard replacement $\sum_k \to [2\Omega/(2\pi)^3] \int d^3k$, and taking the principal value), which gives the Lindhard function $Q_0(q)$:

$$
\begin{aligned}
Q_0(q) &= -\frac{2}{\Omega} \frac{4\pi e^2}{q^2} \sum_{k \leq k_F} \frac{1}{E_{\vec{k}}^0 - E_{\vec{k}+\vec{q}}^0} \\
&= \frac{me^2}{2\pi k_F \hbar^2 \eta^2} \left(1 + \frac{1 - \eta^2}{2\eta} \log \left| \frac{1+\eta}{1-\eta} \right| \right) \quad ; \eta = \frac{q}{2k_F}
\end{aligned}
\qquad (4.33)
$$

and the equation for the induced density becomes:

$$n_{\vec{q}}^{ind} = \frac{2}{\Omega} \sum_{k \leq k_F} \frac{< \vec{k} + \vec{q}|W^0|\vec{k} > + V_{\vec{q}}^{ort}}{E_{\vec{k}}^0 - E_{\vec{k}+\vec{q}}^0} - n_{\vec{q}}^{ind}[1 - G(q)]Q_0(q) \qquad (4.34)$$

This equation is readily solved for $n_{\vec{q}}^{ind}$, and the geometrical structure factor can be factored out:

$$n_{\vec{q}}^{ind} = S(\vec{q}) \frac{1}{1 + Q_0(q)[1 - G(q)]} \frac{2}{\Omega} \sum_{k \leq k_F} \frac{< \vec{k} + \vec{q}|w^0|\vec{k} > + v_{\vec{q}}^{ort}}{E_{\vec{k}}^0 - E_{\vec{k}+\vec{q}}^0} \qquad (4.35)$$

4.4 The total pseudopotential

Adding the ionic and the electronic contributions calculated in the previous sections, one obtains the total pseudopotential, which is thus also proportional to the geometrical structure factor:

$$< \vec{k} + \vec{q}|W|\vec{k} > = S(\vec{q}) < \vec{k} + \vec{q}|w|\vec{k} > \qquad (4.36)$$

where the form factor is written as follows:

$$< \vec{k} + \vec{q}|w|\vec{k} > = < \vec{k} + \vec{q}|\tilde{w}^0|\vec{k} > + w_q^1 \qquad (4.37)$$

with:

$$< \vec{k} + \vec{q}|\tilde{w}^0|\vec{k} > = < \vec{k} + \vec{q}|w^0|\vec{k} > + v_q^{ort} \qquad (4.38)$$

The orthogonalisation hole is then included in the ionic pseudopotential, and both contributions are from now on considered as a single object (the so-called "pseudo atom"). The remaining electronic contribution from the induced density is $\frac{4\pi e^2}{q^2}[1 - G(q)]n_q^{ind}$, and using the expression for the induced density one obtains:

$$w_q^1 = \frac{4\pi e^2}{q^2}\frac{1 - G(q)}{1 + Q_0(q)[1 - G(q)]}\frac{2}{\Omega}\sum_{k \leq k_F}\frac{< \vec{k} + \vec{q}|\tilde{w}^0|\vec{k} >}{E_k^0 - E_{k+q}^0} \qquad (4.39)$$

In Fig. 3 the Heine-Abarenkov pseudopotential for Na including exchange effects (using our expression for $G(q)$), is compared to the pseudopotential with RPA screening. For illustrative purposes, both the approximations with and without the orthogonalisation hole are shown for this material. In Fig. 4 the Heine-Abarenkov pseudopotential (including the othogonalisation hole) is plotted for Al and Si.

With the approximation of Abarenkov and Heine, the orthogonalisation hole enters in the expression for $\tilde{w}^0$ in a fairly simple way. A detailed calculation of the orthogonalisation hole from the core wave functions involves matrix elements $< \vec{k} + \vec{q}|P|\vec{k} >$ [Ref. 1]. But as mentioned before, the difference between this full treatment and the one proposed by Abarenkov and Heine is quite small in practice. And even in principle, the difference between both approaches only involves matrix elements of the form $< \vec{k} + \vec{q}|\Psi_c > < \Psi_c|\vec{k} >$. This means that a different type of Austin-Heine-Sham pseudopotential is considered. Therefore, the difference is immaterial in principle, if one would solve the pseudo Schrödinger equation exactly.

Note however that for a perturbative treatment, the matrix elements of the pseudopotential have to be sufficiently small. It is thus important that in practice the approximation for the orthogonalisation hole is of the same order of magnitude as in the full treatment.

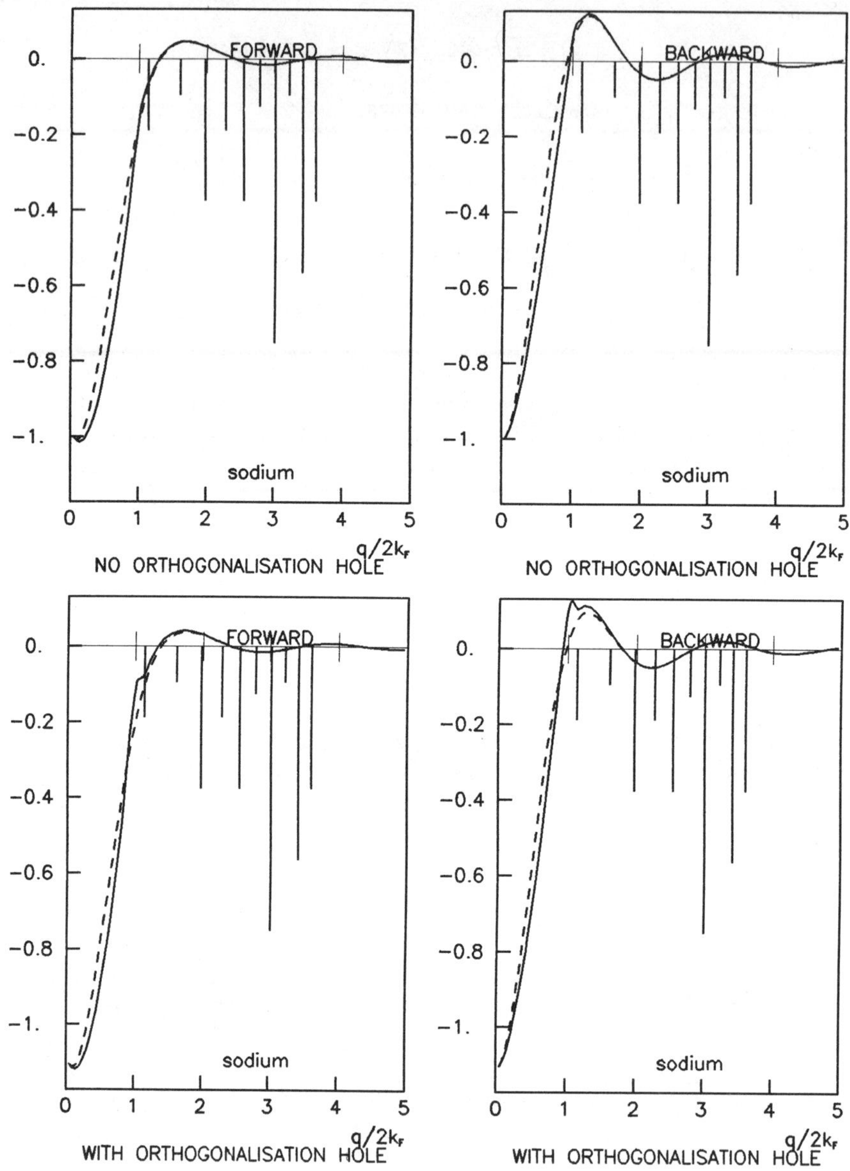

Figure 3. Matrix elements $< \vec{k} + \vec{q}|w_{HA}|\vec{k} >$ for Na

in units $2E_F/3$, for $|\vec{k}| = k_F$.

Forward and backward : $\vec{q}$ parallel resp. antiparallel to $\vec{k}$.
Dashed line: RPA screening; Full line: including dynamical exchange effects.
The vertical lines indicate the positions of the reciprocal lattice vectors,
the length being proportional to the number of lattice vectors.

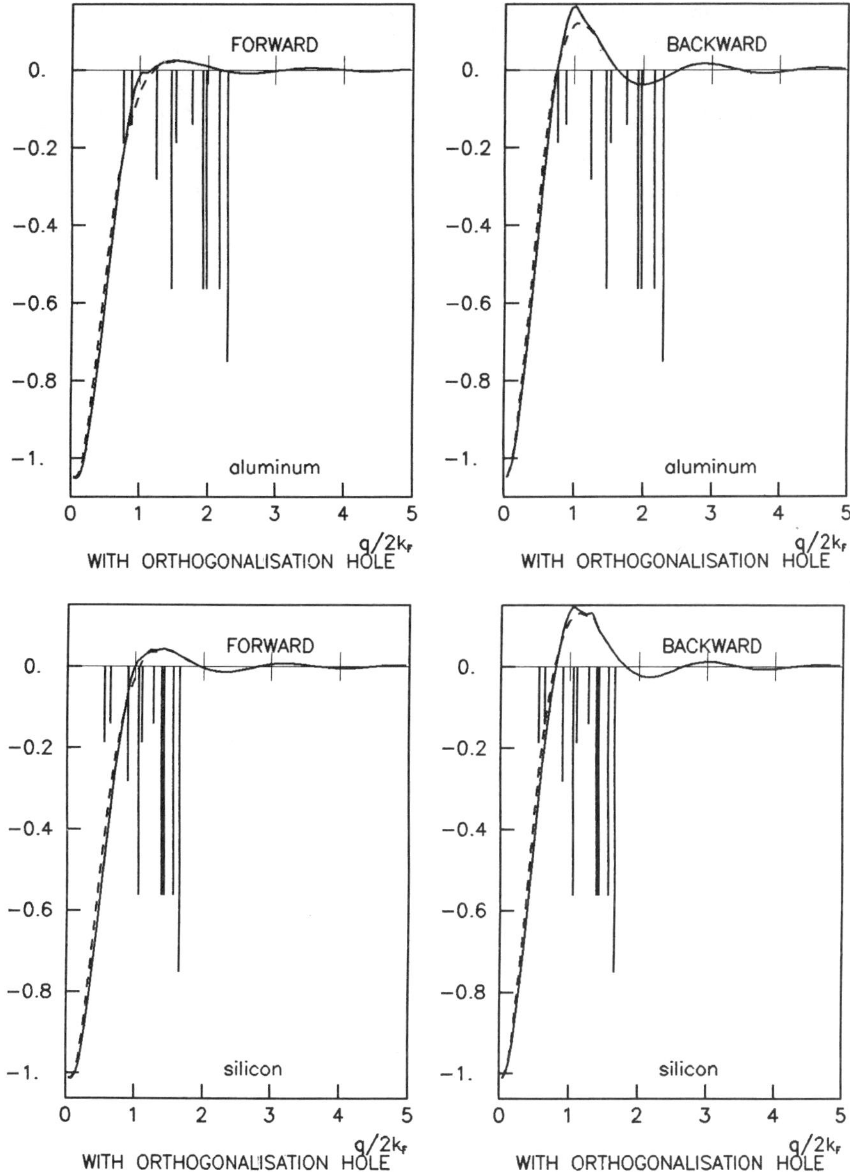

Figure 4. Cfr. previous figure, but for Al and Si.
The pseudopotential without orthogonalisation hole is not shown here.

Again, one should remember that the orthogonalisation hole does not appear in the BHS framework, for which one simply can put $Z^* = Z$. For reference, in Fig. 5 and 6 the BHS pseudopotential matrix elements are shown. The same comments apply in the comparison between the BHS and HA pseudopotential, as were previously given for the ionic contribution.

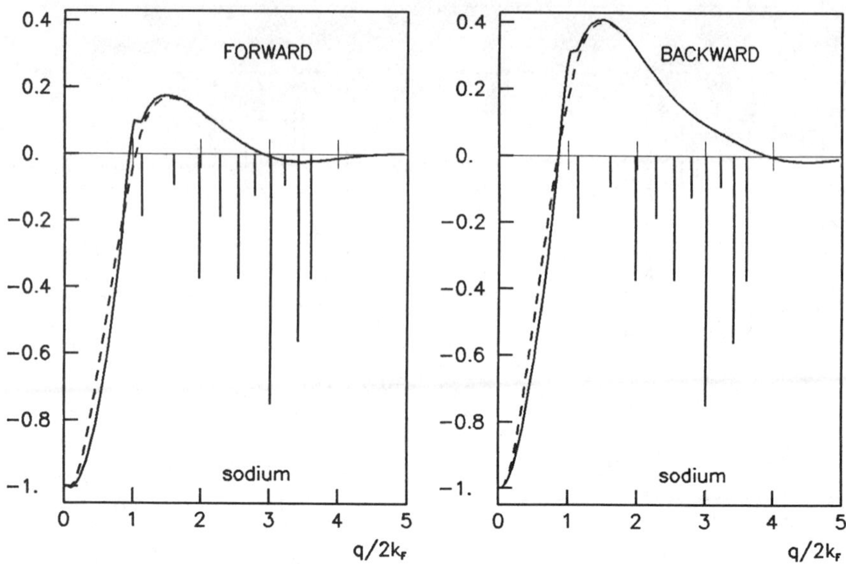

Figure 5. BHS pseudopotential matrix elements for Na.
The same conventions are followed as in the previous figures.

It should also be emphasized that degenerate perturbation theory has to be applied if $E_{\vec{k}}^0 = E_{\vec{k}+\vec{q}}^0$, i.e. if $q^2 = -2\vec{q}.\vec{k}$ with $k \leq k_F$. Because the pseudopotential is proportional to the structure factor, this degenerate perturbation theory is only required if $\vec{q}$ is a reciprocal lattice vector. The condition on $\vec{k}$ and $\vec{q}$ is then the condition for Bragg reflection. Degenerate perturbation theory thus gives a band gap of magnitude $2|< \vec{k} + \vec{q}|w|\vec{k} >|$ at the Brillouin zone boundary.

In the following sections, particularly for the calculation of phonon spectra, the correct small-wave-vector limit of the pseudopotential is required for the internal consistency of the calculations. This limit is easily calculated here. The ionic contribution for $\vec{q} \to 0$, is the Fourier transform of the Coulomb potential of an ion with charge $-Ze$ acting on an electron with charge e. Therefore:

$$< \vec{k} + \vec{q}|w^0|k > |_{\vec{q} \to 0} \to -\frac{4\pi Z e^2}{\Omega_0 q^2} \tag{4.40}$$

With elementary algebraic manipulations one then obtains:

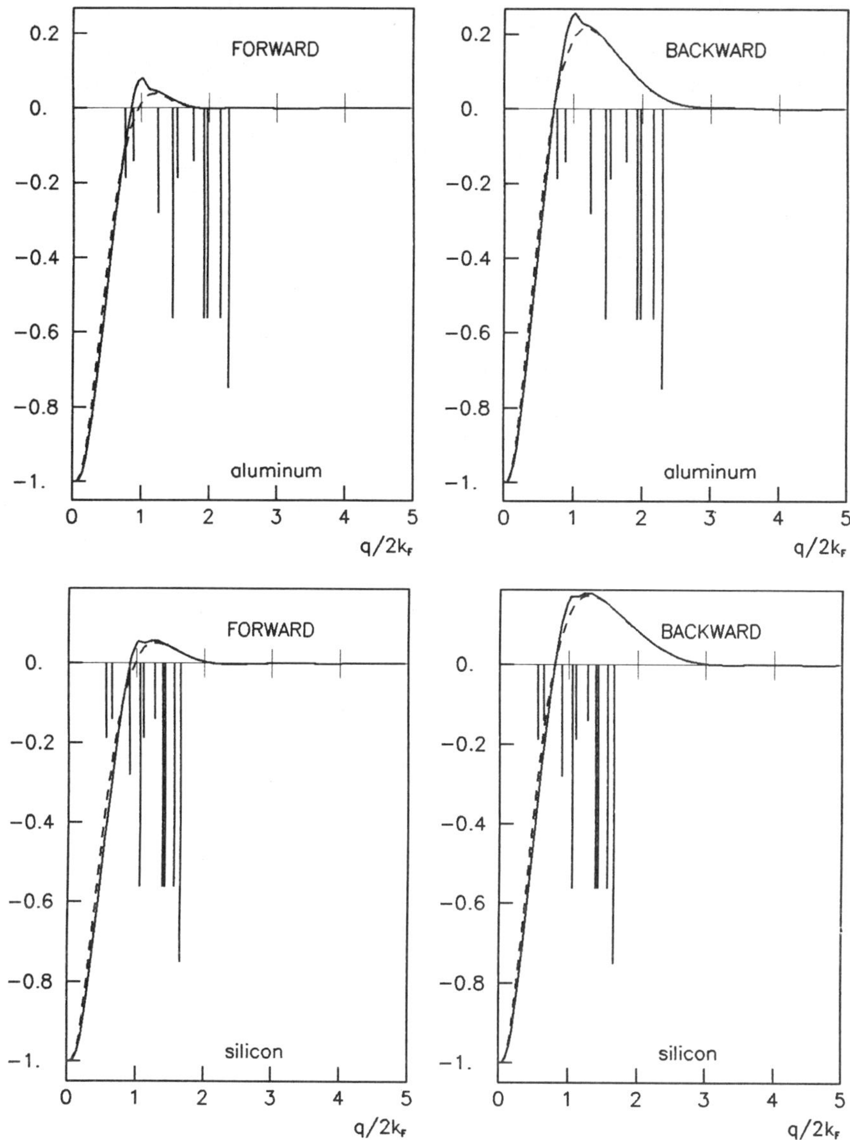

Figure 6. Cfr. previous figure, but for Al and Si.

$$<\vec{k}+\vec{q}|w|k>\big|_{\vec{q}\to 0} \to \frac{<\vec{k}+\vec{q}|w^0|\vec{k}> - \frac{4\pi(Z^*-Z)e^2}{\Omega_0 q^2}\frac{3j_1(qR_c)}{qR_c}[1-G(q)]}{1+Q_0(q)[1-G(q)]}$$

$$(4.41)$$

The exchange and correlation interaction is of short range. Thus, $G(q) \to 0$ for $\vec{q} \to 0$. Using the well known limit for the Lindhard function:

$$Q_0(q)|_{\vec{q} \to 0} \to \frac{k_{FT}^2}{q^2}; \qquad k_{FT}^2 = \frac{4me^2 k_F}{\pi \hbar^2} \qquad (4.42)$$

one thus obtains:

$$< \vec{k} + \vec{q}|w|k > |_{\vec{q} \to 0} \to -\frac{4\pi Z^* e^2}{\Omega_0 k_{FT}^2} = -\frac{Z^*}{Z}\frac{2}{3}E_F \qquad (4.43)$$

It should be noted that in a **local** approximation (i.e. under the assumption that $< \vec{k}+\vec{q}|w^0|\vec{k} >$ is independent of $\vec{k}$), the total pseudopotential becomes:

$$< \vec{k} + \vec{q}|w|k > |_{local} \approx \frac{< \vec{k} + \vec{q}|w^0|\vec{k} > -\frac{4\pi(Z^*-Z)e^2}{\Omega_0 q^2}\frac{3j_1(qR_c)}{qR_c}[1 - G(q)]}{1 + Q_0(q)[1 - G(q)]}$$

$$(4.44)$$

or:

$$< \vec{k} + \vec{q}|w|k > |_{local} \approx \frac{< \vec{k} + \vec{q}|\tilde{w}^0|\vec{k} >}{\tilde{\epsilon}(q)} \qquad (4.45)$$

where $\tilde{\epsilon}(q)$ is the "electron–test-charge" dielectric function of the uniform electron gas, given by:

$$\tilde{\epsilon}(q) = 1 + Q_0(q)[1 - G(q)] \qquad (4.46)$$

This relation illustrates the basic feature that the ionic part of the pseudo-potential and the orthogonalisation hole are screened by the induced charge density. However, since the orthogonalisation hole consists of electrons, its potential also includes an exchange and correlation effect, apart from the Coulomb interaction.

It should be emphasized that the use of the dielectric function of the dielectric function of the electron gas in this treatment, does not mean that this is assumed to be the dielectric function of the solid. In the solid, the response properties are described by the dielectric matrix, even if perturbation theory is applicable. This point will be discussed in the next chapter.

4.5 The Fermi surface

For the calculation of the dielectric matrix to second order in the pseudopotential, we will need the Fermi surface to the same order. For this calculation we already have all the ingredients. The derivation by Wallace[3] will completely be followed, except for the fact that we immediately include the non-locality of the pseudopotential.

As is well known, the Fermi surface to **zero order** is a sphere with radius k_F, determined by the requirement that the total number of electrons equals the number of states inside the Fermi sphere. Using the periodic boundary conditions, one finds that there are $[2\Omega/(2\pi)^3]^{-1}$ states per unit volume in wave number space. (A factor of 2 for the spins is included). The total number of electron states inside a sphere of radius k_F is $[2\Omega/(2\pi)^3]^{-1} 4\pi k_F^3/3$, which equals the total number of electrons ZN. The Fermi wave vector is then obtained from the averaged electron density:

$$\frac{\Omega_0}{Z} = \frac{3\pi^2}{k_F^3} \tag{4.47}$$

The Fermi energy to second order in the pseudopotential is to be found from the corresponding expression for the energy:

$$E_{\vec{F}} = \frac{\hbar^2 F^2}{2m} + \sum_{\vec{q}\neq 0} \frac{|<\vec{F}+\vec{q}|W|\vec{F}>|^2}{E_{\vec{F}}^0 - E_{\vec{F}+\vec{q}}^0} \tag{4.48}$$

where $\vec{F}$ denotes the Fermi wave vector. (Its direction has to be taken explicitly into account). In this expression a term $< \vec{F}|W|\vec{F} >$ has been omitted, since it is a constant, which only shifts the energy scale. (Of course, this shift will be maintained consistently during the whole calculation) For a calculation of the Fermi wave vector to second order, $\vec{F}$ can be replaced in the perturbation term in the right hand side by its zero order value:

$$\vec{k}_F = \frac{k_F}{F}\vec{F} \tag{4.49}$$

The equation for the magnitude of the Fermi wave vector then becomes:

$$|\vec{F}| \approx \sqrt{\frac{2mE_{\vec{F}}}{\hbar^2}} + 2\left(\frac{m}{\hbar^2}\right)^2 \frac{1}{k_F} \sum_{\vec{q}\neq 0} \frac{|<\vec{k}_F+\vec{q}|W|\vec{k}_F>|^2}{q^2 + 2\vec{q}\cdot\vec{k}_F} \tag{4.50}$$

(Note that this means the *magnitude in some given direction*. The requirement that the total number of electrons equals the number of states with energy less than or equal to $E_{\vec{F}}$ means:

$$ZN = \frac{2\Omega}{(2\pi)^3} \int_0^{\vec{F}} d^3k = \frac{2\Omega}{(2\pi)^3}\left(\int_0^{\vec{k}_F} d^3k + \int_{\vec{k}_F}^{\vec{F}} d^3k\right) \tag{4.51}$$

Again, a factor of 2 is included to account for the spin degeneracy. Because

k_F is defined in such a way that the first integral in the right hand side equals ZN, one is left with the requirement:

$$\int_{\vec{k}_F}^{\vec{F}} d^3k \approx k_F^2 \int d^2\Omega_{\vec{F}}(|\vec{F}| - |\vec{k}_F|) = 0 \qquad (4.52)$$

where use has been made of the fact that $|\vec{F}| - |\vec{k}_F|$ is a second order quantity in the pseudopotential, which accounts for the radial part of the integration.

Only the angular integration is left over. Inserting the expression for $|\vec{F}|$ in this equation, one obtains:

$$\sqrt{\frac{2mE_{\vec{F}}}{\hbar^2}} - |\vec{k}_F| \approx -\frac{1}{2\pi}\left(\frac{m}{\hbar^2}\right)^2 \frac{1}{k_F} \int d^2\Omega_{\vec{F}} \sum_{\vec{q}\neq 0} \frac{|<\vec{k}_F + \vec{q}|W|\vec{k}_F>|^2}{q^2 + 2\vec{q}.\vec{k}_F} \qquad (4.53)$$

or:

$$E_{\vec{F}} \approx \frac{\hbar^2 k_F^2}{2m} - \frac{m}{2\pi\hbar^2}\sum_{\vec{q}\neq 0} \int d^2\Omega_{\vec{F}} \frac{|<\vec{k}_F + \vec{q}|W|\vec{k}_F>|^2}{q^2 + 2\vec{q}.\vec{k}_F} \qquad (4.54)$$

From this equation one can calculate the magnitude of the Fermi wave vector in a given direction:

$$\begin{aligned}
|\vec{F}| \approx k_F + 2\left(\frac{m}{\hbar^2}\right)^2 \frac{1}{k_F}\sum_{\vec{q}\neq 0}\Bigg[&\frac{|<\vec{k}_F + \vec{q}|W|\vec{k}_F>|^2}{q^2 + 2\vec{q}.\vec{k}_F} \\
&-\frac{1}{4\pi}\int d^2\Omega_{\vec{F}} \frac{|<\vec{k}_F + \vec{q}|W|\vec{k}_F>|^2}{q^2 + 2\vec{q}.\vec{k}_F}\Bigg]
\end{aligned} \qquad (4.55)$$

where the angular integration is over all possible directions of $\vec{k}_F$. Since the pseudopotential is proportional to the geometrical structure factor, this procedure allows to calculate the Fermi surface in quite complicated structures (provided one has to do with materials for which pseudopotential perturbation theory provides a reasonable description), without substantial numerical complications.

The simplicity of this formula can be appreciated if one considers the local approximation, in which case the angular integrations can be performed analytically (because the pseudopotential matrix elements can be taken out of the integral):

$$|\vec{F}| \approx k_F + \frac{m}{\hbar^2 k_F} \sum_{\vec{q} \neq 0} |S(\vec{q})|^2 |< \vec{k}_F + \vec{q}|w|\vec{k}_F >|^2$$

$$\left[\frac{1}{q^2 + 2\vec{q}.\vec{k}_F} + \frac{m}{2\hbar^2 q k_F} \log|\frac{1 + q/2k_F}{1 - q/2k_F}| \right] \tag{4.56}$$

The logarithmic term cancels the average over all directions of the direction dependent term. This means that the deviations of the Fermi surface from a sphere in a certain direction, are compensated by the deviations in other directions, such that the averaged deviation over all directions is zero.

The most important deviations of the Fermi surface from a sphere, are obviously to be expected from reciprocal lattice vectors in a direction antiparallel to the Fermi wave vector under consideration.

It should be emphasized that a divergence occurs if $\vec{k}_F$ is at the boundaries of the Brillouin zone, namely if $q^2 = -2\vec{q}.\vec{k}_F$. This is a direct consequence of the second order perturbation theory, which fails in this region because of the degeneracy of the unperturbed free electron states. As mentioned above, degenerate perturbation theory has to be applied in this region, and the derivation of the Fermi surface has to start from the secular equation:

$$\left(\frac{\hbar^2 F^2}{2m} - E_{\vec{F}} \right)\left(\frac{\hbar^2 |\vec{F} + \vec{q}|^2}{2m} - E_{\vec{F}} \right) = |< \vec{k}_F + \vec{q}|W|\vec{k}_F >|^2 \tag{4.57}$$

The algebra involved is quite cumbersome, but the interested reader is referred e.g. to Harrison [Ref.1], for a detailed derivation of the Fermi surface, including applications to the more complicated case at two or three intersecting Bragg planes.

We mention a quite important result from Harrison's treatment, namely *that summations over the true Fermi surface only differ to third order in the pseudopotential from summations over the Fermi sphere, in which the singularities near the Bragg reflection planes are disregarded by taking the principal value in the integrals.*

Chapter 5
TOTAL ENERGY AND PSEUDOPOTENTIAL PERTURBATION THEORY

5.1 The energy-wavenumber characteristic[1]

Given the band energies to second order, the total energy is given by:

$$E = \sum_{k \leq k_F} E_{\vec{k}} + \frac{1}{2} \sum_{i \neq j} \frac{Z^2 e^2}{|\vec{R}_i - \vec{R}_j|} - \frac{e^2}{2} \int d^3 r_1 \int d^3 r_2 \frac{n(\vec{r}_1)n(\vec{r}_2)}{|\vec{r}_1 - \vec{r}_2|}$$
$$+ E^{XC}[n(\vec{r})] - \int d^3 r n(\vec{r}) V^{XC}(\vec{r}) \tag{5.1}$$

where the second term accounts for the Coulomb interaction between the ions of charge Z, the third term corrects for the double counting of the Coulomb energy between the electrons in summing the band energies, and the last two terms replace the sum over the one-electron exchange and correlation potentials in the band energies by the exchange and correlation energy.

The evaluation presents some problems of book keeping, which are simplified by temporarily introducing the short hand notation:

$$(n_a)(n_b) \equiv \int d^3 r_1 \int d^3 r_2 \frac{n_a(\vec{r}_1)n_b(\vec{r}_2)}{|\vec{r}_1 - \vec{r}_2|} \tag{5.2}$$

In this notation the energy becomes:

$$E = \sum_{k \leq k_F} E_{\vec{k}} + E^{XC}[n(\vec{r})] - \int d^3 r n(\vec{r}) V^{XC}(\vec{r}) + \frac{e^2}{2}(ion)(ion)' - \frac{e^2}{2}(n)(n) \tag{5.3}$$

where (ion) denotes the distribution of point ions with valence Z at the ion positions, and the prime in the ion-ion energy means that the interaction of an ion with itself should be omitted.

Remembering that the electron distribution consists of three terms (a uniform distribution $n_0 = Z^*/\Omega_0$, orthogonalisation holes which net negative contribution $-(Z^* - Z)$ at each ion position, and induced density oscillations), the Coulomb interaction term $(n)(n)$ can be decomposed into:

$$-\frac{1}{2}(n)(n) = -\frac{1}{2}(n_0 + n^{ort})(n_0 + n^{ort}) - (n_0 + n^{ort})(n^{ind}) - \frac{1}{2}(n^{ind})(n^{ind}) \tag{5.4}$$

The band energy is calculated before, and its contribution is:

$$\sum_{k \le k_F} E_{\vec{k}} = \sum_{k \le k_F} \frac{\hbar^2 k_F^2}{2m} + \sum_{k \le k_F} <\vec{k}|W|\vec{k}> + \sum_{\vec{q} \ne 0} \sum_{k \le k_F} \frac{|<\vec{k}+\vec{q}|W|\vec{k}>|^2}{E_{\vec{k}}^0 - E_{\vec{k}+\vec{q}}^0}$$

$$(5.5)$$

The integral over the kinetic term gives $\frac{3}{5} N Z E_F$. For the contribution of the diagonal elements, one should not use their value $-\frac{2}{3}\frac{Z^*}{Z} E_F$, calculated before, since this limit was based on the electronic pseudopotential contribution to first order. This is not accurate enough for the energy to second order, and one has to study this contribution in some more detail. The pseudopotential consists of a core-electron part, and the potential from the point ions and the valence electrons. The sum over its diagonal elements thus is the sum over the core contribution plus the averaged potential from the ion and the valence electrons. For the latter, the full unperturbed electron distribution $(n_0 + n^{ort})$ is to be used, instead of averaging with the averaged electron density.[*]

$$\sum_{k \le k_F} <\vec{k}|W|\vec{k}> = \sum_{k \le k_F} <\vec{k}|W^{core}|\vec{k}> + \int d^3r [n_0 + n^{ort}(\vec{r})]V^{XC}(\vec{r})$$

$$+ e^2(n_0 + n^{ort})(ion + n^{ort} + n_0 + n^{ind})$$

$$(5.6)$$

where the core term is obtained from the ion pseudopotential by subtracting the point-ion part. The Coulomb interaction term can be combined with the ion-ion interaction term in the total energy and with the double counting correction for the electron interactions. It is a matter of straightforward algebra to show that the only terms left over are:

$$\frac{1}{2}(ion)(ion)' + \frac{1}{2}(n_0)(n_0) + \frac{1}{2}(n^{ort})(n^{ort}) +$$

$$(n_0)(ion + n^{ort}) + (n^{ort})(ion) - \frac{1}{2}(n^{ind})(n^{ind})$$

which is easily recombined into:

$$\frac{1}{2}(ion + n^{ort} + n_0)(ion + n^{ort} + n_0)' + NE^{ort} - \frac{1}{2}(n^{ind})(n^{ind})$$

where E^{ort} denotes the interaction potential of an orthogonalisation hole

---

[*]This argument is rather heuristic. A full derivation in detail is quite involved, but can e.g. be found in Harrison.

at a given site with the point ion at that site, plus half its Coulomb self energy. The first term in this expression is the Coulomb energy of point ions wit valence Z^* (instead of Z) in a uniform compensating background of electrons. The contribution of the last term is expressed in the Fourier transforms of the involved density:

$$(n^{ind})(n^{ind}) = \Omega \sum_{\vec{q} \neq 0} \frac{4\pi}{q^2} n^{ind}_{-\vec{q}} n^{ind}_{\vec{q}} \tag{5.7}$$

For the sake of clarity, let us rewrite the energy, including the results obtained so far (although still rather formal):

$$E = N\frac{3}{5}ZE_F + NE^{ort} + \sum_{k \leq k_F} <\vec{k}|W^{core}|\vec{k}>$$
$$+ NE^{es} + E^{XC}[n(\vec{r})] - \int d^3r\, n^{ind}(\vec{r}) V^{XC}(\vec{r}) \tag{5.8}$$
$$+ \sum_{\vec{q} \neq 0} |S(\vec{q})|^2 \sum_{k \leq k_F} \frac{|<\vec{k}+\vec{q}|w|\vec{k}>|^2}{E^0_{\vec{k}} - E^0_{\vec{k}+\vec{q}}} - N\Omega_0 \sum_{\vec{q} \neq 0} \frac{2\pi e^2}{q^2} n^{ind}_{-\vec{q}} n^{ind}_{\vec{q}}$$

where E^{es} is the energy *per ion* of N ions with valence Z^* at positions $\vec{R}_j$ in a uniform compensating background of electrons.

The electrostatic contribution E^{es} per ion can be calculated by the Ewald-Fuchs procedure which is quite well known. For completeness, the derivation is given in Appendix. The result is:

$$E^{es} = \frac{Z^{*2}e^2}{2}\left[\frac{1}{N}\sum_{i \neq j} f(|\vec{R}_i - \vec{R}_j|) + \frac{1}{\Omega_0}\sum_{\vec{q} \neq 0}|S(\vec{q})|^2\frac{4\pi}{q^2}e^{-q^2/4\eta} - \frac{\pi}{\eta\Omega_0} - 2\sqrt{\frac{\eta}{\pi}}\right] \tag{5.9}$$

where:

$$f(r) = \frac{1}{r}\frac{2}{\sqrt{\pi}}\int_0^{r\sqrt{\eta}} dt\, e^{-t^2} \tag{5.10}$$

and η is an arbitrary constant, determined in such a way that both lattice sums are fastly convergent. By taking the derivative, one readily checks that E^{es} is in fact independent of η. A reasonable choice for the value of η is obtained if one imposes that the structure independent terms give the maximum contribution, leaving the smallest possible contribution to the lattice sums. This gives:

$$\eta = \left(\frac{9\pi}{16}\right)^{1/3} \frac{1}{R_0^2} \qquad \text{where} \quad \frac{4\pi}{3} R_0^3 = \Omega_0 \qquad (5.11)$$

For an estimate of E^{ort}, we assume that the ortogonalisation hole is uniformly distributed over the core volume. The energy can then readily be calculated:[*]

$$E^{ort} = 3(Z^* - Z)\left(\frac{Z}{2} + \frac{Z^* - Z}{5}\right)\frac{e^2}{R_0}\left[\frac{Z}{Z^* - Z}\right]^{1/3} \qquad (5.12)$$

Of course, this contribution is only estimated here. A detailed calculation requires the knowledge of the full core wave functions (and even then quite different values are reported in the litterature). Note however that this contribution is independent of the structure, and thus not too important for the discussions later on.

The only problem left is then the exchange and correlation contribution. As discussed by Hohenberg and Kohn, the exchange and correlation energy is a unique functional of the density, which to second order in the density fluctuations can be written as:

$$E^{XC}[n(\vec{r})] = E^{XC}[n_0] + \int d^3 r \int d^3 x K_{[n_0]}(|\vec{r} - \vec{x}|)\tilde{n}(\vec{r})\tilde{n}(\vec{x}) \qquad (5.13)$$

where $\tilde{n}(\vec{r})$ denotes the non-uniform density contributions. In Fourier expansion, this expression becomes:

$$E^{XC}[n(\vec{r})] = E^{XC}[n_0] + \Omega \sum_{\vec{q} \neq 0} K(q)|n_{\vec{q}}^{ind}|^2 \qquad (5.14)$$

and since the exchange potential is the functional derivative with respect to the density, the Fourier transform of the exchange and correlation potential

---

[*]Consider a uniform electron distribution n within a sphere of radius R_c. The potential energy of an electron for $r < R_c$ is then $v(r) = 2\pi n e^2(R_c^2 - r^2/3)$ (whereas for $r > R_c$ it is given by $4\pi n e^2 R_c^3/3r$). The potential energy of a point ion of valence Z at the center of the sphere is thus given by $-2\pi Z n e^2 R_c^2$, and the self energy of this electron distribution is obtained by integrating $nv(r)/2$ inside the sphere, with the result $16\pi^2 n^2 e^2 R_c^5/15$. Adding both terms, and remembering that for the orthogonalisation hole, the density is $-(Z^* - Z)$ per core volume $4\pi R_c^3/3$, where $Z^* - Z = ZR_c^3/R_0^3$, the self energy expression follows immediately in terms of Z^* and the atomic volume.

is $V_{\vec{q}}^{XC} = 2K(q)n_{\vec{q}}$. In previous sections, we used for this potential the expression $V_{\vec{q}}^{XC} = -(4\pi e^2/q^2)G(q)n_{\vec{q}}$, and the relation between this function $G(q)$ and the factor $K(q)$ in the Hohenberg-Kohn expression for the exchange and correlation energy is thus:

$$K(q) = -\frac{2\pi e^2}{q^2}G(q) \tag{5.15}$$

In the chapter on the dielectric response, it has been argued that

$$\lim_{q \to 0} \frac{4\pi e^2}{q^2}G(q) = -\frac{\partial^2(n\epsilon^{XC})}{\partial n^2} \tag{5.16}$$

For the exchange and correlation terms we thus obtain two contributions, namely from the homogeneous electron density $n_0 = Z^*/\Omega_0$ outside the core region (since inside the core the orthogonalization hole repels the valence electrons), and a contribution from the density fluctuations. The first term thus gives a contribution:

$$
\begin{aligned}
N\Delta^{XC}(n_0) &= N\left(\Omega_0 - \frac{4\pi R_c^3}{3}\right)n_0\epsilon^{XC}(n_0) \\
&= NZ^*\left(1 - \frac{Z^* - Z}{Z}\right)\epsilon^{XC}(n_0)
\end{aligned}
\tag{5.17}
$$

The interaction between the orthogonalisation hole and the induced density in the exchange and correlation energy cancels against the same contribution in the integral over the exchange and the correlation potential. The only term left for the induced density is then its self interaction, for which the energy contribution cancels half of the potential correction. This simply introduces a correction factor $1 - G(q)$ in the Coulomb correction, which was the last term in the expression for the total energy. After combining terms one is left with:

$$E/N = E^0 + E^{es} + E^{bs} \tag{5.18}$$

where E^0 is a structure independent term:

$$E^0 = \frac{3}{5}ZE_F + E^{ort} + \sum_{k \leq k_F} <\vec{k}|W^{core}|\vec{k}> + \Delta^{XC}(n_0) \tag{5.19}$$

and the bandstructure energy E^{bs} per ion is:

$$E^{bs} = \Omega_0 \sum_{\vec{q}\neq 0}\left[|S(\vec{q})|^2 \frac{1}{\Omega}\sum_{k\leq k_F} \frac{|<\vec{k}+\vec{q}|w|\vec{k}>|^2}{E^0_{\vec{k}} - E^0_{\vec{k}+\vec{q}}} - \frac{2\pi e^2}{q^2}[1-G(q)]|n^{ind}_{\vec{q}}|^2 \right]$$

$$(5.20)$$

Inserting the relation between the induced density and the electronic part of the pseudopotential, one again realizes that the structure factor can be factored out, and the structure independent energy-wavenumber characteristic $F(q)$ only depends on the atomic properties:

$$E^{bs} = \sum_{\vec{q}\neq 0} |S(\vec{q})|^2 F(q)$$

$$F(q) = \frac{2\Omega_0}{(2\pi)^3}\left[\int_{k_F} d^3k \frac{|<\vec{k}+\vec{q}|w|\vec{k}>|^2}{E^0_{\vec{k}} - E^0_{\vec{k}+\vec{q}}} - \frac{\pi^2 q^2}{2e^2}\frac{1}{1-G(q)}|w^1_q|^2 \right]$$

$$(5.21)$$

where the integration is limited to the Fermi sphere. Decomposing the pseudopotential in its ionic contribution and its electronic contribution, terms in w^1_q can be collected, and one obtains:

$$F(q) = \frac{2\Omega_0}{(2\pi)^3}\left[\int_{k_F} d^3k \frac{|<\vec{k}+\vec{q}|\tilde{w}^0|\vec{k}>|^2}{E^0_{\vec{k}} - E^0_{\vec{k}+\vec{q}}} + \frac{(2\pi)^3}{4}\frac{|w^1_q|^2}{4\pi e^2/q^2}\frac{1+Q_0(q)[1-G(q)]}{1-G(q)} \right]$$

$$(5.22)$$

which in a local pseudopotential approximation would simply become:

$$F(q) \approx -\frac{\Omega_0}{2}\frac{q^2}{4\pi e^2}|\tilde{w}^0_q|^2 \frac{Q_0(q)}{1+Q_0(q)[1-G(q)]}$$

$$\approx +\frac{\Omega_0}{2}\frac{q^2}{4\pi e^2}|\tilde{w}^0_q|^2\left(\frac{1}{\epsilon(q)}-1\right)$$

$$(5.23)$$

where $\epsilon(q)$ is the "test-charge–test-charge" dielectric function of the uniform electron gas:

$$\epsilon(q) = 1 + \frac{Q_0(q)}{1-G(q)Q_0(q)}$$

$$(5.24)$$

Returning to the full non-local expression, it is more appropriate for numerical purposes to take the orthogonalisation potential out of the integral, since its contribution can be calculated analytically. After some algebra one obtains:

$$F(q) = \frac{2\Omega_0}{(2\pi)^3}\Bigg[\int_{k_F} d^3k \frac{|<\vec{k}+\vec{q}|w^0|\vec{k}>|^2}{E^0_{\vec{k}} - E^0_{\vec{k}+\vec{q}}}$$
$$+ \frac{(2\pi)^3}{4}\frac{q^2}{4\pi e^2}\left(|w^1_q + v^{ort}_q|^2 \frac{1 + Q_0(q)[1 - G(q)]}{1 - G(q)} - \frac{|v^{ort}_q|^2}{1 - G(q)}\right)\Bigg] \qquad (5.25)$$

where the only numerical integration left is over the ionic part of the pseudopotential.

It is quite important to examine the behaviour of $F(q)$ in the limit $\vec{q} \to 0$. The dominant contribution to the ionic pseudopotential is then the point charge potential, including the orthogonalisation hole:

$$<\vec{k}+\vec{q}|\tilde{w}^0|\vec{k}>\big|_{\vec{q}\to 0} \to -\frac{4\pi Z^*e^2}{\Omega_0 q^2} \qquad (5.26)$$

and in this limit the local approximation is correct. Since $Q_0(q)$ behaves as $1/q^2$ for $\vec{q} \to 0$, and $G(q)$ tends to zero, one obtains:

$$F(q)\big|_{\vec{q}\to 0} \to -\frac{2\pi Z^{*2}e^2}{\Omega_0 q^2}$$

As will be clear from the following section, this limiting behaviour is required in order to find acoustical phonons for solids with one atom per primitive cell.

Finally, since part of the electrostatic terms in the energy also involves a lattice sum in reciprocal space, these terms can be combined with the band structure energy. The energy per ion can then be rewritten as:

$$E = E^0 + E^{real} + E^{rec}$$
$$E^{real} = \frac{Z^{*2}e^2}{2}\left[\frac{1}{N}\sum_{i\neq j} f(|\vec{R}_i - \vec{R}_j|) - \frac{\pi}{\eta\Omega_0}\right] \qquad (5.27)$$
$$E^{rec} = \sum_{\vec{q}\neq 0} |S(\vec{q})|^2 H(q) - Z^{*2}e^2\sqrt{\frac{\eta}{\pi}}$$

where:

$$f(r) = \frac{1}{r}\mathrm{erfc}(r\sqrt{\eta}) \qquad (5.28)$$

$$H(q) = F(q) + \frac{2\pi Z^{*2} e^2}{\Omega_0} \frac{e^{-q^2/4\eta}}{q^2} \tag{5.29}$$

This expression will also be used as the starting point for the calculation of the vibration spectrum.

5.2 The vibration spectrum

From the energy expression in the previous section, one can consider ion displacements from equilibrium. Taking these deviations into account to second order, the corresponding hamiltonian is of couse quadratic in the displacements, and can be diagonalized. The eigen frequencies are then found by diagonalizing the dynamical matrix:

$$M\omega_{\vec{Q}}^2 a_{m_1,\alpha}(\vec{Q}) = \sum_{m_2,\beta} D_{m_1,\alpha;m_2,\beta}(\vec{Q}) a_{m_2,\beta}(\vec{Q}) \tag{5.30}$$

where α, β are the Cartesian components of the normal coordinates $\vec{a}_m(\vec{Q})$, with $\vec{Q}$ a vector in the first Brillouin zone. The indices m count the number ν of atoms per primitive cell. The dynamical matrix is derived in appendix, and its explicit expression is:

$$MD_{m_1,\alpha;m_1,\beta}(\vec{Q}) = \chi_{m_1,\alpha;m_1,\beta}(\vec{Q}) - \sum_{m_2} \chi_{m_1,\alpha;m_2,\beta}(\vec{Q} = 0)$$

$$MD_{m_1,\alpha;m_2,\beta}(\vec{Q}) = \chi_{m_1,\alpha;m_2,\beta}(\vec{Q}) \quad \text{if } m_1 \neq m_2 \tag{5.31}$$

where:

$$\chi_{m_1,\alpha;m_2,\beta}(\vec{Q}) = \frac{2}{\nu} \sum_{\vec{G} \neq -\vec{Q}} (\vec{Q}+\vec{G})_\alpha (\vec{Q}+\vec{G})_\beta H(|\vec{Q}+\vec{G}|) e^{i\vec{G}\cdot(\vec{p}_{m_1}-\vec{p}_{m_2})}$$

$$- Z^{*2} e^2 \sum_p \left[e^{-i\vec{Q}\cdot\vec{r}} \frac{\partial^2 f(r)}{\partial r_\alpha \partial r_\beta} \right]_{\vec{r}=\vec{X}_p + \vec{p}_{m_1} - \vec{p}_{m_2} \neq 0} \tag{5.32}$$

where $\vec{X}_p$ denotes the positions of the N/ν Bravais cells, and $\vec{p}_m$ the position of atom m in the primitive cell.

The phonon frequencies are thus obtained by diagonalizing a $3\nu \times 3\nu$ matrix, where ν is the number of atoms per primitive cell. In some symmetry directions, this diagonalization can be performed analytically, and the calculation of the phonon frequencies then only requires some lattice sums. For the ionic contribution, these lattice sums can often even be done analytically. But the numerical evaluation in general is quite simple, so that there is no need to put much effort in finding analytical procedures along specific directions.

Since the sum of the eigenvalues is the trace of the dynamical matrix, one easily derives some simple sum rules. Consider the limit $\eta = 0$, which is allowed since η is an arbitrary constant, introduced for numerical convergence. One readily finds:

$$\sum_{\alpha} \chi_{m_1,\alpha;m_2,\alpha}(\vec{Q}=0) = \frac{2}{\nu} \sum_{\vec{G}\neq 0} G^2 H_{\eta=0}(G) e^{i\vec{G}\cdot(\vec{p}_{m_1}-\vec{p}_{m_2})} \tag{5.33}$$

$$\sum_{\alpha} \chi_{m,\alpha;m,\alpha}(\vec{Q}) = \frac{1}{\nu} Q^2 H_{\eta=0}(Q) + \frac{2}{\nu} \sum_{\vec{G}\neq -\vec{Q}} |\vec{Q}+\vec{G}|^2 H_{\eta=0}(|\vec{Q}+\vec{G}|) \tag{5.34}$$

These relations constitute the necessary ingredients to calculate the trace of the dynamical matrix:

$$\begin{aligned}
\sum \omega_{\vec{Q}}^2 &= \sum_{m,\alpha} D_{m,\alpha;m,\alpha}(\vec{Q}) \\
&= \frac{1}{M} \sum_{m,\alpha} \chi_{m,\alpha;m\alpha}(\vec{Q}) - \frac{1}{M} \sum_{m_1,m_2,\alpha} \chi_{m_1,\alpha;m_2\alpha}(\vec{Q}=0) \\
&= \frac{2}{M} Q^2 H_{\eta=0}(Q) + \frac{2}{M} \sum_{\vec{G}\neq 0} \Bigg(|\vec{Q}+\vec{G}|^2 H_{\eta=0}(|\vec{Q}+\vec{G}|) \\
&\quad - G^2 H_{\eta=0}(G) \sum_{m_1,m_2} e^{i\vec{G}\cdot(\vec{p}_{m_1}-\vec{p}_{m_2})} \Bigg)
\end{aligned} \tag{5.35}$$

In the long wavelength limit, this trace is thus given by:

$$\begin{aligned}
\lim_{\vec{Q}\to 0} \sum \omega_{\vec{Q}}^2 &= \frac{2}{M} \lim_{\vec{Q}\to 0} Q^2 H_{\eta=0}(Q) \\
&\quad + \frac{2}{M} \sum_{\vec{G}\neq 0} G^2 H_{\eta=0}(G) \left(1 - \sum_{m_1,m_2} e^{i\vec{G}\cdot(\vec{p}_{m_1}-\vec{p}_{m_2})} \right)
\end{aligned} \tag{5.36}$$

In the case of one atom per primitive cell, the contribution of the reciprocal sums is zero. Explicitly using the expression for $H(q)$, the requirement to find acoustic phonons thus implies:

$$\lim_{q\to 0} q^2 F(q) = -\frac{2\pi Z^{*2} e^2}{\Omega_0} \tag{5.37}$$

Phonon frequencies in Aluminum.		
	[1,0,0] T L	[1/2,1/2,1/2] T L
Experiment	3.65 6.08	2.63 6.06
Calculated (a)	4.95 8.03	3.32 8.34
(b)	3.93 6.13	3.17 6.32
(c)	4.97 8.10	3.31 8.47
(d)	3.90 6.27	3.31 6.63
(e)	4.78 8.06	3.47 8.42
(f)	3.57 6.31	3.81 6.38

Table 1. Phonon frequencies in 10^{13} rad/sec for Al.
(a) HA-pseudopotential, no orthogonalisation hole, RPA screening;
(b) HA-pseudopotential, no orthogonalisation hole, including exchange;
(c) HA-pseudopotential, with orthogonalisation hole, RPA screening;
(d) HA-pseudopotential, with orthogonalisation hole, including exchange;
(e) BHS-pseudopotential, RPA screening;
(f) BHS-pseudopotential, including exchange.

As already mentioned in the previous section, this condition is indeed fulfilled by the energy-wave number characteristic. In essence, this relation (Eq. 37) establishes the internally consistent treatment of the potential between the ions and the electrons, as compared to the ion-ion interaction.

Again, one should keep in mind that in the framework of the BHS pseudopotentials, the orthogonalisation hole does not enter in the calculations.

In the table, we show some results for the phonon spectrum, as obtained from Eq. 30-32, with the different approximations discussed here, which are compared to the experimental phonon frequencies in Al[2].

5.3 The dielectric matrices[3]

The phonon calculation from the pseudopotential perturbation theory, with the Heine-Animalu and the BHS pseudopotential, raises the question whether the perturbation approach is sufficiently adequate.

Since the dielectric formulation is not based on the weakness of the pseudopotential in the small wave vector region, one might be interested in examining the difference between both approaches for simple metals. Let us therefore calculate the polarizability matrix to second order in the pseudopotential.

In terms of the wave functions of the valence electrons, the polarizability matrix is given by:

$$\Pi(\vec{q}_1, \vec{q}_2) = \frac{1}{\Omega} \sum_{\vec{k}_1, \vec{k}_2}{}' \frac{p_{\vec{k}_1} - p_{\vec{k}_2}}{E_{\vec{k}_1} - E_{\vec{k}_2}} < \Psi_{\vec{k}_1} | e^{-i\vec{q}_1 \cdot \vec{r}} | \Psi_{\vec{k}_2} >< \Psi_{\vec{k}_2} | e^{i\vec{q}_2 \cdot \vec{r}} | \Psi_{\vec{k}_1} >$$

$$(5.38)$$

where $p_{\vec{k}}$ is the occupation number of the state $|\Psi_{\vec{k}} >$. The wave function can be derived from the pseudo wave function, which is given by:

$$|\phi_{\vec{k}} >= |\vec{k} > + \sum_{\vec{q}}{}' a_{\vec{q}}(\vec{k})|\vec{k} + \vec{q} > \tag{5.39}$$

The relation between the energy and the expansion coefficients is:

$$E_{\vec{k}} = E_{\vec{k}}^0 + < \vec{k}|W|\vec{k} > + \sum_{\vec{q}}{}' a_{\vec{q}}(\vec{k}) < \vec{k}|W|\vec{k} + \vec{q} > \tag{5.40}$$

$$a_{\vec{q}}(\vec{k}) = \frac{< \vec{k} + \vec{q}|W|\vec{k} > + \sum_{\vec{q}'}{}' a_{\vec{q}'}(\vec{k}) < \vec{k} + \vec{q}|W|\vec{k} + \vec{q}' >}{E_{\vec{k}} - E_{\vec{k}}^0} \tag{5.41}$$

where $E_{\vec{k}}^0$ denotes the free-particle energy.

The relation $|\Psi_{\vec{k}} >= (1 - P)|\phi_{\vec{k}} >$ then allows to write the wave function as:

$$|\Psi_{\vec{k}} >= |\vec{k} > + \sum_{\vec{q}} c_{\vec{q}}(\vec{k})|\vec{k} + \vec{q} > \tag{5.42}$$

where:

$$c_0(\vec{k}) = - < \vec{k}|P|\phi_{\vec{k}} > \tag{5.43}$$

$$c_{\vec{q} \neq 0}(\vec{k}) = a_{\vec{q}}(\vec{k}) - < \vec{k} + \vec{q}|P|\vec{k} > \tag{5.44}$$

It is useful to normalize these wave functions properly:

$$< \Psi_{\vec{k}_1}|\Psi_{\vec{k}_2} >= \delta_{\vec{k}_1, \vec{k}_2} \tag{5.45}$$

giving the conditions:

$$c_{\vec{k}_1 - \vec{k}_2}(\vec{k}_2) + c^*_{\vec{k}_2 - \vec{k}_1}(\vec{k}_1) + \sum_{\vec{q}} c^*_{\vec{q}}(\vec{k}_1) c_{\vec{k}_1 + \vec{q} - \vec{k}_2}(\vec{k}_2) = 0 \tag{5.46}$$

$$c_0(\vec{k}) + c_0^*(\vec{k}) + \sum_{\vec{q}} |c_{\vec{q}}(\vec{k})|^2 = 0 \tag{5.47}$$

The matrix elements in the polarizability matrix can easily be evaluated:

$$< \Psi_{\vec{k}_1} | e^{-i\vec{q}_1 \cdot \vec{r}} | \Psi_{\vec{k}_2} > = \delta_{\vec{k}_2, \vec{k}_1 + \vec{q}_1} + c_{\vec{k}_2 - \vec{q}_1 - \vec{k}_1}^*(\vec{k}_1) + c_{\vec{k}_1 - \vec{k}_2 + \vec{q}_1}(\vec{k}_2)$$
$$+ \sum_{\vec{q}} c_{\vec{q}}(\vec{k}_2) c_{\vec{k}_2 + \vec{q} - \vec{q}_1 - \vec{k}_1}^*(\vec{k}_1)$$

$$\tag{5.48}$$

Using the normalization condition in the form:

$$c_{\vec{k}_2 - \vec{q}_1 - \vec{k}_1}^*(\vec{k}_1) = -c_{\vec{k}_1 - \vec{k}_2 + \vec{q}_1}(\vec{k}_2 - \vec{q}_1) - \sum_{\vec{q}} c_{\vec{q} + \vec{k}_2 - \vec{q}_1 - \vec{k}_1}^*(\vec{k}_1) c_{\vec{q}}(\vec{k}_2 - \vec{q}_1)$$

$$\tag{5.49}$$

this matrix element becomes:

$$< \Psi_{\vec{k}_1} | e^{-i\vec{q}_1 \cdot \vec{r}} | \Psi_{\vec{k}_2} > = \delta_{\vec{k}_2, \vec{k}_1 + \vec{q}_1} + c_{\vec{k}_1 - \vec{k}_2 + \vec{q}_1}(\vec{k}_2) - c_{\vec{k}_1 - \vec{k}_2 + \vec{q}_1}(\vec{k}_2 - \vec{q}_1))$$
$$+ \sum_{\vec{q}} [c_{\vec{q}}(\vec{k}_2) - c_{\vec{q}}(\vec{k}_2 - \vec{q}_1)] c_{\vec{k}_2 + \vec{q} - \vec{q}_1 - \vec{k}_1}^*(\vec{k}_1)$$

$$\tag{5.50}$$

The matrix element $< \Psi_{\vec{k}_2} | e^{i\vec{q}_2 \cdot \vec{r}} | \Psi_{\vec{k}_1} >$ is obtained from the previous one by taking the complex conjugate, and replacing $\vec{q}_1$ by $\vec{q}_2$.

To second order in the expansion coefficients $c_{\vec{q}}(\vec{k})$, the matrix elements of the plane waves in the polarizability become:

$$< \Psi_{\vec{k}_1} | e^{-i\vec{q}_1 \cdot \vec{r}} | \Psi_{\vec{k}_2} > < \Psi_{\vec{k}_2} | e^{i\vec{q}_2 \cdot \vec{r}} | \Psi_{\vec{k}_1} > \approx$$
$$\delta_{\vec{k}_2, \vec{k}_1 + \vec{q}_1} \delta_{\vec{q}_1, \vec{q}_2}$$
$$+ \delta_{\vec{k}_2, \vec{k}_1 + \vec{q}_1} [c_{\vec{q}_2 - \vec{q}_1}^*(\vec{k}_1 + \vec{q}_1) - c_{\vec{q}_2 - \vec{q}_1}^*(\vec{k}_1 + \vec{q}_1 - \vec{q}_2)]$$
$$+ \delta_{\vec{k}_2, \vec{k}_1 + \vec{q}_2} [c_{\vec{q}_1 - \vec{q}_2}(\vec{k}_1 + \vec{q}_2) - c_{\vec{q}_1 - \vec{q}_2}(\vec{k}_1 + \vec{q}_2 - \vec{q}_1)]$$
$$+ \delta_{\vec{k}_2, \vec{k}_1 + \vec{q}_1} \sum_{\vec{q}} c_{\vec{q} + \vec{q}_1 - \vec{q}_2}(\vec{k}_1) [c_{\vec{q}}^*(\vec{k}_1 + \vec{q}_1) - c_{\vec{q}}^*(\vec{k}_1 + \vec{q}_1 - \vec{q}_2)]$$
$$+ \delta_{\vec{k}_2, \vec{k}_1 + \vec{q}_2} \sum_{\vec{q}} c_{\vec{q} + \vec{q}_2 - \vec{q}_1}^*(\vec{k}_1) [c_{\vec{q}}(\vec{k}_1 + \vec{q}_2) - c_{\vec{q}}(\vec{k}_1 + \vec{q}_2 - \vec{q}_1)]$$
$$+ [c_{\vec{k}_1 - \vec{k}_2 + \vec{q}_1}(\vec{k}_2) - c_{\vec{k}_1 - \vec{k}_2 + \vec{q}_1}(\vec{k}_2 - \vec{q}_1)]$$
$$\times [c_{\vec{k}_1 - \vec{k}_2 + \vec{q}_2}^*(\vec{k}_2) - c_{\vec{k}_1 - \vec{k}_2 + \vec{q}_2}^*(\vec{k}_2 - \vec{q}_2)]$$

$$\tag{5.51}$$

The expansion of the energy denominators to second order gives:

$$\frac{1}{E_{\vec{k}_1} - E_{\vec{k}_2}} - \frac{1}{E^0_{\vec{k}_1} - E^0_{\vec{k}_2}}$$

$$\approx \frac{\sum_{\vec{q}}{}^{'}[< \vec{k}_1|W|\vec{k}_1 + \vec{q} > a_{\vec{q}}(\vec{k}_1) - < \vec{k}_2|W|\vec{k}_2 + \vec{q} > a_{\vec{q}}(\vec{k}_2)]}{(E^0_{\vec{k}_1} - E^0_{\vec{k}_2})^2}$$

$$(5.52)$$

Also the sum over the occupied states has then to be considered over the Fermi surface instead of the Fermi sphere. As shown in the previous section, this correction to second order gives:

$$\frac{1}{\Omega} \sum_{\vec{k}} p_{\vec{k}} f(\vec{k}) = \frac{1}{\Omega} \sum_{|\vec{k}| < k_F} f(\vec{k}) + \frac{k_F^2}{(2\pi)^3} \int d^2\Omega_{\vec{F}} (|\vec{F}| - k_F) f(\vec{k}_F) \quad (5.53)$$

With some algebra, the different contributions in the polarizability matrix can then be combined, giving for the zero order polarizability:

$$\Pi^{(0)}(\vec{q}_1, \vec{q}_2) = \delta_{\vec{q}_1, \vec{q}_2} \frac{1}{\Omega} \sum_{\vec{k}} \frac{p^0_{\vec{k}} - p^0_{\vec{k}+\vec{q}_1}}{E^0_{\vec{k}} - E^0_{\vec{k}+\vec{q}_1}} = -\delta_{\vec{q}_1, \vec{q}_2} \frac{q_1^2}{4\pi e^2} Q_0(q_1) \quad (5.54)$$

where $Q_0(q)$ is the Lindhard function in the free electron gas, discussed before, and $p^0_{\vec{k}}$ are the unperturbed occupation numbers, defining the summations over the Fermi sphere. The first order polarizability is given by:

$$\Pi^{(1)}(\vec{q}_1, \vec{q}_2) = \frac{1}{\Omega} \sum_{\vec{k}} \frac{p^0_{\vec{k}} - p^0_{\vec{k}+\vec{q}_1}}{E^0_{\vec{k}} - E^0_{\vec{k}+\vec{q}_1}} [c^*_{\vec{q}_2-\vec{q}_1}(\vec{k} + \vec{q}_1) - c^*_{\vec{q}_2-\vec{q}_1}(\vec{k} + \vec{q}_1 - \vec{q}_2)]$$

$$+ \frac{1}{\Omega} \sum_{\vec{k}} \frac{p^0_{\vec{k}} - p^0_{\vec{k}+\vec{q}_2}}{E^0_{\vec{k}} - E^0_{\vec{k}+\vec{q}_2}} [c_{\vec{q}_1-\vec{q}_2}(\vec{k} + \vec{q}_2) - c_{\vec{q}_1-\vec{q}_2}(\vec{k} + \vec{q}_2 - \vec{q}_1)]$$

$$(5.55)$$

which is obviously zero for $\vec{q}_1 = \vec{q}_2$. There are thus no diagonal first order corrections to the matrix elements of the polarizability:

$$\Pi^{(1)}(\vec{q}, \vec{q}) = 0 \quad (5.56)$$

For the second order correction to the diagonal matrix elements of the polarizability one obtains after some algebra:

$$\Pi^{(2)}(\vec{q}_1, \vec{q}_1) = \frac{4k_F^2}{(2\pi)^3} \int d^2\Omega_F \frac{|\vec{F}| - k_F}{E_{k_F}^0 - E_{k_F+q_1}^0}$$

$$+ \frac{1}{\Omega} {\sum_{\vec{k}_1, \vec{k}_2}}' \frac{p_{\vec{k}_1}^0 - p_{\vec{k}_2}^0}{E_{\vec{k}_1}^0 - E_{\vec{k}_2}^0} |c_{\vec{k}_1 - \vec{k}_2 + \vec{q}_1}(\vec{k}_2) - c_{\vec{k}_1 - \vec{k}_2 + \vec{q}_1}(\vec{k}_2 - \vec{q}_1)|^2$$

$$+ \frac{1}{\Omega} \sum_{\vec{k}} \frac{p_{\vec{k}}^0 - p_{\vec{k}+\vec{q}_1}^0}{E_{\vec{k}}^0 - E_{\vec{k}+\vec{q}_1}^0} \sum_{\vec{q}} \left(c_{\vec{q}}(\vec{k})[c_{\vec{q}}^*(\vec{k} + \vec{q}_1) - c_{\vec{q}}^*(\vec{k})] + c.c. \right)$$

$$- \frac{1}{\Omega} \sum_{\vec{k}} \frac{p_{\vec{k}}^0 - p_{\vec{k}+\vec{q}_1}^0}{(E_{\vec{k}}^0 - E_{\vec{k}+\vec{q}_1}^0)^2} {\sum_{\vec{q}}}' \bigg(< \vec{k}|W|\vec{k} + \vec{q} > a_{\vec{q}}(\vec{k})$$

$$- < \vec{k} + \vec{q}_1|W|\vec{k} + \vec{q}_1 + \vec{q} > a_{\vec{q}}(\vec{k} + \vec{q}_1) \bigg)$$

$$(5.57)$$

In this expression, the first term comes from the Fermi surface corrections,
the second and the third from the deviation of the wave function from plane
waves, and the fourth from the energy denominators to second order. A
similar calculation can be performed for the off-diagonal matrix elements to
second order, but these are not required for a second order calculation.

For the electron–test-charge dielectric matrix we obtained:

$$\tilde{\epsilon}(\vec{q}_1, \vec{q}_2) = \delta_{\vec{q}_1, \vec{q}_2} - \frac{4\pi e^2}{q_1^2}[1 - G(q_1)]\Pi(\vec{q}_1, \vec{q}_2) \qquad (5.58)$$

which to second order thus becomes:

$$\tilde{\epsilon}^{(0)}(\vec{q}_1, \vec{q}_2) = \delta_{\vec{q}_1, \vec{q}_2}\big(1 + Q_0(q_1)[1 - G(q_1)]\big)$$

$$\tilde{\epsilon}^{(1)}(\vec{q}_1, \vec{q}_1) = 0$$

$$\tilde{\epsilon}^{(1)}(\vec{q}_1, \vec{q}_2) = -\frac{4\pi e^2}{q_1^2}[1 - G(q_1)]\Pi^{(1)}(\vec{q}_1, \vec{q}_2) \qquad (5.59)$$

$$\tilde{\epsilon}^{(2)}(\vec{q}_1, \vec{q}_2) = -\frac{4\pi e^2}{q_1^2}[1 - G(q_1)]\Pi^{(2)}(\vec{q}_1, \vec{q}_2)$$

For the calculation of the dynamical matrix, one needs the inverse test-
charge–test-charge dielectric function:

$$\epsilon^{-1}(\vec{q}_1, \vec{q}_2) - \delta_{\vec{q}_1, \vec{q}_2} = \frac{1}{1 - G(q_1)}[\tilde{\epsilon}^{-1}(\vec{q}_1, \vec{q}_2) - \delta_{\vec{q}_1, \vec{q}_2}] \qquad (5.60)$$

The inversion of the dielectric matrix $\tilde{\epsilon}$ to second order is easily done by
expanding the dielectric matrix and its inverse in the definition of the inverse.

In order to avoid too many indices, we temporarily introduce the notation:

$$\tilde{e}(\vec{q}_1, \vec{q}_2) \equiv \tilde{\epsilon}^{-1}(\vec{q}_1, \vec{q}_2) \tag{5.61}$$

giving:

$$
\begin{aligned}
\delta_{\vec{q}_1,\vec{q}_2} &= \sum_{\vec{q}_3} \tilde{e}(\vec{q}_1, \vec{q}_3)\tilde{\epsilon}(\vec{q}_3, \vec{q}_2) \\
&= \sum_{\vec{q}_3} \tilde{e}^{(0)}(\vec{q}_1, \vec{q}_3)\tilde{\epsilon}^{(0)}(\vec{q}_3, \vec{q}_2) \\
&\quad + \sum_{\vec{q}_3}\left(\tilde{e}^{(1)}(\vec{q}_1, \vec{q}_3)\tilde{\epsilon}^{(0)}(\vec{q}_3, \vec{q}_2) + \tilde{e}^{(0)}(\vec{q}_1, \vec{q}_3)\tilde{\epsilon}^{(1)}(\vec{q}_3, \vec{q}_2) \right) \\
&\quad + \sum_{\vec{q}_3}\left(\tilde{e}^{(2)}(\vec{q}_1, \vec{q}_3)\tilde{\epsilon}^{(0)}(\vec{q}_3, \vec{q}_2) + \tilde{e}^{(1)}(\vec{q}_1, \vec{q}_3)\tilde{\epsilon}^{(1)}(\vec{q}_3, \vec{q}_2) \right. \\
&\quad \left. + \tilde{e}^{(0)}(\vec{q}_1, \vec{q}_3)\tilde{\epsilon}^{(2)}(\vec{q}_3, \vec{q}_2) \right) + \cdots
\end{aligned}
\tag{5.62}
$$

With some elementary algebra, one then obtains:

$$
\begin{aligned}
\tilde{e}^{(0)}(\vec{q}_1, \vec{q}_2) &= \delta_{\vec{q}_1,\vec{q}_2} \frac{1}{\tilde{\epsilon}^{(0)}(\vec{q}_1, \vec{q}_1)} \\
\tilde{e}^{(1)}(\vec{q}_1, \vec{q}_2) &= - \frac{\tilde{\epsilon}^{(1)}(\vec{q}_1, \vec{q}_2)}{\tilde{\epsilon}^{(0)}(\vec{q}_1, \vec{q}_1)\tilde{\epsilon}^{(0)}(\vec{q}_2, \vec{q}_2)} \\
\tilde{e}^{(2)}(\vec{q}_1, \vec{q}_1) &= - \frac{1}{[\tilde{\epsilon}^{(0)}(\vec{q}_1, \vec{q}_1)]^2}\left(\tilde{\epsilon}^{(2)}(\vec{q}_1, \vec{q}_1) - \sum_{\vec{q}_3 \neq 0} \frac{\tilde{\epsilon}^{(1)}(\vec{q}_1, \vec{q}_3)\tilde{\epsilon}^{(1)}(\vec{q}_3, \vec{q}_1)}{\tilde{\epsilon}^{(0)}(\vec{q}_3, \vec{q}_3)} \right)
\end{aligned}
\tag{5.63}
$$

The evaluation of the inverse test-charge–test-charge dielectric matrix is then straightforward:

$$
\begin{aligned}
[\epsilon^{-1}(\vec{q}_1, \vec{q}_2)]^{(0)} &= \delta_{\vec{q}_1,\vec{q}_2}\left(1 - \frac{Q_0(q_1)}{1 + Q_0(q_1)[1 - G(q_1)]}\right) = \delta_{\vec{q}_1,\vec{q}_2}\frac{1}{\epsilon(q_1)} \\
[\epsilon^{-1}(\vec{q}_1, \vec{q}_1)]^{(1)} &= 0
\end{aligned}
$$

$$[\epsilon^{-1}(\vec{q}_1, \vec{q}_2)]^{(1)} = \frac{4\pi e^2}{q_1^2} \frac{\Pi^{(1)}(\vec{q}_1, \vec{q}_2)}{\tilde{\epsilon}(q_1)\tilde{\epsilon}(q_2)}$$

$$[\epsilon^{-1}(\vec{q}_1, \vec{q}_1)]^{(2)} = \frac{4\pi e^2}{q_1^2} \frac{1}{\tilde{\epsilon}(q_1)^2}\bigg(\Pi^{(2)}(\vec{q}_1, \vec{q}_1)$$

$$- \sum_{\vec{q}_3 \neq 0} \frac{4\pi e^2[1 - G(q_3)]}{q_3^2 \tilde{\epsilon}(q_3)} \Pi^{(1)}(\vec{q}_1, \vec{q}_3)\Pi^{(1)}(\vec{q}_3, \vec{q}_1) \bigg) \tag{5.64}$$

where $\epsilon(q)$ and $\tilde{\epsilon}(q)$ are the test-charge–test-charge and electron–test-charge dielectric functions of the homogeneous electron gas. This expansion of the dielectric matrix can be inserted in the electronic contribution to the dynamical matrix, which was given by:

$$D^{el}_{m_1,\alpha;m_2,\beta}(\vec{Q}) = \frac{\Omega_0}{\nu M} \sum_{\vec{G}_1,\vec{G}_2} v^0_{\vec{G}_1-\vec{Q}} v^0_{\vec{Q}-\vec{G}_2} \frac{|\vec{Q} - \vec{G}_1|^2}{4\pi e^2}(\vec{G}_1 - \vec{Q})_\alpha(\vec{G}_2 - \vec{Q})_\beta$$

$$\times\, e^{-i\vec{G}_1\cdot\vec{p}_{m_1}} e^{i\vec{G}_2\cdot\vec{p}_{m_2}}\left(\epsilon^{-1}(\vec{Q} - \vec{G}_1, \vec{Q} - \vec{G}_2) - \delta_{\vec{G}_1,\vec{G}_2}\right)$$

$$-\, \delta_{m_1,m_2} \frac{\Omega_0}{\nu M} \sum_{\vec{G} \neq 0} \nu v^0_{\vec{G}} n_{-\vec{G}} G_\alpha G_\beta e^{-i\vec{G}\cdot\vec{p}_{m_1}} \tag{5.65}$$

where $\vec{G}_1, \vec{G}_2$ are reciprocal lattice vectors, ν is the number of ions per primitive cell, and $\vec{p}_m$ denotes the relative position of the ion m in the primitive cell. Note that this derivation of the dielectric matrix was obtained with a local approximation v^0_q for the ionic pseudopotential, and in this approximation the energy-wave number characteristic was obtained as:

$$F(q) = \frac{\Omega_0}{2} \frac{q^2}{2\pi e^2} |v^0_q|^2\left(\frac{1}{\epsilon(q)} - 1 \right) \tag{5.66}$$

The zero order contribution to the dielectric matrix imposes that $\vec{G}_1 = \vec{G}_2$. One then immediately recognizes the energy-wave number characteristic in the first term of this expression for the dynamical matrix. Furthermore, in the framework of the local ionic pseudopotential, the induced electron density is given by:

$$n_{\vec{G}} = -\frac{G^2}{4\pi e^2} S(\vec{G}) v^0_{\vec{G}}\left(\frac{1}{\epsilon(G)} - 1 \right) \tag{5.67}$$

and thus also the second term in the expression for the dielectric matrix is readily expressed in the energy-wave number characteristic (also writing the structure factor explicitly):

$$
\begin{aligned}
D^{el}_{m_1,\alpha;m_2,\beta}(\vec{Q}) =\ & \frac{2}{\nu M} \sum_{\vec{G}} (\vec{G} - \vec{Q})_\alpha (\vec{G} - \vec{Q})_\beta F(|\vec{G} - \vec{Q}|) e^{-i\vec{G}\cdot(\vec{p}_{m_1} - \vec{p}_{m_2})} \\
& - \delta_{m_1,m_2} \frac{2}{\nu M} \sum_{\vec{G}\neq 0} G_\alpha G_\beta F(G) \sum_m e^{-i\vec{G}\cdot(\vec{p}_{m_1} - \vec{p}_m)} \\
& + \frac{\Omega_0}{\nu M} \sum_{\vec{G}_1 \neq \vec{G}_2} v^0_{\vec{G}_1 - \vec{Q}} v^0_{\vec{Q} - \vec{G}_2} \frac{|\vec{Q} - \vec{G}_1|^2}{4\pi e^2} (\vec{G}_1 - \vec{Q})_\alpha (\vec{G}_2 - \vec{Q})_\beta \\
& \times e^{-i\vec{G}_1\cdot\vec{p}_{m_1}} e^{i\vec{G}_2\cdot\vec{p}_{m_2}} \left[\epsilon^{-1}(\vec{Q} - \vec{G}_1, \vec{Q} - \vec{G}_2)\right]^{(1)} \\
& + \frac{\Omega_0}{\nu M} \sum_{\vec{G}} |v^0_{\vec{G} - \vec{Q}}|^2 \frac{|\vec{Q} - \vec{G}|^2}{4\pi e^2} (\vec{G} - \vec{Q})_\alpha (\vec{G} - \vec{Q})_\beta \\
& \times e^{-i\vec{G}\cdot(\vec{p}_{m_1} - \vec{p}_{m_2})} \left[\epsilon^{-1}(\vec{Q} - \vec{G}, \vec{Q} - \vec{G})\right]^{(2)}
\end{aligned}
\tag{5.68}
$$

The first two terms are precisely the electronic contribution to the dynamical matrix, as obtained from the pseudopotential perturbation theory.

The last two terms, being corrections from the dielectric matrix, are formally of third and fourth order in the pseudopotential. But since perturbation theory is questionable with pseudopotential matrix elements at small wave vector, their contribution should be examined in some more detail.

In the first and second order contributions to the dielectric matrix, only pseudopotential matrix elements appear at reciprocal lattice vectors. If the pseudopotential perturbation for the energy makes sense, these matrix elements can be considered as a perturbation. Consequently, the contributions required are these, where the potential $v^0_{\vec{q}}$ is needed at wave vectors smaller than a reciprocal lattice vector. This is obviously only the case for $\vec{G}_1 = 0$ and for $\vec{G}_2 = 0$ in the first order dielectric correction, and for $\vec{G} = 0$ in the second order dielectric correction. It turns out that for simplicity in the notation, one better writes the first order correction in the dielectric matrix in terms of the first order polarizability.

$$D^{el}_{m_1,\alpha;m_2,\beta}(\vec{Q}) =$$

$$\frac{2}{\nu M} \sum_{\vec{G}} (\vec{G} - \vec{Q})_\alpha (\vec{G} - \vec{Q})_\beta F(|\vec{G} - \vec{Q}|) e^{-i\vec{G}\cdot(\vec{p}_{m_1} - \vec{p}_{m_2})}$$

$$- \delta_{m_1,m_2} \frac{2}{\nu M} \sum_{\vec{G}\neq 0} G_\alpha G_\beta F(G) \sum_m e^{-i\vec{G}\cdot(\vec{p}_{m_1} - \vec{p}_m)}$$

$$- \frac{\Omega_0}{\nu M} \sum_{\vec{G}\neq 0} \frac{v^0_{-\vec{Q}}}{\tilde{\epsilon}(Q)} \frac{v^0_{\vec{Q}-\vec{G}}}{\tilde{\epsilon}(|\vec{Q}-\vec{G}|)} Q_\alpha (\vec{G}-\vec{Q})_\beta e^{i\vec{G}\cdot\vec{p}_{m_2}} \Pi^{(1)}(\vec{Q}, \vec{Q}-\vec{G})$$

$$- \frac{\Omega_0}{\nu M} \sum_{\vec{G}\neq 0} \frac{v^0_{\vec{Q}}}{\tilde{\epsilon}(Q)} \frac{v^0_{\vec{G}-\vec{Q}}}{\tilde{\epsilon}(|\vec{Q}-\vec{G}|)} (\vec{G}-\vec{Q})_\alpha Q_\beta e^{-i\vec{G}\cdot\vec{p}_{m_1}} \Pi^{(1)}(\vec{Q}-\vec{G}, \vec{Q})$$

$$+ \frac{\Omega_0}{\nu M} |v^0_{\vec{Q}}|^2 Q_\alpha Q_\beta \frac{Q^2}{4\pi e^2} \left[\epsilon^{-1}(\vec{Q},\vec{Q})\right]^{(2)}$$

$$(5.69)$$

At this stage, the first and second order corrections to the polarizability have to be evaluated. In the framework of the local pseudopotential, adopted here, we note that the orthogonalisation hole is closely connected to the non-locality of the pseudopotential, as shown by Shaw and Harrison[4]. Therefore, in a local treatment, one should also neglect the orthogonalisation hole, from which it follows:

$$c_{\vec{q}}(\vec{k}) = a_{\vec{q}}(\vec{k}) = S(\vec{q}) \frac{v^0_{\vec{q}}}{E^0_{\vec{k}} - E^0_{\vec{k}+\vec{q}}} \qquad (5.70)$$

The first order polarizability then becomes:

$$\Pi^{(1)}(\vec{Q}, \vec{Q}-\vec{G}) = \frac{2v_{\vec{G}}}{\Omega} \sum_{\vec{k}} p^0_{\vec{k}} \left[\frac{S(\vec{G})}{E^0_{\vec{k}} - E^0_{\vec{k}+\vec{Q}}} \left(\frac{1}{E^0_{\vec{k}+\vec{Q}} - E^0_{\vec{k}+\vec{Q}-\vec{G}}} + \frac{1}{E^0_{\vec{k}} - E^0_{\vec{k}+\vec{G}}} \right) \right.$$
$$\left. + \frac{S(\vec{G})}{E^0_{\vec{k}} - E^0_{\vec{k}+\vec{Q}-\vec{G}}} \left(\frac{1}{E^0_{\vec{k}+\vec{Q}-\vec{G}} - E^0_{\vec{k}+\vec{Q}}} + \frac{1}{E^0_{\vec{k}} - E^0_{\vec{k}-\vec{G}}} \right) \right]$$

$$(5.71)$$

and:

$$\Pi^{(1)}(\vec{Q}, \vec{Q}-\vec{G}) = [\Pi^{(1)}(\vec{Q}-\vec{G}, \vec{Q})]^* \qquad (5.72)$$

Although the algebra involved is quite tedious, the integrals can be done analytically. Using the representation:

$$\frac{1}{ab} = \int_0^1 d\lambda \frac{1}{[a\lambda + b(1-\lambda)]^2}$$

one first considers the angular integrations, for which the following integral
is needed:

$$\int d^2\Omega_{\vec{k}} \frac{1}{1 + \vec{k}.\vec{u}_1} \frac{1}{1 + \vec{k}.\vec{u}_2} = 8\pi \int_{\lambda_1}^{\lambda_2} d\lambda \frac{1}{\lambda^2 - d}$$

$$\lambda_1 = 2k^2 \vec{u}_1.(\vec{u}_2 - \vec{u}_1)$$

$$\lambda_2 = 2k^2 \vec{u}_2.(\vec{u}_2 - \vec{u}_1)$$

$$d = 4k^2(\vec{u}_1 - \vec{u}_2)^2 + 4k^4[u_1^2 u_2^2 - (\vec{u}_1.\vec{u}_2)^2]$$

To obtain this result, only elementary integrations are required, and a simple
translation to eliminate the linear term in λ from the function $a\lambda^2 + b\lambda + c$.
Consider then the first term in the first-order polarizability, using the above
representation for the angular integrations:

$$\int d^2\Omega_{\vec{k}} \frac{1}{E_{\vec{k}}^0 - E_{\vec{k}+\vec{Q}}^0} \frac{1}{E_{\vec{k}+\vec{Q}}^0 - E_{\vec{k}+\vec{Q}-\vec{G}}^0} = -\left(\frac{2m}{\hbar^2}\right)^2 \frac{8\pi}{Q^2(2\vec{Q}.\vec{G} - G^2)} \int_{\lambda_1}^{\lambda_2} \frac{d\lambda}{\lambda^2 - d}$$

$$\lambda_1 = -\frac{8k^2}{Q^2(2\vec{Q}.\vec{G} - G^2)}(\vec{Q}.\vec{G} - G^2)$$

$$\lambda_2 = -\frac{8k^2}{Q^2(2\vec{Q}.\vec{G} - G^2)}\left(\vec{G}.\vec{Q} - \frac{G^2 Q^2}{2\vec{Q}.\vec{G} - G^2}\right)$$

$$d = \frac{64k^2}{Q^4(2\vec{Q}.\vec{G} - G^2)^2}\left(\frac{G^2 Q^2(\vec{G} - \vec{Q})^2}{4} - k^2[G^2 Q^2 - (\vec{Q}.\vec{G})^2]\right)$$

$$(5.73)$$

The substitution $\lambda = -8\mu/[Q^2(2\vec{Q}.\vec{G} - G^2)]$ introduces a new integration
variable μ in which the expression is much simpler:

$$\int d^2\Omega_{\vec{k}} \frac{1}{E_{\vec{k}}^0 - E_{\vec{k}+\vec{Q}}^0} \frac{1}{E_{\vec{k}+\vec{Q}}^0 - E_{\vec{k}+\vec{Q}-\vec{G}}^0} = \left(\frac{2m}{\hbar^2}\right)^2 \pi \int_{\mu_1}^{\mu_2} \frac{d\mu}{\mu^2 - c}$$

$$\mu_1 = k^2(\vec{Q}.\vec{G} - G^2)$$

$$\mu_2 = k^2\left(\vec{Q}.\vec{G} - \frac{Q^2 G^2}{2\vec{Q}.\vec{G} - G^2}\right)$$

$$(5.74)$$

$$c = \frac{k^2}{4}G^2 Q^2(\vec{Q} - \vec{G})^2 - k^4[Q^2 G^2 - (\vec{Q}.\vec{G})^2]$$

A similar analysis can be done for the angular integrations in the remaining three terms of the first order polarizability. It turns out that the only difference in all these terms is in the integration limits. The result is:

$$\Pi^{(1)}(\vec{Q}, \vec{Q} - \vec{G}) = S(\vec{G}) \frac{4 v_G}{(2\pi)^3} \pi \left(\frac{2m}{\hbar^2}\right)^2 \sum_{i=1}^{4} \int_0^{k_F} k^2 \, dk \int_{k^2 a_i}^{k^2 b_i} \frac{dx}{x^2 - sk^2 + tk^4}$$

$$a_1 = a_2 = \vec{Q}.\vec{G} - G^2$$

$$a_3 = a_4 = -\vec{Q}.\vec{G}$$

$$b_1 = -b_3 = \vec{Q}.\vec{G} - \frac{Q^2 G^2}{2\vec{Q}.\vec{G} - G^2}$$

$$b_2 = b_4 = -\vec{Q}.\vec{G} + Q^2$$

$$s = \frac{1}{4} G^2 Q^2 (\vec{Q} - \vec{G})^2 \geq 0$$

$$t = Q^2 G^2 - (\vec{Q}.\vec{G})^2 \geq 0$$

$$(5.75)$$

Consider then the non-trivial integral:

$$J(u) = \int_0^{k_F} k^2 \, dk \int_0^{k^2 u} \frac{dx}{x^2 - sk^2 + tk^4} \tag{5.76}$$

From symmetry it follows that $J(-u) = -J(u)$, which means that in the first order polarizability the terms in b_1 and b_3 cancel against each other, and one is left with:

$$\Pi^{(1)}(\vec{Q}, \vec{Q} - \vec{G}) = S(\vec{G}) \frac{v_G}{2\pi^2} \left(\frac{2m}{\hbar^2}\right)^2 [2J(b_2) - 2J(a_1) - 2J(a_3)] \tag{5.77}$$

To evaluate the integral, we first introduce the integration variable $y = x/k$ instead of x, which will allow to avoid the quadratic behaviour of the integrand in both variables:

$$\begin{aligned}
J(u) &= \int_0^{k_F} k \, dk \int_0^{ku} \frac{dy}{y^2 - s + tk^2} \\
&= \int_0^{u k_F} dy \int_{y/u}^{k_F} \frac{k \, dk}{y^2 - s + tk^2}
\end{aligned}$$

The last integral is elementary in the integration variable k^2, in which the denominator is linear.

$$J(u) = \frac{1}{2t} \int_0^{uk_F} dy \, \log\left|\frac{k_F^2 t + y^2 - s}{y^2 t/u^2 + y^2 - s}\right|$$

$$= \frac{1}{2t} \int_0^{uk_F} dy \left[\log|y^2 + k_F^2 t - s| - \log\left|y^2 - \frac{su^2}{t + u^2}\right| - \log\left(\frac{t + u^2}{u^2}\right)\right]$$

$$= -\frac{uk_F}{2t} \log\left(\frac{t + u^2}{u^2}\right) + \frac{1}{2t} \int_0^{uk_F} dy \left[\log|y^2 + k_F^2 t - s| - \log\left|y^2 - \frac{su^2}{t + u^2}\right|\right]$$

The integrals over y can be done, e.g. with an integration by parts:

$$\int dy \, \log|y^2 + r| = y \log|y^2 + r| - 2 \int \frac{y^2 \, dy}{y^2 + r}$$

$$= y \log|y^2 + r| - 2y + 2r \int \frac{dy}{y^2 + r}$$

$$= y \log|y^2 + r| - 2y + 2\Theta(r)\sqrt{r} \tan^{-1}\frac{y}{\sqrt{r}}$$

$$+ \Theta(-r)\sqrt{-r} \, \log\left|\frac{y + \sqrt{-r}}{y - \sqrt{-r}}\right|$$

By combining terms, one then finds with some elementary algebra:

$$J(u) = \frac{1}{2t}\left[-\sqrt{\frac{su^2}{t + u^2}} \log\left|\frac{uk_F\sqrt{t + u^2} + \sqrt{su^2}}{uk_F\sqrt{t + u^2} - \sqrt{su^2}}\right|\right.$$

$$+ 2\Theta(k_F^2 t - s)\sqrt{k_F^2 t - s} \tan^{-1}\frac{uk_F}{\sqrt{k_F^2 t - s}} \tag{5.78}$$

$$\left.+ \Theta(s - k_F^2 t)\sqrt{s - k_F^2 t} \, \log\left|\frac{uk_F + \sqrt{s - k_F^2 t}}{uk_F - \sqrt{s - k_F^2 t}}\right|\right]$$

This completes the evaluation of the first order polarizability. Note that for $t = 0$ (i.e. for $\vec{G}$ and $\vec{Q}$ parallel), the appropriate limit is:

$$\lim_{t \to 0} J(u) = -\frac{1}{2u}\left(k_F + \frac{u}{2\sqrt{s}}\left(k_F^2 - \frac{s}{u^2}\right)\log\left|\frac{uk_F + \sqrt{s}}{uk_F - \sqrt{s}}\right|\right) \tag{5.79}$$

We evaluated the phonon frequencies, resulting from the dynamical matrix (Eq. 69) including the first order polarizability (Eq. 75, 77, 78). However, for Na and Al, the effect turned out to be of the order of 1 to 2 percent. This is in agreement with earlier results[5], who only found essential higher-order effects in Be. Up to now, we did not evaluate the contribution of the second order polarizability.

Chapter 6

CONCLUDING REMARKS

The main purpose of the present notes was to elucidate in a systematic way the basic concepts, underlying most evolved theories of the electronic structure, dynamics and quantum structural properties of solids. Quite some attention has been paid to the one-electron approximation, which is one of the basic assumptions in actual calculations. Also the dielectric matrix formulation was treated rather in detail, since it enters in many studies of both the static and dynamical properties of solids.

Since the exchange and correlation interaction still forms an important (and in principle unsolved) problem of many-particle physics, the recent evolution in this field has been discussed in the homogeneous electron gas. To the best of our knowledge, similar studies have not been performed yet in real solids, and in particular in semiconductors. In practice, one essentially has to rely on the local-density approximation (which turns out to be quite successful, as will be shown by several lecturers at this ASI). As far as the exchange part in the exchange and correlation interaction is concerned, we have proposed an alternative procedure, going beyond the local-density approximation. At least in the electron gas, this method has shown to give substantial improvements as compared to previous theories (both in explaining the plasmon dispersion, as in realizing internal consistency of the theory).

For illustrative purposes, this exchange treatment was introduced in the pseudopotential perturbation theory, and revealed substantial exchange effects in the phonon dispersion of simple metals (in contrast to Harrison's suggestion[1]).

Nevertheless, the agreement with the experimental data is not as good as obtained by other recent first-principle[2] calculations, also based on the BHS pseudopotentials. This indicates that several corrections still have to be taken into account:

a. In the formalism as we presented so far, no attempt was made to include the correlation effects beyond the dynamical exchange effects. These correlations will in a first stage probably have to be included in the framework of the local-density approximation.

b. Up to now, we did not examine the dependence on the atomic volume. In Al e.g., it has been shown [Ref. 2] that an increase in the atomic volume by 3%, lowers the phonon frequencies by about 10%.

For these reasons, the formalism as presented in these notes, only constitutes a first step, and is intended to indicate some general trends.

Furthermore, for treating semiconductors (and also the stucture of simple metals), one has to leave the second order perturbation theory. The self-consistent pseudopotentials which are then needed, are not discussed in these lecture notes at all. They will be discussed by several lecturers at this ASI, in connection with several other topics, not even mentioned in these notes.

In fact, the self-consistent pseudopotentials form an important subject in our research group, and lies at the basis of the first ab-initio calculation[3] of the phonon spectrum in Si. The material presented here is mainly intended for further use in the self-consistent study of the electronic properties of solids.

Appendix A

ELECTROSTATIC ENERGY OF IONS IN A UNIFORM BACKGROUND

The electrostatic energy *per ion* of point ions with valence Z in a uniform compensating background of electrons is given by:

$$E^{es} = \frac{1}{2N} \sum_{i \neq j} \frac{Z^2 e^2}{|\vec{R}_i - \vec{R}_j|} - \frac{e^2}{2N} \int d^3 r_1 \int d^3 r_2 \frac{n^2}{|\vec{r}_1 - \vec{r}_2|} \qquad (A.1)$$

where $n = Z/\Omega_0$, with Ω_0 the average volume per ion. Note that we used the fact that for a homogeneous background, the magnitude of the attractive term between the ions and the electrons is twice the repulsive term from the background. Since for numerical purposes the sum over the Coulomb interactions is slowly convergent, the Ewald-Fuchs procedure provides a convenient and accurate way for obtaining fast convergence. The basic idea is to write the Coulomb interaction as a sum of two terms:

$$\frac{1}{r} = \frac{2}{\sqrt{\pi}} \frac{1}{r} \int_0^{r\sqrt{\eta}} dt\, e^{-t^2} + \frac{2}{\sqrt{\pi}} \frac{1}{r} \int_{r\sqrt{\eta}}^{\infty} dt\, e^{-t^2} \qquad (A.2)$$

where η is an arbitrary constant, which will be determined to obtain optimal convergence. For the first term, one considers the Fourier expansion:

$$\frac{1}{r} \int_0^{r\sqrt{\eta}} dt\, e^{-t^2} = \sum_{\vec{q}} \frac{2\pi}{\Omega} \frac{e^{-q^2/4\eta}}{q^2} \sqrt{\pi} e^{i\vec{q}\cdot\vec{r}} \qquad (A.3)$$

where Ω is the total volume of the crystal, and the wave vectors $\vec{q}$ satisfy the periodic boundary conditions. Using then the resulting expression

$$\frac{1}{r} = \frac{2}{\sqrt{\pi}} \frac{1}{r} \int_{r\sqrt{\eta}}^{\infty} dt\, e^{-t^2} + \frac{4\pi}{\Omega} \sum_{\vec{q}} e^{i\vec{q}\cdot\vec{r}} \qquad (A.4)$$

one readily recognizes the geometrical structure factor in the electrostatic
energy:

$$S(\vec{q}) \equiv \frac{1}{N} \sum_j e^{-i\vec{q}\cdot\vec{R}_j} \qquad (A.5)$$

and using:

$$\sum_{i \neq j} e^{i\vec{q}\cdot(\vec{R}_i - \vec{R}_j)} = N^2 |S(\vec{q})|^2 - N \qquad (A.6)$$

the ionic contribution becomes:

$$\begin{aligned}
\frac{1}{2N} \sum_{i \neq j} \frac{Z^2 e^2}{|\vec{R}_i - \vec{R}_j|} = {} & \frac{Z^2 e^2}{N} \sum_{i \neq j} \frac{2}{\sqrt{\pi}} \frac{1}{|\vec{R}_i - \vec{R}_j|} \int_{|\vec{R}_i - \vec{R}_j|\sqrt{\eta}}^{\infty} dt\, e^{-t^2} \\
& + \frac{2\pi Z^2 e^2}{\Omega_0} \sum_{\vec{q}} |S(\vec{q})|^2 \frac{e^{-q^2/4\eta}}{q^2} - \frac{2\pi Z^2 e^2}{\Omega} \sum_{\vec{q}} \frac{e^{-q^2/4\eta}}{q^2}
\end{aligned} \qquad (A.7)$$

Converting the sum into an integral ($\sum_{\vec{q}} \to \Omega/(2\pi)^3$ because of the periodic
boundary conditions), the last term can be evaluated, giving $-Z^2 e^2 \sqrt{\eta/\pi}$.

For the background interaction, the same decomposition can be used,
which gives:

$$\begin{aligned}
\frac{1}{N} \int d^3 r_1 \int d^3 r_2 \left(\frac{Z}{\Omega_0}\right)^2 \frac{1}{|\vec{r}_1 - \vec{r}_2|} = {} & \frac{Z^2}{\Omega_0} \int d^3 r\, \frac{1}{r} \\
= {} & \frac{Z^2}{\Omega_0} \int d^3 r\, \frac{2}{\sqrt{\pi}} \frac{1}{r} \int_{r\sqrt{\eta}}^{\infty} dt\, e^{-t^2} \\
& + \frac{4\pi}{\Omega} \sum_{\vec{q}} \frac{e^{-q^2/4\eta}}{q^2} \frac{Z^2}{\Omega_0} \int d^3 e^{i\vec{q}\cdot\vec{r}}
\end{aligned}$$
$$(A.8)$$

Because $\int d^3 r\, e^{i\vec{q}\cdot\vec{r}} = \Omega \delta_{\vec{q},0}$, the last term exactly cancels the divergent term
at $\vec{q} = 0$ in the sum over the ion positions. The remaining integral can
be done (e.g. by partial integration, starting from $r\,dr = dr^2/2$), and one

Table 1. Values of α_{str} for simple lattices

structure		α_{str}
sc		1.760
fcc		1.79172
bcc		1.79186
hcp,	$c/a = 1.4$	1.78664
	1.5	1.78997
	1.6	1.79156
	$\sqrt{8/3}$	1.791676
	1.7	1.79128
	1.8	1.78909

obtains:

$$\int_0^{\infty} r\,dr \int_{r\sqrt{\eta}}^{\infty} dt\,e^{-t^2} = \frac{\sqrt{\pi}}{8\eta} \tag{A.9}$$

Collecting terms, one finds for the electrostatic energy:

$$E^{es} = \frac{Z^2 e^2}{2N} \sum_{i \neq j} \frac{2}{\sqrt{\pi}} \frac{1}{|\vec{r}_i - \vec{r}_j|} \int_{|\vec{r}_i - \vec{r}_j|\sqrt{\eta}}^{\infty} dt\,e^{-t^2} + \frac{2\pi Z^2 e^2}{\Omega_0} \sum_{\vec{q} \neq 0} |S(\vec{q})|^2 \frac{e^{-q^2/4\eta}}{q^2}$$

$$- Z^2 e^2 \sqrt{\frac{\eta}{\pi}} - \frac{Z^2 e^2}{2\Omega_0} \frac{\pi}{\eta} \tag{A.10}$$

Differentiation with respect to η confirms that this expression is indeed independent of η as it should be.

The choice of η determines the convergence of the summations. If one requires that the maximum contribution is given by the structure independent terms, one finds:

$$\eta = \left(\frac{9\pi}{16}\right)^{1/3} \frac{1}{R_0^2} \tag{A.11}$$

where R_0 is the radius of a sphere with volume Ω_0.

For a primitive lattice, the difference between two lattice vectors is still a lattice vector. This reduces the double sum over the lattice positions to a single sum, without the factor $1/N$ in front. One easily shows that the electrostatic energy can be written as:

$$E^{es} = -\alpha_{str} \frac{Z^2 e^2}{2R_0} \tag{A.12}$$

where α_{str} is a dimensionless factor, which only depends on the structure of the lattice. Some values for simple metals are given in Table A.1. These values are comparable to the value 1.8, which is found for a point ion in a uniform electron distribution, contained in a sphere with radius R_0 centered around the ion. For more complicated structures, the value of α_{str} for simple lattices might be quite helpful, since normally the lattice can be decomposed in superpositions of Bravais cells. E.g. for the diamond structure one finds:

$$\alpha_{diamond} = \frac{\alpha_{fcc}}{2^{1/3}} - \frac{\alpha_{sc}}{2} + \frac{\alpha_{bcc}}{2^{2/3}} = 1.671 \qquad (A.13)$$

Also for point vacancies, interstitials and stacking faults the knowledge of this constant is of interest (see e.g. Harrison for some applications), but then one compares energies at different structures, and in many cases this comparison is easier with $\eta = \infty$ or $\eta = 0$. Only for the calculation of the reference structure, the full Ewald-Fuchs procedure is needed.

Appendix B

DYNAMICAL MATRIX BY EWALD-FUCHS METHOD

In this appendix, the Ewald-Fuchs method for the calculation of the dynamical matrix is outlined in quite some detail, provided the structure dependent contribution to the energy per ion is given by:

$$E(\vec{R}_1, \cdots, \vec{R}_N) = \sum_{\vec{q} \neq 0} |S(\vec{q})|^2 H(q) + \frac{Z^{*2} e^2}{2N} \sum_{i \neq j} f(|\vec{R}_i - \vec{R}_j|) \qquad (B.1)$$

where $S(\vec{q})$ is the geometrical structure factor, and:

$$f(r) = \frac{1}{r} \mathrm{erfc}(r\sqrt{\eta}) \qquad (B.2)$$

$$H(q) = F(q) + \frac{2\pi Z^{*2} e^2}{\Omega_0} \frac{e^{-q^2/4\eta}}{q^2} \qquad (B.3)$$

The appearance of the energy-wave number characteristic is not essential for the derivation, but is quite natural from perturbation theory with pseudo-potentials. In the more elaborate treatments, as e.g. the dielectric formulation, the electronic contribution is more complicated, as is outlined in the related section. This contribution has then to be treated in the appropriate way. The main purpose of the Ewald-Fuchs method is to handle the ionic contribution, which is written as in the previous appendix, where η is an arbitrary constant, chosen to obtain optimum convergence for the direct and the reciprocal lattice sum. Denoting the equilibrium positions of the ions by $\vec{R}_j^{\,0}$, and the displacements from equilibrium by $\vec{\xi}_j$:

$$\vec{R}_j = \vec{R}_j^{\,0} + \vec{\xi}_j \qquad (B.4)$$

the Lagrangian corresponding to these displacements in the adiabatic approximation for the electrons is given by:

$$L = \sum_{j} \frac{M}{2}\left(\frac{d\vec{\xi}_j}{dt}\right)^2 - N\sum_{\vec{q}\neq 0}|S(\vec{q})|^2 H(q) - \frac{Z^{*2}e^2}{2}\sum_{i\neq j} f(|\vec{R}_i - \vec{R}_j|) \qquad (B.5)$$

The equation of motion thus becomes:

$$M\frac{d^2\vec{\xi}_j}{dt^2} + N\sum_{\vec{q}\neq 0} H(q)\frac{\partial |S(\vec{q})|^2}{\partial \vec{\xi}_j} + \frac{Z^{*2}e^2}{2}\sum_{k\neq l}\frac{\partial f(|\vec{R}_l - \vec{R}_k|)}{\partial \vec{\xi}_j} = 0 \qquad (B.6)$$

which is easily written as:

$$M\frac{d^2\vec{\xi}_j}{dt^2} + 2i\sum_{\vec{q}\neq 0} H(q)\vec{q}e^{i\vec{q}.\vec{R}_j}S(\vec{q}) + Z^{*2}e^2\sum_{l\neq j}\left[\frac{\vec{r}}{2}\frac{df(r)}{dr}\right]_{\vec{r}=\vec{R}_j-\vec{R}_l} = 0 \qquad (B.7)$$

For the calculation of the phonons in the harmonic approximation, this equation of motion has to be considered to first order in the displacements (which corresponds to a second order calculation in the potential). Expanding the exponential function and the structure factor, the sum in reciprocal space becomes:

$$2i\sum_{\vec{q}\neq 0} H(q)\vec{q}e^{i\vec{q}.\vec{R}_j}S(\vec{q}) \approx 2i\sum_{\vec{q}\neq 0}\vec{q}H(q)e^{i\vec{q}.\vec{R}_j^0}S^{(0)}(\vec{q})$$

$$- 2\sum_{\vec{q}\neq 0}\vec{q}(\vec{q}.\vec{\xi}_j)H(q)e^{i\vec{q}.\vec{R}_j^0}S^{(0)}(\vec{q}) + \frac{2}{N}\sum_{\vec{q}\neq 0}\sum_{l}\vec{q}(\vec{q}.\vec{\xi}_l)H(q)e^{i\vec{q}.(\vec{R}_j^0-\vec{R}_l^0)}$$

$$(B.8)$$

Inserting the explicit expression for the equilibrium structure factor, this term becomes:

$$2i\sum_{\vec{q}\neq 0} H(q)\vec{q}e^{i\vec{q}.\vec{R}_j}S(\vec{q}) \approx 2i\sum_{\vec{q}\neq 0}\vec{q}H(q)e^{i\vec{q}.\vec{R}_j^0}S^{(0)}(\vec{q})$$

$$- \frac{2}{N}\sum_{\vec{q}\neq 0}\sum_{l}\vec{q}\big([\vec{q}.(\vec{\xi}_j - \vec{\xi}_l)]\big)H(q)e^{i\vec{q}.(\vec{R}_j^0-\vec{R}_l^0)} \qquad (B.9)$$

For the contribution of the sum in real space, one observes that

$$\left[\frac{\vec{r}}{r}\frac{df(r)}{dr}\right]_{\vec{r}=\vec{R}_j-\vec{R}_l} = \left[\frac{\vec{r}}{r}\frac{df(r)}{dr}\right]_{\vec{r}=\vec{R}_j^0-\vec{R}_l^0} + \left[[(\vec{\xi}_j-\vec{\xi}_l).\vec{\nabla}_{\vec{r}}]\vec{\nabla}_{\vec{r}}f(r)\right]_{\vec{r}=\vec{R}_j^0-\vec{R}_l^0}$$

$$\text{(B.10)}$$

Since the zero-order terms have to be zero in the equilibrium configuration, the equation of motion to first order in the displacements thus becomes:

$$\begin{aligned}
M\frac{d^2\vec{\xi}_j}{dt^2} = & \frac{2}{N}\sum_{\vec{q}\neq 0}\sum_{l}\vec{q}\big([\vec{q}.(\vec{\xi}_j-\vec{\xi}_l)]\big)H(q)e^{i\vec{q}.(\vec{R}_j^0-\vec{R}_l^0)} \\
& - Z^{*2}e^2\sum_{l\neq j}\left[[(\vec{\xi}_j-\vec{\xi}_l).\vec{\nabla}_{\vec{r}}]\vec{\nabla}_{\vec{r}}f(r)\right]_{\vec{r}=\vec{R}_j^0-\vec{R}_l^0}
\end{aligned}$$

$$\text{(B.11)}$$

Consider now a lattice with ν atoms per primitive cell. If $\vec{X}_p$ denotes the N/ν positions of the primitive cells, and $\vec{\rho}_m$ the relative coordinates of the ions in each primitive cell (with $m = 0, 1, \cdots, \nu-1$), the structure factor $S^{(0)}(\vec{q})$ at equilibrium is zero, except if $\vec{q}$ is a reciprocal lattice vector $\vec{G}$ of the Bravais lattice, where it is given by:

$$S^{(0)}(\vec{G}) = \frac{1}{N}\sum_{p,m}e^{-i\vec{G}.(\vec{X}_p+\vec{\rho}_m)} = \frac{1}{\nu}\sum_{m}e^{-i\vec{G}.\vec{\rho}_m} \qquad \text{(B.12)}$$

The ion displacements, characterized by the two indices p and m, can then be expanded in a Fourier sum:

$$\vec{\xi}_{p,m} = \sum_{\vec{Q}}\vec{a}_m(\vec{Q})e^{i\vec{Q}.(\vec{X}_p+\vec{\rho}_m)} \qquad \text{(B.13)}$$

where $\vec{Q}$ is restricted to the first Brillouin zone. The inverse relation is easily obtained if one multiplies this expansion with $e^{-i\vec{Q}_1.\vec{X}_p}$ and sums over the Bravais cells. This sum is zero, except if $\vec{Q}_1-\vec{Q}$ is a reciprocal lattice vector. Since both vectors are in the first Brillouin zone, this condition means that $\vec{Q}_1 = \vec{Q}$. As a results one obtains:

$$\vec{a}_m(\vec{Q}) = \frac{\nu}{N}\sum_{p}\vec{\xi}_{p,m}e^{-i\vec{Q}.(\vec{X}_p+\vec{\rho}_m)} \qquad \text{(B.14)}$$

Applying this Fourier transform, and looking for the eigen frequencies $\omega_{\vec{Q}}$ of these normal coordinates, one finds:

$$M\omega_{\vec{Q}}^2 \vec{a}_{m_1}(\vec{Q}) = -\frac{\nu}{N}\frac{2}{N}\sum_{\vec{q}\neq 0}\sum_{p_1}\sum_{p_2,m_2}\vec{q}\big([\vec{q}.(\vec{\xi}_{p_1,m_1}-\vec{\xi}_{p_2,m_2})]\big)H(q)$$

$$\times\, e^{i\vec{q}.(\vec{X}_{p_1}+\vec{\rho}_{m_1}-\vec{X}_{p_2}-\vec{\rho}_{m_2})}e^{-i\vec{Q}.(\vec{X}_{p_1}+\vec{\rho}_{m_1})}$$

$$+\frac{\nu}{N}Z^{*2}e^2\sum_{p_1}\sum_{p_2,m_2}{}'\big[[(\vec{\xi}_{p_1,m_1}-\vec{\xi}_{p_2,m_2}).\vec{\nabla}_{\vec{r}}]\vec{\nabla}_{\vec{r}}f(r)\big]_{\vec{r}=\vec{X}_{p_1}+\vec{\rho}_{m_1}-\vec{X}_{p_2}-\vec{\rho}_{m_2}}$$

$$\times\, e^{-i\vec{Q}.(\vec{X}_{p_1}+\vec{\rho}_{m_1})}$$

$$(B.15)$$

where the prime at the last summation means that the term with $p_1 = p_2$ and $m_1 = m_2$ is to be excluded.

Consider then e.g. the term in $\vec{\xi}_{p_1,m_1}$ in the reciprocal sum. The summation over p_2 gives N/ν, provided $\vec{q}$ is a reciprocal lattice vector $\vec{G}$. But since $e^{i\vec{G}.\vec{X}_{p_1}} = 1$, the remaining contribution involves:

$$\sum_{p_1}\sum_{m_2}\vec{\xi}_{p_1,m_1}e^{i\vec{G}.(\vec{\rho}_{m_1}-\vec{\rho}_{m_2})}e^{-i\vec{Q}.(\vec{X}_{p_1}+\vec{\rho}_{m_1})}$$

$$=\frac{N}{\nu}\sum_{m_2}\vec{a}_{m_1}(\vec{Q})e^{i\vec{G}.(\vec{\rho}_{m_1}-\vec{\rho}_{m_2})} \qquad (B.16)$$

For the term in $\vec{\xi}_{p_2,m_2}$ in the reciprocal sum, one observes that the summation over p_1 gives a factor N/ν provided $\vec{q}-\vec{Q}$ is a reciprocal lattice vector $\vec{G}$. One then still has to evaluate:

$$\sum_{p_2}\sum_{m_2}\vec{\xi}_{p_2,m_2}e^{i\vec{G}.\vec{\rho}_{m_1}}e^{-i(\vec{G}+\vec{Q}).(\vec{X}_{p_2}+\vec{\rho}_{m_2})}$$

$$=\frac{N}{\nu}\sum_{m_2}\vec{a}_{m_2}(\vec{Q})e^{i\vec{G}.(\vec{\rho}_{m_1}-\vec{\rho}_{m_2})} \qquad (B.17)$$

where we used the fact that $e^{-i\vec{G}.\vec{X}_{p_2}} = 1$, and the relation between the collective coordinates and the displacement vectors. Combining terms in the reciprocal sum, one then finds:

$$M\omega_{\vec{Q}}^2 \vec{a}_{m_1}(\vec{Q}) =$$

$$-\frac{2}{\nu}\sum_{\vec{G}\neq 0}\sum_{m_2}\vec{G}\big(\vec{G}.\vec{a}_{m_1}(\vec{Q})\big)H(G)e^{i\vec{G}.(\vec{\rho}_{m_1}-\vec{\rho}_{m_2})}$$

(continued)

$$+ \frac{2}{\nu} \sum_{\vec{G}} \sum_{m_2} (\vec{G} + \vec{Q})\big((\vec{G} + \vec{Q}).\vec{a}_{m_2}(\vec{Q})\big)H(|\vec{G} + \vec{Q}|)e^{i\vec{G}.(\vec{\rho}_{m_1} - \vec{\rho}_{m_2})}$$

$$+ \frac{\nu}{N}Z^{*2}e^2 \sum_{p_1} \sideset{}{'}\sum_{p_2,m_2} \Big[\big[(\vec{\xi}_{p_1,m_1} - \vec{\xi}_{p_2,m_2}).\vec{\nabla}_{\vec{r}}\big]\vec{\nabla}_{\vec{r}} f(r)\Big]_{\vec{r}=\vec{X}_{p_1}+\vec{\rho}_{m_1}-\vec{X}_{p_2}-\vec{\rho}_{m_2}}$$

$$\times\, e^{-i\vec{Q}.(\vec{X}_{p_1}+\vec{\rho}_{m_1})} \tag{B.18}$$

For the evalution of the remaining summations over the primitive lattice vectors, the same procedure can be applied identically, if the condition on the position vector $\vec{r}$ is replaced by an integral over $\vec{r}$ via a δ-function, which is used in the representation:

$$\delta(\vec{r} - \vec{R}) = \frac{1}{\Omega} \sum_{\vec{q}} e^{i\vec{q}.(\vec{r}-\vec{R})} \tag{B.19}$$

Proceeding then along the same lines in the (artificially introduced) reciprocal space, one finds:

$$M\omega_{\vec{Q}}^2 \vec{a}_{m_1}(\vec{Q}) =$$

$$-\frac{2}{\nu} \sum_{\vec{G}\neq 0} \sum_{m_2} \vec{G}\big(\vec{G}.\vec{a}_{m_1}(\vec{Q})\big)H(G)e^{i\vec{G}.(\vec{\rho}_{m_1} - \vec{\rho}_{m_2})}$$

$$+\frac{2}{\nu} \sum_{\vec{G}} \sum_{m_2} (\vec{G} + \vec{Q})\big((\vec{G} + \vec{Q}).\vec{a}_{m_2}(\vec{Q})\big)H(|\vec{G} + \vec{Q}|)e^{i\vec{G}.(\vec{\rho}_{m_1} - \vec{\rho}_{m_2})}$$

$$+\frac{N}{\nu}Z^{*2}e^2 \frac{1}{\Omega} \int d^3r \sum_{\vec{G}} \sum_{m_2} e^{-i\vec{G}.\vec{r}}[\vec{a}_{m_1}(\vec{Q}).\vec{\nabla}_{\vec{r}}]\vec{\nabla}_{\vec{r}} f(r) e^{i\vec{G}.(\vec{\rho}_{m_1} - \vec{\rho}_{m_2})}$$

$$-\frac{N}{\nu}Z^{*2}e^2 \frac{1}{\Omega} \int d^3r \sum_{\vec{G}} \sum_{m_2} e^{-i(\vec{G}+\vec{Q}).\vec{r}}[\vec{a}_{m_2}(\vec{Q}).\vec{\nabla}_{\vec{r}}]\vec{\nabla}_{\vec{r}} f(r)$$

$$\times\, e^{i\vec{G}.(\vec{\rho}_{m_1} - \vec{\rho}_{m_2})} \tag{B.20}$$

In the last two terms, one then has to rediscover the δ-function which was introduced earlier. This is easily done in the following way:

$$\sum_p \delta(\vec{x} - \vec{X}_p) = \frac{1}{\Omega} \sum_{\vec{q},p} e^{i\vec{q}.(\vec{x}-\vec{X}_p)} = \frac{N}{\nu\Omega} \sum_{\vec{q}} \delta_{\vec{q},\vec{G}} e^{i\vec{q}.\vec{x}} = \frac{N}{\nu\Omega} \sum_{\vec{G}} e^{i\vec{G}.\vec{x}}$$

$$(B.21)$$

where the previous representation for the δ-function was used, and the fact that the structure factor of the primitive cell only differs from zero at a reciprocal lattice vector. This relation allows to represent the reciprocal sum in the last two terms as a sum over primitive lattice vectors, giving:

$$M\omega_{\vec{Q}}^2 \vec{a}_{m_1}(\vec{Q}) =$$
$$-\frac{2}{\nu} \sum_{\vec{G}\neq 0} \sum_{m_2} \vec{G}\big(\vec{G}.\vec{a}_{m_1}(\vec{Q})\big) H(G) e^{i\vec{G}.(\vec{\rho}_{m_1}-\vec{\rho}_{m_2})}$$
$$+\frac{2}{\nu} \sum_{\vec{G}} \sum_{m_2} (\vec{G}+\vec{Q})\big((\vec{G}+\vec{Q}).\vec{a}_{m_2}(\vec{Q})\big) H(|\vec{G}+\vec{Q}|) e^{i\vec{G}.(\vec{\rho}_{m_1}-\vec{\rho}_{m_2})}$$
$$-Z^{*2}e^2 \sum_{p_2,m_2}{}' \left[\big(e^{-i\vec{Q}.\vec{r}}\vec{a}_{m_2}(\vec{Q}) - \vec{a}_{m_1}(\vec{Q})\big).\vec{\nabla}_{\vec{r}}\vec{\nabla}_{\vec{r}} f(r)\right]_{\vec{r}=\vec{X}_{p_2}+\vec{\rho}_{m_1}-\vec{\rho}_{m_2}}$$

$$(B.22)$$

It is a matter of straightforward algebra to express this equation as a problem of matrix diagonalization:

$$M\omega_{\vec{Q}}^2 a_{m_1,\alpha}(\vec{Q}) = \sum_{m_2,\beta} D_{m_1,\alpha;m_2,\beta}(\vec{Q}) a_{m_2,\beta}(\vec{Q}) \qquad (B.23)$$

where α, β denote Cartesian components, and m_1, m_2 are summation indices over the ν atoms per primitive cell. The explicit expression for the dynamical matrix is more compact if one introduces an auxiliary matrix of the same dimensions:

$$\chi_{m_1,\alpha;m_2,\beta}(\vec{Q}) = \frac{2}{\nu} \sum_{\vec{G}\neq-\vec{Q}} (\vec{Q}+\vec{G})_\alpha (\vec{Q}+\vec{G})_\beta H(|\vec{Q}+\vec{G}|) e^{i\vec{G}.(\vec{\rho}_{m_1}-\vec{\rho}_{m_2})}$$
$$-Z^{*2}e^2 \sum_p \left[e^{-i\vec{Q}.\vec{r}} \frac{\partial^2 f(r)}{\partial r_\alpha \partial r_\beta}\right]_{\vec{r}=\vec{X}_p+\vec{\rho}_{m_1}-\vec{\rho}_{m_2}\neq 0}$$

$$(B.24)$$

One then readily checks that the dynamical matrix is given by:

$$MD_{m_1,\alpha;m_1,\beta}(\vec{Q}) = \chi_{m_1,\alpha;m_1,\beta}(\vec{Q}) - \sum_{m_2} \chi_{m_1,\alpha;m_2,\beta}(\vec{Q}=0)$$
$$MD_{m_1,\alpha;m_2,\beta}(\vec{Q}) = \chi_{m_1,\alpha;m_2,\beta}(\vec{Q}) \quad \text{if } m_1 \neq m_2$$

$$(B.25)$$

For numerical purposes, it is important that the derivatives of $f(r)$ can be calculated analytically:

$$\frac{\partial^2 f(r)}{\partial r_\alpha \partial r_\beta} = \frac{r_\alpha r_\beta}{r^2}\left[\frac{3}{r^3}\mathrm{erfc}(r\sqrt{\eta}) + \left(\frac{6}{r^2} + 4\eta\right)\sqrt{\frac{\eta}{\pi}}e^{-\eta r^2}\right]$$
$$- \delta_{\alpha,\beta}\frac{1}{r^2}\left[\frac{1}{r}\mathrm{erfc}(r\sqrt{\eta}) + 2\sqrt{\frac{\eta}{\pi}}e^{-\eta r^2}\right] \tag{B.26}$$

Concentrating now on the ionic contribution, which is obtained by neglecting the energy-wave number characteristic in $H(q)$, one finds:

$$\chi^{ion}_{m_1,\alpha;m_2,\beta}(\vec{Q}) = \frac{4\pi Z^{*2}e^2}{\nu\Omega_0}\sum_{\vec{G}\neq-\vec{Q}}\frac{(\vec{Q}+\vec{G})_\alpha(\vec{Q}+\vec{G})_\beta}{|\vec{Q}+\vec{G}|^2}e^{-|\vec{Q}+\vec{G}|^2/4\eta}e^{i\vec{G}\cdot(\vec{p}_{m_1}-\vec{p}_{m_2})}$$
$$- Z^{*2}e^2\sum_{p}\left[e^{-i\vec{Q}\cdot\vec{r}}\frac{\partial^2 f(r)}{\partial r_\alpha \partial r_\beta}\right]_{\vec{r}=\vec{X}_p+\vec{p}_{m_1}-\vec{p}_{m_2}\neq 0}$$
$$\tag{B.27}$$

Since η is an arbitrary constant, one might as well consider the limit $\eta \to 0$. In this case, the second contribution simply drops out, and one is left with:

$$\chi^{ion}_{m_1,\alpha;m_2,\beta}(\vec{Q}) = \frac{4\pi Z^{*2}e^2}{\nu\Omega_0}\sum_{\vec{G}\neq-\vec{Q}}\frac{(\vec{Q}+\vec{G})_\alpha(\vec{Q}+\vec{G})_\beta}{|\vec{Q}+\vec{G}|^2}e^{i\vec{G}\cdot(\vec{p}_{m_1}-\vec{p}_{m_2})} \tag{B.28}$$

Of course, this expression is rather slowly convergent numerically, but it is precisely the ionic contribution to the dynamical matrix as found in the section on harmonic phonons from the dielectric approach. The equivalence of both expressions for χ^{ion} allows to evaluate the ionic contribution with a finite value of η. For actual computations, values of the same order as for the calculation of the electrostatic energy assures fast numerical convergence.

REFERENCES

CHAPTER 1

1. J. B. Mann, *Atomic structure calculations, 1: Hartree-Fock Energy Results for Elements Hydrogen to Lawrencium* (Clearinghouse for Technical Information, Springfield, Va.), 1967
    C. F. Fischer, Atomic Data **4**, 451 (1972)

2. See e.g. D. W. Langer, R. N. Euwema, K. Era and T. Koda, Phys. Rev. **B2**, 4005 (1970)

3. P.Hohenberg and W. Kohn, Phys. Rev. **136**, B864 (1964)

4. W. Kohn and L. J. Sham, Phys. Rev. **140**,A1133 (1965)

5. E. P. Wigner and F. Seitz, Phys. Rev. **43**, 804 (1933); **46**, 509 (1934)

6. M. Gell'mann and K. Brueckner, Phys. Rev. **106**,364, (1957)

7. E. P. Wigner, Trans. Faraday Soc. **34**, 678 (1938)

8. G. D. Mahan, *Many-particle Physics, Plenum, 1981*

9. In the next sections, we closely follow D. C. Wallace, *Thermodynamics of Crystals*, J. Wiley, New York, 1972, chapter 5.

10. H. Ehrenreich and M. H. Cohen, Phys. Rev. **115**, 786 (1959)

11. G. Niklasson, Phys. Rev. **B10**, 3052 (1974)

12. P. E. Van Camp, V. E. Van Doren and J. T. Devreese, Phys. Rev. Lett. **42**, 1224 (1979)

CHAPTER 2

1. J. Lindhard, Kong. Danske Vid. Selsk., Mat.-fys. Medd. **24**, No. 8 (1954)
D. Bohm and D, Pines, Phys. Rev. **92**, 609 (1953)
D. Pines and P. Nozières, *The Theory of Quantum Liquids*, Vol. 1, (Benjamin, New York), 1966

2. D. M. Miliotis, Phys. Rev. **B3**, 701 (1971)
P. Zacharias, J. Phys. **F5**, 645 (1975)
H. J. Höhberger, A. Otto and E. Petri, Solid State Commun. **16**, 175 (1975)
P. M. Platzman and P. Eisenberger, Phys. Rev. Lett. **33**, 152 (1974)
P. E. Batson, C. H. Chen and J. Silcox, Phys. Rev. Lett. **37**, 937 (1976)
P. C. Gibbons, S. E. Schnatterly, J. J. Ritsko and J. R. Fields, Phys. Rev. **B13**, 2451 (1976)
S. E. Schnatterly in *Solid State Physics* **34**, eds. H. Ehrenreich, F. Seitz and D. Turnbull (Academic Press, New York), 1979

3. A. J. Glick and R. A. Ferrell, Ann. of Phys. **11**, 359 (1960)

4. J. Hubbard, Proc. Roy. Soc. **A240**, 539 (1957); **A243**, 336 (1958)
D. J. W. Geldart and S. H. Vosko, Can. J. Phys. **44**, 2137 (1966)
L. Kleinman, Phys. Rev. **160**, 585 (1967); **172**, 383 (1968)
K. S. Singwi, M. P. Tosi, R. H. Land and A. Sjölander, Phys. Rev. **176**, 589 (1968)
M. P. Tosi, Il Nouvo Cimento, **1**, 160 (1968)
D. Langreth, Phys. Rev. **181**, 753 (1969)
L. Hedin and S. Lundqvist, in *Solid State Physics* **23**, eds. H. Ehrenreich, F. Seitz and D. Turnbull (Academic Press, New York), 1969
K. S. Singwi, A. Sjölander, M. P. Tosi and R. H. Land, Solid State Comm. **7**, 1503 (1969); Phys. Rev. **B1**, 1044 (1970)
D. J. W. Geldart and R. Taylor, Can. J. Phys. **48**, 155; 167 (1970)
A. W. Overhauser, Phys. Rev. **B3**, 1888 (1971)
P. Vashishta and K. S. Singwi, Phys. Rev. **B6**, 875 (1972)

L. J. Sham, Phys. Rev. **B7**, 4357 (1973)
K. N. Pathak and P. Vashishta, Phys. Rev. **B7**, 3649 (1973)

5. R. Dekeyser, Physica (Utrecht) **38**, 189 (1968)

6. J. C. Kimball, Phys. Rev. **A7**,1648 (1973); **B14**, 2371 (1976)

7. G. Niklasson, Phys. Rev. **B10**, 3052 (1974)

8. A. A. Kugler, J. Stat. Phys. **12**, 35 (1975)

9. F. Brosens, L. F. Lemmens and J. T. Devreese, Phys. Stat. Sol. (b) **74**, 45 (1976)
    F. Brosens, J. T. Devreese and L. F. Lemmens, Phys. Stat. Sol. (b) **80**, 99 (1977)

10. H. Mori, Progr. Theor. Phys. **33**, 423 (1965)

11. G. Mukhopadhyay and A. Sjölander, Phys. Rev. **B17**, 3580 (1978)
    F. Yoshida, S. Takeno and H. Yasuhara, Progr. Theor. Phys. **64**, 40 (1980)
    H. De Raedt and B. De Raedt, Phys. Rev. **B18**, 2039 (1978)

12. J. T. Devreese, F. Brosens and L. F. Lemmens, Phys. Rev. **B21**, 1394 (1980)
    F. Brosens, J. T. Devreese and L. F. Lemmens, Phys. Rev. **B21**, 1363 (1980)
    J. T. Devreese and F. Brosens, in *Proceedings of the 1981 NATO Advanced Study Institute on Electron Correlations in Solids, Molecules and Atoms*, eds. J. T. Devreese and F. Brosens (Plenum Press, New York), 1983

13. W. E. Brittin and W. E. Chappell, Rev. Mod. Phys. **34**, 620 (1962)

14. G. Niklasson, Phys. Rev. **B10**, 3052 (1974)
    G. Niklasson, in *Proceedings of the 1981 NATO Advanced Study Institute on Electron Correlations in Solids, Molecules and Atoms*, eds. J. T. Devreese and F. Brosens (Plenum, New York), 1983

15. S. Schnatterly, in *Proceedings of the 1981 NATO Advanced Study Institute on Electron Correlations in Solids, Molecules and Atoms*, eds. J. T. Devreese and F. Brosens (Plenum, New York), 1983

16. D. N. Tripathy and S. S. Mandal, Phys. Rev. **B16**,231 (1979)

17. F. Toigo and T. O. Woodruff, Phys. Rev. **B2**, 3958 (1970); **B4**, 371 (1971)

18. B. K. Rao, D. N. Tripathy and S. S. Mandal, J. Phys. **F9**, L51 (1979)
    L. F. Lemmens, F. Brosens and J. T. Devreese, Phys. Rev. **B20**, 5398 (1979)

F. Brosens, J. T. Devreese and L. F. Lemmens, J. Phys. **F10**, L27 (1980)
A. Holas and K. S. Singwi, J. Phys. **F11**, L285 (1981)

19. D. J. W. Geldart and R. Taylor, Can. J. Phys. **48**, 155; 167 (1970)

20. A. Holas, P. K. Aravind and K. S. Singwi, Phys. Rev. **B20**, 4912 (1979)

21. F. Brosens and J. T. Devreese, Phys. Rev. **B29**,543 (1984)

22. N. E. Brener and J. L. Fry, Phys. Rev. **B19**, 1720 (1979); **B22**, 2737 (1980)

23. L. Hedin and S. Lundqvist, in *Solid State Physics* **23**, eds. H. Ehrenreich, F. Seitz and D. Turnbull (Academic Press, New York), 1969

24. H. Nachtegaele, F. Brosens and J. T. Devreese, Phys. Rev. **28**, 6064 (1983)

25. K. Utsumi and S. Ichimaru, Phys. Rev. **B23**, 3291 (1981)
S. Ichimaru and K. Utsumi, Phys. Rev. **B24**, 7385 (1981)

26. L. Lantto, P. Pietiläinen and A. Kallio, Phys. Rev. **B26**, 5568 (1982)

27. D. Ceperley, Phys. Rev. **B18**, 3126 (1978)
D. M. Ceperley and B. J. Adler, Phys. Rev. Lett. **45**, 566 (1980)

CHAPTER 3

1. A. O. E. Animalu and V. Heine, Phil. Mag. **12**, 1249 (1965)

2. L. Szasz and G. McGinn, J. Chem. Phys. **47**, 3495 (1967)

3. A. Zunger and M. L. Cohen, Phys. Rev. **B18**, 5449 (1978)

4. A. U. Hazi and S. A. Rice, J. Chem. Phys. **48**, 495 (1968)

5. J. C. Barthelat and P. Durand, Chem. Phys. Lett. **16**, 63 (1972)

6. M. A. Ratner, S. Topiol and J. W. Moskowitz, Chem. Phys. **20**, 1 (1976)

7. A. Redondo, W. A. Goddard and T.C. McGill, Phys. Rev. **B15**, 5038 (1977)

8. K. M. Ho, M. L. Cohen and M. Schlüter, Chem. Phys. Lett. **46**, 608 (1977)

9. W. A. Harrison, in *Pseudopotentials in the Theory of Metals*, (Benjamin, New York, 1966)

10. V. Heine, M. L. Cohen and V. Heine, V. Heine and D. Weaire, in *Solid State Physics* **24**, Eds. H. Ehrenreich, F. Seitz and D. Turnbull (Academic, New York, 1970)

11. M. Schlüter, J. R. Chelikowsky, S. G. Louie and M. L. Cohen, Phys. Rev. **B12**, 4200 (1975)

12. J. A. Appelbaum and D. Hamann, Rev. Mod. Phys. **48**, 479 (1976)

13. S. G. Louie, K. M. Ho, J. R. Chelikowsky and M. L. Cohen, Phys. Rev. **15**, 5627 (1977)

14. G. P. Kerker, K. M. Ho and M. L. Cohen, Phys. Rev. **B18**, 5473 (1978)

15. W. A. Harrison, in *Pseudopotentials in the Theory of Metals*, (Benjamin, New York, 1966)

16. V. Heine, M. L. Cohen and V. Heine, V. Heine and D. Weaire, in *Solid State Physics* **24**, Eds. H. Ehrenreich, F. Seitz and D. Turnbull (Academic, New York, 1970)

17. D. Stroud and N. W. Ashcroft, J. Phys. F **1**, 113 (1971)

18. P. B. Allen and M. L. Cohen, Phys. Rev. **187**, 525 (1969)

19. J. C. Phillips and L. Kleinman, Phys. Rev. **116**, 287 (1975)

20. C. Herring, Phys. Rev. **57**, 1169 (1940)

21. B. J. Austin, V. Heine and L. J. Sham, Phys. Rev **127**, 276 (1962)

22. V. Heine, in *The Physics of Metals. I. Electrons*, J. M. Ziman, ed. Cambridge University Press, New York, 1969, p. 7

23. W. A. Harrison, Phys. Rev. **131**, 2433 (1963)

24. N. W. Ashcroft, Physics Letters **23**, 48 (1966)

25. N. W. Ashcroft, J. Phys. **C1**, 232 (1968)

26. I. V. Abarenkov and V. Heine, Phil. Mag. **12**, 529 (1965)

27. A. O. E. Animalu and V. Heine, Phil. Mag. **12**, 1249 (1965)
    A. O. E. Animalu, Proc. Roy. Soc. A **294**, 376 (1966)

28. W. A. Harrison, *Electronic Structure and the Properties of Solids*, ed. W. H. Freeman, San Francisco, 1980

29. R. Kahn and W. A. Goddard III, J. Chem. Phys. **56**, 2685 (1972)
    C. F. Melius and W. A. Goddard III, Phys. Rev **A10**, 1528 (1974)

S. Topiol, A. Zunger and M. A. Ratner, Chem. Phys. Lett. **49**, 367 (1977)

30. D. R. Hamann, M. Schlüter and C. Chiang, Phys. Rev. Lett. **43**, 1494 (1979)

31. G. B. Bachelet, D. R. Hamann and M. Schlüter, Phys. Rev. **B26**, 4199 (1982)

32. R.W. Shaw and W. A. Harrison, Phys. Rev. **163**, 604 (1967)

CHAPTER 4

1. W. A. Harrison, *Pseudopotentials in the Theory of Metals*, Benjamin, New York, 1966

2. V. Heine, in *Solid State Physics* **24**, eds. H. Ehrenreich, F. Seitz and D. Turnbull (Academic, New York), 1970

3. D. C. Wallace, *Thermodynamics in Crystals*, Ed. J. Wiley, New York, 1972

CHAPTER 5

1. The general outline of this section follows Harrison, Ref. 1 of the previous chapter.

2. R. Stedman and G. Nilsson, Phys. Rev. **145**, 492 (1966)

3. For this derivation we closely follow D. C. Wallace, *Thermodynamics of Crystals*, J. Wiley, New York, 1972

4. R. W. Shaw and W. A. Harrison, Phys. Rev. **163**, 604 (1967)

5. C. M. Bertoli, V. Bortolani, C. Calandra and F. Nizzoli, Phys. Rev. Lett. **31**, 1466 (1973)

CHAPTER 6

1. W. A. Harrison, *Electronic Structure and the Properties of Solids*, W. H. Freeman, San Francisco, 1980, p. 393

2. See e.g. P. K. Lam and M. L. Cohen, Phys. Rev. **25**, 6139 (1982)

3. P. E. Van Camp, V. E. Van Doren and J. T. Devreese, Phys. Rev. Lett. **42**, 1224 (1979)

EXCHANGE - CORRELATION POTENTIAL FOR THE QUASI-

PARTICLE BLOCH STATES OF A SEMICONDUCTOR

W. Hanke, N. Meskini[+] and H. Weiler

Max-Planck-Institut für Festkörperforschung
D-7000 Stuttgart 80
The Federal Republic of Germany

ABSTRACT

A summary is given of recent investigations we have performed on a) the interrelation of many-body perturbation theory for
the non-local and energy-dependent self-energy and the density-
functional theory, b) the single-particle-like excitations in Si
which are calculated within the time-dependent screened Hartree-
Fock (TDSHF) approximation, and c) an analytical energy- and
local-density ($\rho^{1/3}$-) dependent model for the self-energy of nonmetallic systems in general.

We elaborate on recent attempts to derive the local and energy-dependent density-functional potential v_{xc} from the diagrammatic
structure of many-body perturbation theory for the exact exchange-
correlation energy, without explicit recourse to an extremal principle. The local v_{xc} can be related to the nonlocal and dynamic
self-energy Σ obtained from perturbation theory.

We summarize our recent calculations of quasiparticle states
in silicon which aim at a first-principle understanding of the
self-energy corrections in a prototype semiconductor. TDSHF is used
by replacing the exchange operator by a dynamically screened interaction. In contrast to previous calculations in semiconductors the
wave-vector and frequency-dependent two-particle propagator is calculated from first principles and includes local-field and particle-hole (excitonic) effects. The band gap, the valence-band width
and the general features of quasiparticle decay are in good accord
with experiment, though our construction of bare HF states via a
density matrix built from pseudopotential eigenstates puts limits
on this comparison.

On the basis of these numerical results, we outline the construction of an analytic, energy-dependent and local density-dependent model for the self-energy operator. This tight-binding model reproduces the computational results for dynamical self-energies in both the insulator C and semiconductor Si within a few percent across valence and conduction bands. The strength of the self-energy scales with $\rho^{1/3}$. In particular, it is shown how the intrinsically non-local (due to long-range electron-hole polarizations) potential can still approximately be converted via a Lorentz-sphere construction into a local potential. Together with the perturbation-theoretical expression for the "exact" v_{xc} of density-functional theory this tight-binding model is finally used to derive a model expression for v_{xc} of a non-metal.

I. INTRODUCTION

A central problem in one-electron band calculations is the proper inclusion of exchange and energy-dependent, dynamic correlation. From a formal many-particle point of view the justification for an effective single-particle description is contained in the Dyson equation for the one-particle Green's function [1-3]. One approach, which aims at an exchange-correlation potential or self-energy, is provided by the many-body perturbation theory [2,3]. In particular, it relates the single-particle excitations to two-particle (electron-hole) and higher-order excitations [2]. Recently, we have followed this approach [4-6], which has as an important ingredient that the two-particle excitations are directly accessible by experiment [4]. This implies additional control in the approximations of the theory.

A second approach, on which most modern band calculations are based, rests on the density-functional theory of Hohenberg, Kohn and Sham [7,8]. This theory established that all properties of an interacting electron system may be considered as functionals of the ground-state charge density $\rho(\vec{r})$ rather than of an applied external potential $v_{ext}(\vec{r})$. An approximate ground-state energy functional, i.e. the local-density approximation [8], has proved to be of high accuracy for atoms [9], molecules [9] and solids [10,11,12]. The LDA generates one-electron-like eigenfunctions and eigenvalues which are used, without formal justification, to interpret single-particle excitations. Although for metals the agreement between the LDA eigenvalues and experiment is often of satisfying accuracy, LDA typically underestimates band gaps in semiconductors and insulators by 30% to 50% [12]. Sham and Kohn also formulated an LDA for the dynamical self-energy, valid in the limit of slowly varying density [1].

In metals a certain amount of convergence between the many-body perturbation theory and the density-functional theory has been

114

achieved [3]. This is primarily due to the very construction of the
LDA functional within the homogeneous electron-gas limit. On the
other hand, in insulators and semiconductors this is not the case.
Here the localized and inhomogeneous charge distributions seem to
contradict the use of a functional derived from the metallic, homogeneous electron gas.

The purpose of the present review is threefold: Firstly, we
want to address the formal relationship between the density-functional effective one-particle potential and the non-local, energy-
dependent self-energy as derived from the many-body perturbation
theory [2]. To this we elaborate in sec. II on recently introduced
results [13,14]. In particular, the general structure of many-body
theory [2] is used to derive the density-functional potential v_{xc}
without recourse to an extremal principle. We make largely use of
Baym's [2] self-consistent approximations for one- and two-particle
Green's functions. We also derive an expression which has recently
been given by Sham and Schlüter [13] for the gap correction to the
density-functional eigenvalues. This will enable us in sec.IV to
construct an analytic form for the correction, utilizing model results for the self-energy Σ_{xc}.

Secondly, we summarize recent numerical investigations we have
performed on single-particle excitations embedded in the one-particle Green's function of the proto-type semiconductor Si [15]. This
work has the same quantitative objective as our earlier work on the
insulator diamond [5,6]. The self-energy was obtained in the time-dependent screened Hartree-Fock (TDSHF) approximation. It consists in
replacing the bare exchange (HF) operator by a dynamically screened
interaction. To be consistent with the variety of experimental facts
(in particular, optical absorption), also the dielectric function
and thus the two-particle excitations (electron-hole, plasmon) were
treated in TDSHF [16]. This guarantees the above-mentioned additional
control of the various approximations (order of perturbation theory,
wave-functions, energies) by experiment. The correlated bandstructure was derived from a local-orbital representation of the equations
of motion for one- and two-particle Green's functions utilizing in
particular the short-range character of the self-energy [1]. Like in
our previous work on diamond, it is found also for Si that the experimental band gap, the valence-band width and the general features
of quasiparticle decay observed in XPS data can be reproduced by a
first-principles calculation. However, in Si, our approximate construction of the underlying HF states, using a non-local density matrix constructed from LDA wavefunctions, puts limits on this comparison. The self-energy operator scales with energy away from the gap
to a significantly lesser extent than in diamond and is predominantly local in $\vec{r}$-space. On the basis of the Si and our earlier diamond results the Green's function many-body approach is compared
with other recent developments, for example, an LDA approach to the
self-energy [17].

Finally, we discuss in sec. IV an analytical, energy-dependent and local density-dependent model for the self-energy, which we have recently proposed [15]. The model incorporates the general features of self-energy corrections born out by the elaborate computations in both Si and C: due to incomplete screening in nonmetals, the bare particle induces a long-range electron-hole (e-h) polarization, which then sets up the correlation potential felt by the very particle. In tight-binding the polarization can be modelled by a classical dipole sum which, in a continuum approximation, converges into a "local" $\rho^{1/3}$-potential. This extends to non-metals arguments well-known in connection with the exchange hole of metals [19]: the potential acting on the particle in the center of the hole (here the exchange-correlation hole) can either be calculated as the continuum dipole sum outside the spherical region ("Lorentz sphere") or, complementary, by just considering the potential of the uniformly charged sphere surrounded by vacuum. These arguments also illucidate how the intrinsically long-ranged polarization cloud carried by a quasiparticle in a semiconductor or insulator can still be effectively converted into a local effect. Our self-energy formula then models a "LDA" approach to self-energies for systems that bear not the remotest resemblance to a homogeneous electron-gas with short-range screening.

II. THE GREEN'S FUNCTIONS, SINGLE- AND TWO-PARTICLE EXCITATIONS, AND THEIR INTERRELATION WITH DENSITY-FUNCTIONAL THEORY

In this section we first review briefly the formal connection between properties of the one- and two-particle Green's function and elementary excitations. Determining the partition function we then work out interrelations between the Kohn-Sham equation of density-functional theory and the general many-body perturbation theory for the exact exchange-correlation energy.

II. A. The Green's Functions and Elementary Excitations

The Green's function $g(\vec{r},\vec{r}';\omega)$ is a solution of the Dyson equation [1,3]

$$(\omega - h(\vec{r}))g(\vec{r},\vec{r}';\omega) - \int \Sigma(\vec{r},\vec{r}'';\omega)g(\vec{r}'',\vec{r}';\omega)d\vec{r}'' = \delta(\vec{r},\vec{r}').$$

$$(2.1)$$

Here h denotes the Hartree (H) potential and Σ the self-energy or mass operator. A solution of the Dyson equation can be constructed by expanding g in a complete set of left (and right) eigenfunctions $\psi_{n\vec{k}}^{+}$ (and $\psi_{n\vec{k}}$)

$$h(\vec{r})\,\psi_{n\vec{k}}(\vec{r},E)+\int\Sigma(\vec{r},\vec{r}',E)\psi_{n\vec{k}}(\vec{r}',E)d\vec{r}' = \varepsilon_{n\vec{k}}(E)\psi_{n\vec{k}}(\vec{r},E) \quad (2.2)$$

In terms of these functions g is given by

$$g(\vec{r}, \vec{r}';E) = \sum_{n,k} \psi_{n\vec{k}}(\vec{r},E) \psi^{+}_{n\vec{k}}(\vec{r}',E)/(E-\varepsilon_{n\vec{k}}(E)) \qquad (2.3).$$

If the equation, $E-\varepsilon_{n\vec{k}}(E) = 0$, has a solution for some $n,\vec{k}$ and E (i.e. the quasiparticle energy $E_{n\vec{k}}$ with wave vektor k and band index n), then g has a pole or local maximum at $E_{n\vec{k}}$. It then follows from the spectral representation of g, that the N-1 or N+1 particle system has an eigenstate with this energy. In general, the values of E satisfying the quasiparticle condition, $E - \varepsilon_{n\vec{k}}(E) = $ local minimum, are complex, in which case the real part corresponds to a physical energy and the imaginary part to a decay rate. Certain features of the quasiparticle ($E_{n\vec{k}}$) spectrum, like the valence band width, can approximately be related to photoemission(for example, XPS) data [20].

To make now contact with our approach to the quasiparticle states, consider first the Hartree-Fock (HF) approximation. In this

Fig. 1a)   Fig. 1b)

approximation the bare electron or hole (described by g) couples only to particles of the same spin via the Coulomb interaction v (Fig.1a). Beyond HF dynamical correlation effects appear due to the bare particle interacting with electron-hole (e-h) pairs and plasmons. These and higher-order excitations are contained in an expansion of g in the dynamically screened Coulomb interaction $v\varepsilon^{-1}$. The first term in this expansion is schematically indicated in Fig.1b. This is formally expressed in the following coupled chain of integral equations connecting the self-energy Σ, the single-particle Green's function g, the vertex function Γ, and the two-particle Green's function G_3:

$$\Sigma(1,2) = i\, v(1^{+}5)\, \varepsilon^{-1}(53)\, g(14)\, \Gamma(42,3), \qquad (2.4)$$

$$\Gamma(12,3) = \delta(12)\delta(13) + \frac{\delta\Sigma(12)}{\delta g(45)}\, g(46)\, g(75)\, \Gamma(67,3) \qquad (2.5)$$

where "1" denotes the space-time coordinate $(\vec{r}_1 t_1)$, and the repeated numerals are understood to indicate appropriate integration. Σ and Γ are diagrammatically represented in Fig. 2a) and 2b). $v\varepsilon^{-1}$ entering eq. (2.4) is the dynamically screened interaction, which contains the two-particle excitations and which serves as an expansion parameter. In terms of the two-particle Green's function G, ε^{-1} is given by

$$\varepsilon^{-1}(1,2) = \delta(12) - i\, v(13)\, G_4(33;22) \tag{2.6}$$

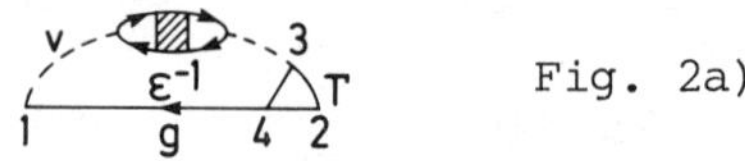

Fig. 2a)

Fig. 2b)

The chain is completed by relating the two-particle Green's function via the Bethe-Salpeter equation to the self-energy

$$G(11';22') = G_o(11';22') + G_o(11';33')\, V(33';44')\, G(44';22') \tag{2.7}$$

Here G_o is the non-interacting e-h propagator

$$G_o(11';22') = g\,(1'2')\, g(21) \tag{2.8a}$$

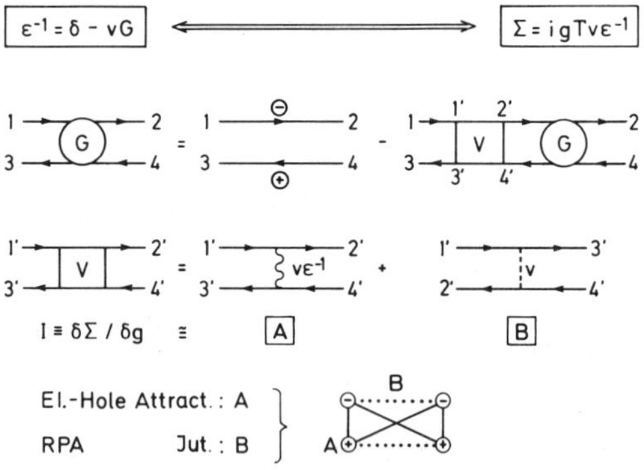

Fig.3

and V the irreducible e-h interaction (see also Fig.3)

$$V(33';44') = -\,\delta\Sigma(3'3/\delta g(4'4) + \delta(33')\,\delta(44')\,v(34) \tag{2.8b}$$

The starting point in this chain of equations is the Hartree (H) approximation, in which case $\Sigma = 0$. An approximate "Hartree" calculation is shown in Fig.4 for Si. This calculation is "approximate" in that it is not a self-consistent H calculation, but instead is based on a self-consistent pseudopotential-LDA calculation [21]. After achieving selfconsistency the LDA exchange-correlation v_{xc} was dropped in the extraction of the eigenvalues E_{nk} and only the Hartree term retained. We note that in this approximation Si is a semi-metal with partial band-overlap.

In the Hartree approximation $\Sigma \equiv 0$ the vertex function Γ is given by $\delta(12)\delta(13)$ (eq.(2.5)). In the next step the self-energy follows from eq.(2.4) with $v\varepsilon^{-1}$ replaced by the bare v

$$\Sigma(12) = iv(12)g(12) \qquad (2.9)$$

This equation, when solved self-consistently (i.e. by realising that g is also determined by Σ through Dyson's eq.(2.1)) gives the Hartree-Fock self-energy and the single-particle HF equation.

The need for dynamical correlation corrections in semiconductors and insulators becomes evident by comparing in Table 1 results

Table 1: HF and LDA Comparison with Experiment

	gap (eV)			valence-band width (eV)		
	HF	LDA	Exp.	HF	LDA	Exp.
MgO	18		7.8			
Si		2.5	3.4		12.0	14.7±0.3
C	15	6.3	7.4±0.2	29	20.4	24.2±1

of various HF calculations [22,23] with experiment: the HF gaps are typically by about a factor two too large. Fig. 4 displays also approximate (using in the density matrix pseudopotential wavefunctions, see sec. III) HF energy bands for Si. The too large HF gap is essentially a result of the fact, that an excitation across the gap corresponds to moving a more or less localized electron (in a bonding-orbital state, see sec. III) with a given spin against the Pauli exclusion principle to a neighbouring unit cell which has already a localized electronic state in this very spin configuration (in the ground state of Si the 4 bonding orbitals per unit cell are doubly occupied). This, of course, oversimplified picture also illustrates the effect of the dynamic correlations: the dynamically screened particles no longer feel the bare exchange interaction, and as a result, the gap is reduced.

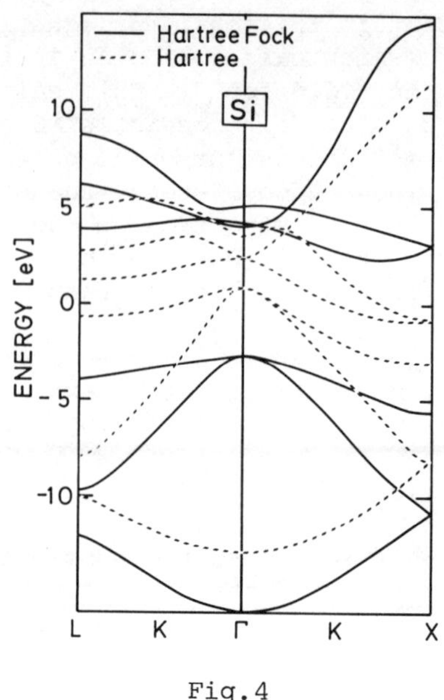

Fig.4

Table 1 and Fig.4 also reveal a general problem of self-energy determinations: the large Fock correction puts the bands much
further away from experiment than the "Hartree" calculation. The
dynamic correlation correction is again large and, to a significant
part, just cancels the Fock correction. Thus, to achieve satisfying accuracy for combined exchange plus correlation, the individual parts have to be calculated with significantly higher accuracy.
This brings us to the question of how the standard methods for calculating one-electron states in solids are related to the quasi-
particle picture.

II. B.   Density-Functional Theory

The local-density approximation (LDA) of Kohn and Sham [8] provides an easily accessible effective one-particle potential for exchange and correlation, which has proved to be of high accuracy
for ground-state properties of atoms [9], molecules [9] and solids [10-12].
Though the theory is extendable to the one-particle Green's function [1], the LDA eigenvalues formally are not justified to be taken
as single-particle excitation energies [1,13,26]. In insulators and
semiconductors the LDA typically underestimates band gaps by 30-
50% [12,24]. We want to investigate to what extent this deviation is
due to the use of a ground-state potential instead of a self-energy
for single-particle excitations.

Consider the pair-correlation function $P(\vec{r})$ in Fig.5. For a rough estimate of the exchange-correlation effects in Σ, we can replace the bare Coulomb interaction v in the Hartree potential h in eq. (2.2) by

$$v(\vec{r}) \rightarrow v(\vec{r}) \, P(\vec{r}) \tag{2.10}$$

As schematically indicated in Fig.5, in H approximation P equals 1 everywhere (Σ_H = 0), in HF $P_x \rightarrow 1/2$ for $\vec{r} \rightarrow 0$ due to the exclusion principle. Including exchange plus correlation $P_{xc} \rightarrow 0$ for $\vec{r} \rightarrow 0$. The corresponding exchange-correlation hole is short-ranged of order $1/k_F(\vec{r})$, where $k_F(\vec{r})$ is a local Fermi vector[1].

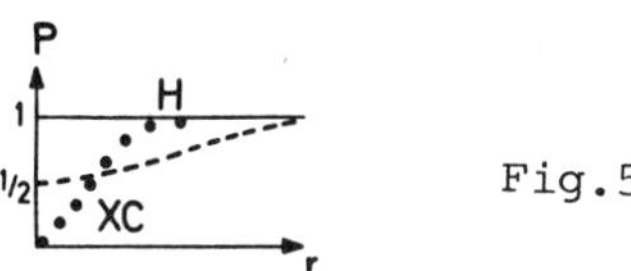

Fig.5

This short-range property of the combined exchange-correlation effects is a basic ingredient of the LDA of Kohn and Sham: from the stationary properties of the ground-state functional E = E[ρ] , Kohn and Sham showed that the density $\rho(\vec{r})$ can be determined exactly from one-particle-like equations solved self-consistently as follows[8] :

$$\{h(\vec{r})+V_{ext}(\vec{r})+v_{xc}(\rho(\vec{r}))\}\psi_{n\vec{k}}(\vec{r}) = \xi_{n\vec{k}}\psi_{n\vec{k}}(\vec{r}) \tag{2.11}$$

with

$$\rho(\vec{r}) = \sum_{n,\vec{k}}^{occ} |\psi_{n\vec{k}}(\vec{r})|^2 \tag{2.12}$$

and

$$v_{xc}(\rho(\vec{r})) = \delta E_{xc}[\rho]/\delta\rho(\vec{r}) \tag{2.13}$$

In the LDA the exchange-correlation portion $E_{xc}[\rho]$ is approximated by[8]

$$E_{xc}[\rho] \cong \int \rho(\vec{r})\varepsilon_{xc}[\rho(\vec{r})]d\vec{r} \tag{2.14}$$

and

$$v_{xc}(\rho(\vec{r})) \cong d[\rho\varepsilon_{xc}(\rho)]/d\rho \tag{2.15}$$

In eq. (2.14) ε_{xc} is the exchange-correlation energy per electron of a uniform electron gas but with a position dependent density $\rho(\vec{r})$.

Eq. (2.11) coupled with eq. (2.15) is commonly used in the determinations of bandstructures, for example, in the calculations for C[25] and Si[24] listed in Table 1.

For the actual quasi-particle excitations $E_{n\vec{k}}$ one has to solve the Dyson equation (2.1) or (2.2) with a non-local energy-dependent self-energy Σ. Sham and Kohn[1] have suggested LDA approximations for Σ, which is also a functional of density. To this they split Σ into a local and non-local component

$$\Sigma(\vec{r},\vec{r}';E) = (V_{ext}(\vec{r})+V_{H}(\vec{r}))\,\delta(\vec{r}-\vec{r}') + M(\vec{r},\vec{r}';E - (V_{ext}(\vec{r}_o)-$$
$$V_{H}(\vec{r}_o))) \tag{2.16}$$

where $V_{ext}(\vec{r}_o)+V_{H}(\vec{r}_o)$ (V_H: Hartree potential) is the electrostatic potential at $\vec{r}_o = 1/2\,(\vec{r}+\vec{r}')$. The simplest approximation replaces M by the mass operator M_{hom} of a uniform electron gas with the local density $\rho(\vec{r})$. In this case an explicit energy dependence near the chemical potential μ (or E_F) can be extracted, i.e.

$$M_{hom}(E) = v_{xc}(\rho) + (E-\mu)(1-m^*(\rho)) + \dots \tag{2.17}$$

where $m^*(\rho)$ is the density-of-states mass of the uniform gas with density ρ. Thus, in the homogeneous electron-gas limit, the mass operator M_{hom} equals the ground-state LDA potential $v_{xc}(\rho)$ of eq. (2.15). The first-order correction in eq. (2.17) is real, and gives in first-order perturbation theory the difference between quasi-particle energies $E_{n\vec{k}}$ and LDA eigenvalues $\xi_{n\vec{k}}$ proportional to the distance from the Fermi energy, $(E-\mu)$. The lowest-order imaginary part of (2.17) can be shown to be of the form $-ic(\rho)(E-\mu)^2$, with $c > 0$.

Wang and Pickett have recently pursued the above LDA approach to the self-energy, incorporating in an empirical manner a direct gap E_g in the single-particle spectrum[17]. They calculate the self-energy operator M_{hom} using the gW approximation[3] (with the vertex $\Gamma \equiv 1$ in eq. (2.4)) with a bare Coulomb interaction v screened by a homogeneous model semiconductor dielectric function. For Si, they find a predominantly energy-dependent correction to the eigenvalues which removes a significant part of the discrepancy for the gaps. If, on the other hand, the metallic homogeneous electron-gas result M_{hom} of eq.(2.17) is used, the direct gap of Si is enhanced by only 5% compared to the LDA eigenvalues $\xi_{n\vec{k}}$ of eq.(2.11)[13]. The Wang-Pickett scheme presents an interesting, pragmatic approach to self-energy corrections. However, it lacks formal justification: it is based on the above intrinsically metallic and homogeneous electron-gas description of the self-energy by Sham and Kohn, mixing it with the perturbation theoretical gW approximation. It, furthermore, forces a gap (of assumed magnitude) into the single-particle and two-particle spectra.

Of course, there exist also other efforts to go beyond LDA, such as "self-interaction corrections" and the weighted (non-local) density approximation (WDA)[8,52]. The former approach tries to correct the incomplete cancellation of the self-interaction terms in the Coulomb and exchange potentials, when the exchange is treated within LDA. The self-interaction correction seems to work well in atoms and in wide-gap (atomiclike) insulators, but its adaption to extended states remains an unsolved problem. Inclusion of non-locality via the WDA has recently been tested for the semiconductor GaAs. It seems to improve the description of the valence bands, but does not resolve the gap problem (i.e. too small a gap) experienced by the simpler local expressions.

In the following section, we want to elaborate on the interrelation between the "exact" local exchange-correlation potential v_{xc} and the non-local and energy-dependent self-energy Σ.

II. C.   Interrelation between the Density-Functional Potential v_{xc} and the Self-Energy Σ of Many-Body Perturbation Theory.

Recently Sham and Schlüter[13] and in somewhat different form Langreth[14], have implemented well-known results from the diagrammatic structure of many-body perturbation theory[2] for the exact exchange-correlation, E_{xc}, to derive the DF potential v_{xc} without explicit recourse to a minimum or extremal principle. We present here our own derivation of this interrelation which is largely based on the self-consistent approximations derived by Baym and Baym and Kadanoff[28] for one- and two-particle Green's functions. In particular, this will enable us in sec. IV to construct approximations to v_{xc}, utilizing explicit forms of the self-energy operator Σ_{xc}.

Let us start by defining the two Green's functions g_o and g by

$$[\omega - h(\vec{r}) - V_{ext}(\vec{r}) - v_{xc}(\vec{r})] g_o(\vec{r},\vec{r}';\omega) = \delta(\vec{r}-\vec{r}') \qquad (2.18)$$

and

$$[\omega - h(\vec{r}) - V_{ext}(\vec{r})] g(\vec{r},\vec{r}';\omega) - \int \Sigma_{xc}(\vec{r},\vec{r}'';\omega) g(\vec{r}'',\vec{r}';\omega) d\vec{r}'' = \delta(\vec{r}-\vec{r}')$$
$$(2.19)$$

The first gives the "unperturbed" one in terms of the self-consistent potential $V_{ext} + v_{xc}$, where v_{xc} is the appropriate density-functional potential from eq. (2.13), i.e. $v_{xc} = \delta E_{xc}/\delta \rho$. On the other hand, the "full" Green's function g is defined as in eq. (2.1) through the non-local and dynamic self-energy operator Σ_{xc}.

From the requirement that both density-functional theory and the many-body perturbation theory give the correct density, i.e.

$$\rho(\vec{r}) = -ig(\vec{r}t,\vec{r}t^+) = -ig_o(\vec{r}t,\vec{r}t^+) \qquad (2.20)$$

an explicit connection between the local v_{xc} and the non-local, energy-dependent Σ_{xc} can immediately be derived. From the Dyson equation $g = g_0 + g_0 \Sigma' g$ and eq.(2.20) it follows that

$$\int \frac{d\omega}{2\pi} \int d\vec{r}_1 d\vec{r}_2 g_0(\vec{r},\vec{r}_1;\omega) \Sigma'(\vec{r}_1,\vec{r}_2;\omega) g(\vec{r}_2,\vec{r};\omega) = 0 \qquad (2.21)$$

where $\Sigma' = \Sigma_{xc} - v_{xc}$. Thus we find

$$\int d\vec{r}' v_{xc}(\vec{r}') \int \frac{d\omega}{2\pi} g_0(\vec{r},\vec{r}';\omega) g(\vec{r}',\vec{r};\omega) = \int d\vec{r}_1 \int d\vec{r}_2 \int \frac{d\omega}{2\pi} g_0(\vec{r},\vec{r}_1,\omega)$$
$$\cdot \Sigma_{xc}(\vec{r}_1,\vec{r}_2;\omega) g(\vec{r}_2,\vec{r};\omega) \qquad (2.22)$$

As already mentioned by Sham and Schlüter, from eq.(2.22) v_{xc} in LDA can easily be obtained. It is also straightforward to obtain the local v_x in the HF approximation: to this only the bare exchange $\Sigma_x \sim vg^x$ (in this case $g_0 \sim g^{HF}$) is kept and g replaced by g_0 in eq.(2.22). This local exchange potential is equivalent to an effective local potential derived by Talman and Shadwick[28] from a minimization procedure. For atoms, such as carbon, neon and aluminium they demonstrated, that this local exchange potential results in total energies, which exceed the HF total energies by less than 0.005%.

The value of the relation (2.22) will become further evident in sec.IV, where we use an explicit model for the Σ_{xc} of an insulator or semiconductor, to derive a model expression for v_{xc} of a non-metal.

Eq.(2.22) as an interrelation between v_{xc} and Σ_{xc}, can be derived by expressing the partition function in density-functional theory and identifying it with the partition function of the general many-body perturbation theory.

To this we follow closely Baym[2] and Baym and Kadanoff[29] in their discussion of perturbation theory for non-uniform systems. They consider non-equilibrium thermal Green's functions at an arbitrary temperature β^{-1} in the presence of an external, in general non-local potential U that acts in the imaginary time interval $0 < it < \beta$. Writing the operators in the interaction representation for the external disturbance, the one-particle Green's function is

$$g(1,1';U) = -i\mathrm{Tr}\{e^{-\beta H}T[S\psi(1)\psi^+(1')]\}/\mathrm{Tr}(e^{-\beta H}T[S]) \qquad (2.23)$$

where H is the unperturbed Hamiltonian and S the S matrix

$$S = \exp\left[-i\int_0^{-i\beta} d2d2'\psi^+(2)U(2,2')\psi(2')\right] \qquad (2.24)$$

Then, due to conservation laws, a function W exists, such that in the imaginary time interval $0 < it < \beta$

$$g(1,1') = \delta W/\delta U(1',1) \tag{2.25}$$

clearly, for the exact g (eq.(2.23)) W is

$$W = \ln \text{Tr}\{e^{-\beta(H-\mu N)} T[\exp(-i\int_o^{-i\beta}\psi^+(1)U(1,1')\psi(1'))]\} \tag{2.26}$$

Thus, in equilibrium, i.e. when $U(1,1') = \delta(t_1-t_1')U(\vec{r}_1,\vec{r}_1')$, W is just the logarithm of the partition function.[1] It is presented by a set of "closed" diagrams [2]. A second, central quantity in Baym's theory is $\emptyset$, a functional of g and v, which generates Σ (note that Σ includes here the H potential from h(r) in eq.(2.19))

$$\Sigma(1,1') = \delta\emptyset/\delta g(1',1) \tag{2.27}$$

If we introduce the four-dimensional matrix notation by defining the following trace of a quantity

$$\text{tr}X = \int_o^{-i\beta}dt\int d\vec{r}X(\vec{r}t,\vec{r}t^+) \tag{2.28}$$

we can alternatively write for eq.(2.27)

$$\delta\emptyset = \text{tr }\Sigma\delta g \tag{2.29}$$

Eq.(2.27) together with the Dyson equation (2.19) (with $V_{ext} \to U$) defines so-called "$\emptyset$-derivable" approximations [2]. $\emptyset$ derivability guarantees then a self-consistent approximation in many-body perturbation theory which, for a given order of approximation, satisfies various conservation laws. For example the particle-hole attraction in eq.(2.9) is given by (Ward identity) $\delta^2\emptyset/\delta g\delta g$.

Relating $\emptyset$ to the quantity W in eq.(2.25) leads to a very useful formula, from which all the equilibrium properties (in which we are interested in) of $\emptyset$-derivable approximations follow. A small change in U induces via (2.29) the change $\delta\emptyset = \text{tr }\Sigma\ \delta g$. On the other hand, from eq.(2.25) $\delta W = \text{tr}(\delta U)g$, and therefore

$$\delta W = \delta\emptyset - \delta(\text{tr}\Sigma g) - \text{tr }[\delta\{g_b^{-1}-U-\Sigma\}]g \tag{2.30}$$

where g_b is the bare Green's function defined by

$$g_b^{-1}(1,1') = (i\partial_t+\nabla^2)\delta(1-1') \text{ with } \delta g_b^{-1} = 0.$$

From the Dyson eq. (2.19) we have $g_b^{-1}-U-\Sigma = g^{-1}$. Thus, with

$-\text{tr}(\delta g^{-1})g = \delta \text{ tr } \ln(-g)$, it follows that

$$W = \emptyset - \text{tr }\Sigma g + \text{tr } \ln(-g). \tag{2.31}$$

In the equilibrium case, eq (2.31) gives the logarithm of the partition function. To set up our interrelation, we can alternatively derive the partition function from density-functional theory [30,31]. At finite temperatures the grand potential depends only on $v_{ext}(r)-\mu$, where μ is the chemical potential. $v_{ext}(r)-\mu$ is uniquely determined by the density $\rho(\vec{r})$, and the functional $\Omega_v[\rho'(\vec{r})]$ takes its minimum value (i.e. the grand potential) at the correct $\rho(\vec{r})$. This then leads to (combining eqs.(168)-(181) of ref. 30)

$$\Omega_{v_{ext}-\mu}[\rho(\vec{r})] = \Omega_o - \frac{1}{2} \int \frac{\rho(\vec{r})\rho(\vec{r}')}{|\vec{r}-\vec{r}'|} - \int v_{xc}(\vec{r})\rho(\vec{r})d\vec{r} + E_{xc},$$

(2.32)

where $E_{xc}[\rho(\vec{r})]$ defines the exchange-correlation contribution to the free energy $E[\rho(\vec{r})]^{30}$, and Ω_o is the grand potential of non-interacting single particle electrons with density distribution $\rho(r)$ and entropy S_s

$$\Omega_o = \sum_i \xi_i \, f(\xi_i-\mu) - T\, S_s[\rho(\vec{r})] - \mu N$$

(2.33)

ξ_i are the eigenvalues of the finite-temperature analogue of the single-particle-like equation (2.11) $f(\xi_i-\mu) = 1/[1+\exp[\beta(\xi_i-\mu)]]$ is the Fermi function.

Making now use of eq.(2.31) for a non-interacting system, we can relate Ω_o to the "unperturbed" Green's function g_o (eq.(2.18)) of density-functional theory:

$$\Omega_o = - \text{ tr ln } (-g_o)/\beta$$

(2.34)

Using again $\rho(\vec{r}) = -ig(\vec{r}t,\vec{r}t^+)$, it, furthermore, follows that

$$-\frac{1}{2} \int \frac{\rho(\vec{r})\rho(\vec{r}')}{|\vec{r}-\vec{r}'|}\, d\vec{r}d\vec{r}' - \int v_{xc}(\vec{r})\rho(\vec{r})d\vec{r} = -\frac{1}{\beta}[\phi_H+\text{tr }v_{xc}g]$$

(2.35)

The β^{-1} (actually $(-i\beta)^{-1}$) factor stems from the time integration in the trace. Thus, identifying the grand potential in eq. (2.32) with the grand potential from many-body perturbation theory in eq.(2.31)

$$\Omega_{v_{ext}-\mu}[\rho(\vec{r})] = \frac{1}{\beta}[-\phi + \text{tr }\Sigma g - \text{tr ln}(-g)],$$

(2.36)

we find $E_{xc}[\rho(\vec{r})]$ from many-body perturbation theory without recourse to a minimum principle

$$E_{xc}[\rho(\vec{r})] = \frac{1}{\beta} \text{tr}\{\ln(g^{-1}g_o) + (\Sigma_{xc}-v_{xc})g\} - \phi_{xc}/\beta$$
$$= \frac{1}{\beta} \text{tr}\{\ln(1-\Sigma'g_o) + \Sigma'g\} - \phi_{xc}/\beta$$

(2.37)

where, again, $\Sigma' \equiv \Sigma_{xc} - v_{xc}$. Like W the basic quantity ϕ_{xc} of Baym's theory is given by a set of "closed" diagrams, i.e. the sum of all exchange and higher order energy diagrams (in terms of v and g)^{2,13}. As already discussed in ref.13, functional derivative of

126

eq.(2.37) yields the eq.(2.22) for the exchange-correlation potential v_{xc} in terms of Σ_{xc}. In fact, the variation of E_{xc}

$$\delta E_{xc} = \frac{1}{\beta} \, \text{tr} \, \{g\delta g_o^{-1} + g_o^{-1}\delta g_o + \Sigma'\delta g\} - \delta\phi_{xc}/\beta \tag{2.38}$$

can independently be made with respect to δn, $\delta g_o(1,1')$, $\delta g(1,1')$
and $\delta g_o^{-1}(1,1')$, each variation resulting in an integral relation
between v_{xc} and Σ_{xc}: For example, variation with respect to $g_o(1,1')$
results in

$$\Sigma'(1',2)g(2,2')g_o^{-1}(2',1) = \text{tr} \, \{v_{xc}[i\frac{\delta n}{\delta g_o(1,1')} - \delta g/\delta g_o(1,1')]\}$$

$$\tag{2.39}$$

However, all relations unfortunately involve the Green's function g_o; this poses an additional (self-consistency) problem, in that
g_o involves eigenstates which can only be calculated when v_{xc} is already known.

A somewhat different formulation has recently been suggested[14]
It again rests on Baym's analysis but expresses the Hartree plus
exchange-correlation free energy as a coupling constant integral

$$E_{xc}\left[\rho(\vec{r})\right] + \frac{1}{2}\int\rho(\vec{r})\rho(\vec{r}')v(\vec{r}-\vec{r}')d\vec{r}d\vec{r}' = \frac{1}{2\beta} \int_o^1 \frac{d\lambda}{\lambda} \, \text{tr} \, \Sigma_\lambda \, g_\lambda$$

$$\tag{2.40}$$

$$= \frac{1}{\beta} \int_o^1 (\frac{\partial\phi_\lambda}{\partial\lambda})_g \, d\lambda$$

where $\lambda = 1$ corresponds to the actual, physical problem. The coupling constant integration can be performed and shown to result in
the Kohn-Sham relation $v_{xc} = \delta E_{xc}/\delta n$ not only for the exact E_{xc}, but
also for any ϕ-derivable approximation to it[14]. The expression for
v_{xc} derives formally from

$$\frac{dv_\lambda(\vec{r})}{d\lambda} = -\int d\vec{r}'\chi_\lambda^{-1}(\vec{r},\vec{r}')(\frac{\partial n(\vec{r}')}{\partial\lambda})_v \tag{2.41}$$

where $v_{\lambda=1}$ is the one-body potential of the physical problem and
$v_{\lambda=0} = v_H + v_{xc}$. χ is the usual density-response function determined
from the Bethe-Salpeter equation (2.7) with the interaction in eq.
(2.9) being just the irreducible particle-hole interaction $V = -\delta\Sigma/\delta g$. This formulation avoids the self-consistency problem of the
previous expression (2.22) for v_{xc}, but it presents a formidably
complicated coupling constant integration in eq.(2.41).

Sham and Schlüter[13] have recently given an expression for the
energy gap of an insulator. It is obtained from the eigenvalues of
the one-particle density-functional equation of eq.(2.11) plus a
finite correction due to the discontinuity of the functional derivative of the exchange-correlation energy. This correction was then

expressed in terms of v_{xc} and the improper self-energy $\tilde{\Sigma}$. We present here a derivation of this correction, which further illucidates the interrelation of density-functional and general many-body perturbation-theory.

From eqs. (2.18) and (2.19) we have the Dyson equation relating g_o and g

$$g = g_o + g_o \Sigma' g \tag{2.41}$$

with

$$\Sigma' = \Sigma_{xc} - v_{xc}^{(-)} . \tag{2.42}$$

$v_{xc}^{(-)}$ is the "exact" exchange-correlation potential to be used in density-functional eq. (2.18) for the N-particle system. Eq. (2.41) can also be written as

$$g = g_o + g_o \tilde{\Sigma} g_o \tag{2.43}$$

with

$$\tilde{\Sigma} = \Sigma' + \Sigma' g_o \tilde{\Sigma} \tag{2.44}$$

The solution g_o can be written as

$$g_o(\vec{x}t,\vec{x}'t') = -i \sum_{n\vec{k}} \psi_{n\vec{k}}(\vec{x}) \psi^*_{n\vec{k}}(\vec{x}') \exp[-i(E_n(\vec{k}) \cdot (t-t'))]$$

$$\cdot \{\Theta(t-t')\Theta(E_n(\vec{k})-\mu) - \Theta(t'-t)\Theta(\mu-E_n(\vec{k}))\} \tag{2.45}$$

We see, that in order to get an increase in density equal $\delta n_+ = |\psi_c|^2$, we must fix the chemical potential μ just below ε_c [13], the minimum of the conduction-band eigenstates of density-functional theory. Therefore, g_o undergoes a change equal to

$$\delta g_o(\vec{x}t,\vec{x}'t') = (+i)\, \psi_c(\vec{x}) \psi^*_c(\vec{x}')\, \exp[-i\varepsilon_c(t-t')] \tag{2.46}$$

or, in Fourier space,

$$\delta g_o(\vec{x},\vec{x}';\omega) = \int e^{i\omega(t-t')} \delta g_o(\vec{x}t,\vec{x}'t')\, d(t-t')$$

$$= 2\pi i \delta(\omega-\varepsilon_c) \psi_c(\vec{x}) \psi^*_c(\vec{x}') \tag{2.47}$$

In fact, from eqs. (2.45), (2.46) it follows

$$\delta n = (-i) \cdot \delta g_o(\vec{x}t,\vec{x}t^+) = (-i) \cdot (i) \psi^*_c(\vec{x}) \psi_c(\vec{x})$$

$$= |\psi_c(\vec{x})|^2 = \delta n_+ \tag{2.48}$$

Now, starting from equation (2.37) giving the exchange correlation energy as a functional of density, i.e.

$$E_{xc}[n] = i \ \text{Tr} \ \{\ln (1-\Sigma' g_o) + \Sigma' g\} - i \ \emptyset_{xc}[n]$$

$$= i \ \text{Tr} \ \{\ln (g^{-1} g_o) + \Sigma' g\} - i \ \emptyset_{xc}[n], \tag{2.49}$$

omitting everywhere a factor $1/\beta$, the change in this energy corresponding to a change $\delta n_+ = |\psi_c(\vec{x})|^2$ in density is:

$$\delta E_{xc}[n] = \text{Tr}\{v_{xc}^{(+)} \ \delta n_+\}$$

$$= i \ \text{Tr}\{g_o^{-1} g[\delta g^{-1} g_o + g^{-1} \delta g_o] + \delta\Sigma' \ g + \Sigma' \delta g\} - i \delta\emptyset_{xc}[n]$$

$$= i \ \text{Tr}\{g\delta g^{-1} + g_o^{-1}\delta g_o + \delta\Sigma' g + \Sigma' \delta g - \Sigma_{xc}\delta g\} \tag{2.50}$$

Using the Dyson equation $g = g_o + g_o \Sigma' g$ and the relation defining $\Sigma_{xc} = \dfrac{\delta\emptyset_{xc}}{\delta g}$, which gives $\delta\emptyset_{xc} = \text{Tr} \ \Sigma_{xc}\delta g$, we can write $\delta E_{xc}[n]$ as (see also (2.38)):

$$\delta E_{xc}[n] = i \ \text{Tr}\{g\delta g_o^{-1} + g_o^{-1}\delta g_o - v_{xc}^{(-)}\delta g\}, \tag{2.51}$$

or,

$$\text{Tr} \ \{v_{xc}^{(+)} \delta n_+ + iv_{xc}^{(-)} \delta g\} = i \ \text{Tr}\{g\delta g_o^{-1} + g_o^{-1}\delta g_o\} \tag{2.52}$$

Replacing g by $g_o + g_o \tilde{\Sigma} g_o$ and δg_o^{-1} by $-g_o^{-1}\delta g_o g_o^{-1}$, we get

$$\text{Tr} \ \{v_{xc}^{(+)} \delta n_+ + iv_{xc}^{(-)} \delta g\} = -i \ \text{Tr} \ \{\tilde{\Sigma}\delta g_o\}, \tag{2.53}$$

From $g(\vec{x}t,\vec{x}t^+) \equiv g_o(\vec{x}t,\vec{x}t^+)$, follows

$$(-i) \delta g(\vec{x}t,\vec{x}t^+) = (-i) \ \delta g_o(\vec{x}t,\vec{x}t^+) = \delta n_+ = \psi_c(\vec{x}) \psi_c^*(\vec{x}) \tag{2.54}$$

Equation (2.53) can then be written as:

$$\text{Tr} \ \{(v_{xc}^{(+)} - v_{xc}^{(-)}) \ \delta n_+\} = -i \ \text{Tr} \ \{\tilde{\Sigma}\delta g_o\} \tag{2.55}$$

or

$$\int d\vec{x}\psi_c^*(\vec{x}) \{v_{xc}^+(\vec{x}) - v_{xc}^{(-)}(\vec{x})\}\psi_c(\vec{x}) = -i \int d1d1'\tilde{\Sigma}(1,1') \delta g_o(1',1^+) \tag{2.55}$$

Using the Fourier representations

$$\tilde{\Sigma}(\vec{x}t,\vec{x}'t') = \int \frac{d\omega}{2\pi} \ \tilde{\Sigma}(\vec{x},\vec{x}';\omega) \ e^{i\omega(t-t')}$$

$$\delta g_o(\vec{x}t',\vec{x}t^+) = \int \frac{d\omega'}{2\pi} \ \delta g_o(\vec{x}',\vec{x},\omega') \ e^{i\omega'(t'-t^+)} \tag{2.56}$$

we arrive at

$$-i \int d1\, d1'\, \tilde{\Sigma}(1,1') \delta g_0(1',1^+) = -i\int d\vec{x}\, d\vec{x}' \int d(t-t') \int \frac{d\omega}{2\pi} \tilde{\Sigma}(\vec{x},\vec{x}';\omega)$$

$$\cdot \int \frac{d\omega'}{2\pi}\, \delta g_0(\vec{x}',\vec{x};\omega')\, e^{i(\omega-\omega')(t-t')}\, e^{-i\omega'\eta}$$

$$\stackrel{\eta \to 0}{=} -i \int d\vec{x} \int d\vec{x}' \int \frac{d\omega}{2\pi}\, \tilde{\Sigma}(\vec{x},\vec{x}';\omega)\, \delta g_0(\vec{x}',\vec{x};\omega)\, e^{-i\omega\eta} \qquad (2.57)$$

Thus, finally we have for the gap correction

$$\Delta = \int d\vec{x}\, \psi_c^*(\vec{x}) \left\{ v_{xc}^{(+)}(\vec{x}) - v_{xc}^{(-)}(\vec{x}) \right\} \psi_c(\vec{x}) = (-i)(2\pi i) \int d\vec{x} \int d\vec{x}' \int \frac{d\omega}{2\pi}$$

$$\tilde{\Sigma}(\vec{x},\vec{x}';\omega)\, \psi_c(\vec{x}')\psi_c(\vec{x})\, \delta(\omega-\varepsilon_c)$$

or

$$\Delta = \int d\vec{x} \int d\vec{x}'\, \psi_c^*(\vec{x})\, \tilde{\Sigma}(\vec{x},\vec{x}';\varepsilon_c)\, \psi_c(\vec{x}') \qquad (2.58)$$

III. QUASI-PARTICLE STATES IN A SEMICONDUCTOR

The above considerations indicate the need for a first-principles calculation of single-particle-like excitations in insulating and semiconducting crystals. From the point of view of density-functional theory, an a-priori determination of the non-local, dynamic self-energy of an insulator or semiconductor gives, via eq. (2.22) the possibility to construct a new functional v_{xc} appropriate for non-metals. At least for a tight-binding model of Σ_{xc} we will construct such a local potential v_{xc} in sec. IV. It will be interesting to confront such a tight-binding model with the commonly used homogeneous electron-gas construction of a LDA functional v_{xc}.

Recently, we have carried out quantitative Green's function determinations of the quasi-particle states of diamond, carefully taking into account the tetrahedral bonding properties [5,6]. The exchange-correlation effects both in the one- and in the two-particle Green's functions were significantly enhanced, when compared with a homogeneous treatment [16]. The mass operator or dynamical correlation was found to strongly scale with energy away from the gap, but to be predominantly local in r-space. Band gaps and widths and the general features of quasi-particle decay, observed in XPS [20] were found to be reproducable by such a first-principle calculation.

Here we want to summarize our recent calculations of single-particle excitations in Si [15], which aim at a similar quantitative understanding of the self-energy corrections in a proto-type semi-conductor.

III. A. The Time-Dependent Screened Hartree-Fock (TDSHF) Approximation

We employ the time-dependent screened HF approximation (TDSHF), which replaces the bare Coulomb interaction v in the HF self-energy (Fig.1) by a dynamically screened interaction (eq.(2.4)) $v\varepsilon^{-1}$:

$$\Sigma_{TDSHF}(\vec{r},\vec{r}';E) = \frac{i}{2\pi} \int dE' e^{iE'\delta} g(\vec{r},\vec{r}';E+E') \int v(\vec{r}-\vec{r}'')$$
$$\varepsilon^{-1}(\vec{r}',\vec{r}'';E')d\vec{r}'' \tag{3.1}$$

The corresponding vertex function Γ derives from the integral equation (2.5) with the kernel

$$I(33';44') = \delta\Sigma(3'3)/\delta g(4'4) \tag{3.2}$$

This irreducible e-h interaction, via a Ward identity, also determines the particle-hole interaction in the Bethe-Salpeter equation (2.7) for the two-particle propagator. Ideally, we should take Σ_{TDSHF} from eq.(3.1) in eq.(3.2), with

$$I(33',44') = -\delta(34)\,\delta(3'4')\,v(3,5)\varepsilon^{-1}(5,3') - (gv\,\frac{\delta\varepsilon^{-1}}{\delta g}) \tag{3.3}$$

The last term on the right-hand side involves an extremely complicated functional integral, because ε^{-1} is a very complex functional of g due to the Bethe-Salpeter equation and its e-h interaction, which again depends on Σ and thus on g. We have estimated the influence of this term both in Si and in C in one- and two-particle propagators and found it negligible. One pragmatic reason for this is the well-known fact that the (static) screened Coulomb e-h attraction (the first term on the right-hand side of eq.(3.3)) alone determines extremely accurately binding energies (Wannier, intermediate range, Frenkel) of excitons[34]. For example, it gives rise to the commonly used effective-mass approximation for Wannier excitons[32]. Here, furthermore, the replacement of the dynamically screened interaction $v\varepsilon^{-1}$ in eq.(3.3), which determines the e-h interaction in the two-particle Green's function G, by a static approximation can be justified[16]. We have also found this approximation in our investigations of the optical absorption spectrum of both C[33] and Si[4] to result in quantitative agreement with experiment both with respect to peak positions and oscillator strengths.

The vertex Γ not only enters the two-particle propagator G but also the self-energy Σ in eq.(2.4). We calculate this self-energy using the lowest-order in the screened dynamical interaction. Again this approximation cannot be rigorously justified in terms of smallness of higher-order terms. The next term in the expansion adds a factor $(2\pi)^{-4}$ ggvε^{-1}dkdω in k-representation. It can be shown to be of order $r_s/(12\pi(1+r_s/2)) \simeq 1/25$ in Si. Nevertheless, it is our opinion that the extension from a static theory of correlations in insulating and semiconducting crystals to a dynamic one is a major step forward, which already in lowest order contains the coupling to the most prominent two-particle interaction, i.e. e-h and plasmons. Furthermore, we emphasize that the dynamical interaction both in C and in Si was calculated for the first time not on a (Penn) model basis but from a priori, including in particular the (local-field) effects due to the charge-density inhomogeneity of the solid. The left-out higher-order terms certainly give numerical changes, but they should not influence the main conclusions (see secs.III. C and IV).

III. B   Calculation of Single- and Two-Particle Propagators

We have developed in previous work [4-6] a practical scheme for calculating the self-energy corrections and quasi-particle states in periodic crystals. One main ingredient of this scheme is the short-range property of Σ in $|\vec{r}-\vec{r}'|$, which has been proven by Sham and Kohn [7] on the basis of general considerations of many-body perturbation theory. As a consequence, $\Sigma(\vec{r},\vec{r}';E)$ will have matrix elements in a set of local orbitals (Wannier, LCAO's, muffin-tin, etc.) appreciably different from zero only about the diagonal, $\vec{r} \equiv \vec{r}'$.

Therefore, by expanding the quasi-particle Bloch function $\psi_{n\vec{k}}(\vec{r})$ of eq.(2.2) in terms of these orbitals $\phi_\nu(\vec{r}-\vec{l})$, localized about the N lattice sites $\vec{l}$,

$$\psi_{n\vec{k}}(\vec{r}) = N^{-1/2} \Sigma_{\nu,\vec{l}} \, C_{\nu n}(\vec{k}) \, e^{i\vec{k}\cdot\vec{l}} \, \phi_\nu(\vec{r}-\vec{l}) \qquad (3.4)$$

one can cast the Dyson integral equation (2.2) into an algebraic eigenvalue problem for the coefficients $C_{\nu n}(\vec{k})$:

$$\Sigma_{\nu'} \; \{\Sigma_{\vec{l}} \, e^{i\vec{k}\cdot\vec{l}} \, [\int\phi_\nu^*(\vec{r}) \, h(\vec{r}) \, \phi_{\nu'}(\vec{r}-\vec{l})d\vec{r} + \int\phi_\nu^*(\vec{r}) \, \Sigma(\vec{r},\vec{r}';E)$$

$$\phi_{\nu'}(\vec{r}'-\vec{l})d\vec{r}d\vec{r}']\} \, C_{\nu'n}(\vec{k}) = \varepsilon_{n\vec{k}} \, C_{\nu n}(\vec{k}) \qquad (3.5)$$

As discussed in sec.II, the time-dependent screened HF approximation (TDSHF) is employed. It consists in complementing the usual time-dependent H approximation (or RPA, i.e. $\delta\Sigma/\delta g \to 0$ in eq.(2.9)) by including the ladder diagrams (or e-h attraction, see the term "B" in the diagrams of Fig.3). Using the local-orbital representa-

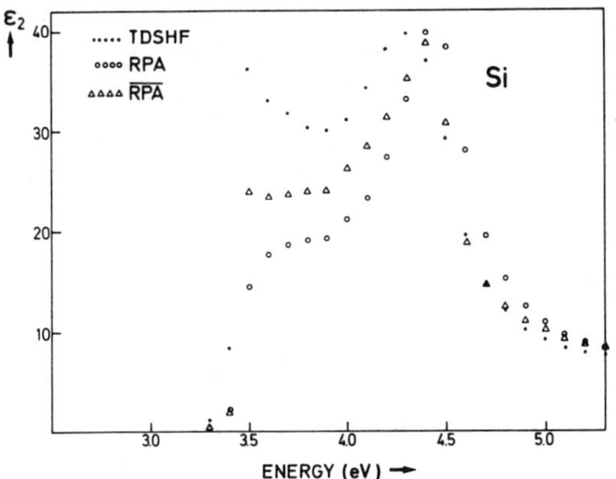

Fig.6: Comparison of the theoretical results for the imaginary part of the dielectric function in Si.

both the energy- and $\vec{r}$-dependence of the screening of semiconductors and insulators. As an example, which tests the ω-dependence, we display in Figs.6-8 the imaginary part of the dielectric func-

tion of eq.(3.4) the Bethe-Salpeter equation (2.7) can be cast in a matrix equation (with the dimension determined by the number of orbital overlaps) and inverted, with G given in matrix notation

$$G = G^O \left\{ 1 - (v - V_{xc}) G^O \right\}^{-1}$$

$$(3.6)$$

where V_{xc} denotes here the e-h attraction and v the RPA Coulomb matrix. We have shown that these many-body effects are essential to even a qualitative description of

tion (optical absorption) as well as the real part in comparison with experiment. (For details, see ref.35) We note, that the RPA local-field effect (denoted by RPA), which contains the Bragg reflections or $\vec{G} \neq \vec{G}'$ matrix elements of the dielectric matrix $\varepsilon(\vec{q}+\vec{G}, \vec{q}+\vec{G}'; \omega)$, shifts absorption structure to higher energies, whereas the combined RPA plus e-h attraction shifts structure (the E_1-structure in Si by about 0.2 eV) to lower energies. The combined many-body effects also enhance the low-energy oscillator strength by up to a factor two, when compared with the non-interacting e-h transition without local-field effects (termed RPA in the figures)

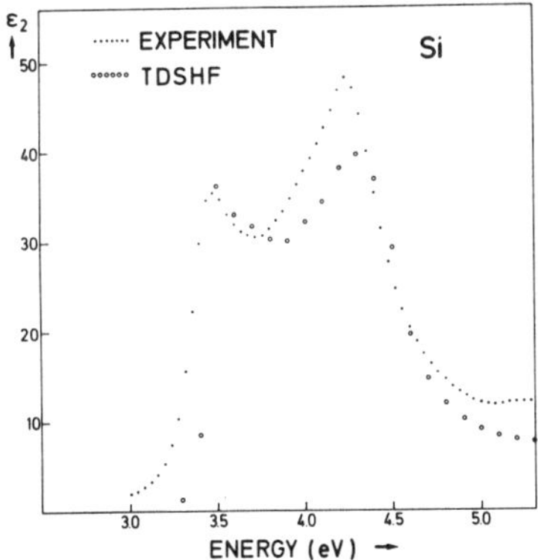

Fig.7: Comparison between theory (TDSHF) and experiment for the imaginary part of the dielectic function of Si.

and result in good accord with experiment. A similar trend of these many-body effects is found in the optical response of insulators

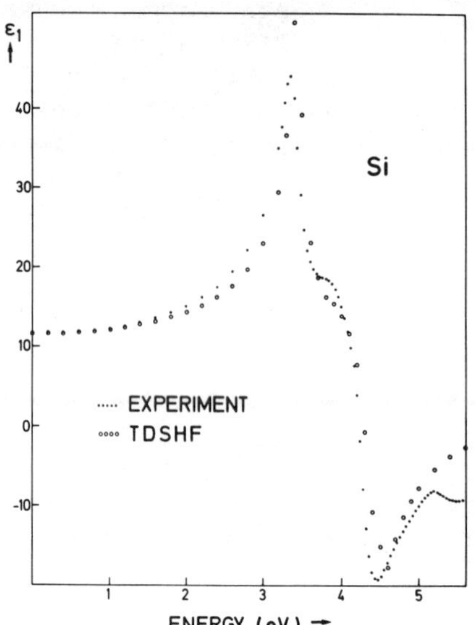

Fig.8: Comparison between theory and experiment for the real part of the dielectric function of Si.

and compound semiconductors[36]. However, the interaction effects are also indispensable in the $\vec{r}$-dependence of the response[35] to a static point-charge. Fig.9 shows the induced charge density of a positive, substitutional point defect in Si in the (110) plane, which contains (semicircle on the left edge) the impurity position and (crosses) the neighboring atoms in this plane. Solid or dashed lines encircle regions of negative or positive charge density, respectively. The point-charge impurity may serve here as an illustration of a "snap shot" of the screening-density profile, induced by a hole as a quasi-particle. We note that the induced charge is strongly anisotropic, with positive and negative oscillators, which are concentrated essentially along the tetrahedral bonds. The screening obviously is radically different (the oscillations are here due to the Bragg reflections and not due to Friedel oscillations like in a simple metal) from a homogeneous electron-gas metal. We will come back to the consequences of this different screening pattern for the exchange-correlation potential shortly.

Expressing finally the self-energy operator Σ in eq.(3.5) in the TDSHF and employing the local representation in eq.(3.6) for the screening propagator, we arrive at the following expression for the matrix elements of Σ' (only the non-HF part)[5]

$$\int \phi_\tau^*(\vec{r}) \; \Sigma'(\vec{r},\vec{r}';E) \; \phi_{\tau'}(\vec{r}'-\vec{t}) \, d\vec{r}\,d\vec{r}' = \frac{i}{2\pi N} \int_0^\infty dE' \sum_{\nu\nu'} \sum_{\vec{m}\vec{m}'}$$

$$\{g_{\nu\nu'}(-\vec{m};E+E') + g_{\nu\nu'}(-\vec{m},E-E')\} \sum_{\vec{q}} e^{i\vec{q}\cdot\vec{m}}$$

$$\cdot W'_{\vec{m}+\vec{m}'\nu\tau;\vec{t}+\vec{m}'\nu'\tau'}(\vec{q},E') \tag{3.7}$$

with (in matrix notation, as in eq.(3.6))

$$W'(\vec{q},E') = v(\vec{q}) \; G(\vec{q},E') \; v(\vec{q}) \tag{3.8}$$

134

and the one-particle Green's function

$$g_{\nu\nu'}(\vec{s}-\vec{s}';E) = N^{-1} \sum_{n} \sum_{k}^{BZ} C_{\nu n}^{(o)}(\vec{k})\, e^{i\vec{k}(\vec{s}-\vec{s}')}\, C_{\nu'n}^{(o)}(\vec{k})/(E-E_{nk}^{(Q)}+$$

$$i\delta\,\mathrm{sign}(E_{n\vec{k}}-E_F))\qquad\qquad(3.9)$$

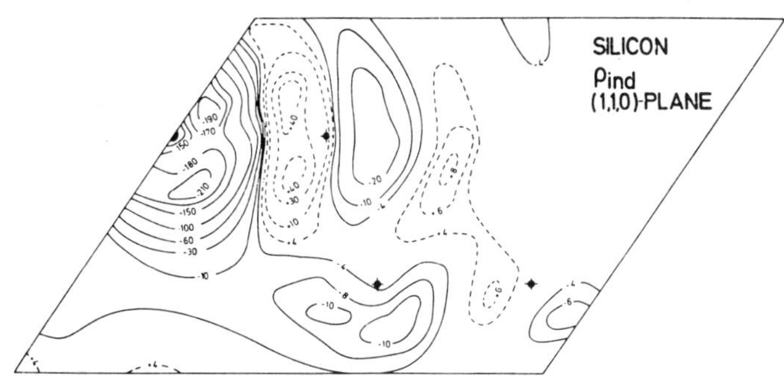

Fig.9: Induced charge density of a positive substitutional point defect in Si in the (110) plane. Numbers multiplied by 10^{-4} give the charge density in atomic units.

Here $C_{\nu n}^{(o)}$ and $E_{nk}^{(Q)}$ denote a zeroth-order approximation for the quasiparticle states. In our Si calculation this zeroth-order approximation was extracted from an empirically fitted pseudopotential bandstructure (see ref.4 and 35). This bandstructure is fitted in terms of a fourth-nearest neighbor (in the fcc-lattice sites) overlap model of bonding and antibonding orbitals as described[4] in our earlier work on optical properties and impurity screening[35]. Also the calculation of the two-particle Green's function is based on this bandstructure and follows closely the impurity studies (for details see in particular, ref.35).

In our previous diamond work[5,6,16] we determined the bare HF part of Σ by making a Slater-Koster fit to an existing HF band calculation, while the correlation part was determined, as in the present Si work, by evaluating the matrix elements in an explicit basis set, which represents a zeroth-order approximation to the actual quasi-particle states.

As discussed in ref.5, the procedure for calculating the HF in C is not completely rigorous, because the density matrix $\rho(\vec{r},\vec{r}')$ that appears in the HF matrix elements should in principle be recalculated as correlations are added. Here the only difference in methodology between our C and Si calculations enters: whereas in C several accurate HF band calculations are available, this is not the case in Si. We, therefore, approximate in Si not only the correlation part by evaluating it in an approximation to the quasi-particle states, but use the same "pseudopotential wave function basis set" also for the non-local density operator $\rho(\vec{r},\vec{r}')$ in the HF matrix elements. The above-mentioned correlations entering $\rho(\vec{r},\vec{r}')$ are incorporated in a consistent manner, but to lowest order in the still required self-consistency circle.

An interesting observation can be added here: if the local orbitals are not constructed from current-conservation criterion as discussed below but instead are adjusted to fit an LDA calculation based on the Kohn-Sham local potential v_{xc}, we can alternatively calculate $\rho(\vec{r},\vec{r}')$ and construct approximate HF bands. These bands were found to agree reasonably well (r.m.s. deviation for valence bands $\sim$1 eV) with the "current conservation" HF results. This lends some further justification to our approximate HF construction: at least for atoms and molecules [37,38] it is well-known that the orbitals of a self-consistent X_α calculation with $\alpha \approx 2/3$, i.e. the Kohn-Sham value, form a very good approximation to the actual HF orbitals.

Now to the definition of the basis set of eq.(3.4). This basis function set has already been used in ref.35 to calculate the screened interaction W' of eqs.(3.7, 3.8) and has been justified and described in all detail there: we take the empirical pseudopotential energies $E(n,\vec{k})$ as the above zeroth-order approximation $E^{(0)}(n,\vec{k})$ to the actual quasi-particle energies. The basis functions are then constructed by expanding them in a Gaussian basis. The Gaussian coefficients are extracted from the criterion of current conservation, i.e. a Ward identity. On the one hand, this just guarantees the required consistency in evaluating actual operators (current and density) between local basis functions. On the other hand, our previous work on the screening W' (or ε^{-1}) demonstrated that this basis set determination results in quantitative agreement with a variety of experimental data [4,35]. So, clearly, this is a reasonable starting point for the quasi-particle wave functions and energies.

The current conservation guarantees consistency in the approximation for both one (g)- and two-particle (W')-propagators in eq. (3.7). Just as in our previous C work, we therefore use the same basis set as an approximation for the actual quasi-particle states everywhere in eq.(3.7). Of course, as discussed above, this procedure is approximate in that it is a pragmatic "short-cut" in the self-consistency circle for the quasi-particle states.

III. C.   Results for the Quasi-Particle Energies in Si

Fig. 10 displays our results for the quasi-particle band-structure in Si for two symmetry directions in an absolute energy scale. The full lines give our HF results. The HF approximation gives a direct gap of 6.8 (eV) compared to 3.3-3.4 (eV) in optical experiments, and a valence-band width of 13.5 (eV) compared to 12-13 (eV) in photoemission data [19] (see also Table II). The dashed lines (TDSHF) are the dynamically correlated bands, where the two-particle Green's function contains both RPA (including local-field effects) and e-h attraction effects. The direct gap is reduced to 3.5 (eV) and the valence-band width to 12.9 (eV) by the dynamical correlation. Both values are in good accord with the experimental data. Our numerical results clearly demonstrate that the self-energy correction is essentially $\vec{k}$-independent. Thus, $\Sigma(\vec{r},\vec{r}';E)$ is rather short-ranged and local across a given band. It displays a significantly weaker energy dependence from the top to the bottom of the valance bands, when compared with our earlier diamond results [5]

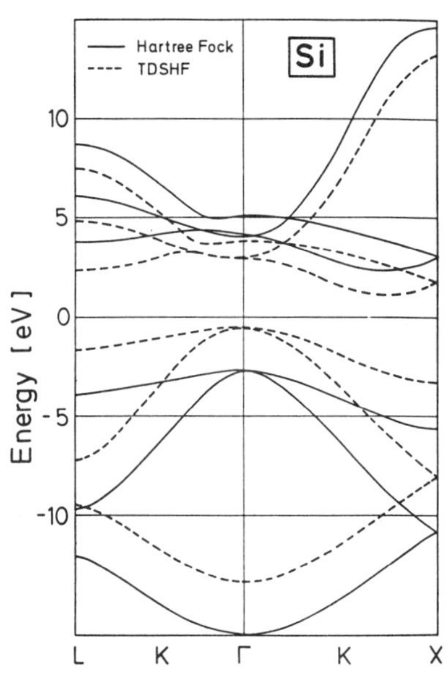

Fig.10: Quasi-particle bands in Si.

To gain deeper insight into the physical significance of various steps in the calculation, consider the summary presented in Table II.

The "TDSHF" column gives the full calculation. The column "E-H" resorts to a calculation, where the plasmon pole is chopped off at 14 (eV) in the dynamically screened interaction W' of eq. (3.7). Again, as in our previous diamond calculation, we find the gap practically unchanged. However, the plasmon-type correlations, which are predominatly of long-range nature and therefore essentially non-local, introduce a shrinking of 0.1 (eV) of the valence-band width (in C :1.25 (eV)). But still, also for the band width, the dominating self-energy correction stems from the e-h excitations, which bring the HF value of 13.5 (eV) down to 13.03 (eV).

An approximation for Σ, which has been suggested by Hedin [2] is the so-called Coulomb-hole plus screened-exchange (COHSEX) method.

It is based on splitting up the contour integration in eq.(3.1) into

$$\Sigma = \{\text{Residues of } g \cdot W(\omega - E_{nk})\} + \{g(\omega_{pole}) \cdot \text{Residues of } W\}$$

(3.10)

where ω_{pole} denotes the poles or elementary excitations of ε^{-1}. In the usual form the COHSEX approximation further assumes that ε^{-1} or W is independent of frequency (see, in particular, Fig.11 and the discussion in the sec.IV). This amounts to taking $\omega \equiv E_{nk}$ in eq. (3.10), in which case Σ becomes independent of ω, i.e.

$$\Sigma(1,2) = -\rho(1,2) \, W(1,2;\omega{=}0) + 1/2 \, \delta(1,2) \, [W(1,2;\omega{=}0) - v(1,2)]$$

(3.11)

where ρ is again the density matrix.

In this form it is evident that the first term on the right-hand side gives the screened exchange. The second term is obviously the potential from the "Coulomb-hole" in the charge density around a given electron. The COHSEX approximation has been used for Si in connection with various Penn-type model dielectric functions [39-42].

We can gain some insight into the validity of the COHSEX approximation from Table II and Fig. 11. The band gap is found practically unchanged, when compared with the full TDSHF calculation. However, the valence-band width comes out to be 13.33 (eV) and thus is only slightly renormalized compared with HF (13.5 eV).

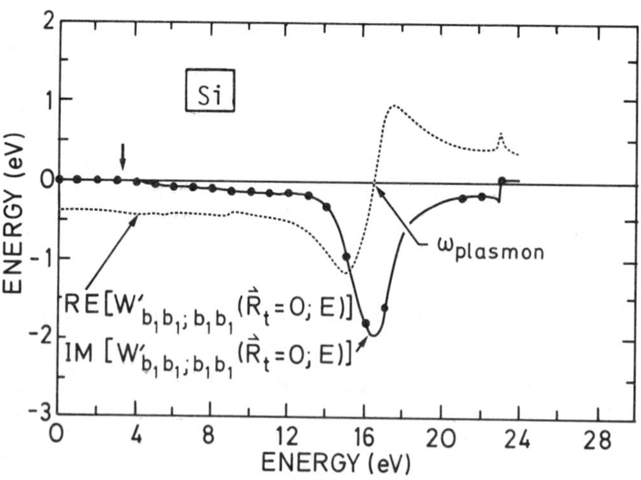

Fig.11: Real and imaginary part of the screened interaction in Si.

138

We found similar discrepancies for the higher conduction bands. Fig.11, containing the largest matrix element of the screened interaction as function of ω, tells us that the assumption in COHSEX of an ω-independent real part of W'(and ImW'=0) is very reasonable to about 8 (eV); from there on we have dynamic screening effects to include, which becomes obvious near the plasmon pole.

This suggests, that an energy-independent approximation works best at energies around the gap, as it is implied in the very construction of the COHSEX approximation, and as it has already been remarked by Kane [43]. Moving down through the valence or conduction bands to higher energies, the energy-dependence of Σ becomes progressively more important, corresponding to an increasing weight of the plasma resonance. These conclusions support some of the findings of a non-local energy-dependent pseudopotential calculation [44].

We would like to stress here that most of the above trends have already been established in our previous diamond calculation [5,6]. However, there exist also some important differences: one example is furnished by the RPA calculation for Σ, leaving out the e-h attraction in the screened interaction W.

In contrast to C, where in the RPA both the gap and band-width increase by about 1 (eV) compared to the full TDSHF calculation, we find in Si gap and band-width practically unchanged. This reduction in the influence of e-h interaction effects in the two-particle excitations on the one-particle spectrum is to be expected: when we compare the screening pattern in Si of the point charge in Fig.9 with the induced charge pattern in the insulator diamond (ref.35), we find that the essentially one-dimensional flow (along bonding channels) of screening charge in C is much more spread out and becoming three-dimensional in Si. As a result e-e interactions are not confined to these "one-dimensional" channels. This implies a significantly larger influence of exchange plus correlation in C giving rise to e-h attraction.

Finally, let us consider in Fig.12 our findings for the decay rate of the quasi-particles (hole in the valence bands) in Si, and again compare with our C results. First, we note that below a threshold $Im(E_{n\vec{k}}) = 0$ the lifetime $\{2\,ImE_{n\vec{k}}\}^{-1}$ is infinite. The mechanism for the decay of the quasi-holes is the creation of e-h pairs. Thus, the radiationless transition to a final state with more than one hole in the valence bands is an Auger process, that can be single,

Table II: Quasi-Particle Energies

	exp	H	HF	TDSHF	LDA	E-H	COHSEX
E_{gap} (eV)	3.4	1.3	6.8	3.5	2.9	3.5	3.5
E_{val} (eV)	12.5	14.7	13.5	12.9	12.8	13.03	13.33

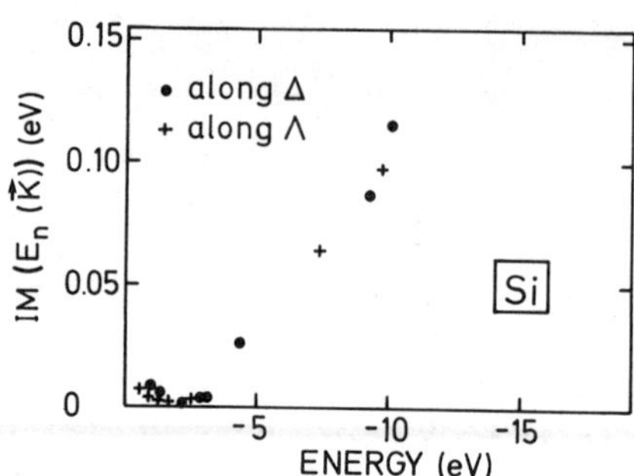

Fig.12: Quasi-particle decay in Si.

double etc. From the elementary treatment of Auger transitions we should find this threshold at essentially the gap. An additional shift of about 2 (eV) stems from the fact that the first peak in the valence density of states has its half-maximum at about -1 (eV) [5,6]. Indeed, our calculations reveal a sharp rise beyond this threshold. This in accordance with empirical broadening factors, which are neccessary to relate valence-band photoemission spectra to theoretical line shapes of the electronic density of states [19]. The main difference to our C results is the by about a factor of 10 smaller value for the $\mathrm{ImE}_{n\vec{k}}$ at the bottom of the valence band. This clearly is a consequence of the significantly smaller Coulomb correlation in Si.

IV. AN ANALYTICAL, ENERGY- AND LOCAL-DENSITY-DEPENDENT MODEL FOR SELF-ENERGY

We make now use of our numerical results for the self-energy operator in C and in Si to derive a simple, analytic model. It will in particular elucidate how an intrinsically long-ranged and non-local e-h polarization giving rise to the dynamical correlation in a non-metal can still approximately be cast into a local effect. We then use the relation (2.22), which expresses the Kohn-Sham potential v_{xc} in terms of Σ_{xc}, to derive a model expression for v_{xc}. This model for v_{xc} should work not in the commonly used

140

homogeneous electron gas (metallic) limit, but in the opposite
tight-binding of a large-gap insulator.

IV.A. Model for Σ_{xc}

We start from the TDSHF expression for the matrix elements of
Σ' (the non-HF part) in eqs. (3.7) and (3.8). As already mentio-
ned, Σ' is short-ranged in $|\vec{r}-\vec{r}'|$, a fact which can be proven em-
ploying graphical considerations, and which has been checked numerically by Hedin[2] for the electron gas. That this short-range property holds also for non-metals, has already been shown in our
diamond work[5] and is again displayed in Fig. 13a):

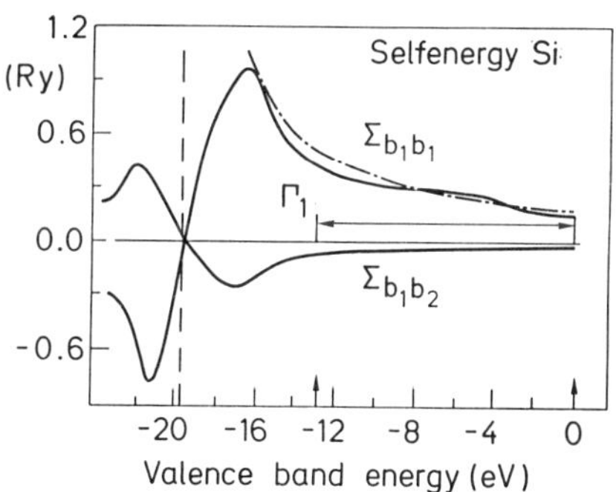

Fig.13a: Valence-band self-energy matrixelements (real part). Dash-
dotted line gives analytical model results (see sec. IV).

The short-range character of $\Sigma(\vec{r},\vec{r}';E)$ results in a drastic reduction of the valence-band matrixelement, if both bonding orbitals
b_1 and b_2 are located on the same lattice site $\vec{l}=o$, but one orbital (b_2) is turned into a different tetrahedral direction. That
this short-range property holds also for the conduction-band matrixelements is shown in Fig. 13b).

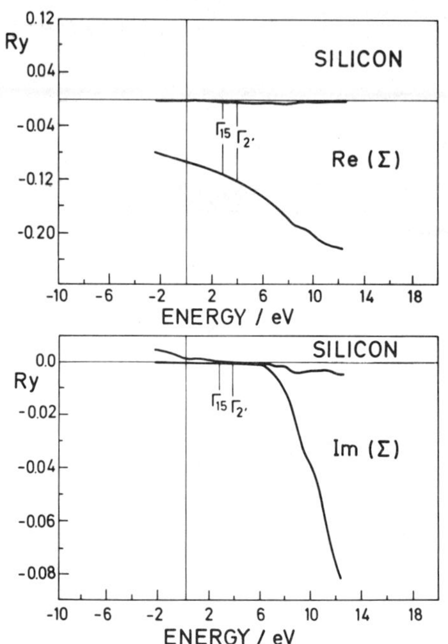

Fig. 13b: Real (upper part) and imaginary part of conduction-band
self-energy matrixelements $\Sigma_{a_1 a_1}$.

The reason for this short-range behavior is due to a rather
different energy dependence displayed by the one-and two-particle
propagators (see detailed discussion in ref. 5): the overall shape
of the different matrixelements of W' is quite similar to the largest matrix element $W'_{b_1 b_1; b_1 b_1}(\vec{t}=o,E)$ shown in Fig. 11 . It gives
the induced part of the dynamically screened potential between two
pairs of bonding orbitals centered at the same site and pointing
along the same bond [111]. Here, we have introduced the notation

$$W'_{\nu_i \tau_i; \nu_j \tau_j}(\vec{t},E) = \frac{1}{N} \sum_{\vec{q}}^{BZ} e^{-i\vec{q}\cdot\vec{t}} W'_{o\nu_i\tau_i; o\nu_j\tau_j}(\vec{q},E) \qquad (4.1)$$

142

The reason behind the similarity in overall shape of the matrix
elements is that it is controlled by the screening matrix
$G^O(1-(v-V_{xc})G^O)^{-1}$ which is common to all of them. Both real and
imaginary parts of W' are approximately constant over the energy
range of the valence bands and show strong variations only near
the plasmon pole ω_p. On the other hand, both real and imaginary
parts of the matrix elements of the one-particle properties
$g_{oo'}(-m;\omega)$ display strong oscillatory behavior for $m \neq o$ (see Fig.14
and 15). Performing the convolution integrals in eq. (3.7) and noting that, by overlap arguments only the terms with $m+m' = t+m'=o$
with the bonds ν parallel to $\tau,\tau'=\nu'$ contribute significantly leads
to the observed short-range property of Σ'.

In this on-site approximation the matrix elements (3.7) become

$$\langle\phi_{b_i}(\vec{r})|\Sigma'(\vec{r},\vec{r}';E)|\phi_{b_j}(\vec{r}')\rangle = \frac{i}{2\pi} \, dE' \int_0^\infty [\overline{g}_{b_i b_j}^{(v)}(\vec{t}=o;E,E')$$

$$\cdot W'_{b_i b_i b_j b_j}(\vec{t}=o,E') + \overline{g}_{a_i a_j}^{(c)}(\vec{t}=o;E,E')W'_{a_i b_i a_j b_j}(\vec{t}=o;E')] \quad (4.2)$$

for the valence bands, where b_i stands for bonding orbitals with
$i = 1,2,3,4$ and $\overline{g} = g(E+E')+g(E-E')$. A similar expression holds
for the conduction-band matrix elements. In Si, very much like in
C (see Figs. 3 and 4 of ref. 6), the bb-bb matrix elements of W'

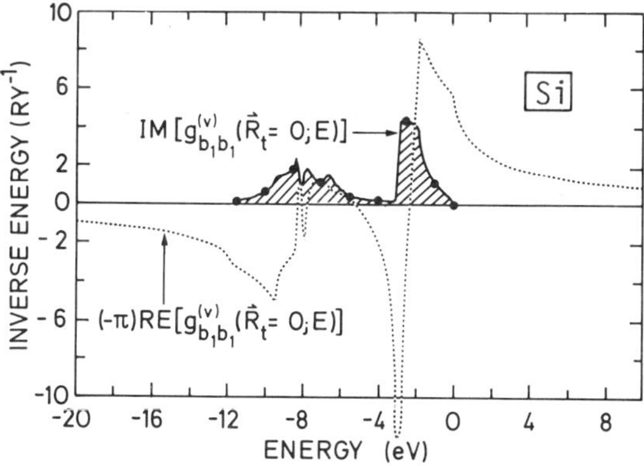

Fig. 14: Real and imaginary parts of the one-particle propagator
($\vec{t}$=o in eq. (4.1)).

are about an order of magnitude larger than the ba-ba matrixele-
ments. This stems from the fact that, as the wavevector $|\vec{q}+\vec{G}| \to o$,
the e-h form factors in W' have a finite limit in the first case
but not in the second[6]. Thus, we can leave out the second term on
the r.h.s. of eq. (4.2) and, furthermore, consider only the overlap terms with $b_i = b_j$.

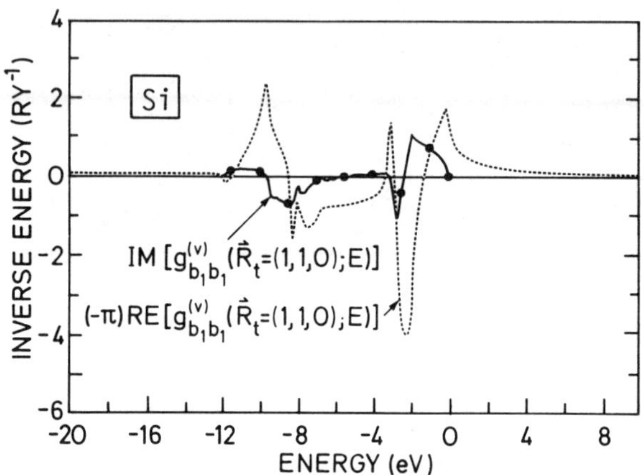

Fig. 15: Real and imaginary parts of the one-particle Green's
         function ($\vec{t}$=(1,1,O) nearest-neighbor, in eq. (4.1))

The interplay of the screening propagator W' with the one-
particle Green's function $g^{(v)}$ is shown in Fig. 16 for the numerical values of W' and g in the case of diamond. We note, in particular, that for small valence-band energies around the gap, the
self-energy matrix-element is determined by convoluting the real
part of W' (which is essentially constant across the valence bands)
with the imaginary part of $g^{(v)}$ (which is the weighted density of
states). Fig. 16 then gives a pictorial insight into the validity
of the screened exchange approximation (which is just given by
$\int$Img$\cdot$ReWdE') and the (energy-independent) COHSEX approximation as
discussed in sec. III. It is also clear from this figure, why both
approximations can only work near the gap: for energies E further
into the valence band also Im W' has to be convoluted with Re $g^{(v)}$,
and the energy dependence of W' becomes increasingly important.
From eqs. (3.7), (3.8) and (4.1) W' can be written in the local-
orbital basis as follows

$$W'_{ob_1 b_1;ob_1 b_1}(\vec{t}=o;E) = \frac{1}{N} \sum_{\vec{q}}^{BZ} e^{-i\vec{q}\cdot\vec{t}} W'_{ob_1 b_1;ob_1 b_1}(\vec{q},E)$$

144

$$= \frac{1}{N} \sum_{\vec{q}}^{BZ} v_{ob_1b_1;l\nu\mu}(\vec{q}) \; G_{l\nu\mu;l'\nu'\mu'}(\vec{q},E) \; v_{l'\nu'\mu';ob_1b_1}(\vec{q},E) \qquad (4.3a)$$

where, as in eq. (3.6)$_4$, the screening matrix $G(\vec{q},E)$ is given by (in matrix notation):

$$G(\vec{q},E) = G^{O}(\vec{q},E) \; \{ \; 1-[\; v(\vec{q}) - V_{xc}(\vec{q})] \; G^{O}(\vec{q},E)\}^{-1} \qquad (4.3b)$$

$G^{O}(\vec{q},E)$ is the RPA irreducible polarizability in local representation

$$G^{O}_{l\nu\mu;l'\nu'\mu'}(\vec{q},E) = \frac{1}{N} \sum_{\vec{k}}^{BZ} e^{i(\vec{k}+\vec{q})(\vec{l}-\vec{l}')}$$

$$\cdot \sum_{n_1,n_2} c^{*}_{\nu n_2}(\vec{k}) C_{\mu n_1}(\vec{k}+\vec{q}) \; \frac{2 \; [f_{n_1}(\vec{k}+\vec{q}) - f_{n_2}(\vec{k})]}{E_{n_1}(\vec{k}+\vec{q})-E_{n_2}(\vec{k})-E-i\eta} \; c_{\nu' n_2}(\vec{k}) c^{*}_{\mu' n_1}(\vec{k}+\vec{q})$$

$$(4.4)$$

$f_n(\vec{k})$ being the occupation number of the quasi-particle state $|n\vec{k}\rangle$. The expansion coefficients $c_{\nu n}(\vec{k})$ (defined in eq. (3.4)) and the band eigenvalues are the same as those to be utilized in eq. (3.9) for the one-particle Green's function $g_{\nu\nu'}(\vec{s}-\vec{s}';E)$.

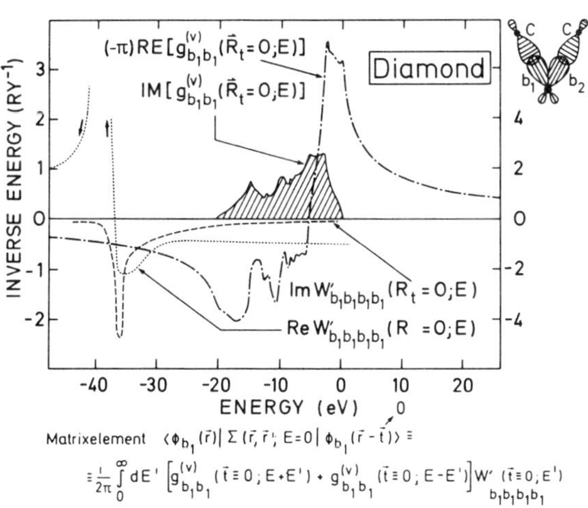

Fig.16:

As in eqs. (3.6) and (3.8) v in eq. (4.3) is the Fourier transform of the matrix of the bare Coulomb potential between pairs of local orbitals:

$$v_{\vec{l}\nu\mu,\vec{l}'\nu'\mu'}(\vec{q}) = \sum_{\vec{m}} e^{-i\vec{q}\cdot\vec{m}} \langle \phi_\nu^*(\vec{r}-\vec{m}) \phi_\mu(\vec{r}-\vec{l}-\vec{m}) | v(\vec{r}-\vec{r}') | \phi_{\nu'}^*(\vec{r}') \phi_{\mu'}(\vec{r}'-\vec{l}') \rangle$$

In our model eq. (4.3) is further simplified by assuming that the dominant $\vec{q}$-dependence in the BZ-integration stems from the Coulomb matrices v, and that we may put $\vec{q}\equiv o$ in the e-h propagator $G(\vec{q},E)$. This assumption has been shown in refs. 5,6 and 35 to be very reasonable for the irreducible propagator G^o. Here, we extend it also to the denominator $1-[v(\vec{q})-V_{xc}(\vec{q})] G^o(\vec{q},E)$ in eq. (4.3).

Our next approximation which we have to introduce to derive an explicit expression for G^o cannot directly be justified by comparison with our numerical result in Si or C. It consists in implementing a two- (flat) band model, where only a bonding (or p-type) band with average energy E_{val} ($\equiv E(x_4)$, in crystals with the diamond structure) and an anti-bonding (s-type) band with energy E_{cond} ($\equiv E(x_1)$) is retained. The choice of E_{val} and E_{cond} is dictated by the requirement, that they determine the Penn gap $E_{cond}-E_{val}\equiv E_g$, i.e. the main optical transition (see refs. 4 and 33). In an insulator like LiF they are chosen so as to determine the main p-s transition as revealed by the optical absorption data (see below).

This tight-binding two-band model simplifies the one-particle propagators in eq. (3.9) to

$$g_{\nu\nu'}(\vec{s}-\vec{s}';E) = \delta_{\nu\nu'}/(E-E_n(\vec{k}) + i\delta[E_n(\vec{k})-E_F]), \qquad (4.6)$$

where $E_{n=val}(\vec{k}) = E(x_4)$ for $g^{(v)}$ and $E_{n=cond}(\vec{k}) = E(x_1)$ for $g^{(c)}$. For the irreducible two-particle propagator G^o in eq. (4.4) we have

$$G^o_{\vec{l}\nu\mu;\vec{l}'\nu'\mu'}(\vec{q},E) \cong - \frac{4E_g}{E_g^2-\omega^2} \delta_{\vec{l},o} \delta_{\vec{l}',o} \delta_{\nu\nu'}\delta_{\nu\mu}\delta_{\nu'\mu'} \qquad (4.7)$$

and for the interacting G

$$G_{\vec{l}\nu\mu,\vec{l}'\nu'\mu'}(\vec{q},E) \cong \frac{4E_g}{\omega^2-\omega_p^2} \delta_{\vec{l},o}\delta_{\vec{l}',o} \delta_{\nu\nu'}\delta_{\nu\mu}\delta_{\nu'\mu'} \qquad (4.8)$$

In contrast to G^o, G in fact is not diagonal in the bonding indices ν and ν', and so the expression (4.7) presents a diagonal approximation. The plasmon pole $\omega=\omega_p$ approximates the pole structure in $\{1-[v(\vec{q}=o)-V_{xc}(\vec{q}=o)] G^o(\vec{q}=o,\omega)\}^{-1}$ with $\omega_p = \{E_g^2 + [v(\vec{q}=o)-V_{xc}(\vec{q}=o)].4E_g\}^{1/2}$.

With these simplifications, the valence-band self-energy takes finally the form

$$\Sigma^{(v)}(\vec{r},\vec{r}';E) = - \sum_{b_i,l} W(\vec{r},\vec{r}';\overline{E}_{val}-E)\,\phi_{b_i}(\vec{r}-\vec{l})\,\phi^*_{b_i}(\vec{r}'-\vec{l})$$

$$+ \frac{2E_g}{\omega_p} \sum_{b_i,l,a_i} c_i(\vec{r}-\vec{l})\,c_i(\vec{r}'-\vec{l}) \left\{ \frac{\phi_{b_i}(\vec{r}-\vec{l})\phi^*_{b_i}(\vec{r}'-\vec{l})}{\overline{E}_{val}-\omega_p} + \right.$$

$$\left. \frac{\phi_{a_i}(\vec{r}-\vec{l})\phi^*_{a_i}(\vec{r}\perp\vec{l})}{\overline{E}_{cond}-\omega_p} \right\} \qquad (4.9)$$

where b_i (a_i) denote bonding (anti-bonding) orbitals and c the Coulomb matrix element

$$c_i(\vec{r}-\vec{l}) = \int \phi^*_{b_i}(\vec{r}_1-\vec{l})\,v(\vec{r}-\vec{r}_1)\,\phi_{a_i}(\vec{r}_1-\vec{l})\,d\vec{r}_1 \qquad (4.10)$$

The first term in eq. (4.9) presents the screened exchange, with

$$W(\vec{r},\vec{r}';\overline{E}_{val}-E) = v(\vec{r}-\vec{r}') + \frac{4E_g}{(\overline{E}_{val}-E+i\eta)^2-\omega_p^2} \sum_{l,\nu} c_\nu(\vec{r}-\vec{l})c_\nu(\vec{r}'-\vec{l})$$

$$(4.11)$$

The second term in eq. (4.9) is the Coulomb-hole contribution. From this equation, together, with the corresponding expression for $\Sigma^{(c)}$, we can write down the bonding matrix elements

$$\langle\phi_{b_i}(\vec{r})|\,\Sigma'(\vec{r},\vec{r}',E)\,|\phi_{b_i}(\vec{r}')\rangle = - V^2_{b_ib_i} \frac{4E_g}{(E(x_4)-E+i\eta)^2-\omega_p^2}$$

$$+ \frac{2E_g}{\omega_p}\left[\frac{V^2_{b_ib_i}}{-\omega_p-E+E(x_4)} + \frac{V^2_{b_ia_i}}{-\omega_p-E+E(x_1)}\right] , \qquad (4.12)$$

and similarly, for the anti-bonding matrix element:

$$\langle\phi_{a_i}(\vec{r})|\Sigma'(\vec{r},\vec{r}';E)|\phi_{a_i}(\vec{r}')\rangle = - V^2_{b_ia_i} \frac{4E_g}{(E(x_4)-E+i\eta)^2-\omega_p^2}$$

$$+ \frac{2E_g}{\omega_p}\left[\frac{V^2_{b_ia_i}}{-\omega_p-E+E(x_4)} + \frac{V^2_{a_ia_i}}{-\omega_p-E+E(x_1)}\right] \qquad (4.13)$$

In eqs. (4.9) through (4.13) valence-band energies are defined as positive and conduction-band energies as negative, with the zero of energy being, for example, the top of valence bands. In eqs. (4.12 and (4.13) the first and second terms on the r.h.s. give again screened exchange and Coulomb-hole contributions, respectively.The matrix element $V^2_{b_i b_i}$ is defined as

$$V^2_{b_i b_i} = \sum_{\vec{1},j} \left| \iint \phi^*_{b_i}(\vec{r})\, \phi_{b_i}(\vec{r})\; v(\vec{r}-\vec{r}')\phi^*_{b_j}(\vec{r}-\vec{1})\phi_{a_j}(\vec{r}'-\vec{1})\,d\vec{r}d\vec{r}' \right|^2$$

(4.14)

In the matrix element $V^2_{b_i a_i}$ the bonding orbital $\phi^*_{b_i}(\vec{r})$ in eq. (4.14) is replaced by the anti-bonding orbital $\phi^*_{a_i}(\vec{r})$, and in $V^2_{a_i a_i}$ additionally $\phi_{b_i}(\vec{r})$ replaced by the corresponding anti-bonding orbital $\phi_{a_i}(\vec{r})$. The matrix elements $V^2_{b_i a_i}$ is in Si about a factor 10^3 and in C a factor 10 smaller than $V^2_{b_i b_i}$ or $V^2_{a_i a_i}$, and will therefore be omitted in our model.

The physical content of $V^2_{b_i b_i}$ or $V^2_{a_i a_i}$ is obvious: the particle (with density proportional to $(\phi^*_{b_i}\phi_{b_i})$ or $(\phi^*_{a_i}\phi_{a_i})$) induces an e-h polarization at lattice sites $\vec{1}$, which in turn sets up the dynamical correlation potential felt by the same particle.

The $(\vec{1}=o)$-term in $V^2_{b_i b_i}$ vanishes because of bonding-antibonding symmetry. The remaining lattice sum can further be approximated, by expanding the Coulomb interaction into a multipole series.

Keeping only the dipole term we have

$$V^2_{b_i b_i} \stackrel{\sim}{=} \sum_{\vec{1},j} e^4 \frac{(\vec{p_j}.\vec{1})^2}{|\vec{1}|^6}$$

(4,15)

where

$$\vec{p_j} = \int \phi^*_{b_j}(\vec{r})\, \vec{r}\, \phi_{a_j}(\vec{r})\,d\vec{r}$$

(4.16)

The lattice sum in (4.15) can be evaluated, (for fcc lattice: $\sum_{\vec{1}} 1/1^4 = (\sqrt{2}/a_o)^4 .8\pi$) yielding

$$V^2_{b_i b_i} \cong 8\pi e^4\, p^2/\Omega_o \left(\frac{4}{3a_o}\right) \cong 8\pi e^4 p^2/\Omega_o \left(\frac{2}{3}^{1/3} <\rho>^{1/3}\right) \qquad (4.17)$$

In the dipole expansion the same results holds also for $V_{a_i a_i}$.

In eq. (4.17) $\cos^2(\vec{p}\cdot\vec{l})$ has been averaged to 1/3 and the average density $<\rho> = 4(2/a_o)^3$ been introduced. The dipole matrix element $p=|\vec{p}|$ is determined in the tight-binding basis[4] and put equal to $(a_o/4)(3/2)^{1/2}$ here. Towards the plane-wave ($E_g \to o$) limit it has to be replaced by $2\pi/(a_p mE_g)$, in which case eq. (4.12) turns into a plasmon-pole formula,[46] as discussed in the homogeneous electron gas by Lundqvist[46] and Hedin and Lundqvist[2].

The proportionality of the correlation potential in eq. (4.16) to $<\rho>^{1/3}$ points to an important physical aspect: although intrinsically long-ranged, the correlation potential of an insulator or semiconductor can still be converted to a "local" potential. This conversion becomes very transparent from an alternative consideration of eq. (4.15). The potential acting on a particle in the center of a sphere (the "Lorentz cavity", which plays here roughly — see the remark concerning ref. 48, below — the role of the exchange-correlation hole) can either be calculated as the continuum dipole sum outside the spherical region or, complementary, by just considering the potential of a uniformly charged sphere with density $<\rho>$, surrounded by vacuum. This illucidates how the long-ranged, non-local e-h polarization effect in a non-metal can still be cast into a local effect, with the latter being in the spirit of Slater's $<\rho>^{1/3}$ exchange model for metals.[47]

Sterne and Inkson[48] have recently proposed a similar construction of an extreme tight-binding model for Σ, where the exchange-correlation hole is directly identified with the Lorentz sphere. This results (apart from a different expression used for p and the screening ε^{-1} in a factor of 2 difference to our $V^2_{b_i b_i}$ and thus also to our self-energy matrixelements (see the comparison with the extensive numerical Σ-calculation in Figs. 13 and 17.

Identifying $<\rho>^{1/3}$ in eq. (4.16) as being due to an $\vec{r}$-dependent local potential, we can construct correlation self-energies, which act locally on Bloch states $\psi_{val,\vec{k}}(\vec{r})$ and $\psi_{cond,\vec{k}}(\vec{r})$, respectively (again omitting the terms of order $V^2_{b_1 a_1}$)

$$\Sigma'_{val}(\vec{r},E) = \rho^{1/3}(\vec{r})\; e^2 \left(\frac{2\pi}{3}\right)^{1/3} \frac{8\pi p^2 e^2}{3\Omega_o} \frac{2E_g}{\omega_p(\omega_p + \bar{E}_{val} - E)} \qquad (4.18a)$$

and

$$\Sigma'_{cond}(\vec{r},E) = -\rho^{1/3}(\vec{r})\, e^2 \left(\frac{2\pi}{3}\right)^{1/3} \frac{8\pi p^2 e^2}{3\Omega_o} \frac{2E_g}{\omega_p(\omega_p - E_{cond} + E)} \qquad (4.18b)$$

Eq. (4.18a) gives a fairly accurate account of our numerical values in Si and C for the dynamical correlation in the valence band. Fig. 13 compares the model results in Si with the TDSHF numerical self-energy matrix element $\Sigma_{b_i b_i}$ across the valence band. Our model places the pole in Σ at $E = E_{val} + \omega_p$, in accordance with the a-priori calculations. On the other hand, the model overestimates the a-priori numerical self-energy corrections in the conduction band both in C and Si (by about 20%, clearly a short-coming of the extreme tight-binding model for the more extended conduction wave functions.

The analytical form in (4.18a) gives a simple scaling of Σ' with energy. In Fig. 17, we have further approximated it by a linear form for diamond. With the parameter, $E(x_4) = \bar{E}_{val} = -.53$, $E(x_1) = E_{cond} = .39$, $E_g = .92$, $\omega_p = 2.43$ (all in a.u.), taken from photoemission and optical experiment[20], and $<\rho>^{1/3} = .47$ [a.u.] the linear form results

$$\Sigma_{b_i b_i}(E) = -0.16\, E + 4. \quad [eV] \qquad (4.19)$$

which, when added in Fig. 17 to the HF energies, describes quite well our numerical valence-band energies (from ref. 5). Formula (4.18a) is also in good agreement with variational results employing the Gutzwiller Ansatz in diamond.[18] However, eqs. (4.18) for Σ_{val} and Σ_{cond} also work as it should from its tight-binding construction for a large-gap insulator like LiF. Here, with the para-

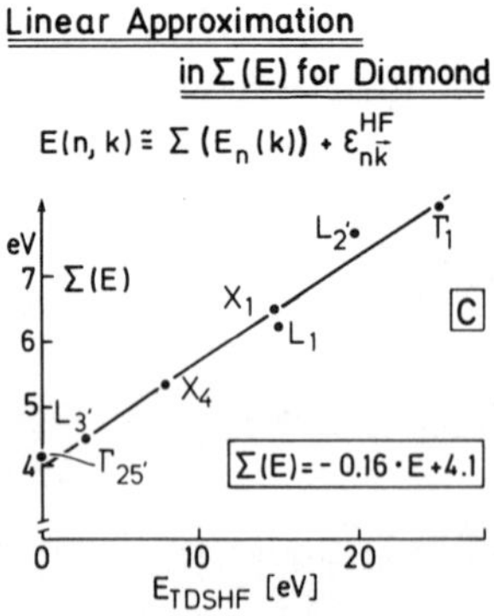

Fig. 17: Quasi-particle model versus a-priori calculation (from ref. 5)

meter p=1, $E_g=1$, $\omega_p=1.66$ and $\langle\rho\rangle^{1/3}=0.38$ [a.u.][49] a reduction of the bare HF band gap of 23 [eV] by 9.8[eV] results, due to the combined dynamical correlation in eqs. (4.18). This is in good accord with the experimental $E_g=13.6$ [eV].

As a final step, one can also evaluate the bare non-local HF in extreme tight-binding. Converting this self-exchange for the occupied states into a dipole form, and again introducing the average density $\langle\rho\rangle$, we can construct an exchange self-energy which, as in eqs. (4.18a) and (4.18b) for the correlation, acts locally on valence-band Bloch states, i.e.

$$\Sigma_{val}^{HF} \simeq -\rho^{1/3}(\vec{r})\ e^2\left(\frac{2\pi}{3}\right)^{1/3} \tag{4.19a}$$

and for the conduction bands,

$$\Sigma_{cond}^{HF} \simeq -V_{b_1 a_1} \tag{4.19b}$$

where

$$V_{b_1 a_1} = \iint \phi_{a_i}^*(\vec{r})\phi_{b_i}(\vec{r})\ v(\vec{r}-\vec{r}')\ \phi_{a_i}^*(\vec{r}')\ \phi_{b_i}(\vec{r}')\ d\vec{r}d\vec{r}' \tag{4.20}$$

Here, a couple of remarks are in order. First of all, the above extreme tight-binding model for the bare exchange can only work in large-gap insulators with well-localized electronic states. This is due to the fact that the unscreened exchange is longer-ranged than correlation and thus samples many more overlap terms in a smaller-gap system (like Si), with more spread -out orbitals. Therefore, it does not come as a surprise, that in diamond Σ_{val}^{HF} gives (to first order) 8.6 (eV) and Σ_{cond}^{HF} 2.1 (eV) (from ref. 33) which, when added to an approximate LDA-Hartree (constructed like our LDA-Hartree bands in Si, see sec. II) gap of 4.3[50] gives a reasonable "HF" gap of 14 (eV). However, in Si, the valence band Fock correction Σ_{val}^{HF} alone is already 5.62 (eV), and thus here the simple HF model clearly fails. This contrasts the correlation part of our self-energy model, which seemingly works both in small- and larger-gap non-metals. We note further that, in tight-binding, the local form of exchange for the valence band (4.19a) is by a factor $(2\pi^2/9)^{1/3} = 1.3$ larger than the Kohn-Sham value. On the other hand, the tight-binding exchange for the conduction band ($-V_{b_1 a_1}$)

is both in C and in Si (0.2 (eV)) significantly smaller than the valence exchange. This difference reflects the well-known fact that the exchange charge density is one for occupied but zero for unoccupied orbitals.

Exchange models based on the nearly-free electron model have recently been used by Sham and Schlüter[13] and by Horsch[50]. When combined with LDA-Hartree bands the approximate HF bands seem to agree reasonably well with abinitio HF-bands for C and some rare-gas solids.[50] This agreement is certainly surprising in that in these solids the electronic states are much more tight-binding than nearly free electron like.

IV. B Tight-Binding Model for the Density-Functional v_{xc} and the resulting Gap Correction.

Taking for the matrix elements of our model potential in eqs. (4.18) and (4.19) again the extreme tight-binding limit, with

$$\langle \phi_{b_i}(\vec{r}-\vec{l}) \mid \Sigma(\vec{r},\vec{r}';E) \mid \phi_{b_i}(\vec{r}-\vec{l}')\rangle = \delta_{ij}\,\delta_{\vec{l}\vec{l}'}\,\Sigma_{val}(E) \qquad (4.21a)$$

and

$$\langle \phi_{a_i}(\vec{r}-\vec{l}) \mid \Sigma(\vec{r},\vec{r}';E) \mid \phi_{a_i}(\vec{r}'-\vec{l}')\rangle = \delta_{ij}\,\delta_{\vec{l}\vec{l}'}\,\Sigma_{cond}(E) \qquad (4.21b)$$

and utilizing the interrelation (2.22) between the Kohn-Sham potential v_{xc} and the self-energy Σ_{xc}, we can derive a tight-binding model for v_{xc}. It constracts the commonly used homogeneous electron-gas construction.

To this we approximate the full, interacting one-particle propagator g in (2.22) by the density-functional g_o and assume the tight-binding overlap as in eq. (4.21) also for v_{xc}^o. After some straightforward manipulations, we arrive at the implicit formula for v_{xc}:

$$[V_{xc}(\vec{r}) - v_{xc}^{val}]\,\rho(\vec{r}) = \frac{-iE_g}{2}\Big\{ \rho(\vec{r})\int_{-\infty}^{0} \frac{\Sigma_{val}(\omega)\frac{d\omega}{2\pi}}{[\omega-E_{val}-i\eta]^2}$$

$$+ \sum_{i,l}|\phi_{a_i}(\vec{r}-\vec{l})|^2\int_{0}^{\infty}\frac{\Sigma_{cond}(\ )\frac{d\omega}{2\pi}}{[\omega-E_{cond}+i\eta]^2}\Big\} \qquad (4.22)$$

where v_{xc}^{val} denotes the valence-band matrixelement in the local basis, as in eq.)4.21a). In deriving (4.22) we have made use of the completeness relation

$$\delta(\vec{r}-\vec{r}') = \sum_{i,l}[\phi_{b_i}^{*}(\vec{r}-\vec{l})\phi_{b_i}(\vec{r}'-\vec{l})+\phi_{a_i}^{*}(\vec{r}-\vec{l})\phi_{a_i}(\vec{r}'-\vec{l})] \qquad (4.23)$$

and the ground state density

$$\rho(\vec{r}) = \sum_{i,l} \phi^*_{b_i}(\vec{r}-\vec{l}) \, \phi_{b_i}(\vec{r}-\vec{l}) \qquad (4.24)$$

Within the same approximation we can derive also an explicit expression for the gap correction Δ to the density-functional eigenvalues (see sec. II). Δ is found to be

$$\Delta = (\Sigma_{cond}(\varepsilon_c) - v_{xc}^{cond}) / [1 - (\Sigma_{cond}(\varepsilon_c) - v_{xc}^{cond})/(\varepsilon_c - \overline{E}_{cond})] \qquad (4.25)$$

To obtain a value for Δ in Si and C, let us simply use the Kohn-Sham exchange for v_{xc} with $v_{xc} = -e^2 \frac{3}{2}(\frac{3}{\pi})^{1/3} \rho^{1/3}$. Replacing in Σ_{cond} and v_{xc}^{cond} the matrixelements of $\rho^{1/3}$ again by $\langle\rho\rangle^{1/3}$ as defined in eq. (4.17), gives, with eq. (4.18b)

$$\Sigma_{cond}(\varepsilon_c) - v_{xc}^{cond} = - V_{b_1 a_1} - V_{b_1 b_1}^2 \frac{2E_g}{\omega_p^2} + e^2 (\frac{3}{\pi})^{1/3} \langle\rho\rangle^{1/3}$$

$$= \begin{array}{l} 11.2 \ (eV):C \\ 8.7 \ (eV):Si \end{array} \qquad (4.26)$$

Here, the first and second terms on the r.h.s. are again the bare exchange and correlation contribution to Σ_{cond}. The Si and C numbers are derived with the values for $V_{b_1 a_1}$ (C:0.14; Si: 0.014 (a.u.)), $V_{b_1 b_1}^2$ from eq. (4.17) (C:1,12;Si:0.48 (a.u.)) and ω_p, E_g and $\langle\rho\rangle^{1/3}$ as listed in the previous section. With the LDA values for the minimum conduction band energy ε_c (C:5,45 (eV))[25], Si: 0.65 (eV)[24]) and $\overline{E}_{cond}$ (C:7.45 (eV)[5], Si: 1.3 (eV)[4,24], we have finally for the gap correction

$$\Delta = \begin{cases} 1.6 \, (eV): \ C \\ 0.6 \, (eV): \ Si \end{cases} \qquad (4.27)$$

Both values describe surprisingly well the difference between the a-priori calculated LDA gaps and the experimental gaps both in (=1.1(eV)[25], 1.8(eV)[11] and in Si (0.6 (eV)-0.9 (eV)[24]). This is due to the fact, that the energy difference in eq. (4.25) is about a factor 6(C) to 13 (Si) larger than the energy difference $(\varepsilon_c - \overline{E}_{cond})$; so Δ is of the order of 1(eV), a result which holds also for other tetrahedrally bonded semiconductors like Ge (Δ=0.7 (eV)[51] or GaAs (Δ=0.5 -1(eV))[52]. In evaluating eq. (4.25) for Δ the same zero of energy has to be used for the density-functional energy scale (ε_c) and the quasi-particle energy scale (E_{cond}). The exact Kohn-Sham bandstructure predicts exactly the center of the fundamental energy gap relative to the ideal vacuum level (whose position is not affected by the presence of a physical surface), and so does the quasi-particle bandstructure. Therefore, we have taken as the common zero the center of the fundamental gap. This amounts to replacing $\varepsilon_c \rightarrow \varepsilon_c/2$

and $\overline{\overline{E}}_{cond} \to \overline{E}_{cond} - E_o/2$, where E_o is the fundamental gap of the quasi-particle bands.

Formula (4.25) for the gap correction can also be tested in an insulator[53] like LiF. With the value for $\rho=0.055$, $E_g = 1$,, $\omega_p =1.66$, $p=1$, $\varepsilon_c=10.6$ $V_{b_1 a_1}=V_{2p,2s} = 0.15$ (all in a.u.) and $\overline{E}_{cond}=1.6$ (a.u.), from optical absorption experiment[43], we find $\Delta=2.3$ (eV). The a-priori Δ is 3(eV) for Slater ($\alpha=1$) exchange and 3.8(eV) for $\alpha=\frac{2}{3}$[53]. The value of the conduction-band exchange $V_{2p,2s}$ has been calculated with atomic 2p(F)- and 2s(Li)-orbitals and, thus is probably larger than that calculated with the actual Wannier functions of the crystal. A smaller value of $V_{2p,2s}$ would further enhance Δ.

V. SUMMARY

Through a calculation of quasiparticle states in silicon, based on an accurate description of the two-particle Green's function, we have aimed at a first-principles understanding of selfenergy corrections, both with respect to non-locality in $\vec{r}$-space and energy dependence in a prototype semiconductor. The single-particle-like excitations, which are found to be in good accord with the experimental data, are governed by a predominantly local potential. It scales with energy away from the gap, however to a significantly less extent than in the insulator C. A simple analytical tight-binding model is discussed for the selfenergy operator. It illucidates the basic differences in the quasiparticle properties of metals and non-metals and reproduces the accurate numerical results for the dynamical correlation both in a covalent semiconductor and insulator (our previous calculations for C) within a few percent. New insights for the applicability and also the limitation of much used one-electron schemes in particular, the density-functional theory have emerged. To this we implemented general results from the many-body perturbation theory for the exact exchange-correlation energy, to derive the density-functional effective one-particle potential v_{xc} without use of a minimum or extremal principle. This enabled us, for example, to construct a simple, analytic approximation to the gap correction to the density-functionsl eigenvalues.

References

+ Faculté des Sciences, Departement de Physique, Campus Universitaire, Belvédère, Tunis, TUNISIA.

1. L.J. Sham and W. Kohn, Phys. Rev. <u>145</u>: 561 (1966)
2. G. Baym, Phys. Rev. <u>127</u>: 1391 (1962)

3. L. Hedin and S. Lundqvist, Solid State Phys. Vol. 23 (Academic N.Y.) 1969 pp.1, and references therein; L. Hedin, Phys. Rev. 139: A796 (1965)

4. W. Hanke and L.J. Sham, Phys. Rev. Lett. 43: 387 (1979); Phys. Rev. B21: 4656 (1980)

5. G. Strinati, H.J. Mattausch and W. Hanke, Phys. Rev. Lett. 45: 290 (1980), Phys. Rev. B25: 2867 (1982)

6. W. Hanke, G. Strinati and H.J. Mauttausch, in "Recent Developments in Condensed Matter Physics", Vol. 1, ed. by J.T. Devreese (Plenum, 1981), p.p. 263

7. P. Hohenberg and W. Kohn, Phys. Rev. 136: B864 (1964)

8. W. Kohn and L.J. Sham, Phys. Rev. 140: A1133 (1965); for recent developments see also O. Gunnarson, M. Jonson and B.I. Lundqvist, Phys. Rev. B20: 3136 (1979), and A.R. Williams, U.v. Barth in "Theory of the Inhomogeneous Electron Gas" ed. by S. Lundqvist and N.H. March (Plenum, 1983)

9. O. Gunnarson and R.O. Jones, Phys. Scr. 21: 394 (1980)

10. V.L. Moruzzi, J.F. J.anak and A. R. Williams, in "Calculated El-Electronic Properties of Metals" (Pergamon, Oxford, 1978)

11. D. Glötzel, B. Segall and O.K. Andersen, Sol. State Comm. 36: 403 (1980)

12. M.T.Yin and M.L. Cohen, Phys.Rev. B26: 5668 (1982)

13. L.J.Sham and M. Schlüter, Phys. Rev. Lett. 51: 1888 (1983)

14. D.C. Langreth, to be published

15. W. Hanke, T.Gölzer and H.J. Mattausch, Sol. State Comm. (1984), and Phys. Rev. B to be published

16. W. Hanke, H.J. Mattausch and G. Strinati, in "Electron Correlations in Solids,Molecules and Atoms", ed. by J.T. Devreese and F. Brosens (plenum, 1983), p.p. 289

17. C.S. Wang and W.E. Picklett, Phys. Rev. Lett. 51: 597 (1983)

18. S. Horsch, P. Horsch and P. Fulde, Phys. Rev. B29: 1870 (1984)

19. J. C. Slater, Phys. Rev. 81: 385 (1951)

20. L. Ley, M. Cardona and R.A. Pollak, in "Photoemission in Solids" ed. by L. Ley and M. Cordona (Springer, Berlin 1979),p.11

21. This LDA "Hartree" calculation was provided by C. Kunc; independently very similar results were found by D. Glötzel.

22. HF results for MgO: S.T. Pantelides, D. J. Mickish and A.B. Kunz Phys. Rev. B10: 5203 (1974), and B10: 2602 (1974)

23. HF results for C:A. Mauger and M. Lannoo, Phys. Rev. B15:2324 (1977)

24. LDA for Si:D.R. Hamann, Phys. Rev. Lett.42:662 (1979)

25. LDA for C: A. Zunger and A.J. Freeman, Phys. Rev. B15:5049(1977)

26.J.P.Perdew and M. Levy, Phys. Rev. Lett. 51:1884 (1983)

27.J.P. Perdew and A. Zunger, Phys. Rev.B23:5048 (1981), and references therein; R.E. Heaton, J.G. Harrison and C.C. Lin, Phys. Rev. B28:5992 (1983)

28.J.D. Talman and W.F. Shadwick, Phys. Rev. A14:36 (1976)

29.G. Baym and L.P. Kadanoff, Phys. Rev. 124:287 (1961)

30.W. Kohn and P. Vashista, in "Theory of the Inhomogeneous Electron Gas", ed. by S. Lundqvist and N.H. March, (Plenum, N.Y., 1983) eq. (168)

31.N.D. Mermin, Phys. Rev. $137A$:1441 (1965)

32.L.J. Sham and T.M. Rice, Phys. Rev. 144:708 (1966)

33.W. Hanke and L.J. Sham, Phys. Rev. B12:4501 (1975)

34.W. Hanke in "Festkörperprobleme XIX", Adv. Sol. St. Phys.(Vieweg, (1979) p.p.43

35.H.J. Mattausch, W. Hanke and G. Strinati, Phys. Rev. B27:3735 (1983)

36.N. Meskini, H.J. Mattausch and W. Hanke, Sol. St. Comm.48:807 (1983.)

37.J.C. Slater, in "The Self--Consistent Field for Molecules and Solids", Vol. 4 (Mc-Graw-Hill, 1974) p.p.26

38.J.W.D. Conolly, "The X$_\alpha$ Method", Modern Theoret. Chem. 7 G.A. Segal ed. (Plenum, 1979); Ch.4

39.E.O. Kane, Phys. Rev. B4:1910 (1971)

40.W. Brinkman and B. Goodman, Phys. Rev. 149:597 (1966)

41.N.O. Lipari and W.B. Fowler, Phys. Rev. B2:3354 (1970)

42.J. Bennett and J.C. Inkson, J. Phys. C10:987 (1977)

43.E.O. Kane, Phys. Rev. B5:1493 (1972) and C11:2017 (1978)

44.J.R. Chelikowsky and M.L. Cohen, Phys. Rev. B10:5095 (1974)

45.M. Cardona, in "Atomic Struct.a. Propert. of Solids" (Academic, 1972)

46.B.I. Lundqvist, Phys. Condens. Matter 6;193 (1967)

47.J.C. Slater, Phys. Rev. 81:385 (1951)

48.P. Sterne and J.C. Inkson, J. Phys. C17:1497 (1984)

49.R.N. Euwema, G.G. Wepfer, G.T. Surralt and D.L. Wilhite, Phys. Rev. B9:5249 (1974)

50.P. Horsch, to be published; and K.P. Bohnen, private communication

51.C.S. Wang, B.M.Klein, Phys. Rev. B24:3393 (1981)

52.F. Manghi, G. Riegler, C.M. Bertoni, C. Calandra, G.B. Bachelet Phys. Rev. B28:6157 (1983)

53.A. Zunger, A.J. Freeman, Phys. Rev. B16:2901 (1977)

54.From "Physics Data" 8-1 (1977), DESY Hamburg, edited by E.E. Koch

AB-INITIO CALCULATION OF THE PHONON FREQUENCIES IN COVA-
LENT SEMICONDUCTORS USING THE DIELECTRIC SCREENING METHOD

J.T. Devreese[°], P.E. Van Camp and V.E. Van Doren

University of Antwerp (RUCA), Groenenborger-

laan 171, B-2020 Antwerpen, Belgium

I. INTRODUCTION

The theory of the lattice dynamics of covalent semiconductors has been developed in the late sixties and
early seventies [1]. The essence of this theory is the
description of the density distribution of the valence
electrons between the ions at their arbitrary and instantaneous positions. Originally this part of the electron
distribution due to deviations of the ions from their
equilibrium positions is derived from linear response
theory. Subsequently phonon dispersion curves can be
obtained from this <u>dielectric screening method</u>. From
1972 the present authors started their efforts to calculate phonon frequencies using this method. A presentation of linear response and dielectric screening theory
is given in these proceedings in the paper by J.T.
Devreese and F. Brosens.

About ten years ago it was proposed by R.M. Martin
and D.J. Chadi to calculate phonon frequencies directly
from the difference between the total energy of the
crystal in the equilibrium configuration and in a dis-

[°] Also at: Department of Physics, University of Antwerp
(UIA), B-2610 Antwerpen-Wilrijk, Belgium and University of Technology, Eindhoven, The Netherlands.

torted state. The total energy difference method leads
to one frequency for that particular crystal state which
is called a frozen phonon. This method is extensively
described in the articles by R.M. Martin, K. Kunc and
S. Louie in these proceedings.

In the dielectric screening method the electron
density response due to the motion of the ions around
their equilibrium positions is calculated in first order
perturbation theory. The potential energy of the crystal
for an arbitrary configuration of the ions is expanded
to second order in the ionic displacements from equilibrium. The expansion coefficients of the second order
term form a matrix. The Fourier transform of this matrix
is the dynamical matrix whose eigenvalues yield the
phonon frequencies. The dynamical matrix has an ionic
and electronic part. The electronic part can be expressed in terms of the electron density response matrix and
of the ionic potential. This method has the advantage
over the total energy difference method that the phonon
frequencies for any arbitrary wave vector can be calculated without additional difficulties. Furthermore in
this method the acoustic sum rule is automatically
satisfied as a consequence of the way the dynamical
matrix is derived. However the dielectric screening
method is limited to harmonic phonons.

The present authors have used the dielectric screening method for all their ab-initio computations of the
macroscopic dielectric constant and phonon frequencies of
Si and Ge. For the calculation of the electron energies
and wave functions needed in the expression of the electron density response matrix they apply the local density
approximation in the Hamiltonian. The advantage and
shortcomings of this approximation are treated at length
in the papers by J.T. Devreese, R. Martin, K. Kunc, S.
Louie, A. Baldereschi and R. Resta in these proceedings.

For the electron-ion potential appearing in the Hamiltonian known pseudopotentials have been used. The Hartree
and exchange-correlation part of the crystal potential
are obtained self-consistently from this ionic pseudopotential. The same ionic pseudopotential is used in
the electronic part of the dynamical matrix. Moreover
the electron density response matrix has been derived in
the same local density approximation as the one used in
the Hamiltonian. This procedure guarantees that the
same exchange-correlation effects are taken into account
in the calculation of the electron wave functions and
energies and in the expression of the electron density
response matrix. The present authors found for the
first time that both the consistencies, i.e. in the
potentials and in the exchange-correlation, are crucial
in the dielectric screening method.

Previously the present authors did not perform the
summation over the conduction bands in the polarizability
matrix but instead approximated this summation by means
of a moment expansion. Results of this approximation
for the phonon dispersion curves of Si have been published in the literature [2]. However in their present work
presented in these proceedings the polarizability matrix
is evaluated by means of a straightforward summation
over all the conduction bands obtained from diagonalization of the Hamiltonian matrix.

Also in their present work, the authors do not take
the lattice parameters of Si from experiment but instead
find its value from minimalization of the total crystal
energy with respect to ionic displacements.

Finally in this paper as well as in previous work
the authors were able to satisfy the generalized
acoustic sum rule. In particular it is found that the
term for the zero reciprocal lattice vector is zero to
within 10^{-4}.

II. THEORY

In the dielectric screening method the total electronic energy is expanded in a Taylor series with respect to the ionic displacements. The relationship between the externally applied potential δV^{EXT} and the induced charge density $\delta\rho$ is given by linear response theory as

$$\delta\rho = \chi \; \delta V^{EXT} \tag{1}$$

This equation defines the density response matrix χ. Eq. (1) is written in matrix notation, i.e. χ is a matrix and $\delta\rho$, δV^{EXT} are column vectors. The polarizability matrix $\tilde{\chi}$ is defined by:

$$\delta\rho = \tilde{\chi} \; \delta V^{IND} \tag{2}$$

i.e. the linear relationship between induced charge density and induced potential. In Fourier space this matrix was first given by [3]

$$\tilde{\chi}(\vec{q},\vec{G},\vec{G}') = \frac{1}{\Omega} \sum_{\ell m} \frac{\eta_\ell - \eta_m}{E_\ell - E_m} <\ell| e^{-i(\vec{q}+\vec{G})\vec{r}} |m> \; .$$

$$\cdot \; <m| e^{i(\vec{q}+\vec{G})\vec{r}} |\ell> \tag{3}$$

A derivation of this expression can be found in the lectures "Basic Concepts in Dielectric Response and Pseudopotentials" by J.T. Devreese in this volume. Ω is the crystal volume, $\vec{G}$ ($\vec{G}'$) are reciprocal lattice vectors and $\vec{q}$ is the phonon wave vector.

The dielectric matrix ε depends on the particular form of the one-electron equations used. In the present work the local density approximation [4] was employed resulting in the following form of the dielectric matrix [5]:

$$\varepsilon = 1 - v_c \; \tilde{\chi} (1 - v_{xc} \; \tilde{\chi})^{-1} \tag{4}$$

where v_c is the Fourier transform of the Coulomb potential and v_{xc} is given by

$$v_{xc}(\vec{q},\vec{G},\vec{G}') = \frac{1}{\Omega} \int d\vec{r} \int d\vec{r}' \; e^{i(\vec{q}+\vec{G})\vec{r}}$$

$$\cdot \; \frac{\delta V_{xc}}{\delta\rho} \; e^{-i(\vec{q}+\vec{G}')\vec{r}'} \tag{5}$$

Here V_{xc} is the exchange-correlation energy (see ref.[4]).

If one uses the Slater form of the exchange-correlation energy, i.e. [6]

$$V_{xc}(\rho) = -\alpha \; \frac{3}{2} \; (\frac{3}{\pi} \rho)^{1/3} \tag{6}$$

then Eq. (5) simplifies to

$$v_{xc}(\vec{q},\vec{G},\vec{G}') = -\alpha \; (\frac{3}{\pi})^{1/3} \; \frac{1}{\Omega} \int dr \; \rho^{-2/3}(\vec{r}) \; . \; e^{-i(\vec{q}+\vec{G}')\vec{r}} \tag{7}$$

This means that in this approximation the exchange-correlation corrections to the dielectric matrix are independent of the phonon wave vector q.

The density response matrix χ (Eq. (1)) is related to the inverse dielectric matrix ε^{-1} through

$$\varepsilon^{-1} = 1 + v_c \; \chi \tag{8}$$

and is calculated by straightforward inversion of ε.

The expressions (7) and (8) have been used to calculate the macroscopic dielectric function of Si [5], Ge [7] and C and α-Sn [8].

The macroscopic dielectric function of Silicon in the Hartree and in the Slater approximation is given in figure 1.

In figure 2 the macroscopic dielectric function in the Hartree and Slater approximations is given for Germanium. For a comparison of our inverse dielectric function of Germanium with other calculations see the paper by K. Kunc in these proceedings.

The electronic contribution to the dynamical matrix can be written as follows:

$$D_{ij}^{en}(\vec{q};ab) = \sum_{GG'} X_{ij}(ab;\vec{q},\vec{G},\vec{G}')$$

$$- \delta_{ab} \sum_{c} X_{ij}(ac;\vec{q},\vec{G},\vec{G}') \tag{9}$$

with

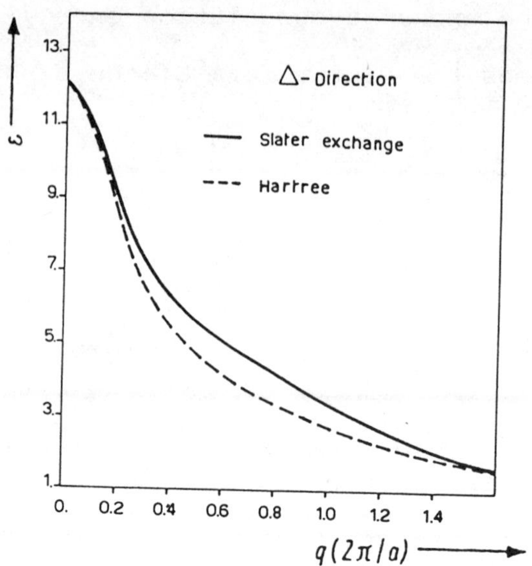

Figure 1: Macroscopic dielectric function of Si in the
Δ-direction. Full line: Slater exchange ($\alpha=1$);
broken line: Hartree approximation. The horizontal axis is in units of $\frac{2\pi}{a}$ [5] .

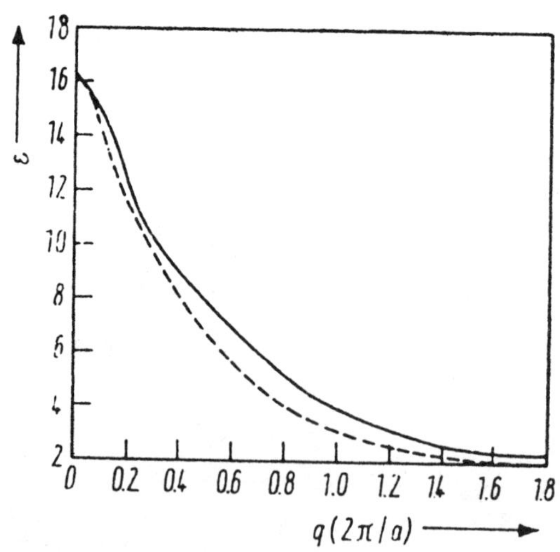

Figure 2: Macroscopic dielectric function of Ge in the
Δ-direction. Full line: Slater exchange ($\alpha=1$);
broken line: Hartree approximation. The horizontal axis is in units of $\frac{2\pi}{a}$ [7] .

162

$$X_{ij}(ab;\vec{q},\vec{G},\vec{G}') = \frac{1}{v_c\sqrt{M_aM_b}} \; (\vec{q}+\vec{G})_i \; V_a(\vec{q}+\vec{G})$$

$$e^{i\vec{G}\vec{R}_a} \; (\vec{q},\vec{G},\vec{G}') \; e^{-i\vec{G}'\vec{R}_b} \; V_b(\vec{q}+\vec{G}')\cdot(\vec{q}+\vec{G}')_j \qquad (10)$$

where V_a (V_b) is the Fourier transform of the electron-ion potential of the a-th ion in the unit cell with masses M_a (M_b).

A derivation of Eqs. (9) and (10) can be found in the lectures "Basic Concepts in Dielectric Response and Pseudopotentials" by J.T. Devreese in this volume.

Diagonalization of the dynamical matrix yields the phonon frequencies for the wave vector q under consideration. It should be noted that phonon frequencies of an arbitrary wave vector can be calculated. This is contrary to the total energy method described in the lecture notes of S. Louie, K. Kunc and R.M. Martin in this volume.

The generalized acoustic sum rule:

$$\lim_{q\to o} \sum_{a\vec{G}} (\vec{q}+\vec{G}) \; \chi(\vec{q},\vec{G},\vec{G}') \; V_a(\vec{q}+\vec{G}) \; e^{-i\vec{G}\vec{R}_a} = \vec{G}'\rho(\vec{G}')$$

$$(11)$$

establishes a relation between the density response matrix $\chi(\vec{q},\vec{G},\vec{G}')$ and the electronic charge density $\rho(\vec{G}')$. As such it is an important property for the internal consistency of the calculation. Physically the sum rule guarantees that the effective ionic potential is invariant under a uniform translation. In the present work the relation given in Eq. (11) for zero reciprocal lattice vector $\vec{G}'$ is satisfied to within 1 part in 10^4.

III. DISCUSSION

III.1. Computational Procedure

a. Electron-ion potential. If we treat the core electrons and the nucleus as one entity, the electron-ion potential is the potential experienced by an electron due to a Si^{4+} ion. Several forms of this potential can be found in the literature (see e.g. refs. [9], [10]). In the work described here (see ref. [12]) we used the

Schlüter-Chelikowsky-Louie-Cohen potential [9] given by:

$$V(\vec{q}) = \frac{4\pi Z V_1}{q^2} (\cos V_2 q + V_3) e^{V_4 q^4} \tag{12}$$

with (in au)

$$V_1 = 1.5432 \qquad V_3 = -0.3520$$

$$V_2 = 0.7907 \qquad V_4 = -0.01807$$

Although strictly speaking $V(\vec{q})$ given in Eq. (12) is not an ab-initio potential (it was fitted to experimental valence and conduction band energies) it was used in this stage of the work. Recent tests performed by the present authors using the Topp-Hopfield potential [10] indicate that the overall features of the phonon frequencies calculated with both potentials are similar. A detailed discussion of results using different potentials will be published.

b. Electronic band structure. In the local pseudo-potential formalism the Kohn-Sham equations are replaced by (in au)

$$(- \frac{1}{2} \nabla^2 + V_p(\vec{r})) \psi_{kn}(\vec{r}) = E_{kn} \psi_{kn}(\vec{r}) \tag{13}$$

where V_p is the local pseudopotential. Since both the pseudopotential V_p and the dynamical matrix depend on the electron-ion potential a self-consistent pseudo-potential band calculation must be performed. The computational procedure is as follows:
1. Determine the ionic potential. In the present work for Silicon Eq. (12) has been used.
2. Choose an empirical starting pseudopotential (e.g. the one given in ref. [11]).
3. Diagonalize the Hamiltonian matrix in a plane wave basis. Typically 137 to 150 plane waves are used by the present authors while additionally some 150 plane

waves are treated in second order perturbation theory.
Next construct the charge density

4. Subsequently calculate the Hartree and exchange-
   correlation potentials. Together with the ionic
   potential the total pseudopotential can then be constructed.

5. Replace the starting pseudopotential by the new one
   and repeat the calculation, i.e. compute a new charge
   density. The iteration cycle is stopped if this new
   charge density differs by less than a predefined
   amount from the previous charge density. Convergence
   is normally reached with less than ten iterations.

Schematically the computational procedure can be summarized in the flow diagram given below.

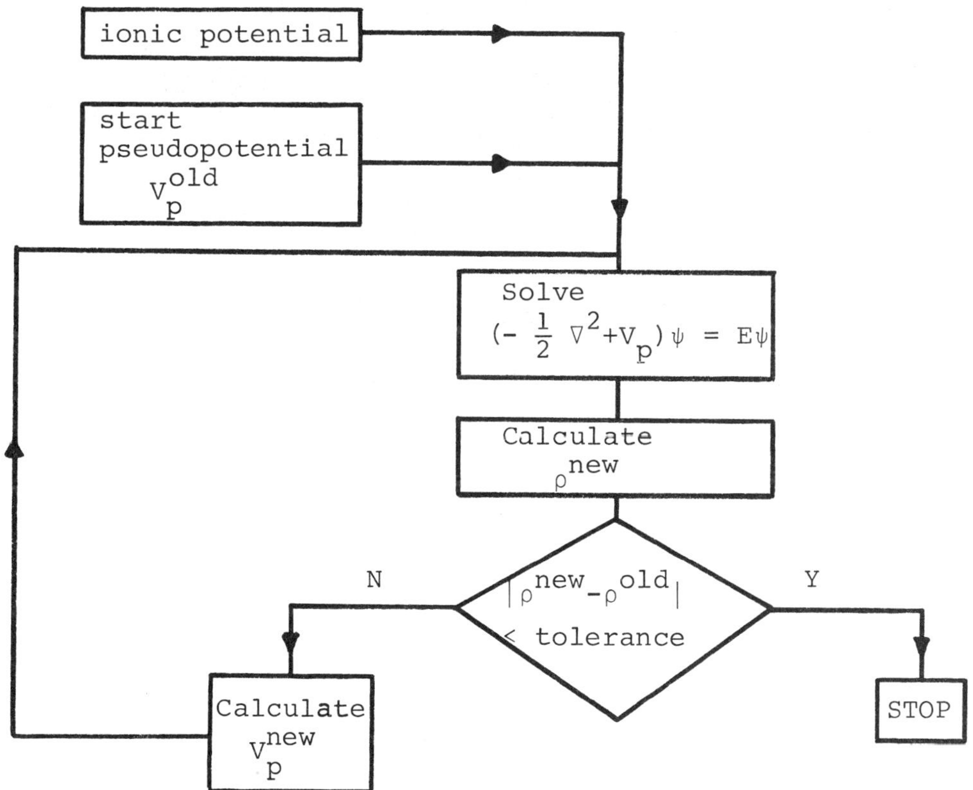

Figure 3: Flow diagram of the self-consistent band
calculation.

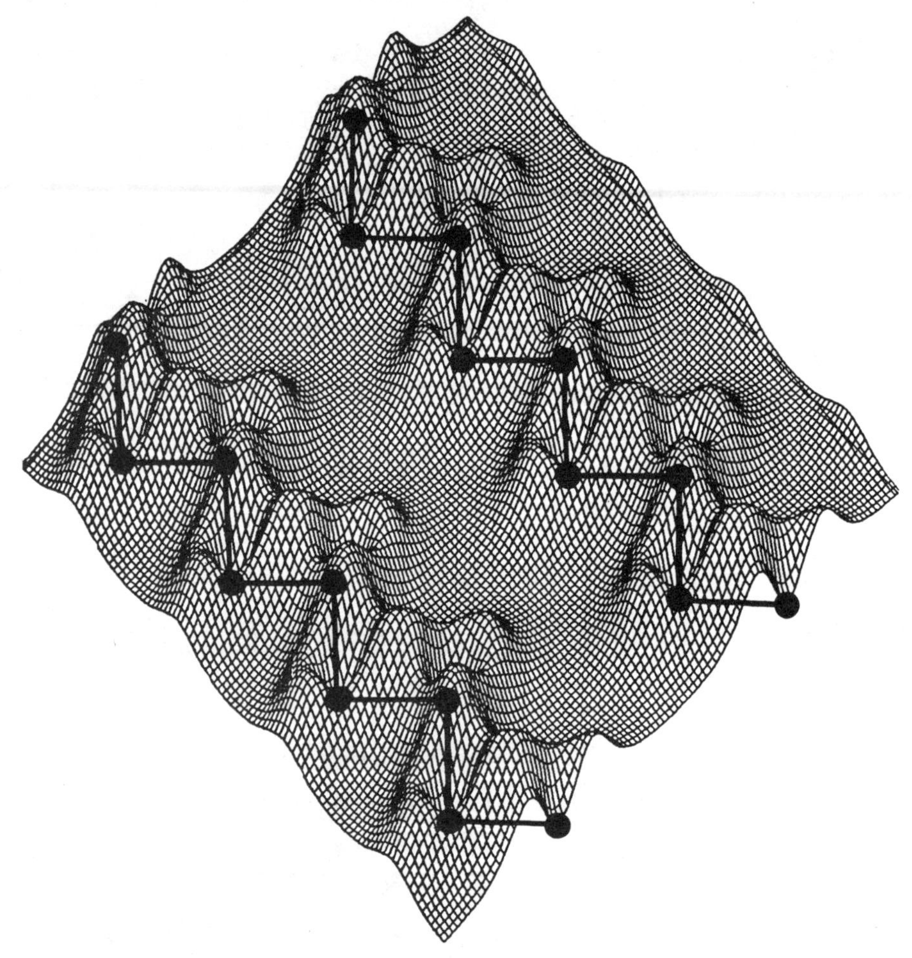

Figure 4: Charge density of Silicon in the [1,1,0] plane (in units of electrons per atom) together with the atoms and bonds.

It should be noted that the final pseudopotential is independent of the chosen starting pseudopotential. A
good starting potential merely decreases the number of
iteration cycles.

A three-dimensional plot[*] of the final charge density in the [1,1,0] plane is displayed in figure 4. Also
shown are the atoms in this plane together with the
bonds.

c. Total energy. In the local density approximation
the total crystal energy can be written as

$$E_T = \sum_{kn} E_{kn} - \frac{1}{2} \int dr \int dr' \frac{\rho(\vec{r})\ \rho(\vec{r}')}{|\vec{r}-\vec{r}'|}$$

$$+ \int d\vec{r}\ \rho(\vec{r})\ [\varepsilon_{xc}(\rho) - v_{xc}(\rho)] \qquad (14)$$

The crystal is free of stress if the total energy E_T is
minimal at the lattice constant used. Therefore E_T is
minimalized with respect to the lattice constant. For
Silicon a value of 5.373 Å is found by the present
authors to be compared with the experimental value [13]
of 5.429 Å. All calculations discussed in the following
were performed using this calculated lattice constant.

d. Convergence of the sum over the conduction bands.
Since the calculation of the sum over the conduction
bands in Eq. (3) is rather time consuming we performed
runs using different numbers of terms in this sum.

Table I: Phonon frequencies of Si in terms of the number
         of conduction bands taken into account together
         with the experimental values (in THz).

number of con- duction bands	Γ	TO(X)	LOA(X)	TA(X)
50	19.71	18.49	15.01	4.02
100	19.04	17.65	14.69	4.33
146	18.74	17.28	14.36	4.38
Experiment [14]	15.5	13.9	12.3	4.5

[*] It is a pleasure to thank Dr. F. Brosens for develop-
ping the computer drawing.

As can be seen from Table I the convergence was found to
be relatively slow, and one has to use all conduction
bands calculated. The number of conduction bands of
course is determined by the number of plane waves used in
the basis in which the electron wave functions are expanded. Using all calculated conduction bands means
that the first order wave function is expanded in exactly the same number of plane waves as the zeroth order
wave function.

e. Convergence of the phonon frequencies in terms
of the dimension of ε. In table II results for the
phonon frequencies are given for different dimensions of
the dielectric matrix $\varepsilon(\vec{q},\vec{G},\vec{G}')$. In all these calculations 146 conduction bands were used.

It is clear from Table II that the convergence of
the phonon frequencies with respect to the dimension of
$\varepsilon(\vec{q},\vec{G},\vec{G}')$ is relatively slow. To include effects of
higher reciprocal lattice vectors one can add a diagonal
"tail", i.e. a diagonal part, to the dielectric matrix
[12]:

$$\varepsilon(\vec{q},\vec{G},\vec{G}') = (1 + \frac{16\pi\ \rho(\vec{o})}{|\vec{q}+\vec{G}|^4})\ \delta_{GG'} \qquad (15)$$

with $\rho(\vec{o})$ the zeroth Fourier component of the charge
density. The sums over reciprocal lattice vectors in
the dynamical matrix are then performed till convergence.

Table II: Phonon frequencies of Si in terms of the
dimension of ε together with the experimental
values (in THz).

dimension $\varepsilon(\vec{q},\vec{G},\vec{G}')$	Γ	TO(X)	LOA(X)	TA(X)
89	22.40	19.31	17.41	6.11
113	18.18	16.75	13.97	3.68
137	18.74	17.28	14.36	4.38
169	18.77	17.22	14.39	4.75
181	18.67	17.12	14.33	4.55
Experiment [14]	15.5	13.9	12.3	4.5

Table III: Phonon frequencies of Si in terms of the
           dimension of H together with the extrapolated
           values and the experimental values (in THz). In
           all cases all conduction bands calculated were
           used.

dimension H	Γ	TO(X)	LOA(X)	TA(X)
137	18.89	17.50	14.51	4.39
150	18.74	17.28	14.36	4.38
169	18.54	17.07	14.20	4.39
193	18.46	17.05	14.29	4.51
Extrapolation	17.10	14.93	12.81	4.16
Experiment [14]	15.5	13.9	12.3	4.5

<u>f. Convergence in terms of the dimension of H</u>. In
table III results for the phonon frequencies are given
for different dimensions of the Hamiltonian matrix H.

As can be seen from the table the convergence is
relatively slow. Therefore, as a first approximation,
an extrapolation is made based on the hypothesis that
the phonon frequencies are inversely proportional to
the number of plane waves used in the Hamiltonian. The
results of this extrapolation based on the values obtained with 137 and 150 plane waves are also shown in Table
III.

2. Results

The results for the phonon frequencies of Silicon
obtained by the present authors are shown, for wave
vectors in Γ-, Δ- and Σ-directions, in figure 5. In
this calculation both the tail in the dielectric matrix
and the extrapolation in the Hamiltonian were used. A
meaningful overall agreement with experiment is obtained.
The discrepancy, averaged over all calculated branches,
is 14%. Generally the acoustic branches are in better
agreement with experiment than the optical modes.

Recently the method described in this paper has also
been applied to Germanium. The ionic potential used is the
Topp-Hopfield form [10] with parameters determined by
the Germanium ion. The calculated lattice constant is
5.517 Å to be compared with the experimental value of

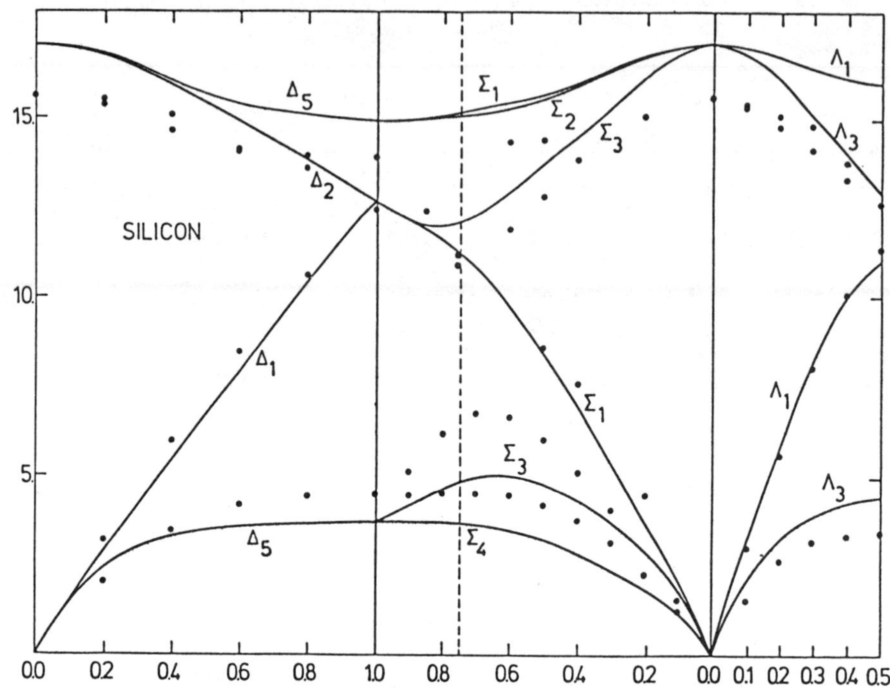

Figure 5: Calculated phonon frequencies of Silicon in
Δ-, Λ- and Σ-directions (in THz). The experimental data [14] are also shown.

5.652 Å [13]. In table IV results of the phonon frequencies are given together with the experimental values.

The calculation of the complete phonon spectrum of Germanium is in progress. The method described in the present paper was programmed for a vector computer (C.D.C. Cyber-205). Executing the same runs on a serial computer would take a prohibitively long computer time. Details of the vectorization of the phonon programs will be published elsewhere [15].

Table IV: Phonon frequencies of Ge (in THz) together
          with the experimental values.

	Γ	TO(X)	LOA(X)	TA(X)
calculated values	9.90	9.40	7.68	2.70
experiment [14]	9.13	8.26	7.21	2.40

IV. CONCLUSIONS

The calculations of the phonon dispersion relations
using the dielectric screening method presented here lead
to the following conclusions:
1. Minimalization of the total crystal energy with respect to the lattice parameters yields a lattice constant which guarantees a stress-free crystal. It is
   found that in particular the TA-mode is very sensitive
   to compatibility between lattice constant and total
   energy.
2. It is found that the consistency between the electron-
   ion potential and the total pseudopotential is essential in order to satisfy the acoustical sum rule. If
   this sum rule is not fulfilled the crystal will be
   unstable against shear forces. In the present work
   it is found that the term of the generalized acoustical sum rule for the zeroth reciprocal lattice vector
   is satisfied to within 10^{-4}.
3. Consistency between the exchange-correlation approximations used in the Hamiltonian and in the linear
   response theory has been realized. The importance of
   different exchange-correlation approximations is
   presently under investigation.
4. The present authors showed that it is very important
   to use a basis consisting of a sufficiently large
   number of plane waves. The summation over the conduction bands in the polarizability matrix should be
   performed using all bands calculated. In this way
   the first order wave functions are expanded in exactly the same basis as the one used for the expansion
   of the unperturbed wave function.

ACKNOWLEDGMENT

Part of the numerical work was performed on a
C.D.C. Cyber-205 with a grant from the "Supercomputer

Project" 1983-84 of the N.F.W.O. (National Fund for Scientific Research, Belgium).

REFERENCES

1. L.J. Sham, Phys. Rev. 188, 1431 (1969).
   R.M. Pick, M.H. Cohen and R.M. Martin, Phys. Rev. B2, 910 (1970).
   F.A. Johnson, Proc. Roy. Soc. A310, 79 (1969); ibid. A310, 89 (1969); ibid. A310, 101 (1969).
   D.C. Wallace, "Thermodynamics of Crystals", J. Wiley, New York (1972).
2. P.E. Van Camp, V.E. Van Doren, J.T. Devreese, Inst. Phys. Conf. Ser. 43, 685 (1979).
   P.E. Van Camp, V.E. Van Doren, J.T. Devreese, Phys. Stat. Sol. (b) 93, 483 (1979).
   P.E. Van Camp, V.E. Van Doren, J.T. Devreese, Phys. Rev. Lett. 42, 1224 (1979).
   J.T. Devreese, P.E. Van Camp, V.E. Van Doren, Int. J. Quant. Chem. 18, 317 (1980).
3. H. Ehrenreich, M.H. Cohen, Phys. Rev. 115, 786 (1959).
   S. Adler, Phys. Rev. 126, 413 (1962).
   N. Wiser, Phys. Rev. 129, 62 (1963).
4. W. Kohn, L. Sham, Phys. Rev. 136, B864 (1964).
5. P.E. Van Camp, V.E. Van Doren, J.T. Devreese, Phys. Rev. B24, 1096 (1981).
6. J.C. Slater, Phys. Rev. 81, 385 (1951).
7. P.E. Van Camp, V.E. Van Doren, J.T. Devreese, Phys. Stat. Sol. (b) 110, K133 (1982).
8. P.E. Van Camp, V.E. Van Doren, J.T. Devreese, unpublished results.
9. M. Schlüter, J. Chelikowsky, S. Louie, M.L. Cohen, Phys. Rev. B12, 4200 (1975).
10. W. Topp, J. Hopfield, Phys. Rev. B7, 1295 (1973).
11. M.L. Cohen, T. Bergstresser, Phys. Rev. 141, 789 (1966).
12. P.E. Van Camp, V.E. Van Doren, J.T. Devreese, in "Ab-Initio Calculation of Phonon Spectra", eds. J.T. Devreese, V.E. Van Doren, P.E. Van Camp, Plenum, New York (1983).
    P.E. Van Camp, V.E. Van Doren, J.T. Devreese, Phys. Rev. B25, 4270 (1982).
13. J. Donahue, "The Structure of Elements", Wiley, New York (1972).
14. G. Nilson, G. Nelin, Phys. Rev. B6, 3777 (1972).
15. P.E. Van Camp, J.T. Devreese, in "Proceedings of the Supercomputer Applications Symposium", to be published (1984).

I.2. Total Energy Methods

CURRENT IDEAS AND METHODS FOR CALCULATION

OF GROUND STATE PROPERTIES OF SOLIDS

Richard M. Martin

Xerox PARC
3333 Coyote Hill Road
Palo Alto, CA 94304, USA

I.   INTRODUCTION

The purpose of my contribution to this NATO Institute is to provide background for understanding the progress that has been made recently in the field of calculation of the properties of solids in their electronic ground state. I will attempt to give my perspective on the fundamental concepts, the most significant steps in the development of current methods, important results of calculations, and directions for future research.   Although this is not meant to be a comprehensive review, I will attempt to describe the breadth of current work, giving examples of diverse methods and applications.   This paper is organized to discuss the basic many-body problem in Section II, the techniques for *ab initio* solutions in Sections in Sections III and IV, and results of calculations in Section V. The notation and perspective that is presented here builds upon the published work by D. J. Chadi, H. Wendel, K. Kunc, O. H. Nielsen, and myself.[1-11] Other papers in this proceedings by Kunc[12] and Nielsen[13] will present further insights into the general theory and specific results.

What are interesting ground state properties?   The static quantities which describe a state of matter at $T = 0$ are its ground state properties. These include the equilibrium structure of a molecule or solid, the average electronic charge density, the total energy, stability of different phases at $T = 0$, and so forth. The changes in the ground state of the matter in the presence of static external influences determine the static compressibility, elastic constants, dielectric suceptibilities, and other linear and non-linear response functions. In addition to these true ground state properties, the dynamics of the ionic motions in most solids can be considered to be on a much slower time scale than the electronic excitations and can be treated by requiring that the electrons follow the ions adiabatically, always in their ground state for every configuration of the ions.[14-16] Thus within this approximation, which is *very good* in most solids, the theory of the electronic ground state suffices to describe the much larger class of properties including phonon dispersion curves, zero-point motion, thermal population of vibrational states, anharmonicity, low-frequency dielectric behavior, etc. These are, of course, all quantities of physical importance and all significant parts of the fabric of our understanding of condensed matter.

A major theme of current work, which I wish to stress here, is that: 1) these static and low-frequency dynamical properties should be considered together as *unified properties* of the electronic ground state, and 2) it is worthwhile to construct a *unified* theory capable of accurate calculation of these properties.

This is not to imply that other aspects of condensed matter are less interesting. On the contrary, a well-defined theoretical framework makes *more fascinating* those cases which are exceptions to the usual rules. The reader is challenged to seek out the exceptions and the places where the present approach is not sufficient.

What is the historical background? Because the structure of matter is one of the basic fields of research in physics, it has been one of the core areas of theoretical physics. While it is not possible to give a history, or even to mention the many fundamental contributions, it is important to recognize the extent to which the work discussed here rests upon the ingenious creations of previous decades. In particular, the reader is referred to the paper[17] by Wigner and Seitz in 1955, who summarize their own work and that of others in understanding the theory of cohesion in metals, and to the multi-volume work of Slater.[18]

Why is this area active at the present time? I believe that there are three principal developments which have led to the recent growth of work on first-principles calculations of ground state properties of matter. The first is the enormous increase in computational power. This has made it possible to carry out calculations on real materials in interesting situations with sufficient accuracy that there can be meaningful detailed comparison with experimental measurements. In addition, it has provided new tools for attacking the fundamental many-body problem. The second important development is the density functional method for electron exchange and correlation.[19-22] The density functional has made it feasible to calculate

176

the ground state energy and charge density including these many-body effects with remarkably accurate results for real solids. This is the starting point for almost all current first-principle calculations of total energies of solids. Finally, there have been significant new developments in experimental techniques and materials preparation that are making it possible to probe the structure of matter in ways never before realized. One advance is the ability to create high pressures and explore the properties of matter over a wide range of densities.[23] This is an almost ideal experimental tool to provide information that can be compared directly with current theoretical calculations. In addition, there are new fields of research in which the atomic structures play a dominant role, yet are largely unknown experimentally. Perhaps the best examples are surfaces of semiconductors, where theoretical predictions have stimulated experiments and led to sweeping changes in understanding of the nature of the surfaces.[24] The synergism of theory and experiment is providing exciting stimulation of both fields and a changing relation of theory to experiment.

What is the recent progress? Since the mid-1970's there has been a qualitative change in the ability to predict accurately structures and related ground state properties of solids without using any information from experiment. This is leading to a changing relation of theory and experiment in which the theory can play a more predictive role in real materials. Recent short reviews of the results of the theoretical work are given in Refs. 6,10, and 25-29. The results include calculations of lattice constants of crystals; elastic, dielectric, and piezoelectric constants; phonon frequencies; pressures for transitions between different phases; and structures of complex crystals. For a large number of cases these are in accurate agreement with experiments. Furthermore other quantities not known experimentally have been predicted, such as energies of non-equilibrium metastable phases;[30,31] the energy of a solid along a continuous path connecting stable phases;[31,32] eigenvectors of phonons;[4,8,9,12] non-linear elastic properties;[11,13] and structures of surfaces[24] and interfaces.[5]

The recent progress is, however, not solely in terms of realistic results. From a purely theoretical point of view, the prediction of the structures and related properties of solids is the basic problem of condensed matter physics: the understanding of the cooperative behavior of systems of $\sim 10^{24}$ strongly-interacting electrons and nuclei. Although current work is focused upon the cooperative behavior of the quantum electronic system, the reader is challenged to foresee the interesting physics problems that can arise from the possibilities for phase transitions, non-linear dynamics, non-equilibrium effects, thermodynamics, and so forth, in the coupled system of nuclei and electrons.

Why emphasize solids and molecules? The essential point for our purposes is that a vast range of interesting ground state problems can be cast in terms of a limited number of possible structures for a molecule, a bulk solid, the surface of a solid, etc. The goal of the present work is to develop a comprehensive theoretical framework for studying these structures and testing them for stability both locally and (to a limited extent) globally. It is important to consider solids and molecules together because it is often the case that different first-principles methods can be tested and compared in small molecules.[26,33-36] This is one of the significant areas of overlap of theoretical chemistry and physics and is a meeting ground for the different approaches to the many-body problems. Although the same principles may apply to the liquid states of matter just as well as to molecules and the solids, the essential problems in liquids involve the statistical mechanics of many different configurations. We will attempt to cast the general principles in forms applicable to liquids but we will not address the physics of liquids.

What are the most important goals for this area of theoretical physics? As background for selecting the appropriate future goals, it is worthwhile to recall the comment made by Wigner and Seitz in 1955:[17]

"If one had a great calculating machine, one might apply it to the problem of solving the Schrödinger equation for each metal and obtain thereby the interesting physical quantities, such as cohesive energy, the lattice constant, and similar parameters. Presumably, the results would agree with experimentally determined quantities and nothing vastly new would be gained from the calculation. It would be preferable, instead, to have a vivid picture of the behavior of the wave functions, a simple description of the essence of the factors which determine cohesion, and an understanding of the origins in the variations in the properties from metal to metal. . . "

The goal for future work should not be only to reproduce experiments or calculate properties not known experimentally. The greatest challenges are to generate new ideas and concepts to gain greater insight into the nature of the many-body problem, to bring unity to the understanding of the properties of solids, and to stimulate further experimental and theoretical research into new fields.

II.   GENERAL THEORY

It is important to start our discussion at a very general point to establish fundamental theorems and to encompass different theoretical approximations and methods. The Hamiltonian describing the interacting system of nuclei and electrons is given by

$$H = \sum_i p_i^2/2m_i + \sum_{i>j} V_{ij} \qquad (1)$$

where $p_i = i \nabla_i$ is the momentum operator, V_{ij} is the Coulomb interaction, and the indices i and j denote *both* nuclei and electrons. The total energy is given by

$$E_{tot} = \langle\psi| H |\psi\rangle/\langle\psi|\psi\rangle \qquad (2)$$

and the ground state is given by the variational principle

$$\delta E_{tot} / \delta\psi^* = 0, \qquad (3)$$

which leads to the many-body Schrödinger equation

$$H |\psi\rangle = E_{tot} |\psi\rangle. \qquad (4)$$

The ground state wavefunction ψ and energy E determine the properties of the system at T = 0. In condensed matter with macroscopic numbers of particles, the wavefunction may describe either a quantum liquid or a solid, with the nuclei and electrons adopting a fixed structure for long periods of time. The existence of a stable structure defines a solid in which the application of moderate forces or stress leads to reversible distortions of the structure, in which the energy E_{tot} increases above its minimum value. It is important to emphasize that this holds in the general case where the nuclei are treated quantum mechanically, including their zero-point motion. For small numbers of particles the system may form molecules with well-defined structures. In all cases, the homogeneity and isotropy of space leads to translation and rotation invariance of the energy.[14-16] We will not be concerned with rigid translations or rotations, and we will only consider cases in which external forces and stress satisfy the condition of zero total force and torque.

Although Eq. (4) is the basic equation for all the following work, it is most useful first to express properties other than the total energy that can be calculated from ψ. In later sections we will discuss approaches to the (approximate) solution of Eq. (4) and the results of calculations.

<u>First Derivatives of the Energy: Generalized Forces</u>

The variational property of the total energy makes it possible to determine directly from the ground state wavefunction not only the energy, but also derivatives of the energy with respect to any parameters in the Hamiltonian. The basic idea is often presented as the celebrated "Hellmann-Feynman theorem" for forces.[37-44] Because this is of such broad significance for the theory, we will derive here the general formulas following the concise formulation given in Refs. 37-39 and discussed in Ref. 7. A more complicated explicit derivation of forces is given in Refs. 41 and 42. For any constant λ in the Hamiltonian, the "theorem" states that the derivative of the total energy can be evaluated from the ground state wavefunction and the explicit dependence of H upon λ:

$$dE_{tot}/d\lambda = \langle\psi|\, dH/d\lambda\, |\psi\rangle \tag{5}$$

This follows because of the variational condition (3). Although the wave function ψ changes with λ, the effect of that change upon the energy is zero to first order. The constant λ may be a true constant, such as the electron mass or charge, or it may be a quantity which is held fixed in the determination of the ground state but which can be varied physically, such as the position of an atom. Each type of variation has useful consequences.[39]

The well-known Hellmann-Feynman theorem for the force on a nucleus results from choosing λ to be the position of a nucleus R_I. If ψ is the exact wave function calculated holding R_I fixed, then the force on nucleus I is

$$F_I = -dE_{tot}/dR_I = -\langle\psi|\, dH/dR_I\, |\psi\rangle, \tag{6}$$

Since the only explicit dependence of H upon R_I is in the electron-nucleus interaction, the operator dH/dR_I involves only that interaction and does not involve *explicitly* the electron-electron interations or electron kinetic energies. In the case of Coulomb interactions, the force in (6) depends only upon the charge density, as is described eloquently by Feynmann[38] and reviewed by Deb.[39] Although non-local ion-electron interactions lead to slightly more complex forms[42] involving the wavefunctions directly, the basic idea is nonetheless exactly analogous. The force defined by (6) may be viewed as the negative of the external force which must be applied to keep R_I in equilibrium. Because Eq. (6) is a special case of the results derived much earlier by Ehrenfest[45] from the equation of motion of the momentum operator, we shall refer to (6) as the "force theorem."

Other types of variations lead to useful expressions as well. For example, the first-order change in energy due to changes in the atomic potential can be derived using Eq. (5). One can consider, for example, the change in energy due to actual changes of atoms, such as replacing an atom in an alloy. (In fact, the exact change in energy for a finite change in potential can be derived from the "integral Hellman-Feynman theorem" discussed in Ref. 39.) Similarly, one can define displacements of regions of the charge density surrounding an atom to derive the force on the nucleus by an alternative form. There are many such examples, and the general idea of using the variational principle to calculate change in energy are very useful. Perhaps the most explicit recent discussion is in the body of work[27,46,47] in which this idea is sometimes referred to as the "Andersen force theorem." Although that work has been restricted to the local density function (LDF) approximation, and the calculational procedures have been tailored to the LDF self-consistent equations, the ideas have more general validity. A review of the specific LDF derivation and the rationale behind the applications is given in Ref. 27.

The variational property of the total energy is also of great value in the calculation of the energy itself. In particular, Eq. (3) shows that the error in the energy is always second order in the difference of an approximate wave function from the exact one. This variational property has been used effectively to reduce the numerical calculations in many cases. Specific utilization of this idea can be found in the work of Wendel and Martin,[2] who found approximate energies for Si starting from potentials given by free-electron screening; Anderson and co-workers,[46] who consider differences in total energies of metals in different close-packed structures, using potentials transferred from one solid environment to another; Chellikowsky and Louie,[48] who showed that the density functional expressions for the total energy of diamond can be solved very accurately using only a superposition of atomic charge densities to construct their approximate potentials; and Gordon and Kim,[49] who have developed an entire framework for calulating properties of closed shell rare gas or ionic solids, using only the superposition of charge densities of atoms or ions. The use of the variational property is discussed in more detail in Ref. 7.

The most important consequence of the variational property, however, may be in the effects of the many-body correlations. With the exception of some notable cases mentioned below, almost all calculations make *uncontrolled* approximations in treating (or ignoring) correlations. The above reasoning leads to the conclusion that only a second order error in the energy is made when the exact correlated many-body wave function is approximated by a simpler function. Presumably, this is very important among the reasons why approximate methods, such as Hartree-Fock, or local density functional, can nevertheless provide very good results for the ground state energy.

The Stress Theorem

The choice of the parameter λ as a scale of length in the physical system leads to a different class of theorems, which are of particular significance for our present work. As shown by Fock,[50] the uniform scaling of the system in all dimensions leads to the quantum virial theorem[11,13,18,41,50-53]

$$3P\Omega = \sum_i \left\langle p_i^2/m_i + \sum_{j \neq i} r_i \nabla_i V_{ij} \right\rangle. \tag{7}$$

This may be derived[11,13,50] from (1) by replacing $r_i \rightarrow (1+\lambda)r_i$ and $p_i \rightarrow (1-\lambda)p_i$ and differentiating in the limit $\lambda \rightarrow 0$. The usual form of the virial theorem, with the left-hand side of (7) replaced by zero, applies for an isolated system. However, if the system is in a fixed external potential (which is not scaled), then P is the pressure exerted upon the system by the external forces and Ω is the volume of the system. The virial theorem is analogous to the force theorem in that it gives an expression for the external pressure (i.e. the force conjugate to the volume) in terms of the internal operators of the Hamiltonian. Unlike the force theorem, however, the virial theorem involves the kinetic energies and the interactions of *all* particles – nuclei *and* electrons.

A generalization of the virial theorem to all components of the stress tensor $\sigma_{\alpha\beta}$ has been given in Refs. 11 and 53 and termed the "stress theorem" by Nielsen and Martin.[11] The derivation follows exactly the scaling argument of Fock[50] but with an anisotropic scaling $r_{i\alpha} \rightarrow (\delta_{\alpha\beta} + \varepsilon_{\alpha\beta})r_{i\beta}$, where $\varepsilon_{\alpha\beta}$ is the strain tensor and α, β denote Cartesian coordinates. The variational principle leads to the stress theorem

$$\sigma_{\alpha\beta}\,\Omega = -dE_{tot}/d\varepsilon_{\alpha\beta} = \sum_i \left\langle p_{i\alpha}p_{i\beta}/m_i + \sum_{j \neq i} r_{i\beta} \nabla_{i\alpha} V_{ij} \right\rangle, \tag{8}$$

which gives all components of the stress tensor in terms of diagonal and off-diagonal kinetic and virial operators. The reader is cautioned that in the limit of large systems, the factor of r in the virial expressions (7) and (8) is in fact *not well-defined*. In order to avoid the difficulties caused by the presence of the positive coordinate r in Eq.(8), it is essential to reexpress the virial in terms of *relative* coordinates. For two-body central interactions $\Phi\left(|r_i - r_j|\right)$, the stress theorem can be written[11]

$$\sigma_{\alpha\beta}\,\Omega = \sum_i \left\langle p_{i\alpha}p_{i\beta}/m_i + \sum_{j \neq i} (r_i - r_j)_\beta\, \Phi'_{ij}\left(|r_i - r_j|\right)(r_i - r_j)_\alpha \right\rangle, \tag{8a}$$

where $\Phi'(x) = d\Phi(x)/dx$. This form is manifestly symmetric and is well-defined for macroscopic systems with short range forces. In the case of

Coulomb interactions, the long range at the interaction requires that the sums over positive and negative charges be treated carefully. This is the same problem as for the energy itself and will not be discussed further here.

The interpretation of the terms in the stress theorem are discussed in more detail in Ref. 13 and in papers to be published in Ref. 11. The basic idea is that the kinetic term is the quantum analogue of classical gas theory. It represents momentum p_α transferred by a flux p_β/m. The diagonal $\alpha\alpha$ terms are always positive, i.e., expansive terms. The virial term is associated with potential forces in the α direction between particles displaced in the β direction. The diagonal terms are an energy density and must be negative (attractive) for system which is in equilbrium ($\sigma_{\alpha\alpha} = 0$) in order to cancel the expansive kinetic term. The off-diagonal terms may, of course, have either sign.

The stress theorem is very useful in large macroscopic systems. In particular, since the structure of a crystal is completely specified by the size and shape of the unit cell and positions of the atoms in the unit cell, the force and stress theorems give *all* the generalized forces conjugate to these variables. If the stress theorem is expressed in terms of relative coordinates and proper care is taken of the Coulomb interactions, then the stress theorem, Eqs. (8) and (8a), give the total stress in terms of the intrinsic bulk Hamiltonian and wave functions in the bulk of the crystal. Thus, the stress and force are sufficient to construct the complete equation of the state of any crystal. In general, the more complex and the lower the symmetry of the crystal, the more useful are the forces and stresses in the calculations. This is further described by Nielson in this proceedings[13] and later in the present paper.

Higher-Order Derivatives

Derivatives of the energy of order higher than the first are of particular importance for structural properties of solids. For example, the second-order derivatives, with respect to nuclear displacement, define the force constants. These are sufficient to determine the exited vibrational states of the solid within the harmonic approximation. Higher order derivatives define anharmonic constants, and so forth. Analytic expressions for each derivative can be derived by expanding the Hamiltonian, H, and wavefunction, ψ, to the desired order.[15,16] Since this is well-known perturbation theory, we can easily recognize the fact that, for all orders greater than the first, the variation in ψ must be retained. Because it involves steps that are significantly different from the calculation of the energy, force, or stress, the evaluation of second- and higher-order derivatives has traditionally developed as a rather separate area with its own techniques and approximations.[15,16,54 – 57]

The perturbation approach is well-adapted to many problems such as the calculation of harmonic phonon frequencies in a solid, the dielectric response, etc. Although it requires the development of additional computational machinery, once that machinery is in hand, it can provide a wealth of information. A very succesful use of the perturbation theory has been the development of very accurate and intuitively appealing expressions for structures, phonons, etc., in simple metals by perturbation theory on the free electron gas.[54] This has been extended in recent years to the much more difficult problem of phonons in semiconductors. There are severe computational problems in many methods, but the experience is growing and leading to impressive results.[55-57] For example, phonon frequencies and phonon dispersion curves of Si have been calculated by inversion of the dielectric matrix in reciprocal space[54-55] and by transformation to localized orbitals,[56] which is a physically motivated form that greatly reduces the size of the computations.[56] A recent example is the calculation of both bulk and surface phonon dispersion curves using a simplified tight binding electronic Hamiltonian.[57]

On the other hand, the so-called "direct" method[6] attacks the problems of calculating the higher derivatives by the simple procedure of calculating the energy, forces, and stress for solids with atoms displaced by finite amounts. From these calculations, force constants and elastic constants to all orders can be calculated using exactly the same methods and approximations as for the equilibrium structure. The great advantage of the direct method is that many situations—simple crystals, distorted crystals, surfaces, molecules, etc.—can all be treated on the same basis. Work using the direct procedure has led to the "unified" theory of structure of solids, which will be discussed in the later part of the present paper. The disadvantage of the direct approach is that it is limited to simpler situations than could be considered in the perturbation methods. Comments on the comparison of perturbation vs. the direct approach have been made by the present author in a previous paper.[7]

Beyond the Variational Principle

Because the central idea in this section has been the variational principle, it is doubly important to emphasize its limits. In his book[59] "Basic Notions of Condensed Matter Physics," P.W. Anderson terms the naive application of the variational principle as "Quantum Chemists' Fallacy No. 1." The substance of the comment is that ideas which are completely sufficient in small systems can go astray in condensed matter. In particular, in the limit of a large system there may be vanishing overlap between the true ground state and a trial state that appears sensible. A clever choice of the nature of the state at the beginning can lead to the solution, whereas no amount of brute force will arrive at the answer starting from a wrong point. Cases in point include the development of a collective order parameter at a phase transition, the Kondo problem, and the (presently-not-understood) Kondo and Anderson Lattice problems.

III.   SOLUTIONS OF THE MANY-BODY PROBLEM

Few exact solutions exist for Hamiltonians which involve many particles. There is, nevertheless, a class of one-dimensional problems that are exactly solvable using the Bethe Ansatz.[60]   Among these are the Hubbard model[61] for electrons moving on sites in 1-d and interacting only on the same site, and the Anderson and Kondo models for interacting electrons as an impurity in a 3-d metal.[62]   The solutions provide bona fide tests for approximate methods and may even apply in certain real solids. Perhaps the most direct application to a real solid has been to cerium, where the unusual properties of the ground state energy of the Kondo problem has been suggested to cause the well-known volume phase transition.[63]

For finite numbers of particles there exist a number of methods to find solutions for the ground state energy which are variational approximations to the exact solution and which can be made to achieve numerically very nearly the exact energy.   The most common general proceedure is by straightforwardly diagonalizing the Hamiltonian in the space of all possible "configurations" of the electrons in the Hartree-Fock orbitals.   These "configuration interaction" calculations[33,64] require the diagonalization of very large matrices which grow rapidly to astronomical sizes as the number of particles increase.   Nevertheless, these are extremely important for small systems like molecules,[33] extrapolations to large numbers can be very valuable.[64] The ideas of the renominalization group[65] provide a way to derive mathematically defined recursions that lead to exact descriptions of the lowest-energy excitations for infinite systems.

Statistical Monte Carlo techniques have recently been adapted to the calculation of the ground state wavefunction and energy of quantum systems of finite numbers of particles.[66,67]   These can provide correlated wavefunctions, which are rigorous upper bounds on the ground state energy.   Probably the most difficult aspect of the calculations at present is including the antisymmetry of Fermions.   One method has been to use fixed nodes in the wavefunctions provided by approximate solutions, but to allow all other degrees of freedom.[67]   Although the methods have been applied to various idealized many-body Hamiltonians,[65] the most important calculations for our purposes have been for interacting electrons in a uniform background and for condensed hydrogen.

One of the important applications of the quantum Monte Carlo calculations has been for the interacting electron gas in a uniform positive background, commonly termed "jellium." Such calculations describe the entire density range from the interacting liquid at high density to the Wigner crystallization of electrons at low density. They provide direct numerical values for the ground state energy, the corelation functions, etc., and are a benchmark for other more analytic treatments of exchange and correlation. The calculations by Ceperly and Alder[67] have provided the best current values of energy of the uniform electron gas, which are now used in the local density functional calculations described in the following section. One possibility for future work may be a systematic study of the inhomogeneous electron gas, to provide a new basis for going beyond the local density approximation.

The Monte Carlo calculations have also been reported hydrogen, which is the simplest of all condensed matter.[68] The ability to carry out such computer simluations of real many-body problems in condensed matter is one of the major breakthroughs, which may lead to entirely new approaches to the theory of condensed matter. The results for the phase transition from the molecular solid to the metallic monatomic solid are of great interest at the present time and are discussed in Section V.

IV.   DENSITY FUNCTIONALS

The approach which is by far the most widely used at the present time for the calculation of the total energy and other ground state properties of matter is the density functional method. The fundamental advance in this area was made by Hohenberg, Kohn, and Sham.[19-21] These authors demonstrated that the exact ground state energy of the many-body quantum-mechanical system is a unique functional of the charge density $n(r)$. The proof is elegant and it rests upon their demonstration that for any inhomogeous charge density, there is a unique potential which has that density as its ground state. Since that time there have been many important contributions, e.g., the extension to spin density density functionals; the calculation of the functional for a free electron gas using Monte Carlo techniques; and analysis of the nature of the corrections in the excited states. Discussion of recent contributions can be found in the book "Theory of the Inhomogeneous Electron Gas"[21] and the reader is referred there for in-depth discussions.

For our purposes, there are three important aspects of the density functional method. First, it is in principle exact and provides an alternative to the direct treatment of the full many-body Hamiltonian discussed above. It is therefore relevant to establish rigorous expressions for other physical quantities, such as the stress, in the density functional formalism. Second, for any functional, variational solutions of the equations satisfy all the properties required to derive the requisite theorems for force, stress, and other derivatives.   Third, there are local approximations to the exact functional which are both simple to use and are remarkably accurate for many properties of the electronic system in normal states of matter: atoms, molecules, and condensed phases. Although the overview provided in this section is restricted to the local-density-functional (LDF) approximation, the reader is encouraged to keep in mind that there is a corresponding exact proceedure and to be alert for places where the LDF forms must be generalized to encompass an exact non-local functional.

The local-density-functional (LDF) expression for total energy of a system of electrons and fixed ions (or nuclei) consists of the kinetic energies of non-interacting electrons, the interactions of the electrons with the ions, the average Hartree electron-electron repulsion, exchange and correlation energy, and the ion-ion interactions. The basic expression is given in Ref. 20 and has been recast in many references, including Refs. 2, 7, 11, 21, 25 – 27, 42 and others discussed below. The LDF makes the ansatz that the difficult part of this problem, the exchange and correlation, is a local function of the charge density n(r). Since the function is assumed to be the same for all systems, it follows that it is the same as in a free-electron gas of the corresponding uniform density. Thus, once one has a good expression for the total energy of an interacting electron gas, then the LDF expressions are completely specified.

Here we give a short review of the LDF equations and methods of solution. In the notation of Wendel and Martin[2] and Martin and Kunc,[7] the energy may be written

$$E_{tot}[n] = E_{II} + \int d^3r\, V_I(r)\, n(r) + F[n], \tag{9}$$

where E_{II} is the ion-ion interaction energy and V_I is the bare potential seen by an electron due to the ions. The functional $F[n]$ contains all kinetic and interaction terms for the electrons and it is very useful to separate the non-local Hartree $E_H[n]$ and single particle kinetic energy $T_s[n]$, leaving a short-range exchange-correlation functional $E_{xc}[n]$:

$$F[n] = E_H[n] + T_s[n] + E_{xc}[n] \tag{10}$$

where

$$E_H[n] = \tfrac{1}{2} \int d^3r\, d^3r'\, n(r)\, |r - r|^{-1}\, n(r') \tag{11}$$

and T_s is given in terms of independent wave functions,

$$T_s[n] = -\tfrac{1}{2} \sum_i \langle \psi_i | \nabla^2 | \psi_i \rangle, \tag{12}$$

with

$$n(\mathbf{r}) = \sum_i |\psi_i(\mathbf{r})|^2. \qquad (13)$$

Note that the total energy is *not* given literally in terms of a functional of the charge density as it is, e.g., in the Thomas-Fermi approximation. Instead, the kinetic energy is given by the form (12) in terms of the wave functions.

The local density functional (LDF) approximation is to assume that $E_{xc}[n]$ is a local function of n

$$E_{xc}[n] = \int d^3r \; n(\mathbf{r}) \; \epsilon_{xc}(n(\mathbf{r})) \qquad (14)$$

where $\epsilon_{xc}(n)$ is a function that is determined in calculations on a free-electron gas of uniform density n. Within the local (LDF) approximation for $E_{xc}[n]$, the solution of the equations for the ground state energy and charge density may be derived from the variational equation for E_{tot} ,

$$\delta E_{tot}/\delta \psi_i^*(\mathbf{r}) = \delta E_{tot}/\delta n(\mathbf{r}) = 0, \qquad (15)$$

subject to the constraint $\int d^3r \; \delta n(\mathbf{r}) = 0$ for charge conservation. This variational condition leads to single-particle Hartree-like equations for the wave functions $\psi_i(\mathbf{r})$,

$$[-\tfrac{1}{2}\nabla^2 + \varphi(\mathbf{r})] \; \psi_i(\mathbf{r}) = \epsilon_i \psi_i(\mathbf{r}), \qquad (16)$$

with the self-consistent field given by

$$\varphi(\mathbf{r}) = V_I(\mathbf{r}) + \int d^3r' \; |\mathbf{r}-\mathbf{r}'|^{-1} \; n(\mathbf{r}') + \mu_{xc}(n(\mathbf{r})) =$$
$$= V_I(\mathbf{r}) + \varphi_H(\mathbf{r}) + \mu_{xc}(n(\mathbf{r})), \qquad (17)$$

where φ_H is the Hartree field and

$$\mu_{xc}(n) = d(n\epsilon_{xc}(n))/dn. \qquad (18)$$

Equations (13), (16), and (17) define a self-consistent problem in which each ψ_i is a solution of a single-particle Schrödinger equation in which the potential φ is a function of all the occupied wavefunctions $\psi_i(\mathbf{r})$. The solutions of these equations have features in common with of all solutions of self-consistent equations, and we can identify six essential steps:

1) choice of the function $\epsilon_{xc}(n)$,
2) fixing the positions of nuclei and the interaction potentials V_I,
3) solution of the differential equation (16) for a given $\varphi(\mathbf{r})$,
4) summation over filled states to find $n(\mathbf{r})$ and a new $\varphi(\mathbf{r})$,

188

5) iteration to arrive at self-consistency

6) optional evaluation of the total energy, forces, and stress.

Each of these steps involves choices which are important for the entire problem and we will discuss aspects which the author considers to be particularly pertinent for present work.

1) <u>LDF Functional</u>. In a first-principles LDF calculation, the function $\epsilon_{xc}(n)$ is not an adjustable function - it is defined to be the xc energy of a homogenous electron gas. At present, there are a number of suggested forms for $\epsilon_{xc}(n)$, which are justifiable approximations to the (unknown) exact function. One simple form which has had especially widespread use is the Wigner interpolation form[69]

$$\epsilon_{xc}(n) = 0.44/(r_s + 7.8), \qquad (19)$$

in atomic units, where $r_s = (3/4\pi n)^{1/3}$ is the Wigner-Seitz radius, which was chosen to reproduce properly the high-density (uncorrelated) limit and the low-density Wigner crystal energy. The function which is now widely accepted as the most accurate current result is the energy calculated numerically for a large range of densities by Ceperly and Alder,[67] using the Monte Carlo method discussed earlier. An analytic fit to these results by Perdew and Zunger[70] is often used at present. Other forms are described in Refs. 21 and 71. An important point is that the different forms for ϵ_{xc} in current use are very similar. There has been limited investigation of the consequences of the effects of different functions upon solid state properties.[72]

2) <u>Potential V_I</u>. All current work assumes that the nuclei are rigidly fixed at positions R_I, and the energy is minimized only with respect to electronic degrees of freedom. We will consider only this case here, but we note that this is an unnecessary asymmetry in the way nuclei and electrons are treated. It would be more elegant to formulate the theory including the nuclei as massive quantum particles and to minimize the total energy with respect to *all* degrees of freedom. Zunger[73] has suggested that the positions of the nuclei, treated as classical particles, be included in the variational scheme as well, but this has not been used in actual calculations, to the knowledge of the author.

Once the nuclei are fixed, the the exact potential $V_I(r)$ is simply a sum of Coulomb interactions $V_I(r) = \sum_I v_I(r - R_I)$. We note, however, that the equations can be transformed to pseudopotentials and pseudo-wavefunctions with essentially arbitrary accuracy for the ground state properties which do not involve the core electrons. The theoretical justification for this introduction of a first-principle pseudopotential rests upon either 1) a transformation of the Hamiltonian to include orthogonalization to the core states, or 2) re-expressing the problem in terms of phase shifts of the valence electron states outside the core. The latter is used in the construction of the norm-conserving, *ab initio* non-local psuedo-potentials.[74-77]. The procedure for defining such *ab initio* non-local potentials has been put forth by by Hamann, Schülter, and Chiang[74] and by Kerker.[75] Starting from theoretical calculations for atoms, they have defined model potentials (different for each angular momentum l) that have correct eigenvalues and wavefunctions outside a central region, smooth wavefunctions inside the central region, and are "norm-conserving." The effect of both the core states and of the rapidly varying strong potential are included in an effective way in the potential. This is a very general procedure and such potentials can even be defined for nuclei like hydrogen. The point is, that these potentials permit accurate calculations of the *changes* in the outer parts of the valence wave functions that occur in molecules or solids, separate from the problem of describing the deeply bound cores. A set of potentials spanning the entire periodic table has been published,[76] and there exist softer potentials, derived by the same methods.[77] These have been used in many calculations, as discussed below.

3)   <u>Solutions of the Schrodinger Equation</u>.   The most difficult computational step in the problem is the solution of the Schrödinger equation. Fortunately, this is a problem of such significance that many different, powerful procedures have been developed.[78] For the purposes of the present directions of research, the most significant difference between the various methods is the division into those which intrinsically depend upon the assumption of a high-symmetry environment of each atom, and those which make no assumptions on the symmetry.

Methods in which the symmetry plays a crucial role from the outset generally utilize a spherical approximation to simplify the problem. These methods include the Korringa, Kohn, Rostocker (KKR) scattered wave theory,[79] augumented plane wave[80] (APW), linear combinations of muffin-tin orbitals[81-82] (LMTO), and its descendant, the augumented spherical wave (ASW).[83] The great strength of these methods is that they permit inclusion of all core electrons in a way no more difficult than in an atomic calculation. In addition, for materials like close-packed metals, where the total charge density is nearly spherically symmetric in each Weigner-Seitz cell, this greatly simplifies the calculations.[81-83] The theoretical techniques for self-consistent calculation of the total energy, using the spherical muffin-tin approximation, were given by Liberman,[84] Janak,[85] and others,[86,87] and applied within the KKR method by Moruzzi, Williams, and Janak.[25]

The power of these simplifications can be seen by considering the limiting case, where the Wigner-Seitz cell itself is assumed to be spherical. Then the entire problem for the solid can be reduced to atomic-like calculations in which the only effect of the periodicity of the solid is in the boundary conditions on the spherical cell. Such methods may be classified, in general, as the atomic sphere approximation (ASA).[81] In a one-atom-per-cell solid, this gives the total energy directly. Because the cells are spherical and neutral, the long range Coulomb interactions explicity cancel, and one is left with only interactions within the single spherical cell.[81-85] In ionic solids, the only complication is that there is more than one sphere, and the charges on the spheres lead to a simple Madelung term.[81-83] This great simplicity also leads to natural atom-like (or ion-like) expressions, from which it is easy to identify the cohesion energy as the difference from real atoms or ions.[81-85] Within this approximation, it is natural to work with the equation in real space as given above. We note that although the spherical approximations provide enormous imprcvements in efficiency, they suffer from the uncontrolled nature of the approximation and the difficulty in treating low symmetry systems. It will be an interesting area for future research to test these methods and provide simple extensions to lower symmetry problems.

At the opposite extreme is the direct solution of the Schrödinger equation in a periodic crystal by Fourier decomposition. Calculation of band structures using plane wave bases is, of course, well known.[78] This is feasible only for "soft" potentials, whose Fourier transform decreases rapidly as a function of the reciprocal lattice vector, G. Because the nuclear Coulomb potential *cannot* be treated in this way, this method is directly applicable in first-principle methods *only* if one can use some soft *ab initio* pseudo-potential, as discussed above, or if one explicitly transforms to an orthogonalized plane wave (OPW) form.[88] Although this procedure is restricted to some form of pseudopotential, it nevertheless is one of the most important techniques, because the Fourier decomposition makes it possible to describe arbitrary forms for the potentials, and charge densities in thc regions bctwccn the atoms when bonding occurs. This is the reason why it has been possible to apply the plane wave basis set to such a large variety of problems and attempt to construct a *unified* theory of ground state properties.[6,10,28]

For the total energy, useful explicit reciprocal-space expressions were given by Ihm, Zunger and Cohen[42], who also covered non-local potentials. The formulas were elaborated by Yin and Cohen,[89,90] and Nielsen and Martin.[11] The expression for the total energy E_{tot} (in atomic units where $m_e = \hbar = e = 1$) per unit cell volume Ω is:

$$E_{tot}/\Omega = \sum_{k,G,i} |\psi_i(k+G)|^2 \; \tfrac{1}{2}(k+G)^2 \; + \; \tfrac{1}{2}\cdot 4\pi \sum'_{G} |n(G)|^2/G^2 \; +$$

$$+ \sum_{G} \varepsilon_{xc}(G)\,\rho(G)^* + \sum'_{G,\tau} S_\tau(G)\, V^L_\tau(G)\, n(G)^* \qquad\qquad (20)$$

$$+ \sum_{k,G,G',i,l,\tau} S_\tau(G) \; \Delta V^{NL}_{l,\tau}(k+G,k+G') \; \psi_i(k+G)\, \psi_i(k+G')^* \; +$$

$$+ \; \Omega^{-1} \Big(\sum_\tau \alpha_\tau\Big)\cdot\Big(\sum_\tau Z_\tau\Big) \; + \; \Omega^{-1}\, \gamma_{Ewald}$$

where k extends over the Brillouin zone, G are reciprocal lattice vectors, ψ is the wave function, i denotes the occupied states for a given k, n is the electronic charge density, Σ' denotes a sum excluding $G=0$, τ labels the atoms in the unit cell, $S_\tau(G)$ is the structure factor, V^L is a local (l-independent) potential, l labels the angular momentum ($l=0,1,2,...$), ΔV^{NL} is a non-local (l-dependent) correction superiposed on V^L, α_τ denotes the average non-Coulombic part[42] of V^L, Z_τ is the ionic charge, and γ_{Ewald} the Madelung energy of point ions in a constant neutralizing background.[91]

This is simply the Fourier transform of Eq. (9) - (14), except that the ionic potential is allowed to be non-local and the ion-ion term is divided into an Ewald term for point charges in a uniform background, and average non-Coulombic terms. The non-local terms and the average non-Coulombic terms are essential for *ab initio* ionic pseudopotentials.[74-77] The Ewald term is collected in the standard way to avoid divergences.[91,92] Let us emphasize that it contains not only ion-ion terms but also it contains the "background," which is, in fact, often the largest electron-electron and electron-ion term! These latter terms are functions only at the volume, i.e., they contribute only to pressure, and not to any changes at constant volume.

There is a significant difference in the way the terms are grouped together in Eq. (20) from that done in the ASA forms.[81-85] In the latter, the assumption is made that the true charge density can be divided into spherically symmetric regions around each atom. This allows the equations to be simplified to involve only intrasphere terms plus simple Madelung terms that involve only the *total* charge of the spheres. This often gives Madelung terms *much* smaller than those appearing in Eq.(20). For example, in NaCl the ASA Madelung term involves the self-consistently determined charges on the spheres (which is $\sim\pm1$). On the other hand, in Eq.(20) the term γ_{Ewald} is defined to be the Madelung energy of charge $+1$ and $+7$ in a uniform background of total charge -8. The fact that in the final solution the electrons really are not uniform but concentrated near the Cl site must be included in the $G\neq0$ potential terms in (20). This is an example which shows clearly the large cancellations that can occur between the γ_{Ewald} and the $G\neq0$ electronic terms. This makes it difficult in general to find simple interpretations for the total energy in this form.

The advantage of Eq.(20) is that it is exact and applies to all crystals with no assumptions. This is very important in many-problems such as the phonon energies discussed below. It would be very valuable to find a fulfilling way given to reformulate (20) to make it easier to interpret and still maintain its exactness.

There are, in addition, other types of band structure techniques which can combine the advantages of the methods discussed thus far. Although these other thechniques are, in general, more difficult computationally and have been used less extensively thus far in solids, their power is such that they are certain to be developed more fully in the future. One class of techniques use localized basis sets, which may be linear combinations of atomic orbitals[78,93] or Gaussian orbitals.[33,94-97] The latter are particularly effective, even though the Gaussian form is not physically motivated, because the three- and four-center integrals can be done analytically. Explicit expressions for the total energy[96,97] and pressure[97] have been derived. This has widespread use in molecular calculations where Gaussians are utilized very effectively.[33] Localized functions combined with plane waves, termed "mixed basis"[43,98] can go even further in providing efficient basis sets.

There has also been a recent extension of the APW method to include arbitrary variations in the potential. In this "full-potential linear" (FLAPW) method,[99] the sphere in the APW method becomes merely a convenient surface for matching wave functions—both inside and outside the sphere, the potential is allowed to have all non-spherical components, and the equations is solved with essentially arbitrary accuracy. This can provide a major avenue to be able to treat core electrons, d and f states, and low-symmetry situations simultaneously. It has been applied only in limited cases thus far.[99]

It may be noted that there are difficulties associated with the highly efficient choices of basis sets. In general, they are chosen differently to represent each situation. An example of the consequences is the calculation of forces using the Hellmann-Feynman idea. If the basis set changes with the position of the atom (as in cases where a localized function is centered on an atom), then this introduces extra terms into expressions, such as Eq. (6). These are well known in the chemistry literature.[39,40] and are being incorporated into current solid state theoretical work.[43,44]

4) <u>Sum Over All Filled States</u>. The sum over filled states in condensed matter is in fact an integral over a continuous variable, which is the momentum k in a crystal. Judicious choices of how to approximate the

integral by a finite sum can have great benefits in reducing the computation required and improving the accuracy. This problem is very different for metals and for insulators.

For metals one must integrate up to the Fermi energy (which must be found self-consistently). Since the Fermi surface defining the occupied states in k space bears no simple relation to the Brillouin zone, no unique choice of k points will be optimum for different problems (or even for the different iterations in the same problem!). One can choose to either select a k-point set optimized for a particular type of band structure, or to find a general way to utilize calculations on a "universal" grid. The first choice was made, e.g., by Lam and Cohen[100] for Al. They utilized a radial grid to take advantage of the nearly parabolic bands and to determine the Fermi surface accurately along only a few directions. However, the latter approach is needed for non-simple metals. The tetrahedral interpolation method[101,102] is very powerful in this regard, but it does require careful use to properly interpolate when bands are nearly degenerate. A very simple Gaussian weighting procedure has been used by Ho, Fu and Harrnon[103] in which the sum over k points is weighted by a Gaussian $\exp(-A(\epsilon_i - \epsilon_F)^2)$. This has the great virtue of simplicity, and has been used by them for very complex bands in distorted transition metals. Recently, Needs and coworkers[32,104] have used this procedure for equally difficult cases, such as the semi-metal As,[104] in which there are small pieces of Fermi surfaces in different parts of the zone, which change as a function of pressure.

For insulators, however, the integrals over filled bands can be carried out very efficiently using the "special points" of Baldereschi[105] and Chadi and Cohen[106] and cast in a more systematic form by Monkhorst and Pack[107]. This defines sets of points which are rational fractions of reciprocal lattice vectors and which form increasingly accurate grids. Accurate calculations can often be done with very small numbers of points, e.g., 2 or 10 inequivalent points in diamond-structure materials, or one point in complex crystals and "supercells".[1-13] This great simplification of the problem is, in fact, the reason why it has been possible to carry out the calculations of small energy differences in semiconductors with far less numerical problems than would be the case with any other method known to the author. In this sense, the special points idea is the clever trick that has made it feasible to actually carry out the "unified" calculations of structures, phonons, etc., envisioned in the Introduction.

5) Self-Consistency. The methods to achieve self-consistency are to a large extent independent of the other choices in solving the LDF equations. They are, in fact, a part of general problems encountered in the numerical analysis, and arise in many problems in phyisics. The simplest procedure, which is often used, is to construct the potential at the n + 1

194

step, φ_{n+1} as a linear combination of the previous "input" potential $\varphi_n^{in} = \varphi_n$, and the "output" potential, φ_n^{out}, constructed using Eq. (13) and the output charge density,

$$\varphi_{n+1} = \alpha\varphi_n^{in} + (1-\alpha)\varphi_n^{out}. \tag{21}$$

There are many cases in which convergence is rapid, i.e., α can be chosen small enough ($\sim\frac{1}{2}$), and this simplest approach has been considered adequate. These cases are ones, such as Si in the diamond structure,[11,13] in which the potential φ is dominated by the ionic potentials. The simplest improvement, letting α vary for different parts of the potential, can also be incorporated in this framework.[13]

There are, however, situations where the achievement of self-consistency requires deeper insights into the problem. This general problem can be effectively illustrated by large cell calculations on semiconductor surfaces and interfaces.[5,24] As discussed in Ref. 7, the long range Coulomb interactions lead to an extreme sensitivity of φ to small changes in the charge density. An error in the input potential φ^{in} is expected to lead to an error $\sim(\epsilon(G)-1)$ times as large in φ^{out}. For G = reciprocal lattice vector of an ordinary crystal, this gives no problem. but for large supercells $\epsilon\sim10$ for the smallest G's and the magnitude of the error in the output is an order of magnitude larger than in the input! Nevertheless, the direction of the change indicates the path toward the self-consistent answer. In representative cases $\alpha\sim0.9$ gives convergence, but in a slow, unsatisfying way. In this situation, it was shown [5,7] that one could use the fact that φ approaches the asymptotic limit φ_∞ exponentially, i. e., $(\varphi_n - \varphi_\infty)\propto\exp(-An)$. By using the values of φ_n for these or more iterations, one can determine the coefficient A and predict the desired asymptotic value φ_∞.

Another difficulty that the change in $\varphi_n(r)$ or its Fourier transform $\varphi_n(G)$ is a function of $\varphi^{out} - \varphi^{in}$ at all r' or G'. This means that the α in Eq. (21) should in fact be a matrix. It is clear[7] that the needed matrix is in fact closely related to the dielectric matrix. Several authors have implemented approximate calculations of the dielectric matrix in order to improve convergence.[108]

There is a numerical method which incorporates both of the above aspects of the problem into a single general procedure, called Broyden's method.[109] The essence of this method is to use some simple version of Eq. (21) for the first iteration, but, in all succeeding iterations, to construct each new potential φ using knowledge of the change in all components of φ for several previous iterations. This effectively generates the dielectric matrix relations, and performs the extrapolation to infinity of all components of the potential. An illustration of the increase in speed of convergence using Broyden's method is given by Nielsen.[13]

6) <u>Energy, Forces, and Stress</u>. In the LDF method, the equations are derived by the variational principle applied to the total energy. In the resulting equations it is the charge density of the ground state that plays the key role and it is the charge density which is made consistent. It is not necessary to evaluate the total energy—indeed before 1970, the primary emphasis in calculations was upon the eigenvalues or the "bands" that resulted. As we have seen, however, it is the total energy and other ground state properties such as the force and stress that are the proper objects of LDF calculations.

Although the expressions for the total energy have been given by Kohn and Sham,[21] it is necessary to make these more specific and to carry out the calculations accurately to achieve our goals. Different expressions have been discussed above. Of course, one can also compute the derivitives. The force theorem (6) applies directly to the LDF formulas. There is no special expression for the density functional, because the force is independent of the electron-electron interactions. All that is needed is the charge density and the wavefunctions for the non-local parts. Stress, on the other hand, is very different, because the electron-electron terms enter. For calculations in reciprocal space using Eq. (20) for the energy, it is straightforward to differentiate (20) directly within respect to $\epsilon_{\alpha\beta}$ to obtain the needed expressions. This is done in Ref. 13.

It is worth pausing to see the simplicity of the way the LDF enters the stress. The essence of the LDF is that the xc energy density is a purely local property depending only upon the charge density at each point. It follows that in the LDF the effect of the xc contributions is to add a pressure at each point, which is the same as the xc pressure in a free electron gas of that density. The total xc pressure is just the integral of local xc pressure. Furthermore, there is *no* contribution of xc terms to shear stress in the LDF. This is true for all local functions, whereas non-local functions (such as the Hartree energy) contribute to shear. This simple form has been used by a number of authors[47,84] in calculations of the pressure. The formulas given in those papers apply only to pressure within the spherical approximation. More general expressions and discussions are given in Refs. 11 and 13.

V. RESULTS

It is not possible here to describe, or even list, all the important *ab initio* calculations of total energies and other ground state properties. We can only choose subjectively those which exemplify best the ideas and the capabilities of the techniques which have been described. I will attempt to select a few results which represent the thresholds of new developments and those which represent current theoretical knowledge of classes of materials.

Hydrogen

There is a central position in the understanding of condensed matter occupied by hydrogen.[110] This simplest form of condensed matter can be viewed as a two-component Fermion system of protons and electrons in which each component can undergo separate and/or correlated transitions. At low pressures, the protons and electrons condense into a molecular system. At extremely high pressures, theory argues convincingly that condensed H is a two-component Fermi liquid.[110] At some intermediate pressure, H is thought to transform to a monatomic metallic solid, and it may become metallic even in the molecular phase. Theoretical and experimental studies of condensed hydrogen are among the most fascinating problems in condensed matter because of the special role of hydrogen in the universe, and because of the possible interesting properties that it may have. For example, it has been proposed that metallic H has a superconducting transition temperature an order of magniture higher than any currently known material.[111,112]

Density functional calculations have been carried out for metallic H with several fixed lattice structures.[112-114] Probably the most surpising result is that a simple cubic lattice is more stable than close-packed lattices for a range of densitites.[112] However, no calculations have yet treated the monatomic metal and the molecular solid on a comparable basis to determine the transition pressure accurately. Current estimates[111] of the transitions pressures range between 1 and 10 Mbar,[115] generally intriguingly above the current experimental range of $\simeq 1-2$ Mbar. It would be very valuable for there to be more complete density functional calculations in the near future.

The quantum Monte Carlo calculations for Fermions have also been applied to a quantitative calculation of the phase diagram of condensed hydrogen.[68] Calculations of the ground state have been carried out as a function of the average volume, treating both the electrons and protons quantum-mechanically, i.e., including the proton zero-point motion. There is only one intrinsic limitation of the method—the finite number of particles which can be included in the many-body ground state. Since the

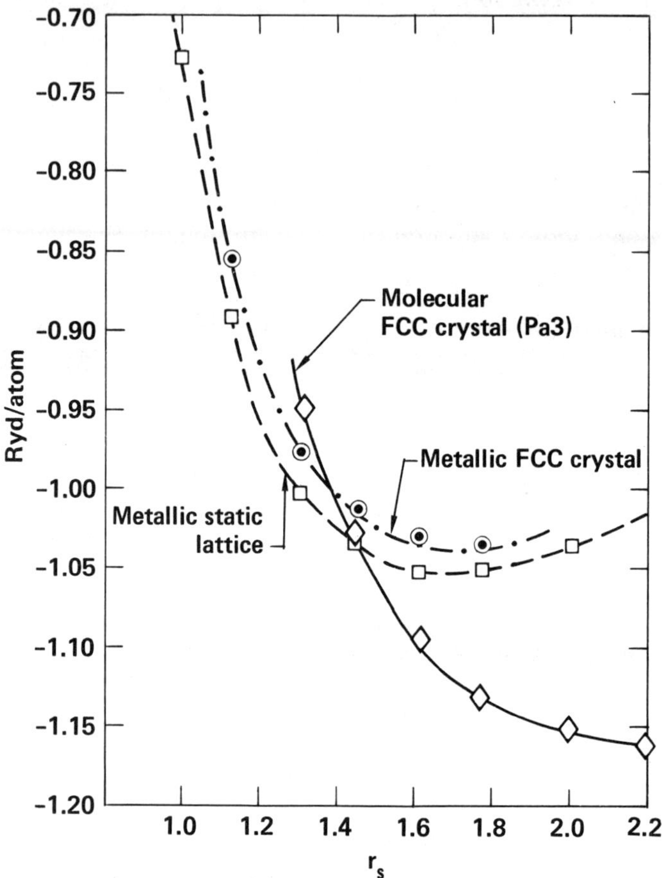

Fig. 1. The phase diagram of condensed hydrogen computed by the quantum Monte Carlo methods. Here r_s is the average proton separation. The calculations treated both electrons and protons quantum mechanically except that the protons are required to have a average fcc structure in the metallic phase and the centers of the molecules are restrained in the molecular phase. The molecular curve agrees well with experiment to 0.5 Mbar and the transition is predicted to occur around 5 Mbar. (Figure courtesy of D. M. Ceperley and B. J. Alder).

calculations of Ref. 68 include 108 atoms, this is expected to be sufficient for most properties. The most difficult problem—the Fermion character of the wave functions—has been handled by an approximate "fixed node" method with additional node "relaxations," and it is argued that this is not a major problem quantitatively.[68,115] Finally, the calculations to date have restrained the lattice symmetry by requiring the protons (or the molecular centers) to remain near the predetermined lattice sites.

The results shown in Fig.1 give the total energy of both phases versus the average particle separation and shows that the molecular phase should transform to the close-packed lattice at a radius of about 1.3 a_0. The calculated pressure at this point is ~5 Mbar.[116] The molecular crystal curve agrees well with experimental data which extends to a pressure of ~0.5 Mbar or ~1.5 a_0. The metallic curve is a rigorous upper bound on the energy, and the results support work done by other methods, such as the density functional calculations. The difference between the two curves for the metallic phase is the zero-point energy, which is clearly an important effect in this quantum solid. This is, in the author's opinion, a major step that represents the most rigorous calculation ever done for a real condensed matter system. It remains to be seen how well it can be applied to more complex problems.

<u>Simple Metals</u>

The elements which do not have partially filled d or f shells are simple in the sense that: 1) the electronic problem can be stated rather simply, 2) analytic approximations to the wavefunctions and Fermi surfaces can be expected to be adequate for the essential aspects, and 3) the known structures are typically simple close-packed lattices. With this in mind, they are nearly ideal embodiments of the fundamental many-body problems of the interacting electron Fermi liquid and have been studied in classic works, such as that of Wigner and Seitz.[17] These simple solids continue to serve as prototypes for each new, "improved" method.

The simple metals are the solids for which the most direct comparisons can be made between the different methods of calculation. In addition, they provide *stringent* tests. Because the energy differences between different structures are small, they are exceedingly difficult to calculate by any of the general methods that is not tailored specifically to these problems. This may be seen simply from the kinetic part of the total energy. If the electrons were free, this would be just an integral of k^2 over the filled Fermi sea. As mentioned in the previous section, the accurate calculation of this simple integral by one of the universal sampling techniques is just as difficult as for complex cases.

Table 1. The lattice constant (a), bulk modulus (B), and the static lattice
cohesive energy (E_{coh}) of Al found in various calculations and
compared to experiment.

	a (a.u.)	B(kbar)	E_{coh}(Ry/atom)
Lam & Cohen (Pseudo., Ref. 100)	7.59	715	−0.268
Liberman (KKR, Ref. 84)	7.65	780	−0.245
Janak et al. (KKR, Ref. 113)	7.59	801	−0.285
Williams et al. (ASW, Ref. 83)	7.60	870	−0.298
Ross & Johnson (APW, Ref. 36)	7.79	940	−0.30
Boettger & Trickey, (LCGTO, Ref. 117)	7.65	968	−0.235
Experiment	7.60	794	−0.248

The comparison of different methods can be illustrated by a recent
paper by Boettger and Tricky[117] on Al. They have carried out calculations
using linear combinations of Gaussian-type orbitals (LCGTO) and compared
with many different calculations and with experiment, as is reproduced in
Table 1. The primary conclusion is that careful calculations using any of
the methods are in remarkably good agreement. For example, the
LCGTO[117], with all core electrons and localized basis states gives the
transition pressure for the fcc to bcc phases to be 3.28±.09 Mbar, whereas
the earlier *ab initio* pseudopotential plane wave basis calculation[100] gave 2.9
Mbar.

Perhaps the most interesting of the recent results for the simple metals
is the identification of broad trends with some surprising results and
predictions. One type of trend is the variation across the periodic table.
For example, Moriarty and McMahon,[119] have examined the trend in the
third row Na, Mg, Al, and Si. In this sequence at high pressures, Si behaves
like a simple metal and is just a continuation of the trends of transition
between the close packed structures. Reference 120 also illustrates the
overall agreement of LMTO[120] and pseudopotential methods[89] for the close
packed phases of Si. The most surprising result to me, however, is that
simple metals do not become simpler under pressure. Instead, they become
more complex as d bands become occupied.[119] The "simple" metals are

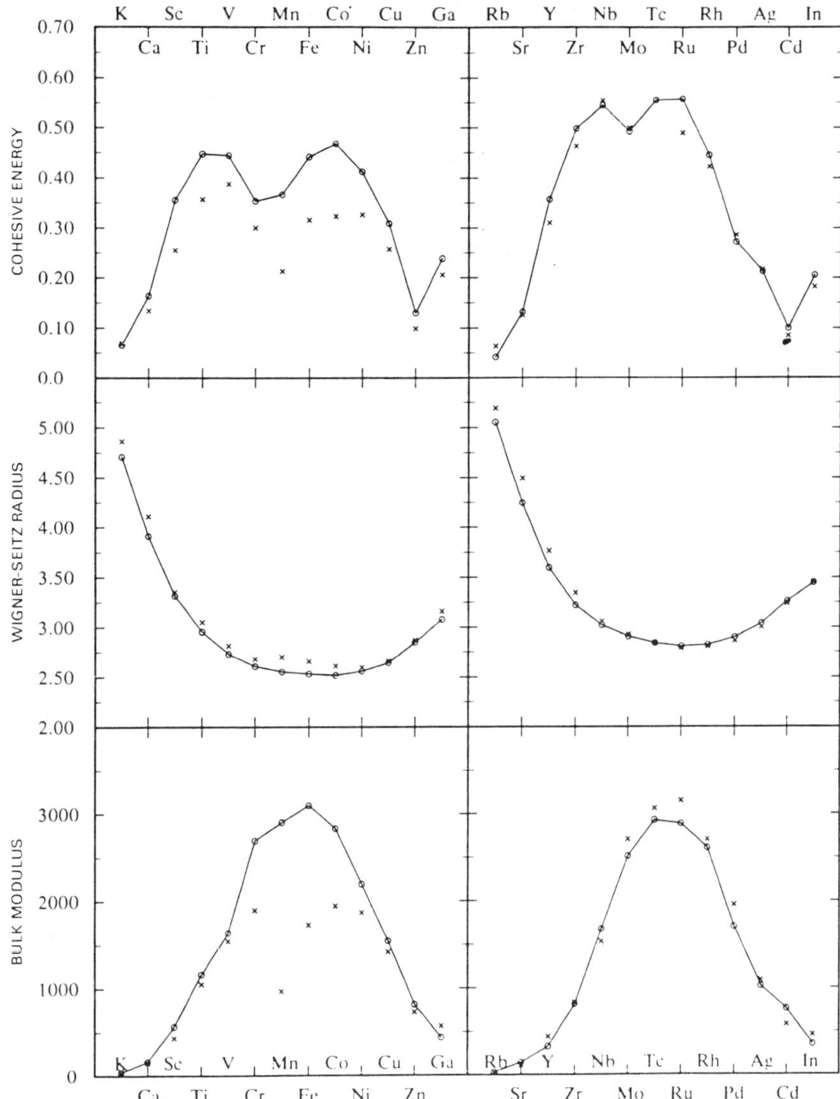

Fig. 2. Calculated properties of the 3d and 4d transition metals-cohesive energy, lattice constant, and bulk modulus- compared with experiment (crosses). This represents a milestone in the development of the methods to calculate E_{tot} with sufficient accuracy to find these quantities. (From refs. 25 and 123, figure courtesy of V. Moruzzi)

predicted to become d band metals and undergo a series of transitions similar to these in the transition metals (when the sequence occurs as a function of atomic number) and even to go to such distorted structures as the Sm structure.[121] For Rb and K, the calculations predict complete s→d transfer by ~600 Kbar (which is within present experimental capability) and transition to the Sm structure at ~3 Mbar (within the realm of possibilities). Thus, even "simple" metals go through complex intermediate structures before they reach the ultimate bcc phase.

Rare Gases

The total energy as a function of pressure of rare gasses presents problems which are technically quite different. Here, I just want to mention that the same APW and LMTO-type calculations have also been done for these materials to derive equations of state.[122] At high densities the LDF method is expected to be as good for rare gases as for other materials. However, we note that the van der Waals interaction is not included in any LDF – it is a non-local correlation effect – and this longest range part of the interaction of rare gas atoms will not be given properly. Probably the greatest interest has been to use pair potentials thus derived from *ab initio* calculations, and to use them in theories of melting, etc. It would take us too far afield to consider the work here.

Transition Metals

One of the most influential steps in the course of recent calculations occured in 1977 when Janak, Moruzzi, and Williams[25,123] published their calculations of the ground state properties of the entire set of third and fourth row transition metals. This has since been incorporated in a book[25] including calculations for essentially all elemental metals. Using the muffin tin approximation and KKR method, they found the energies, lattice constants, and bulk moduli of all the metals, in addition to other properties like the bands, Fermi surfaces, magnetic enhancements, etc. A well known figure of the results is reproduced here as Fig. 2. We can see from the comparison of the points (connected by solid lines) with the experimental crosses that the results of these *ab initio* calculations are very good. The greatest discrepancies are in the middle of the series and originate in the strong magnetic character and Hund's rule coupling in the atoms and solids. The work convincingly demonstrated the ability to handle the enormous total energies and derive the tiny energy differences accurately enough to derive the lattice constants and bulk moduli.

Among the great number of other calculations on transition metals, I would like to mention the work of Ho, Fu, and Harmon as particularly significant for our purposes.[43,103,124] They have developed the mixed basis method[98] to the stage where it is accurate and feasible for transition (and

202

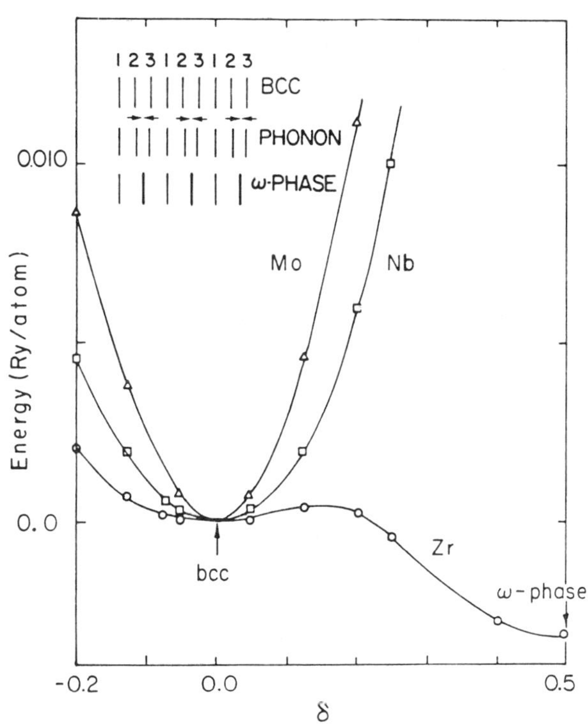

Fig. 3. Total energy of Nb, Zr, Mo as a function of the displacement δ that is the order parameter for the transition to the ω-phase, where each second and third plane collapse to a single plane as indicated. The curvature at δ = 0 gives the harmonic phonon frequency and the second minimum of δ = 0.5 shows the stability of the ω-phase in Zr. (From Ref. 103, figure Courtesy of B. N. Harmon)

rare earth) metals, and they have also used APW methods in complementary ways. The point is that from the APW-type methods the core states can be included and high symmetry problems solved accurately, whereas the mixed-basis pseudopotential methods have made it possible to deal with lower symmetry problems like phonons and displacive phase transitions.

The calculation of the energies describing the famous ω-phase transition can illustrate their results. The transition occurs in Zr, and is a collapse of each second and third (111) planes in the bcc structure into a single plane of double density! This is a very unusual transition and has been the object of much study. In Figure 3 is shown the energy as a function of the displacement δ corresponding to the order parameter of the transition. In this calculation, the authors have found: 1) The phonon frequency (curvature at small displacement). The frequency of this LA mode, 2/3 of the way to the BZ, is high in Mo and only slightly reduced in Nb. It is much lower in Zr but it does not vanish. 2) The complicated changes and crossings of energy bands that occur for larger displacements. This leads to changes in the occupied states which are important in the stability of the crystal. 3) The second minimum in the energy, which becomes the absolute minimum in the energy of Zr and in the ω-phase. and 4) The contributions of the various bands to the stabilization of the ω-phase.

One of the useful points of this work is the way the force, calculated from the force theorem, has been utilized to understand the physics. The force is a simple sum over contributions from the eigenstates of the LDF Schrödinger equation. (This is a property of the exact problem, and for the LDF as well, because the force can be derived only from the charge density.) This has been presented in Ref. 103 in the form of a force-weighted density of states, which contains the interesting information that there is a change in sign for the states near the Fermi energy. If the Fermi level is changed, (e.g., a simple picture of alloying), the tendency to distort changes. The authors have used this to give a physical picture for why there is the tendency toward the ω-phase in Zr, less in Nb, and not in Mo. To the knowledge of the present author, this is the only such calculation of a force-weighted density. This could be very useful in calculations for other materials, as well.

<u>Semiconductors</u>

The class of solids with covalent bonding is perhaps the one for which there have been the most extensive recent developments using the LDF. Although there are many informative and useful simple concepts and empirical relations for covalent bonds, an accurate first-principles description of a covalent solid or molecule cannot be achieved using any simple perturbation arguments based upon free atoms or ions. One must have a rather complete theoretical procedure that is capable of describing the non-spherical charge density, the importance of the kinetic energy in the bands, the strong interactions of the electrons with the nuclei or ion cores, and the exchange and correlation in the bonds. The LDF theory using *ab initio* pseudopotentials[74-77] and plane wave basis states has proved capable of putting such calculations on a firm basis and predicting properties of the covalent solids which were far beyond the reach of *ab initio* calculations only a few years ago. The methods are described in Refs. 2, 7, 11, 42, and 89 and a representative fraction of the work using this approach can be found in Refs. 2 – 13, 24, 28 – 32, 58, 89, 90, 96, and 125 – 140.   In this section we will review a few of the results with emphasis upon those in which I have been involved. This is a primary subject of this conference and others will also cover these subjects.[12,13,130]

Let us note at the outset that the other methods which can handle non-sperical charge densities, such as FLAPW,[99] Gaussian[77,95-97] and mixed basis,[48,98,103] can be applied to these problems as well. Although they have not been so extensively applied, a number of important results will be mentioned.

From the point of view of the present author, the step which led to the curent way of approaching these problems was the work of Wendel and Martin.[2] They chose to set up a calculational procdure which was expressly designed to deal with low symmetry situations, such as the calculation of the small energy difference when the atoms are displaced in a phonon mode.[2] The goal was to be able to apply the same ideas in a *unified* way to other low symmetry problems like the interface between two solids, surfaces, and defects. These have now been realized to a large extent. This work has been focussed upon semiconductors because in those systems there are questions of current interest which could not be answered with previous theoretical methods, such as the very stability of the. open covalent structures, the nature of the surface reconstructions,[24] and interface properties,[5] which are not known experimentally. The same ideas are applicable to metals and wide-gap insulators, but the consequences are of greatest interest for the semiconductors.

The next step that made it possible to do accurate realistic calculations was the construction of *ab initio* non-local pseudopotentials.[74-77] This made it worthwhile to do the calculations with great accuracy and compare in minute detail with experiment. Such calculations have been done for many cases, especially by Yin and Cohen.[89,90,126,127] With the introduction of new ideas like the calculation of stress,[11,13] it has been possible to carry the calculations to even greater precision.[11,13,30,32,134] At the same time has proceeded the development and use of techniques[3-10,12,125,135-137] for exploring new phenomena and new physics problems without such need for detailed accuracy. Although this work may not involve the most accurate *ab initio* pseudopotentials, nevertheless it can use simpler potentials to construct *ab initio* theories of the ground state properties of solids.

Crystal Structures of Covalent Solids

The most extensive investigation of the stable structures of the group IV, III-V and II-VI crystals has been carried out by the researchers associated with the University of California at Berkeley. (See Refs. 28, 21, 48, 58, 59, 90, 125–129 and the paper of S. G. Louie in the present proceedings.[130]) These calculations use a plane wave basis set and the expressions for the energy, given by Ihm, Zunger, and Cohen[42] and discussed above. The more recent calculations have used pseudopotentials defined following the prescriptions for *ab initio* potentials given in Ref. 74. A basic result of this work is that the equilbrium structure is diamond, zinc blende, or wartzite. All other structures are much higher in energy in agreement with the basic experimental fact of the pervasiveness of these tetrahedral structures. The total energy calculated for several volumes and fit to a Murnaghan[141] equation of state was used to determine the total energy, equilibrium lattice constant a_0, bulk modulus B, and dB/dP. The result is that the equilbrium lattice constants are within ~1% of experiment bulk moduli ~5–10% from experiment, and dB/dP is reasonable but is only determined to within a factor of two from the numerical fit. This is an important body of work which has had great influence upon the course of calculations for this class of solids.

The accuracy of the density functional calculations for Si has been tested in considerable detail by Hochshuh,[131] who tested the convergence in the number of plane waves, the number of special points in the integration, and the way in which the self-consistent potential is calculated. He found general agreement with other published results, most closely with the work of Nielsen and Martin[111] who used both energy and pressure in the calculation of the equilbrium lattice.

The other structures studied in this work were β-tin, NaCl, CsCl, and simple metallic structures—fcc, bcc, hcp, and sc (simple cubic). For all the semiconductors, these phases were found to be stable only at high pressure. Since pressure has become an important experimental tool in studies of solids, the calculations of the stability of the different phases and the transition pressures is of particular interest. For Si and Ge, the calculations of Yin and Cohen[126] showed that the transitions should be first to the β-tin structure and then at a much higher pressure to a close-packed structure. The transition pressure to the β-tin structure was found to be close to experiment (~100 kbar for Si compared to the most accepted value of ~125 kbar). It is indeed a new level of theoretical ability to calculate the

206

energies of an open covalent structure and a distorted metallic structure with sufficient accuracy that the energy difference agrees with experiment to within $<\sim 0.05$ eV, which is required to achieve the above accuracy in the transition pressure.

The ionic III-IV materials have a particularly fascinating behavior under pressure because there is a competition between an ionic NaCl structure and a metallic (β-tin-like) structure. GaAs is on the borderline as was shown by the calculations of Froyen and Cohen.[31] Although they could not definitively select the order of the structures, their results are important because they show the range of possibilities. Work such as this can form a basis for enabling understanding of what is now a very uncertain experimental situation.

Among the most provocative results are those for carbon. Yin and Cohen[127] showed that for C, unlike Si or Ge, the simple metallic phases are at high energy and prefer to have high volumes, comparable to that of diamond. Thus it requires enormous pressures to force C to be stable in such structures. A very surprising result was found that, among the simple metallic structures, C transforms first to the simple cubic structure. Furthermore this occurs at an extremely high transition pressure of 23 Mbar. This is very different from previous estimates based upon intrapolations of Si and Ge, and it gives a new perspective on the nature of carbon. The reason for the difficulty in forcing carbon to be metallic was attributed to the lack of d-electrons in the occupied states of C.[127]

Other methods of calculation have also been applied to these problems. In particular, the calculations of total energy in localized basis sets has been developed by several groups.[48,77,93,96,133-136] The ASA-LMTO methods have also been extended to covalent materials by adding empty spheres in the void regions.[121,137] Among the accomplishments has been the calculation of lattice constants of materials allowing the core states to vary with volume.[77,96,121,137] The localized bases make it possible to consider more complex problems, such as defects in crystals.[133-136]

<u>Elastic Properties of Covalent Crystals</u>

It is most natural to consider next the elastic properties of the crystal because this is the dependence of the total energy upon the size and shape of the unit cell. As was discussed above, it is a great advantage to use the direct calculation of stress from the "stress theorem." This has been applied to accurate calculations of the elastic properties of the crystals Si, Ge, and GaAs,[11,13] using methods which have general applicability to all solids.

The simplest quantity to calculate is the pressure, in which case the stress theorem reduces to the well-known virial theorem. Since the pressure is the analytic derivative of the energy with respect to volume, this provides two pieces of information on the energy vs. volume curve from one self-consistent calculation. From only a few calculations near the equilibrium lattice, it is possible to obtain the lattice constant (when $P = O$), the bulk modulus $B = -dP/dV$, and the higher derivatives with much greater accuracy than using energy alone. Furthermore, there is no need to fit to a heuristic equation of state. The comparison of energy versus pressure in calculations is given in Ref. 11.

The real value of the stress calculations, however, is for lower symmetry problems, where several independent pieces of information can be found from a single self-consistent calculation. The simplest illustration given in Ref. 11 is the strain ε_{xx} of a Si crystal along a cubic axis. The two independent stresses, σ_{xx} and σ_{yy}, can be calculated for any magnitude of the strain ε_{xx}. In the linear region $C_{11} = \sigma_{xx}/\varepsilon_{xx}$ and $C_{12} = \sigma_{yy}/\varepsilon_{xx}$. The variation of σ with ε gives a set of third and fourth order elastic constants. The second and third order constants agree well with experiment. The fourth order terms are not known experimentally and the results of Ref. 11 are predictions to be tested.

Simultaneous calculation of force and stress are very valuable in cases where the positions of the atoms in the unit cell are not determined by symmetry. This happens for the diamond and zinc blende structures for ε_{xy} strains. This is a well-known problem and a parameter ζ has been defined to quantify the change of the structure under ε_{xy} strains. This problem and its resolution is discussed in detail by Nielsen in this conference.[13] The basic points are:

1) From only two calculations of force and stress one can determine the optic mode frequency (the restoring force fo the internal displacements), the internal strain parameter ζ, and the elastic constant. This may be compared with a very arduous task of fitting multiple parameter curves if one has only the energy, as was done by Harmon, et al.[96]

2) The resulting frequency and elastic constants are in good agreement with experiment.

3) The value of ζ differs significantly from the values determined by analysis of difficult x-ray measurements. Although many tests have been done to examine possible reasons for the difference, there is no explanation and it may suggest a reexamination of the experiments.

Crystal Structures with Degrees of Freedom

In the previous section we have summarized the use of calculations of energy, force, and stress to determine simple crystal structures and the restoring forces and stresses for distorted structures. In this section we turn to the case of crystal structures containing degrees of freedom – the shape of the unit cell or the positions of the atoms in the unit cell. In these cases it is even more useful to have forces and stresses to determine the equilibrium structure (where all forces and stresses vanish). Furthermore, in the search for the equilbrium structure, one automatically finds forces and stresses for distorted structures, i.e., one calculates phonon frequencies, elastic constants, etc.

The general rule is: the more degrees of freedom, the more valuable is the calculation of analytic derivatives for force and stress.

Let us now turn to some very recent results which suffice to show two points: 1) There are interesting surprises in the structures of solids under pressure, and 2) new developments in the theory, such as the calculation of stress, are very valuable tools for the investigation of stability of crystals.

New calculations[32] of the structural energies of Si were stimulated by the discovery of a new metallic phase of Si under pressure.[142,143] The β – tin structure, which is the well-known[144] high pressure phase of Si that has been studied theoretically,[126] is found to be stable only over a small range of pressures. At only a slightly higher pressure ($\sim$130-160 kbar) Si is found to transform again to a new phase with a *simple hexagonal* structure containing one atom in the unit cell. This structure has not been found in any other element at any pressure to the knowledge of the author. However, it has been observed in some alloys of tin and the In-Bi system.[145] Theoretical calculations have ben done very recently by Needs, et al.,[32] to test the stability of this new structure. The calculations have also led to results beyond the possibilities of experiments and have found surprising conclusion that there is near-stability of an entire continuum of structures.

The simple hexagonal (sh) structure is characterized by each atom having 6 neighbors in the basal plane at a distance a and 2 neighbors above and below the plane at a distance c. The coordination is effectively 8 since $(c/a)_{sh} \sim 1$. This structure is very different from all the covalent structures because it has 3 and 4 fold rings of atoms with $60°$ and $90°$ bond angles. The existence of this structure is an important finding for our knowledge of the properties of group IV elements since it establishes that these geometries exist at low energy and may be important in cases such as surfaces, the liquid, etc. The competing β – tin structure has a centered tetragonal unit cell with a 2 atom basis and a single parameter $(c/a)_\beta$. In the experimental structure there are 4 nearest and 2 next nearest neighbors at a slightly greater distance giving ~6-fold coordination. Thus Si exhibits an increase in effective coordination number from $4 \to 6 \to 8$ with increasing pressure.

In Ref. 32 calculations of the total energy and stress were performed with an *ab initio* norm-conserving pseudopotential,[77] using the Wigner approximation to the exchange and correlation energy.[69] The resulting Schrödinger equation was solved self-consistently in momentum space at 40 points in the irreducible part of the Brillouin zone for β – tin and 60 points for the sh structure. This is a large number which was found to be necessary for accurate calulations for both metallic structures. Plane waves of up to 12 Ryd in energy were included in the basis set, which is very similar to the previous work on β-tin. With these parameters the convergence in the energy differences between phases was estimated to be ~0.01 eV per atom.

Calculations were performed at 8 volumes for β-tin and 9 volumes for the sh structure. At each volume two preliminary calculations were performed with different c/a ratios close to the minimum in energy. By linear extrapolation of the anisotropic part of the stress ($\sigma_{zz} - \sigma_{xx}$ where z is in the c direction and x is in the a direction) it was possible to rapidly find the c/a ratio at which the anisotropic stress is zero. This is the desired equilibrium c/a ratio at which the crystal is in equilrium with an externally applied pressure. With only the linear extrapolation, the residual anisotropy in the stress was typically less than 2 kbars. Thus the true equilibrium structure for each phase was found at each volume. The total energies from calculations performed at the predicted minima are plotted in Fig. 4 together with a curve for diamond Si of comparable accuracy.

The total energies were fitted to Murnaghan's equation of state[141] and are shown as solid lines in Fig. 4. The transition pressure from diamond to β – tin of 70 kbars is rather lower than the experimental value of ~125 kbars and also somewhat lower than the value of 99 kbars from the calculations of Ref. 126. The value of 125 kbars is not within the numerical uncertainty of our calculation. This difference may be due to the use of the local density approximation in our calculation. We also note that their is a wide variation in reported transition pressures[143] and that hysteresis and temperature dependence may complicate the comparison between theory and experiment.

The β – tin to sh transition is calculated to occur at ~143 kbars. This is in excellent agreement with the experimental value of between 130 and 160 kbars.[142,143] However, we must note that if the sh curve were shifted in energy by ~0.01 eV per atom relative to β-tin the transition pressure would change by ~50 kbars. One result of these calculations using stress is that the c/a ratio is found systematically and variations in the ratio with pressure can be determined accurately. From these calculations the volume dependence of the c/a ratio is found to be appreciable in the sh structure, and we find that it increases with the application of pressure in agreement with Ref. 142. In comparison to the experimental c/a ratio of 0.94 close to the transition, the theoretical value is very close, c/a ~0.957. On the other

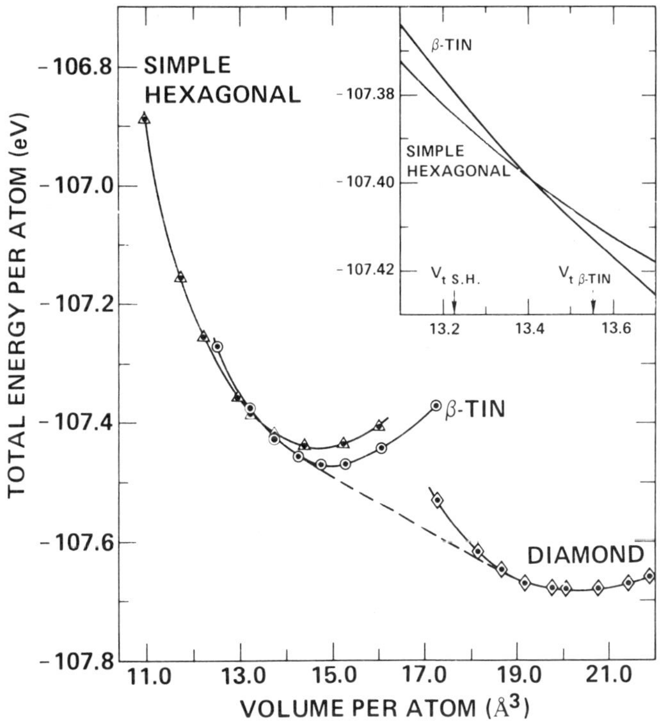

Fig. 4. Calculated phase diagram of silicon including the diamond, β-tin, and the simple hexagonal (sh) structure, which was discovered recently (in high pressure experiments, Refs. 142 and 143). The c/a ratios of both metallic phases are calculated at each point shown by zeroing the shear stress. The β-tin to sh transition pressure is 143 Kbar, in good agreement with experiment. (from Ref. 32)

hand, the c/a ratio does not change so much in the β-tin phase. Close to the transition the theoretical result is $(c/a)_\beta \sim 0.525$, which is below the experimental value of 0.552. This difference may indicate a significant discrepancy with experiment.

The reason for the closeness in energy of the two phases is that there is a simple relation between the β-tin and sh structures, as shown in Ref 145. This is illustrated in Fig. 5. The β-tin stucture is shown in an unconventional way with the c-axis into the page. The two atoms in the cell can be considered as the corner and a face atom. By moving the atoms in the faces of the cell by c/4 in the c direction of β-tin, the lattice has only one atom per cell with vertical rows of atoms as shown by the dashed circles. Finally, only small strains of the cell like those shown are required to make the lattice simple hexagonal. The symmetry of the intermediate structures obtained by such a continuous transition is orthorhombic.[145]

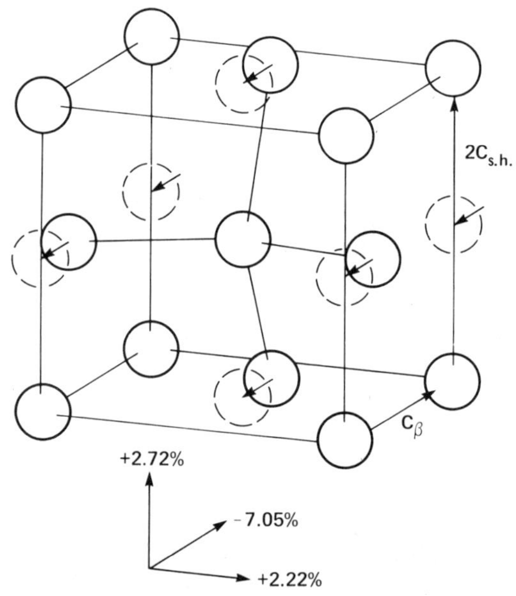

Fig. 5. The relation of the β-tin and simple hexagonal (sh) structures. The solid circles show the β-tin structure. The dashed circles and the changes in dimensions of the cell shown give the sh structure. Calculations along the transition path show that an entire continuum of structures have essentially the same enthalpy at the transition pressure. (from Ref. 32)

One of the most interesting results of Ref. 32 is the exploration of the structures which are not accessible experimentally, for example, along a possible transition path between the stable structures. The total energy was calculated at 5 points along a path between the theoretical transition structures of β-tin and sh, as described in Fig 5. We have arbitrarily chosen each of the 4 parameters describing the path (3 dimensions of the cell and the position of the atoms in the faces of the cell) to change linearly between their values at the end points. Since the transition occurs under conditions of constant external pressure the appropriate energy along the transition path is the sum of the internal energy E and the work done against the external forces $P_t \Delta V$, where P_t is the transition pressure and ΔV is the change in volume. The enthalpy $H = E + P_t \Delta V$ is the same at the end points of a transition between structures at constant pressure. The intermediate values indicate whether there are stable or metastable phases intervening or if there is a large energy barrier to the transition. Along this path H was found to be constant within the convergence of the energy differences between phases of ~0.01 eV per atom.[32] We conclude that if a barrier is present it is smaller than 0.01 eV and that the continuum of structures along the path are potentially stable phases.

In Ref. 32 there are speculations on the possible consequences of the result that near the transition Si is very soft and has almost constant enthalpy for a range of structures between β-tin and simple hexagonal. We have calculated the frequency of the optic phonon shown in Fig. 5 in the β-tin phase to be 87 cm^{-1} at the transition, i.e., a restoring force smaller than that of the optic mode in ordinary Si by a factor of $(520/87)^2 \sim 36$. This low frequency suggests a large temperature dependence of the transition. Futhermore, the soft phonon and the proximity to the transition could enhance superconductivity. Previous studies at high pressure which have shown that Si is a large critical field superconductor with $T_c \sim 7$ K. It would be interesting now to carry out the experiments very near the transition.

The sh structure has not been found in Ge up to applied pressures of ~500 kbars.[142] This large difference between Ge and Si could be explained by only a small change to the energy versus volume curves for Ge from those calculated for Si. If the sh curve was shifted to 0.06 eV higher energy the theoretical transition pressure would shift to over 500 kbars. In Ref. 32 it was suggested that such a shift might be due to the differing amount of d character of the wavefunctions in the two structures in a way similar to the discussion in Ref. 126 for other structures. The d occupation in the sh phase near the transition was found to be ~6% higher than that in β-tin. This difference makes it *less* favorable for Ge to be in the sh structure than for Si. This is because Ge has a filled 3d core the d pseudopotential is more repulsive in Ge than in Si. However, no calculations have been done for Ge in the sh structure.

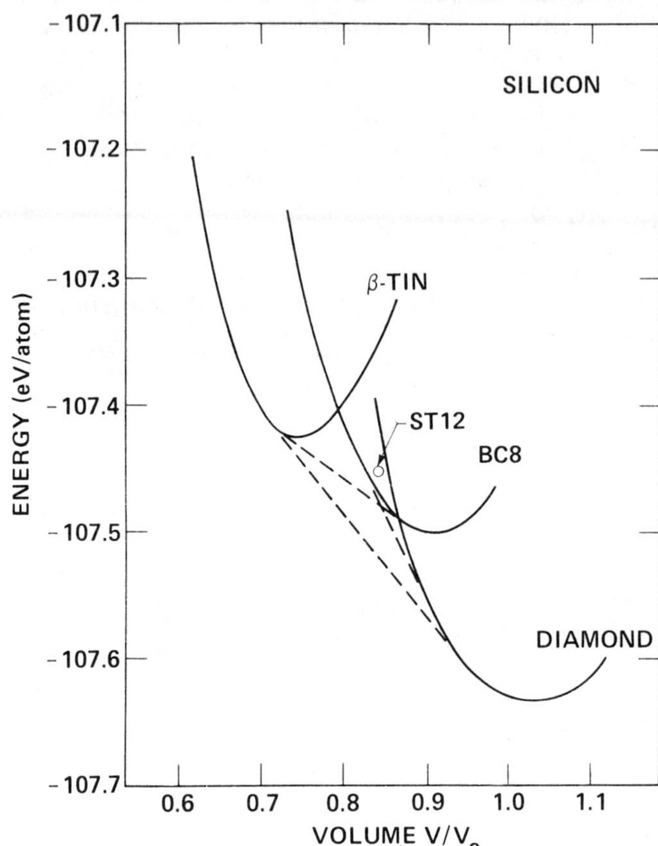

Fig. 6. It is known that Si and Ge can exist in complex tetrhedral structures such as BC8 and ST12 which are created by pressure and are metastable at P = 0. (Refs. 30,144,146). We have carried out calculations that show that these complex phases are indeed metastable at all pressures in Si, since the energy is above the content of diamond – β-tin. The calculated BC-8 structure is determind by zero force and is in good agreement with experiment. (from Ref. 30)

214

The elements Si and Ge are also known to exist in complex crystal structures, BC8 (body-centered cubic with 8 atoms per cell) and ST12 (simple tetragonal with 12 atoms per cell), which can be formed under pressure and are metastable at zero pressure.[144,146] This family of structures has densities intermediate between diamond and the more-closely-packed metallic structures, which is achieved by distorted tetrahedral coordination and connectivity topologically different from diamond. The distortions of the angles are comparable to those thought to be appropriate for the amorphous materials and these structures are often considered as models for the amorphous covalent structures.[147] The phase diagrams of group IV elements therefore have a rich variety of possibilities and a complete theoretical study must include these dense, covalent structures. We have carried out *ab initio* density functional calculations[30] for Si and C in the complex structures compared to the diamond and metallic phases considered in previous work.[89,127] The calculations on these difficult structures were done almost as accurately as those for the simpler structures.[32,89,127] In our calculations for BC8 the structure was fully relaxed, i. e., the internal degree of freedom x was adjusted to zero the force. The more difficult ST12 structure is partially relaxed. Yin has independently found similar results.[148]

In Fig. 6 is shown the energy vs. volume for Si in the diamond, BC8, and β-tin structures, determined by fitting the Murnaghan equation of state[141] to energies calculated at 7, 5, and 8 volumes in each phase, respectively.[30] The equilibrium BC8 structure was found to have $V/V_o = 0.903$ ($V_o = 20.024$ A^3 is the experimental Si volume) and $x = 0.1022$, in satisfactory agreement with the experimental results,[146] $V/V_o = 0.912$ and $x = 0.1003 \mp 0.0008$. The calculated minimum energy for BC8 is 0.13 eV/atom higher than for diamond. The common tangent between diamond and BC8 has a pressure of 107 Kbar, at which point the enthalpies of BC8 and diamond become equal. However, the tangent for the transition to the β-tin structure is at a lower energy and thus we find that the BC8 structure is *metastable* for Si. Although it is not a true equilibrium phase at any pressure, the closeness to the equilibrium line is probably the reason it can be formed under non-equibrium release of pressure.[146] Calculation of the force as a function of the parameter x also leads to a predicted frequency for the Γ_1 Raman mode of 410 cm^{-1} compared to the measured value[149] of 416±2 cm^{-1}. Ref. 30 found BC8 Si to be a semimetal, within the LDA, with bands touching at the H point in the Brillouin zone and a small Fermi surface caused by small overlap of bands.

We have also investigated the stability of C in diamond, BC8 and another structure (MSG) discussed in Ref. 30. In Fig. 7 is shown the enthalpy $H = E + PV$ for C in the region of the transition. Since P ia given by the stress theorem, H may be calculated directly. The main point is that *for*

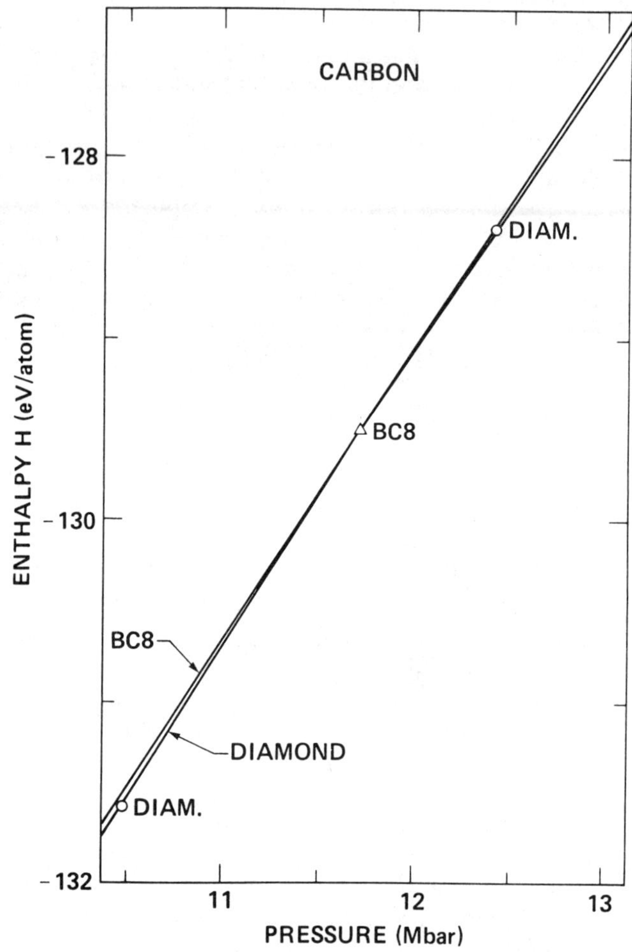

Fig. 7. For carbon the BC8 phase is predicted to be more stable than all previously considered simple metallic phases (Ref. 127). Here we show that calculated enthalpy $H = E + PV$ vs. P gives a transition from diamond to BC8 at ~12 Mbar. This may set the limit of stability of diamond. (from Ref. 30).

216

a range of pressure BC8 carbon is more stable than any of the previously tested simple metallic phases.[127] This is a new upper limit on the stability on diamond of ~12 Mbar which is very high, but much lower than the previous limit of 23 Mbar.[127]

Dynamical Vibrations of Crystals

The final step in the unified theory of the ground electronic state of solids is the calculation of the restoring forces for any of the degrees of freedom of the nuclei in the solid. For small displacements from equilbrium, this may be considered in the harmonic approximation and described completely in terms of the phonon eigenfrequencies and eigenvectors. This is an area of the theory of solids which was studied very much in previous years but suffered from the situation that there was no firm theoretical approach to calculate these properties from first-principles (except for simple metals) and the impossibility to determine uniquely the desired relation of forces to displacements from the experimental information on the eigenfrequencies.

In the last few years the *ab initio* calculations have been carried to the point where the desired relations of forces and displacements can be computed accurately, making no approximations other than the LDF for the electron-exchange and correlation. The basic ideas and a great number of results are described in the paper in this proceedings by Kunc.[12] I will not repeat the information that he is presenting. I will merely summarize a few points and direct the reader to his paper[12] for further information.

The two current ways of carrying out the "direct" *ab initio* calculations of energies or forces have come to be known as the "frozen phonon" method and the force constant method. In the former approach, the atoms are displaced in a given phonon mode and the phonon frequency is calculated either form the energy $E = (1/2)kx^2$ or the force $F = -kx$. In this case both E and F provide exactly equivalent information and $\omega^2 = k/M$, where M is a mode mass. This requires that the eigemode be known in advance. Frozen phonon calculations have prooved very successful and have been applied to many modes in Si,[2,13,58,96,125] Ge,[13,58] GaAs,[4,8,13] C,[48,127] and Se.[132] We note that there is no limitation to small displacements and anharmonic terms have been calculated.[2,4,8,13,48,58]

A different approach has been shown to be capable of providing more extensive information. In the supercell force constant approach,[9,58] the idea is to displace a single atom (on a single plane of atoms) and to carry out a self-consistent calculation to determine the change in electronic charge density at all points due to the displacement of the single atom or plane. By use of the force theorem, from this one calculation can be found many independent forces. If a plane of atoms is displaced, this provides planar force constants which are sufficient to determine an entire phonon dispersion curve—both eigenfrequencies and eigenvectors. Furthermore, it gives directly the forces in real space which is the very quantity needed to describe graphically the nature of the forces and the bonding. Calculations have been carried out to date for GaAs[9,12] Ge,[12] and Si.[58] The most extensive results are for Ge.[12] The case of GaAs is particularly interesting because of the presence of long range forces which appear in infrared activity and the LO-TO splitting.[9,12] These are described in more detail by Kunc.[12]

Just a few years ago there was no theory which could even come close to calculation of the phonon dispersion curves of a crystal like GaAs from first principles. This is indeed a very large improvement in the ability of the theory to make realistic calculations with no input from experiment. There are remaining many problems that go beyond the limited examples given here. These include defects,[133-136] piezoelectric constants,[138] dielectric properties,[139] electronic deformation potentials,[13,140] etc.

VI. CONCLUSIONS

The primary conclusion of my contribution to this NATO Institute is that theoretical calculations have reached the point where there can be accurate *ab initio* calculations of many ground state properties of solids. This is leading to the emergence of a more unified understanding of these properties and to a new relation of theory to experiment – a relation in which theory can be more predictive and in which there can be a more stimulating interaction between theory and experiment. This thesis has been supported by discussion of recent theoretical developments and calculations, and I have attempted to point out a few of the places where new understanding and improvements may lead in the future.

The general theory of the ground state properties of solids was discussed in Section II. The primary results of that section were the formulation of the total energy, forces on atoms, and the macroscopic stress tensor, in terms of the many-body ground state wavefunctions describing the nuclei and electrons. The expressions for force are those well-known as the Hellmann-Feyman theorem. The expressions for stress, derived in Ref. 11, were termed the "stress theorem" since they are generalizations of the virial theorem for pressure. Together, the total energy, force and stress provide complete information on the equation of state of a solid, the relative stability of different phases and (within the adiabatic approximation) the complete dynamics of the vibrations of solids.

Recent developments of the theory were discussed in Sections III and IV, which are devoted respectively to direct solutions of the many-body problems and to the local density functional method. In the former category, an important recent development is the quantum Monte Carlo methods[67,68] which simulate the solution of a finite number of interacting particles. This provides a new approach to the study of matter and is being applied to certain important problems in condensed matter. The method which has been responsible for almost all other recent developments is the density functional method. This is outlined in Section IV, where we have described current methods to solve the self-consistent local density functional equations.

In Section V were described selected results of recent work. The ones which I want to emphasize in particular are: 1) The calculation of the phase diagram of molecular and metallic phases of hydrogen under pressure using the quantum Monte Carlo methods for Fermions.[68] This is an important new development, which has also been applied to certain other light molecules and solids.[116] Extension to other systems is certainly an exciting possibility for future work 2) The calculation of the cohesive energies, lattice constants, and bulk moduli of the entire series of transition metals,[25] which demonstrated convincingly the technical ability to carry out accurate theoretical calculations. This has been extended to some other properties, such as phonons and phase transitions. 3) The calculations on semiconductor solids, in which the author has been involved, demonstrate the ability to treat low symmetry solids with covalent bonding and to calculate a great number of properties in a unified manner. The determination of crystal structures, lattice constants, and phase transitions was considered and our most recent calculations[30,32] on new structures were described in some detail. These are informative examples of calculations on very different structures and the use of forces and stresses to determine equilibrium crystal structures. The calculation of elastic properties using the stress theorem was described briefly. More information and the results of calculations are given by Nielsen in these proceedings.[13] The calculation of phonon frequencies and eignevectors has been carried out for semiconductors to a far greater extent than for other solids. The present theoretical work has shown that many different phonon frequencies can be calculated with general agreement with experiment of order a few percent. The consideration of ionic crystals has led to effective charges, dielectric properties and piezoelectricity, etc. This work is providing entirely new *ab initio* techniques in the theory of lattice vibrations and is described in these proceedings by Kunc.[12] There has not been time or space to discuss the many other important contributions to the calculations of total energy and forces in crystals, surfaces, defects, alloys, molecules, and other structures of matter.

For the future, the continued development of computational methods and ideas in condensed matter physics will certainly lead to new capabilities of theoretical research. There are many possibilities suggested by the current state of the field and, hopefully, many exciting developments which cannot be foreseen at the present time. I believe that even more important than the detailed results will be the continued stimulation of new theoretical concepts and well-chosen experiments. By this synergetic interaction of computational abilities, conceptual developments, and experimental investigations, the "great calculating machine" envisioned by Wigner and Seitz[17] can indeed lead to increased understanding of the nature of condensed matter.

ACKNOWLEDGEMENT

I am very pleased to acknowledge the contributions of my coworkers in this endeavor: R. Biswas, D. J. Chadi, K. Kunc, R. J. Needs, O. H. Nielsen, C. Van de Walle, and H. Wendel. We have benefitted greatly from many useful conversations with W. C. Herring.

This work was supported in part by the Office of Naval Research under ONR Contract No. N00014-82-C-0244.

REFERENCES

1. D. J. Chadi and R. M. Martin, Solid State Commun. 19, 643 (1976).
2. H. Wendel and R. M. Martin, Phys. Rev. Lett. 40, 950 (1978); Phys. Rev. B 19, 5251 (1979).
3. R. M. Martin and K. Kunc, Phys. Rev. B 24, 2081 (1981).
4. K. Kunc and R. M. Martin, Phys. Rev. B 24, 2331 (1981).
5. K. Kunc and R. M. Martin, Phys. Rev. B 24, 3445 (1981).
6. R. M. Martin and K. Kunc, J. de Physique 42, C6-617 (1981) (review).
7. R. M. Martin and K. Kunc, in *Ab Initio Calculation of Phonon Spectra*, ed. by J.T. Devreese (Plenum, 1983), p. 49.
8. K. Kunc and R. M. Martin, in *Ab Initio Calculation of Phonon Spectra*, ed. by J.T. Devreese (Plenum, 1983), p. 65.
9. K. Kunc and R. M. Martin, Phys. Rev. Lett. 48, 406 (1982).
10. K. Kunc and R. M. Martin, in Proc. 16th Int. Conf. on the Physics of Semiconductors, Physica 117B and 118B, 511 (1983) (review).
11. O. H. Nielsen and R. M. Martin, Phys. Rev. Lett. 50, 697 (1983), and to be published.
12. K. Kunc, these proceedings.
13. O. H. Neilsen, these proceedings.
14. M. Born and K. Huang, <u>Dynamical Theory of Crystal Lattices</u> (Oxford University Press, Oxford, 1954).
15. L. J. Sham, Phys. Rev. 188, 1451 (1969).
16. R. M. Pick, M. H. Cohen, and R. M. Martin, Phys Rev. B 1, 910 (1970).

17. E. P. Wigner and F. Seitz, Solid State Physics, ed. by F. Seitz and D. Turnbull (Academic Press, New York, 1955), Vol. 1, p. 97.

18. J. C. Slater, Quantum Theory of Molecules and Solids (McGraw-Hill, New York, 1963, 1967) vol. 1, 3.

19. P. Hohenberg and W. Kohn, Phys. Rev. 136 (1964) B864.

20. W. Kohn and L. J. Sham, Phys. Rev. 140 (1965) A1133, and L. J. Sham and W. Kohn, Phys. Rev. 145 (1966) B561.

21. For reviews of the density functional ideas and techniques, see Theory of the Inhomogenous Electron Gas, ed. by S. Lundquist and N. H. March, (Plenum Press, New York, 1983).

22. Recent work to extend the LDF to excited states includes C. S. Wang and W. E. Pickett, Phys. Rev. Lett. 51, 597 (1983); J. P. Perdew and M. Levy, Phys. Rev. Lett. 51, 1884 (1983); and L. J. Sham and M. Schülter, Phys. Rev. Lett. 51, 1888 (1983).

23. See, for example, the review of high pressure experiments using the diamond cell by A. Jayaraman, Rev. Mod. Phys., 55, 65 (1983).

24. See, for example, the review by J A. Appelbaum and D. R. Hamann, Rev. Mod. Phys. 48, 479 (1976), and recent papers by K. C. Pandey, Phys. Rev. Lett. 47, 1913 (1982); J. E. Northrup and M. L. Cohen, J. Vac. Sci. Technol. 21, 333 (1982); Phys. Rev. Lett. 49, 1349 (1982); and to be published.

25. V. L. Moruzzi, J. F. Janak and A. R. Williams, Calculated Electronic Properties of Metals (Pergamon NY, 1978) and refs. therein

26. A. R. Williams and U. von Barth, in Ref. 19, p. 189.

27. V. Heine, in Solid State Physics, ed by F Seitz and D. Turnbull (Academic Press, New York, 1980), Vol. 35, p. 1.

28. M. L. Cohen, Proc. 15th Int. Conf. on the Physics of Semiconductors, J. Phys. Soc. Japan 49, 13 (1980) (review).

29. See papers in Ab Initio Calculation of Phonon Spectra, ed. by J. T. Devreese (Plenum, 1983).

30. R. Biswas, R. M. Martin, R. Needs, and O. H. Nielsen, to be published in Phys. Rev.

31. S. Froyen and M. L. Cohen, Phys. Rev. 28, 3258 (1983).

32. R. Needs, R. M. Martin, and O. H. Nielsen, to be published.

33. For work on molecules, see, for example, M. M. Franci, et. al., J. Chem. Phys. 77, 3654 (1982) and references therein.

34. O. Gunnarsson, J. Harris, and R. O. Jones, J. Chem. Phys. 67, 3970 (1977).

35. M. M. Goodgame and W. A. Goddard, III, Phys. Rev. Lett. 48, 135 (1982).

36. B. Delley, A. J. Freeman, and D. E. Ellis, Phys. Rev. Lett. 50, 488 (1983).

37. H. Hellmann, Einführung in die Quantenchemie, (Deuticke, Leipzig, 1937), p. 61 and 285.

38. R. P. Feynman, Phys. Rev. 56, 340 (1939) and undergraduate thesis, MIT, unpublished.

39   See the review, "The Force Concept in Chemistry" by B. M. Deb, Rev. Mod. Phys. <u>43</u>, 22 (1973).

40.   P. Pulay, Mol. Phys. <u>17</u>, 197 (1969).

41.   J. C. Slater, J. Chem. Phys. <u>1</u>, 687 (1933).

42.   J. Ihm, A. Zunger, and M. L. Cohen, *J. Phys.* C <u>12</u>, 4409 (1979); erratum <u>13</u>, 3095 (1980).

43.   K.-M. Ho, C.-L. Fu, and B. N. Harmon, Phys. Rev. <u>28</u>, 6687 (1983).

44.   M. Weinert, to be published.

45.   P. Ehrenfest, Z. Phys. <u>45</u>, 455 (1927).

46.   O. K. Anderson, H. L. Skreiver, H. Nohl, and B. Johansson, Pure and Appl. Chem. <u>52</u>, 97 (1979) and A. R. MacIntosh and O. K. Andersen, <u>Electrons of the Fermi Surface</u>, ed. by M. Springford (Cambridge Press, Cambridge, 1979).

47.   D. G. Pettifor, Commun. Phys. <u>1</u>, 141 (1976), J. Chem. Phys. <u>69</u>, 2930 (1978).

48.   J. R. Celikowsky and S. G. Louie, Phys. Rev. B <u>29</u>, 3470 (1984).

49.   R. G. Gordon and Y. S. Kim, J. Chem. Phys. <u>56</u>, 3122 (1972).

50.   V. Fock, Z. Phys. <u>63</u>, 855 (1930).

51.   M. Born, W. Heisenberg and P. Jordan, Z. Phys. <u>35</u>, 557 (1926); E. Schrödinger, Ann. d. Phys. <u>82</u> 265 (1927).

52.   W. Pauli, in *Handbuch der Physik*, Band XXIV, 1. Teil, p. 83-272.

53.   A. G. McLelland, Am. J. Phys. <u>42</u>, 239 (1974); J. Phys. C <u>17</u>, 1 (1984).

54.   For discussion at the second-order perturbation expansions, see the papers in these proceedings by J. T. Devreese, A. Baldereschi, and R. Resta, and Refs. 15 and 16. For review of calculations of harmonic force constants see, for example, S. K. Sinha, Crit. Rev. Solid Stage Sci. <u>3</u>, 273 (1973).

55.   P. E. Van Camp, V. E. Van Doren, and J. T. Devreese, Phys. Rev. Lett. <u>42</u>, 1223 (1929); Phys. Rev. B <u>25</u>, 4270 (1982); and to be published.

56.   C. Falter, M. Selmke, W. Ludwig, and W. Zierau, J. Phys. C <u>17</u>, 21 (1984), and references therein.

57.   D. C. Allan and E. Mele, to be published.

58.   M. T. Yin and M. L. Cohen, Phys. Rev. B <u>25</u>, 4317 (1982).

59.   P. W. Anderson, <u>Basic Notions of Condensed Matter Physics</u>. (Benjamin/Cummings Publishing Co., Menlo Park, CA, 1984).

60.   H. Bethe, Z. Phys. <u>71</u>, 205 (1931).

61.   E. H. Lieb and F. Y. Wu, Phys. Rev. Lett. <u>20</u>, 1445 (1968).

62.   A. M. Tsvelick and P. B. Weigmann, J. Phys. C <u>20</u>, 1707 (1982); N. Andrei, K. Furuya, and J. H. Lowenstein, Rev. Mod. Phys. <u>55</u>, 331 (1983), and references therein.

63.   J. W. Allen and R. M. Martin, Phys. Rev. Lett. <u>49</u>, 1106 (1982).

64.   R. Jullian and R. M. Martin, Phys. Rev. B <u>26</u>, 6173 (1982).

65.   K. G. Wilson, Rev. Mod. Phys. <u>67</u>, 773 (1975).

66. J. E. Hirsch and D. J. Scallapino, Physics Today **36**, 44 (1983) and references therein.
67. C. M. Ceperley, Phys. Rev. B **18**, 3126 (1978); C. M. Ceperley and B. J. Alder, Phys. Rev. Lett. **45**, 566 (1980).
68. C. M. Ceperley and B. J. Alder, Physica **108B**, 875 (1981), and to be published.
69. E. P. Wigner, Phys. Rev. **46**, 1002 (1938).
70. J. Perdew and A. Zunger, Phys. Rev. B **23**, 5048 (1980).
71. L. Hedin and B. I. Lundquist, J. Phys. C **4**, 2064 (1971); O. Gunnarsson and B. I. Lundquist, Phys. Rev. B **13**, 4274 (1976).
72. See for example Ref. 120.
73. A. Zunger, private commun.
74. D. R. Hamann, M. Schlüter and C. Chaing, Phys. Rev. Lett. **43**, 1494 (1979).
75. G. Kerker, J. Phys. C **13**, L189 (1980).
76. G. B. Bachelet, D. R. Hamann and M. Schlüter, Phys. Rev. B **26**, 4199 (1982).
77. G. B. Bachlet, H. S. Greenside, G. A. Baraff, and M. Schluter, Phys. Rev. B **24**, 4745 (1981).
78. A general discussion of band structure methods is given by N. W. Aschroft and N. D. Mermin, Solid State Physics (Saunders College, Philadelphia, 1976); A review is given in Computational Methods in Band Theory, P. M. Marcus, J. F. Janak, and A. R. Williams, editors (Plenom Press, New York, 1971).
79. J. Korringa, Physica **13**, 392 (1947) and W. Kohn and N. Rostocker, Phys. Rev. **94**, 1111 (1954).
80. J. C. Slater, Phys. Rev. **51**, 846 (1937); T. L. Loucks, Augmented Plane Wave Method, (W. A. Benjamin, Menlo Park, CA, 1967).
81. O. K. Andersen, Phys. Rev. B **12**, 3060 (1975).
82. H. L. Skriver, Muffin Tin Orbitals and Electronic Structure (Springer-Verlag, Berlin, 1984) and references therein).
83. A. R. Williams, J. Kübler, and C. D. Gelatt, Jr., Phys. Rev. B **19**, 6094 (1979).
84. D. A. Liberman, Phys. Rev. B **2**, 244 (1970) and B **3**, 2081 (1971).
85. J. F. Janak, Phys. Rev. B **9**, 3985 (1974).
86. M. Ross and K. W. Johnson, Phys. Rev. B **2**, 4709 (1970).
87. F. W. Averill, Phys. Rev. B **6**, 3637 (1972).
88. W. C. Herring, Phys. Rev. **57**, 1169 (1940).
89. M. T. Yin and M. L. Cohen, Phys. Rev. B **25**, 7403 (1982).
90. M. T. Yin and M. L. Cohen, Phys. Rev. B **26**, 3259 (1982).
91. K. Fuchs, Proc. Roy. Soc. London A **151**, 585 (1935).
92. Note that there are many restrictions in the derivation, e. g., there can be no macroscopic field in the energy for the periodic solid.
93. A. Zunger and A. J. Freeman, Phys. Rev. B **15**, 5049 (1977).
94. H. Sambe and R. H. Felton, J. Chem. Phys. **62**, 1122 (1975).

95.   P. Feibelman, J. A. Appelbaum, and D. R. Hamann, Phys. Rev. B $\underline{20}$, 1433 (1979).

96.   B. N. Harmon, W. Weber, and D. R. Hamann, Phys. Rev. B $\underline{25}$, 1109 (1982).

97.   J. C. Boettger and S. B. Trickey, Phys. Rev. B $\underline{29}$, 6425 (1984).

98.   S. G. Louie, K. M. Ho and M. C. Cohen, Phys. Rev. B $\underline{19}$, 1774 (1979).

99.   M. Weinert, E. Wimmer, and A. J. Freeman, Phys. Rev. B $\underline{26}$, 4571 (1982).

100. P. K. Lam and M. L. Cohen, B$\underline{24}$, 4224 (1981).

101. J. Rath and A. J. Freeman, Phys. Rev. B11, 2109 (1975).

102. O. Jepsen, and O. K. Anderson, Solid State Commun. $\underline{9}$, 1763 (1971).

103. K.-M. Ho, C.-L. Fu, and B. N. Harmon, Phys. Rev. $\underline{29}$, 1575 (1984).

104. R. J. Needs and R. M. Martin, to be published.

105. A. Baldereschi, Phys. Rev. B7, 5212 (1973).

106. D. J. Chadi and M. L. Cohen, Phys. Rev. B$\underline{8}$, 5747 (1973).

107. H. J. Monkhorst and J. D. Pack, Phys. Rev. B$\underline{13}$, 5188 (1976).

108. K. M. Ho, J. Ihm, and J. D. Joannopoulos, Phys. Rev. B $\underline{25}$, 4260 (1982).

109. Broyden's method is discussed in Ref. 13.

110. E. Wigner and H. B. Huntington, J. Chem. Phys. $\underline{3}$, 764 (1935).

111. See discussion and references in A. E. Carlsson and N. W. Ashcroft, Phys. Rev. Lett. $\underline{50}$, 1305 (1983).

112. B. I. Min, H. J. F. Jansen and A. J. Freeman, to be published, and references therein.

113. V. T. Rajan and C. W. Woo, Phys. Rev. B $\underline{18}$, 4048 (1978).

114. S. Chakravarty, J. H. Rose, D. Wood, and N. W. Ashcroft, Phys. Rev. B $\underline{24}$, 1624 (1981).

115. E. Ostgaard, Phys. Lett. $\underline{45A}$, 371 (1973).

116. B. J. Alder and D. M. Ceperly, private communication.

117. J. C. Boettger and S. B. Trickey, Phys. Rev. B $\underline{29}$, 6434 (1984).

118. J. F. Janak, V. L. Moruzzi, and A. R. Williams, Phys. Rev. B $\underline{12}$, 1257 (1975).

119. J. A. Morriarity and A. K. McMahon, Phys. Rev. Lett. $\underline{48}$, 809 (1982).

120. A. K. McMahan, M. T. Yin, and M. L. Cohen, Phys. Rev. B $\underline{24}$, 7210 (1981).

121. A. K. McMahan, to be published.

122. See, for example, M. Ross and A. K. McMahon, Phys. Rev. B $\underline{21}$, 1658 (1980), and references therein.

123. V. L. Moruzzi, A. R. Williams and J. F. Janak, Phys. Rev. B $\underline{15}$, 2584 (1977).

124. K.-M. Ho, C.-L. Fu, and B. N. Harmon, to be published.

125. J. Ihm and M. L. Cohen, Phys. Rev. B$\underline{21}$, 1527 (1980); B$\underline{23}$, 1576 (1980).

126. M. T. Yin and M. L. Cohen, Phys. Rev. B$\underline{26}$, 5668 (1982)

127. M. T. Yin and M. L. Cohen, Phys. Rev. B$\underline{24}$, 6121 (1981); Phys. Rev.

Lett. $\underline{50}$, 2006 (1983); Phys. Rev. B$\underline{29}$, 6996 (1984).

128. M. Y. Chou, P. K. Lam and M. L. Cohen, Solid State Comm. $\underline{42}$, 861 (1982).

129. A. Zunger, Phys. Rev. B$\underline{21}$, 4785 (1981).

130. S. G. Louie, these proceedings.

131. E. Holzschuh, Phys. Rev. B$\underline{28}$, 7346 (1983).

132. D. Vanderbilt and J. D. Joannopoulos, Phys. Rev. B$\underline{27}$, 6296, 6302, and 6311 (1983).

133. G. A. Barraff, and M. Schluter, Phys. Rev. Lett. $\underline{41}$, 892 (1978); J. Bernholc, N. O. Lipari, and S. T. Pantelides, Phys. Rev. Lett. $\underline{41}$, 895 (1978).

134. G. A. Barraff, and M. Schluter, and G. Allan, Phys. Rev. Lett. $\underline{50}$, 739 (1984).

135. Y. Bar-Yam and J. D. Joannopoulos, Phys. Rev. Lett. 52, 1129 (1984).

136. R. Car, P. J. Kelly, A. Oshiyama, and S. T. Pantelides, Phys. Rev. Lett. $\underline{52}$, 1814 (1984).

137. D. Glotzel, B. Segall, and O. K. Anderson, Solid State Commun. $\underline{36}$, 403 (1980).

138. J. B. McKitterick, Phys. Rev. B$\underline{28}$, 7384 (1983).

139. K. Kunc and R. Resta, Phys. Rev. Lett. $\underline{51}$, 686 (1983); R. Resta, these proceedings.

140. F. Pollack, these proceedings.

141. F. D. Murnaghan, Proc. Natl. Acad. Sci. U.S.A. $\underline{30}$, 244 (1944).

142. H. Olijnyk, S. K. Sikka and W. B. Holzapfel, Phys. Lett. $\underline{103}$A, 137 (1984).

143. J. Z. Hu and I. L. Spain, to be published.

144. J. Donohue, *The Structures of the Elements* (Wiley, New York 1974).

145. B. C. Giessen, Advan. X – Ray Analysis $\underline{12}$, 23 (1969).

146. R.H. Wentorf, Jr. and J.S. Kaspar, Science $\underline{139}$, 338 (1963);J.S, Kaspar and S.M. Richards, Acta.Cryst. $\underline{17}$, 752 (1964).

147. J.D. Joannopoulos and M.L. Cohen, Solid State Physics $\underline{31}$, 71 (1976).

148. M. T. Yin, to be published.

149. R. J. Kobliska, S.A. Solin, M. Selders, R.K. Chang, R. Alben, M.F. Thorpe and D. Weaire, Phys. Rev. Lett. $\underline{29}$, 725 (1972).

RECENT RESULTS IN SEMICONDUCTOR DYNAMICS

BY AB INITIO 'DIRECT' APPROACH

Karel Kunc

Cavendish Laboratory, University
of Cambridge, England
CNRS, Tour 13, 4 pl. Jussieu
75230 Paris-Cedex 05, France *

1. INTRODUCTION

Of all the physical characteristics of solids, the dynamical
properties give a rather complete description of various aspects of
the electronic ground state: elasticity, phonon frequencies,
dispersion, phase transformations, anharmonicity - they are all
derived from the properties of interatomic bonds. Therefore it
seems only natural to attempt to trace the origins of semiconductor
dynamics back to the behavior of electrons, which ultimately
reduces to electron - electron and electron - nuclei interactions.
These are the starting point of "ab initio" theories.

Microscopic theory of lattice dynamics studies the response of
electron-ion systems to displacements of nuclei; the <u>"direct"</u>
<u>method</u> generally means an approach in which the undistorted crystal
and the crystal with displacements are treated from the very
beginning as two distinct and <u>unrelated</u> systems. The "direct"
treatment of phonons is an alternative to the classical
perturbation approach, in which phonons are viewed as a small
perturbation of the ground state to be treated by linear response
theory. This method, based on the inverse dielectric matrix, is
explained in this Volume in the lecture-notes by J. T. Devreese, R.
Resta and A. Baldereschi.

---

* Permanent address

The direct approach to lattice dynamics starts from the idea of a "frozen phonon" as illustrated in Fig. 1.1: A wave propagating through the crystal imposes upon the amplitudes of the vibrating ions a definite displacement pattern to which the electron clouds instantaneously adapt (adiabatic approximation). It is possible to imagine _freezing_ the vibrations (like a snapshot), and to consider the displaced atoms as a new crystal structure: one of lower symmetry and higher energy. The frequency of the phonon in question (together with anharmonic terms) is then evaluated by comparing the total energy of the crystal in equilibrium with that of the crystal with displacements. The displacement pattern is chosen to correspond to the phonon under consideration.

The idea seems obvious. We could have traced it back to 1970 – although its origins are probably even older: "If we can find the change in energy due to the static distortion of the lattice by the 'frozen' lattice vibration, to second order in _x_, the calculation

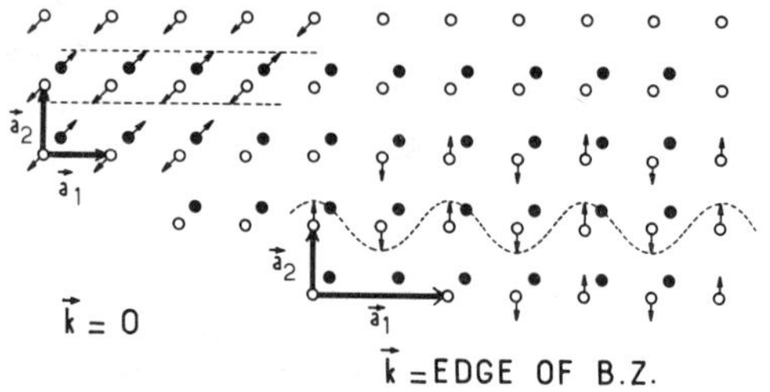

Fig. 1.1:   Schematic representation of a "frozen phonon" corresponding to wavevector k=0 and to a k at an edge of the Brillouin zone. It is possible to imagine freezing the vibrations (like a snapshot) and to consider crystal with displacements as a new crystal structure. The energy of the phonon in question is then evaluated "directly", as the difference of energies of structures with and without displacements. For k=0 the "new" structure has the same translational symmetry as in the undistorted crystal. For k at the Brillouin zone boundary the atoms from neighboring cells vibrate with opposite phases and the translational symmetry is recovered by choosing a unit cell which is double the usual elementary one. This "direct" approach requires the total energy to be reliable to within <0.01 %.

228

of the vibration frequency follows trivially" wrote V. Heine and D. Weaire in Ref. 1 (p. 366). The idea was purely academic in 1970, a kind of "Gedanken-calculation". The evaluation of the total energy of the system of cores and all valence electrons forming a semiconducting crystal - be the configuration displaced or not - is a fundamental conceptual problem which has proved an insurmountable difficulty until recently. Although the main tool we are using presently, the Density Functional[2] (DF), had existed since 1966, it was not realized for some time that the "direct" pathway might be tractable and that the energy could be evaluated with sufficient precision.

The first numerical attempt[3] dates from 1976, but the energy-evaluation based on the empirical tight-binding method proved to be too approximate. It showed, nevertheless, that the frozen phonon energies are not necessarily lost in the <u>roundoff</u> errors of the calculation; it was the physics of the total energy evaluation that was too crude. The essential step was accomplished in 1978 by Wendel and Martin[4] who showed that the Local Density Functional approximation[2] for exchange and correlation leads to correct frozen phonon energies in Silicon. These calculations[4] were not yet self-consistent, and the phonon frequencies were not predicted with extreme precision. However, they were sufficiently exact to confirm that the "direct" approach to dynamical properties of solids was viable: it only needed to be applied with <u>self-consistent</u> charge densities if reasonable accuracy were to be expected. This was done by three different groups independently in 1980-82 (Refs. 5-7). By using a realistic non-local pseudopotential, the energies of principal phonons in Si were predicted very accurately[5], with errors of order 1 - 3 %. The work on GaAs[6] has provided the next step: the treatment of long-range forces, dealing with eigenvectors which are not completely predicted by symmetry, anharmonicity and consequences of its pressure variation. A calculation based on first-principle LCAO[7] (rather than on pseudopotentials) predicted the phonon frequencies in Si with 3 - 10 % errors, and verified the validity of the frozen-core approximation. This justified the use of pseudopotentials in other work. Further references are given in Section 4.5. We note that the <u>elastic</u> properties of <u>metals</u>, involving problems of the same difficulty as frozen phonons, have been studied[8] by DF since 1978.

The above work demonstrated that the LDF approximation is precise enough to be useful in "direct" approaches to lattice dynamics, and that its accuracy is sufficient - in fact extreme. Among all the recent applications of DF to different substances in various situations (atoms, molecules, crystals, surfaces, interfaces, etc.) the frozen phonon calculations have so far been the most stringent test for the validity of the DF method itself.

The frozen phonon approach is limited to only a small number
of high-symmetry phonons. Another direct approach was devised[9-11]
which provides complete phonon dispersion $\omega(\vec{k})$ along an entire
direction $\vec{k}$ - which still has to be a high-symmetry direction. It
is based upon the realization that a monochromatic plane-wave
propagating through a periodic structure does not destroy the
crystal's translational symmetry in the direction perpendicular to
phonon propagation. This method is based on evaluation of force
constants from first principles, by using the Hellmann-Feynman
theorem, and it relies on the same DF as the total energies in the
frozen phonon approach.

The fundamentals of the DF method are discussed in detail
elsewhere in this volume[12]; the present lecture notes start where
those of R. M. Martin ended: the method provides us with 1) energy
of the unit cell; 2) forces on atoms and 3) stress over a unit
volume. Only those details of the method that are specific to our
present applications are summarized in Section 2 . The successive
steps leading to dynamical properties - static equilibrium, frozen
phonon method - are then explained in Sections 3 and 4; the topic
of frozen phonons is treated only briefly in these notes, because
an adequate text already exists: detailed material completing
Section 4 is to be found in Ref. 13.

The central theme of these lectures is the ab initio force
constant method as explained in Section 5 on Germanium and
generalized to polar crystals in Section 7. Section 6, in which
the electric fields in polar crystals are discussed, is not just an
introduction to Section 7 but is also a starting point for a fully
general treatment of dielectric properties within DF, which is the
most recent of the applications of this theory. The underlying
ideas for this approach are briefly recalled in Section 8, and are
discussed in some detail also in the articles by A. Baldereschi and
R. Resta in this volume.

These notes are based on our work on GaAs, and most of the
procedures used in the "direct" method are illustrated on this
prototype. We find the example of a polar crystal more appealing
than the simple Si or Ge, where several interesting problems either
do not exist (macroscopic electric fields, effective charges) or
are trivially dealt with by symmetry (eigenvectors).

2. TOTAL ENERGY CALCULATIONS

A general discussion of the method we are using for the
evaluation of the total energy - the Local Density Functional - is
presented in full detail in another paper in this volume (Ref. 12).
In this Section we summarize the details of its application to our
particular cases of GaAs and Ge. The starting point is the ionic

pseudopotentials of Ga and As[14,15], and the essential approximation consists in choosing the density functional, i.e. in assuming the local form $\alpha \times 3/2 \ (3n(\vec{r})/\pi)^{1/3}$ with $\alpha = 0.8$ for the exchange operator. Dealing with valence electrons only requires the assumption that the cores are rigid (Frozen Core Approximation).

The pseudopotentials of Ga and As are plotted in Fig. 2.1 in reciprocal space. Fig. Al.1 in Appendix shows their variation in direct space, and compares them with the full atomic potentials. For Germanium we choose a potential which is an <u>average</u> of those for Ga and As. [The main advantage of this choice (over the Ge pseudopotential given in Ref. 15 and previously used in Ref. 9) is that the predicted equilibrium lattice constant is close to the experimental value (-1.6 % error). This is crucial for all lattice dynamical calculations. The convergence properties with number of plane waves are expected to be similar to those of GaAs.]

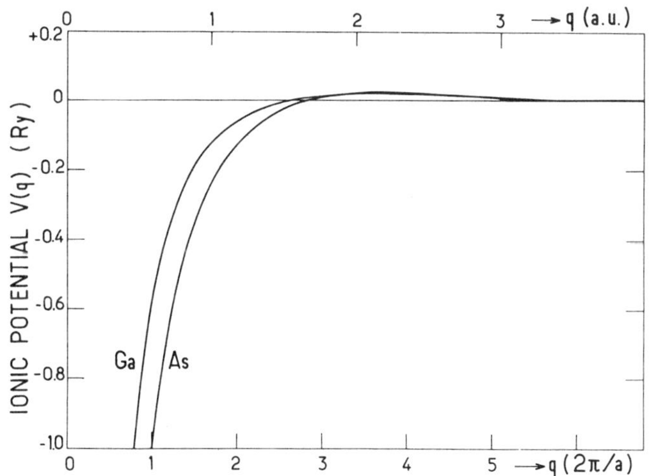

Fig. 2.1:   Local ionic pseudopotentials used throughout this work. Their variation in reciprocal space is given by the formula $V^{ion}(q) = a_1 \ [\cos(a_2 q) + a_3] \ \exp(a_4 q^4)/q^2$ and the coefficients a_1 through a_4 were determined in Ref. 15 from energy bands of GaAs. The alternative units of q displayed at the bottom assume a=5.65 A.

Both pseudopotentials were generated in Refs. 14,15 so as to reproduce the valence bands of GaAs; no adjustment whatsoever was made to fit any phonon properties. Although more precise norm-conserving pseudopotentials have been introduced recently[16-18], with excellent results in subsequent calculations, the work presented in these lectures is based on <u>local potentials</u>. This choice gives us the freedom of working on much larger supercells than the presently available non-local potentials would allow, which is crucial for developing the concept of planar force constants and vibrations of atomic planes – the central theme of these lectures. The disadvantage of this choice is generally lower precision of the numerical results. This turns out to be only a minor problem, however, because the only situation where the non-local potentials convincingly demonstrate their superiority is the TA(X) frozen phonon where the cancellations of different contributions made the calculation particularly sensitive to all aproximations and the slightest imprecision. Our choice of the 'soft-core' <u>local</u> pseudopotential does not affect the understanding of the basic mechanisms behind the electronic screening in lattice vibrations that is the ultimate goal of these lectures.

All the calculations are performed in momentum space and (unless otherwise stated) plane waves with kinetic energy up to 9.15 Ry are included in the expansions of the wave functions. Only those with kinetic energy $\leq$ 2.55 Ry are dealt with exactly, the remaining ones are treated by Löwdin perturbation theory[19] up to second order. This corresponds to approximately 21 + 125 waves when working with the two-atoms cells, 43 + 240 when working with the doubled (four atoms) unit cells, 85 + 500 on quadrupled cells, etc. Two to five special $\vec{k}$-points are used for Brillouin zone integration (corresponding to $(q_1 q_2 q_3) = (222)$ in the notation of Ref. 20) and a fast Fourier transform on a grid of 8 x 8 x 8 to 8 x 8 x 64 points in real space is used to evaluate the Fourier expansions of $n(\vec{r})^{1/3}$. The self-consistency of all components of the potential is achieved, in general, to within less than 0.05 mRy.

3. STATIC EQUILIBRIUM

The first step in any dynamical calculation is to establish static equilibrium: one has to verify, before imposing any displacements, that the ab initio method chosen correctly describes the starting point, and that the undistorted structure is in static equilibrium. This is usually done by calculating the total energy at various lattice constants $\underline{a}$, i.e. for a crystal under (positive or negative) compression (Fig. 3.1). The minimum of the function $E^{tot}(a)$ determines the equilibrium lattice constant a_o, the second derivative at a_o yields the bulk modulus

$$B \equiv -V \, dp/dV \qquad (3.1a)$$

$$= V \, d^2 E^{tot}/dV^2 \qquad (3.1b)$$

and the first question one then asks is how well the calculated a_o and B agree with experiment.

We notice in Fig. 3.1 that the function $E^{tot}(a)$ is rather flat at its minimum, so that a precise determination of the equilibrium

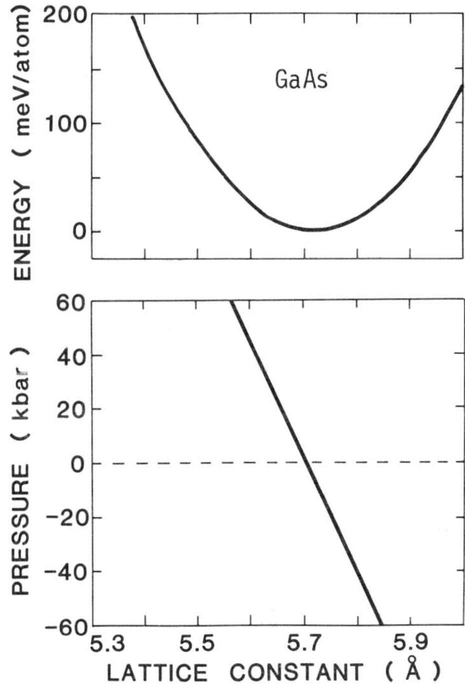

Fig. 3.1: Determination of static equilibrium (lattice constant a_o and bulk modulus B) from calculations of total energy and pressure.

lattice constant requires calculating $E^{tot}(a)$ for a number of values $\underline{a}$; the resulting a_o can still exhibit an appreciable numerical uncertainty. The problem is even more acute in calculation of the second derivative. An elegant way[21] to avoid the numerical difficulties is to use the $\underline{\text{stress theorem}}$ and to find the static equilibrium from variation of $\underline{\text{pressure}}$.

It is explained in Ref. 12 how the self-consistent[22] LDF calculation yields the total energy, forces on atoms and $\underline{\text{average}}$ $\underline{\text{macroscopic stress}}$ σ_{xy}. Knowledge of forces is of little use for establishing static equilibrium because, in crystals as symmetric as 0_h or T_d, the site-symmetry makes the net force on each atom zero, at $\underline{\text{any}}$ volume. But knowledge of the external $\underline{\text{pressure}}$ $p \equiv$ $-1/3$ $(\sigma_{xx} + \sigma_{yy} + \sigma_{zz})$, needed for maintaining the lattice constant at the nonequilibrium value $a \neq a_o$, is very useful because the static equilibrium can also be defined by the relation $p(a) = 0$; the bulk modulus is then found from its definition (3.1a) rather than from (3.1b). We can see from Fig. 3.1 that numerical errors in both finding a root and calculating the slope will be small, and that only a modest number of self-consistent data is needed: even with only two values of $p(a)$ the equilibrium is found precisely enough.

If the calculations shown schematically in Fig. 3.1 were to extend over a large range of lattice constants, the simple linear relation for $p(a)$ would have to be replaced by some $p(V)$ equation of state: the Murnaghan equation[23]

$$E^{tot} = \frac{B_o V}{B_o'} \left[\frac{(V_o/V)^{B_o'}}{B_o' - 1} + 1 \right] + \text{const} \qquad (3.2)$$

is one of the better known ones. An equivalent $p(V)$ expression can then be derived for $p \equiv -dE/dV$. (B' is the pressure derivative of the bulk modulus B_o and V_o the equilibrium volume of the unit cell corresponding to the minimum of energy.)

All Fourier series have to be made $\underline{\text{finite}}$ when performed numerically; the choice of the number of waves used in any calculation is a compromise between the $\underline{\text{computational effort}}$ and the $\underline{\text{errors}}$ caused by the truncation; they are difficult to estimate and one usually resorts to numerical testing. An example of a convergence test for a_o, B – calculated from $p(a)$ as described above (Fig. 3.1) – is shown in Tab. 3.1; the behavior of truncation errors is typical for many similar situations. Whereas the absolute values of both pressure and energy vary considerably with increasing number of waves, the a_o, B calculated from them evolve only slowly. Apparently a large part of the truncation error is systematic. Detailed convergence tests for the different potentials used can be found e.g. in Ref. 24; so far the most detailed study

234

Table 3.1: Determination of static equilibrium in GaAs — test of convergence with number of plane waves. The equilibrium lattice constant a_o is found from the condition $p(a_o) = 0$, the bulk modulus is $B = -V_o(dp/dV)$. (Local potential, Slater exchange $\alpha=0.8$, 2 special points, B from a linear fit, a_o from a quadratic one.)

cutoffs			pressure (kbar)			equilibrium		deviation from the "best" calculated result	
E_1 E_2 (Ry)		typical number of waves	a=5.4 Å	a=5.65 Å	a=5.9 Å	a_o (Å)	B (Mbar)	a_o	B
3	6	26 + 50	+147.044	−14.629	−99.318	5.621	0.925	−1.4 %	−1.1 %
3	9	26 + 115	+141.614	−12.318	−94.247	5.625	0.892	−1.4 %	−4.6 %
6	9	76 + 65	+149.354	−3.300	−85.276	5.643	0.891	−1.0 %	−4.7 %
6	12	76 + 134	+166.895	+13.589	−71.389	5.679	0.910	−0.4 %	−2.7 %
6	18	76 + 315	+181.958	+24.768	−63.019	5.704	0.939	+0.04 %	+0.4 %
9	18	141 + 250	+179.907	+23.404	−63.822	5.701	0.934	−0.02 %	−0.1 %
9	27	141 + 576	+180.490	+23.873	−63.451	5.702	0.935	≡ 0 %	≡ 0 %
Experiment[a]:						5.653	0.769		
% difference:						+0.9 %	+22 %		

(a) Ref. 35

of all systematic errors introduced[25] in the self-consistent algorithms has been done by Holzschuh[25].

After checking the static equilibrium on predictions of a_o, B, the next logical step would be to check the crystal structure. In fact, the function $E^{tot}(V)$ can be recalculated assuming another crystal structure, and the relative positions of different "parabolas" in Fig. 3.1 would then show which of the hypothetical crystal structures has the lowest energy. Such calculations are more than just a check; they become a source of valuable new information. Different structures turn out to be energetically favorable at different pressures, and complete phase diagrams have been obtained in this way for several substances[24,26-30]; new high-pressure structures were predicted even before being found experimentally. This is summarized in the lectures of S. G. Louie in this Volume.

4. FROZEN PHONONS

4.1. The Concept

Having checked the calculated static equilibrium (Section 3), we can start displacing the atoms and evaluate the energy-differences between distorted and undistorted structures. The frequency of the "frozen phonon" defined by the selected displacement pattern is related to its energy by

$$\tfrac{1}{2}\,\omega^2 \sum_\kappa M_\kappa \left|\vec{u}(\kappa)\right|^2 = E(\,u{\neq}0\,) - E(\,u{=}0\,), \qquad (4.0.1)$$

with κ running over all atoms of the unit cell or supercell. We notice that the left-hand side of (4.1.1) is the energy of harmonic oscillator, and thus care has to be taken that on the right-hand side also, only the harmonic part of the total energy increase is calculated.

We are encountering a feature that is inherent in all "direct" approaches to lattice dynamics: as the total energy can only be calculated with finite displacements u, the harmonic terms appear intertwined with the anharmonic contributions, and we have to treat all the expansion terms simultaneously from the very beginning. In addition to the frozen phonon frequencies, we then obtain detailed information on the anharmonicity of the mode in question, data which are difficult to find by other means, both theoretical and experimental. In some cases, it can be verified that the displacement $\vec{u}$ is small enough and does not give rise to any noticeable anharmonic effects. With most displacement patterns, however, the total energy has to be evaluated for several magnitudes of the displacement (typically 5 to 25 values of u): $\Delta E^{tot}(u)$ is then represented by a polynomial in u, truncated at a

236

sufficiently high power of $\underline{u}$, and only its (dominant) harmonic part retained for (4.1.1). Unless the anharmonic terms are the main goal of the investigation, one usually finds that only the <u>lowest order</u> of anharmonicity is numerically important: cubic or quartic, according to the symmetry of the mode.

How large should the finite displacements $\underline{u}$ chosen for calculating $E^{tot}(u)$ be? – The order of magnitude of interest in lattice dynamics is essentially the mean square displacement $\underline{u}_o$, and the range of $\underline{u}$'s in a frozen phonon calculation should include these values – if possible: it is clear that $\underline{u}$ also has to be sufficiently large so that the self-consistent algorithms converge to a state which is (numerically) discernable from the undistorted configuration. In GaAs the typical values are $\bar{u}_o = 0.04$ Å for zero-point vibrations and $\bar{u}_o = 0.1$ Å at room temperature; energies might become numericaly uncertain for $u < 0.01$ Å.

4.2 Zone Center: TO(Γ) mode

The displacement pattern of the TO(Γ) mode in GaAs is shown in Fig. 4.2.1 and the corresponding total energy $E^{tot}(u)$, calculated for various displacements $\underline{u}$, is given in Fig. 4.2.2. The variation

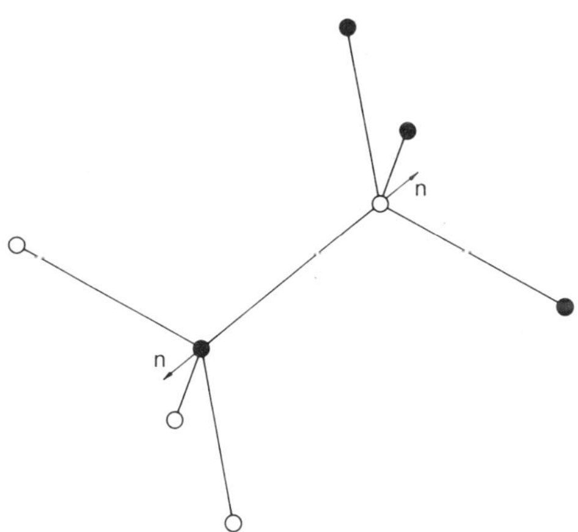

Fig. 4.2.1: Displacement pattern used for evaluation of energies of the TO(Γ) mode. Only the two atoms of the basis are shown displaced, it is understood that all ● are shifted by the same amount, as well as all o in the opposite sense. The "outward" displacements $\underline{u}$, tending to extend the bond, are defined as <u>positive</u> throughout this work.

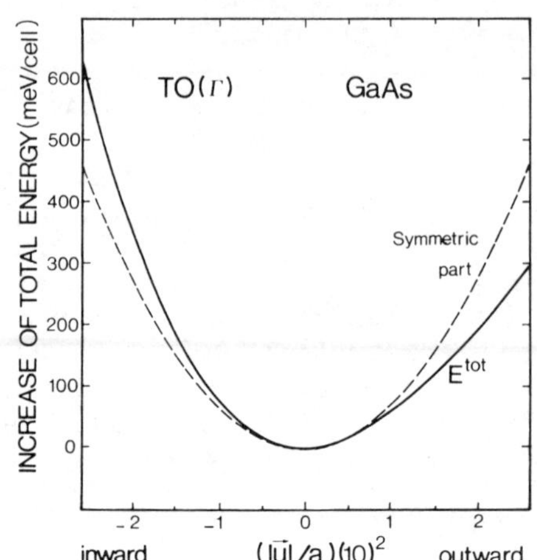

Fig. 4.2.2:   Total energy of the crystal with TO(Γ) mode (Fig. 4.2.1) calculated as a function of displacement $\underline{u}$ (solid line). The considerable asymmetry of the solid curve, showing that it is easier to extend the Ga-As bond (u>0) than to compress it (u<0), reveals a strong cubic anharmonicity of this mode; the "symmetrized" curve $1/2[E(+u)+E(-u)]$ (broken line) is the harmonic part of the phonon energy.

is rather asymmetric - it is easier to extend the bond (u>0) than to compress it (u<0) - which reveals the presence of a strong cubic anharmonicity. By dealing with the "symmetric" part, $1/2[\ E^{tot}(u) + E^{tot}(-u)\ ]$, we can isolate the even-power terms (quadratic, quartic,...), although by the quartic term they are becoming negligible. The resulting harmonic frequency is given in Tab. 4.1: 1.2 % disagreement with experiment.

The most important aspects of the frozen phonon approach have been thoroughly explained already in Ref. 13, and the calculated frequencies of principal frozen phonons in GaAs are given, for illustration, in Tab. 4.1; the reader is recommended to look for all details at pp. 75 - 93 of Ref. 13. The following Section is a summary of the most important conclusions:

238

Table 4.1: Calculated phonon frequencies and the mode Grüneisen
    parameters $\gamma \equiv -(d\omega/\omega) / (dV/V)$ compared with experiment. All
    eigenfrequencies correspond to the <u>harmonic</u> part of the
    expansion $E^{tot}(u)$, i.e. to vibrations with $u \to 0$. All
    frequencies are in THz. (From Ref. 6.)

	TO(Γ)	LO(X)	LA(X)	TO(X)	TA(X)
ν (calc)	8.29	7.55	7.20	7.94	1.87
ν (expt)	8.19[a]	7.22[b]	6.89[c]	7.56[b]	2.41[c]
% diff.	+1.2%	+4.6%	+4.5%	+5.0%	−22%
γ (calc)[d]	1.42	0.91	1.11	1.71	−3.88
γ (expt)	1.39	?	?	1.73	−1.62

(a) Ref. 31 (at 4 K)
(b) Ref. 32 (at 300 K)
(c) Ref. 32 (at 95 K)
(d) Ref. 33

4.3. Summary of Ref. 13

1. The treatment described above can be repeated with modified
lattice constant - and eq. (4.1.1) then yields the pressure
dependence of phonon frequencies (Grüneisen parameter); see Tab.
4.1 for results.

2. For dealing with phonons with $\vec{k}$ at the Brillouin zone
boundary, the unit cell has to be doubled so that the system
recovers translational symmetry.

3. In the LO(X) and LA(X) modes only one sublattice vibrates,
the other is at rest; the higher frequency (LO(X)) was obtained
with Ga vibrating, only the As-sublattice moves in LA(X).

4. When the displacement pattern is not predicted by group
theory, the "direct" approach can be used to find not only
eigenfrequencies but also eigenvectors. The values u(Ga)/u(As) =
0.9 in TO(X) and 1.2 in TA(X) in GaAs were obtained.

5. Comparison of eigendisplacements calculated for TO(X) and
TA(X) in GaAs with the results of 7 phenomenological models (e.g.
"shell model") has shown that their predictions are extremely
scattered, even if the eigenfrequencies $\omega(\vec{k})$ are nearly identical,

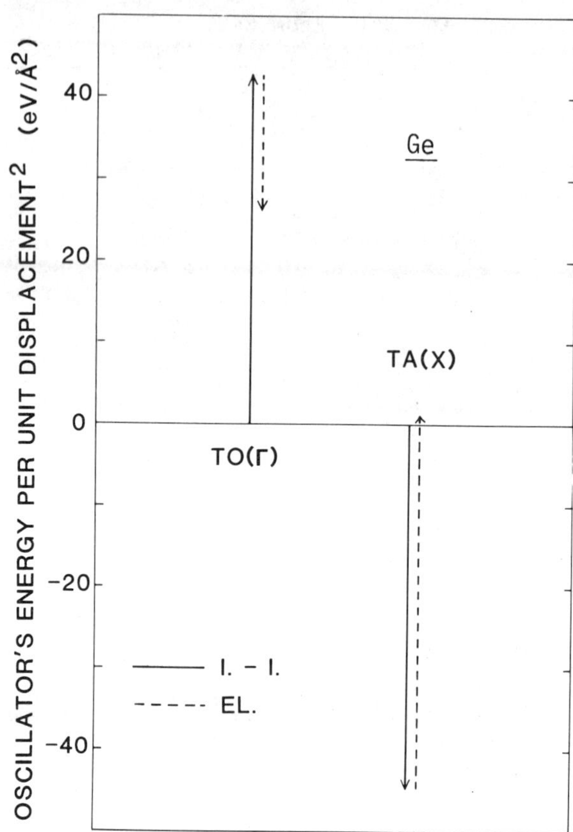

Fig. 4.4.1: Composition of the phonon energies in Ge from the ion-ion energy and "electronic" energy. The ion-ion contribution corresponds to the energy of positive core charges in a uniform negative background (γ^{Ewald}), the electronic term is the sum of all contributions involving electrons: electron-ion energy, kinetic energy of electrons, Hartree and exchange-correlaton terms. In TO(Γ) the ion-ion energy increases with displacement, i.e. the structure is "stabilized" by the ion-ion interactions, which are weakened by the electronic terms; with respect to the TA(X) distortion, the structure is "destabilized" by the electrostatic interactions and "stabilized" by the electronic ones. This behavior of TA(X) is typical of covalent crystals.

240

and in excellent agreement with experiment. One shell model[32] was found to be in good, and one[34] in rough agreement with ab initio calculations.

6. The anharmonic terms can be fairly large: for TO(Γ) in GaAs the cubic and quartic energies at room temperature (u=0.1 Å) are respectively 26 % and 3 % of the harmonic energy. In TA(X) the anharmonic terms at room temperature total 37 % of the harmonic energy.

7. The symmetry of the TA(X) mode forbids odd-power terms in the expansion of $\Delta E^{tot}(u)$, which thus takes the form $Au^2+Bu^4+Cu^6$; the value of C is small but cannot be neglected: as B was found to be negative, the presence of a positive Cu^6 term is dictated by the requirement of structural stability.

8. The relative anharmonicity of the TA(X) mode in GaAs (C/A in the expansion above) rapidly increases with pressure. A soft-mode induced first-order phase transition to orthorhombic structure was predicted.

9. Large relative errors in the calculated TA(X) frequency (see Tab. 4.1) come from nearly complete cancellation between the ion-ion and electronic terms; here we have approached the limits of reliability of the approximations presently used (viz. local potentials and Slater exchange) - but not those of the DF- or frozen phonon method: using the more precise norm-conserving potentials, TA(X) have been obtained with excellent precision on Si, Ge in Ref. 5,37,38. Note that the <u>sign</u> of the Gruneisen parameter for TA(X) is correctly predicted (Tab. 4.1), even if the absolute value is off: TA(X) is a <u>soft</u> mode, its frequency decreases with pressure.

10. The LO(Γ) mode cannot be treated in the same manner as the TO(Γ) frozen phonon, because of the presence of a macroscopic electric field. The method proposed in Ref.6 is the ab initio evaluation of effective charges that determine the LO-TO splitting and will be discussed in detail in Section 6.

4.4. Composition of Phonon Energies in Ge

For the sake of comparison with the alternative force constant approach, which will be explained in Section 5, we have also performed in Ref. 39 a calculation of frozen TO(Γ) and TA(X) phonons in Germanium, using the same class of local potentials (average of the Ga and As ones) and the same number of waves. The result[39] is $\nu = 9.33 \pm 0.01$ THz for TO(Γ) and 1.607 ± 0.02 THz for TA(X) (experiment[36]: 9.12 and 2.40 THz). Various terms composing the phonon energy are displayed graphically in Fig. 4.4.1: the ion-ion energy is largely cancelled by the "electronic energy" (i.e. the sum of the kinetic energy of electrons, Hartree term, exchange-correlation and electron-ion energy).

In TO(Γ) the ion-ion energy <u>increases</u> with displacement, which means that the structure is "stabilized" by the ion-ion contribution and "destabilized" by the electronic one; in TA(X) the terms have opposite signs: the electronic contribution "stabilizes" and the ion-ion energy "destabilizes" the structure. These two types of behavior are representative of other modes[4]: the phonon energy comes systematically from competition between the two types of interactions; most phonons belong to the first category, the TA modes in semiconductors to the second one. (In ordinary metals, all phonons have the composition of the "TO(Γ)-type" of Fig. 4.4.1, with electronic screening even more sizable.) As for the electronic contribution itself, we note that the (largest) electron-ion term is partly cancelled by the Hartree and kinetic energy; the exchange-correlation contribution has the same sign as the electron-ion one and is sizeable only in TO and TA modes at the zone boundary.

4.5. Literature Guide

Several calculations applying the "frozen phonon" concept to different substances have accumulated in the past 3 - 4 years. Besides the work on Si[5] already mentioned in Section 1, phonons in Ge[37,38] and Al[40] were studied as well, with the aid of the norm-conserving pseudopotentials. Ref. 41 demonstrates the use of forces, instead of energies, in frozen phonon calculations. (This will still be explained in Section 5.1.) In the context of static calculations (phase diagrams), S. Froyen evaluated also the TO(Γ) frequency in GaAs[28] (7.84 THz), and energies of the frozen phonons at Γ and X in NaCl[30].

Lattice vibrations in Si were also studied by the first-principle LCAO method in Ref. 7; the frozen core approximation (which is implicitly used in all pseudopotential calculations) was verified by an all-electron calculation[7]. The same technique was later applied to phonons in transition metals (Nb, Mo, Zr) in Ref. 42 and 43. This method, as well as the most important results achieved, are summarized in a recent review[44].

5. FORCES AND PLANAR FORCE CONSTANTS

5.1. Forces and Frozen Phonons

All the reasoning of Section 4 proceeded in terms of energies; this is the quantity most readily accessible by the DF scheme(s).

The theories of small vibrations[45-47] of mechanical systems are
usually formulated in terms of energy-functions: Lagrangians or
classical Hamiltonians. At an even more elementary level - in
systems with one degree of freedom - one gets by with the simplest
picture of <u>restoring force</u>

$$\vec{F} = - k\, \vec{U} \quad ; \quad \omega^2 = k / M \tag{5.1.1}$$

which still holds when $\vec{F}$ turns out to be a <u>generalized</u> force. The
force-picture, appealing in its simplicity, is fully equivalent to
the energy representation in Classical Mechanics.

It is explained elsewhere in this Volume[12] that the
self-consistent charge density and ionic potential corresponding to
a given configuration of atoms not only determine the system's
total energy - but also provide the <u>forces</u>[48,49] acting on all atoms
through the theorem of Hellmann-Feynman (HF). The details of
application in the context of DF, and using the concept of the
pseudopotential, are discussed elsewhere[50,12]; it will be examined
in this Section how we can benefit from this tool and how the
additional information is to be efficiently used.

Returning to the simple displacement pattern of the TO(Γ)
phonon (Fig. 4.2.1), we find that the restoring forces calculated
from self-consistent solutions of the HKS equations are correctly
oriented, as required by (5.1.1), and have values $|\vec{F}(\kappa)| = (7.522 \pm
0.008)\, \sqrt{3} \times 10^{-5}$ dyn for $\vec{u} = 0.005$ (111) $\underline{a}$ and $|\vec{F}(\kappa)| = (10.81 \pm
0.01)\, \sqrt{3} \times 10^{-5}$ dyn for $\vec{u} = -0.005$ (111) $\underline{a}$; as in Section 4.2, the
reason for the two values being considerably different is in a
strong anharmonicity.

We note that in each calculation for every selected $\vec{u}$, we are
getting two values for the force: F(Ga) and F(As); as F(Ga) =
-F(As) must hold, we could estimate the error bars due to an
imperfect self-consistency. In order to eliminate the <u>cubic
anharmonicity</u>, one proceeds as in the case of energies and averages
the two results above; in fact, $E^{tot} - E_0 = Au^2 + Bu^3$ implies $\vec{F} =
-2A\vec{u} - 3B|\vec{u}|^2$, or $|\vec{F}| = 2\,A|\vec{u}| \pm 3\,B\,|\vec{u}|^2_0$, where $\pm$ signs correspond
to outward/inward directions of $\vec{u}$. Taking M as the <u>reduced</u> mass of
the unit cell and $\vec{U}$ the <u>relative</u> displacement of the two atoms, the
frequency $\nu(TO(\Gamma)) = 8.276 \pm 0.004$ THz calculated from eq. (5.1.1)
is about 1 % higher than the value 8.19 THz obtained with the same
number of plane waves through energies - which can be considered as
numerical uncertainty of the algorithm.

Similar calculations in terms of forces are reported in Ref.
41 for modes at Γ, X and L in Si. (No comparision with equivalent
energy calculations was given.) We note that in the above
calculation using forces, the self-consistent solutions
corresponding to the <u>undisplaced</u> configuration was not needed -

because we know in advance that in undisplaced configurations (even beyond equilibrium, e.g. under hydrostatic pressure) the force on each atom is zero, on grounds of tetrahedral symmetry. This certainly represents some economy compared to calculations in terms of energy (two self-consistent calculations instead of three) – which could even become slightly more pronounced in the case of more complicated patterns; e.g. if the problem of coupled modes in Section 4.3.4 is solved in terms of forces, only two self-consistent iterations corresponding to S_1 and S_2 modes are needed, instead of four: neither the equilibrium configuration nor the combined $S_1 \pm S_2$ patterns are of any use. Nevertheless, the advantage is only slight and, as long as the HF theorem is applied in its present form, i.e. to <u>high symmetry</u> displacement patterns which correspond to phonon eigenmodes, it is merely a matter of convenience or taste whether a description in the language of forces or energies is preferred: both paths are <u>physically</u> equivalent, because both are based on the same functional of charge density; eventual (small) differences can only arise from insignificantly different <u>numerical</u> implementation.

In the following Subsections, it will be shown how the HF theorem becomes a powerful tool when applied to <u>low-symmetry</u> displacement patterns.

5.2. Force Constants

In the previous Subsection we did not pay much attention to the spring-constant <u>k</u>, defined as the negative of the restoring force per unit displacement <u>U</u> ; yet this quantity represents the essential information on the simple harmonic oscillator, and is more pertinent to its bonding properties than the eigenfrequency itself. In more complicated systems of atoms and displacements (such as solids), this role is played by the <u>interatomic force constants</u> $\Phi_{\alpha\beta}(\ell\kappa;\ell'\kappa')$ meaning negative of the force acting on atom $(\ell\kappa)$ when atom $(\ell'\kappa')$ is given a unit displacement

$$-F_\alpha(\ell\kappa) = \sum_{\ell'\kappa'\beta} \Phi_{\alpha\beta}(\ell\kappa;\ell'\kappa')\, u_\beta(\ell'\kappa') \qquad (5.2.1)$$

(see e.g. Ref. 51, eq. (2.1.14)). The main task of all ab initio theories of lattice dynamics has been to calculate either force-constants $\Phi_{\alpha\beta}(\ell\kappa;\ell'\kappa')$ – or their Fourier transforms $C(\vec{k})$, the "dynamical matrix", that has for eigenvalues the vibrational frequencies of normal modes. The question we are addressing in this subsection is whether the "direct" approach can lead, in complicated systems, to the force constants Φ, in a way similar to that in which it provided the "spring-constant" <u>k</u> for the simple isolated eigenmode in Section 5.1.

244

Consider the displacement pattern shown in Fig. 5.2.1. In a large supercell, the atom A is displaced by a small amount $u_x(A)$, and self-consistent solutions for charge densities supply the forces on all sites; the one acting on atom B could immediately be translated into $\Phi_{\alpha 1}(B,A)$ - if A were the only displaced atom in the crystal. Unfortunately, the probe atom will "feel" all the other displacements at A', A'', etc. - so that only some meaningless combinations

$$\sum_i \Phi_{\alpha 1}(B,A^i) \tag{5.2.2}$$

might be obtained. The presence of other displaced atoms A^i is dictated by translational symmetry, and cannot be avoided - as long as the <u>supercells</u> are used for performing the electronic calculation. The force constants Φ fall off slowly with distance A-B, so that there is little chance that the supercell could be made sufficiently large to make only one term non-negligible in the sum (5.2.2). Also, attempts at calculating <u>different</u> sums of the type (5.2.2), by choosing different displacement patterns inside the supercell, would lead to systems of linear equations for Φ, which would be numerically ill-defined - even if no roundoff errors were adversely affecting the quality of the calculated data.

The solution was found in calculating <u>meaningful</u> combinations of the type (5.2.2), by conveniently choosing the supercell and the displacements.

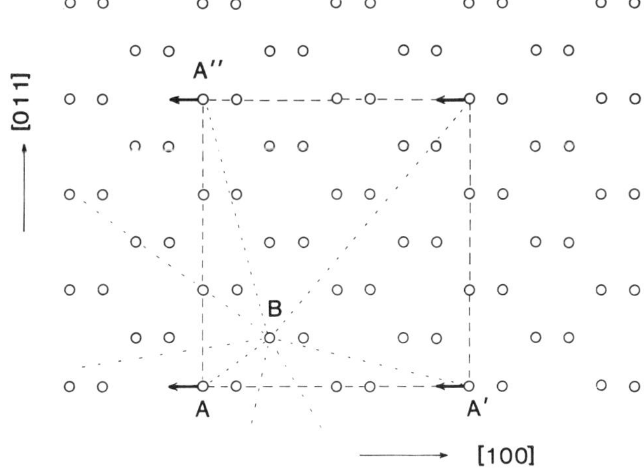

Fig. 5.2.1: The individual inter<u>atomic</u> force constants $\Phi_{\alpha\beta}(\ell\kappa;\ell'\kappa')$ cannot be determined from calculations on supercells.

5.3. Ab initio Planar Force Constants

The ab initio planar force constant approach to lattice dynamics of <u>homopolar</u> semiconductors was proposed in Ref. 9, and generalized to <u>heteropolar</u> compounds in Ref. 10. A recent detailed study[39] applies the method to Germanium. Procedures similar to the original work[9], which was restricted to homopolar substances, were independently applied in Ref. 11 to Silicon. The method is based upon the realization that a monochromatic plane-wave propagating through a periodic structure does <u>not</u> destroy the crystal's translational symmetry in the direction perpendicular to the phonon propagation. This means that in a crystal through which an arbitrary plane wave of wave-vector $\vec{k}$ propagates, the atoms of every plane perpendicular to $\vec{k}$ vibrate in phase; with planes moving as rigid units, the crystal vibrations can be described as those of a linear chain.

Two examples of such chains are given in Fig. 5.3.1. The atomic planes are connected by <u>interplanar</u> force constants k_n and, for symmetric choices of propagation direction, the equations of motion of the "linear chain" lead to a simple secular equation 2 x 2, whatever the range of the forces; different sets $\{k_n\}$ are then needed for different choices of polarization and of propagation direction. For choices less symmetric than $\vec{k} \parallel$ [100] or $\vec{k} \parallel$ [111], the size of the secular determinant may become slightly larger: e.g. 4 x 4 for $\vec{k} \parallel$ [110] and the "mixed" polarization; it cannot exceed, however, 6 x 6 in Ge, even for the off-symmetric choices of $\vec{k}$.

In order to evaluate the planar force constants ab initio, we choose a "long" tetragonal supercell like that in Fig. 5.3.2a, which repeats the elementary unit cell <u>m</u>-times along the direction of propagation [100]; displacement pattern is then chosen such that an <u>entire plane</u> of atoms undergoes a small (longitudinal or transverse) displacement $\vec{u}$. After the self-consistent charge densities are calculated within the DF framework, and the Hellmann-Feynman theorem has supplied the forces on the atoms, the <u>interplanar force constants</u> k_n are found as the negative of the force acting on an atom in the plane <u>n</u> per unit displacement of the plane <u>0</u>:

$$- F(n) = k_n \, u_o \tag{5.3.1a}$$

or, in analogy with the more conventional notation[51],

$$- F(\ell\kappa) = K(\ell\kappa;\ell'\kappa') \, u(\ell'\kappa') . \tag{5.3.1b}$$

Different sets of force-constants are obtained for longitudinal and transverse vibrations; for a different propagation direction we have to start from a different unit cell - such as the one in Fig.

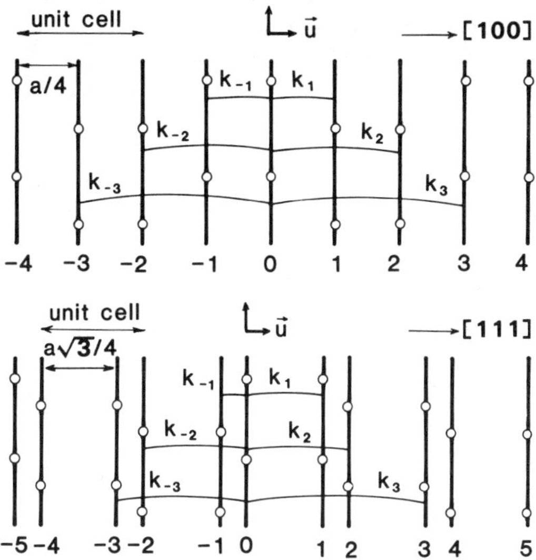

Fig. 5.3.1:   Linear chains representing the vibrations in Ge
propagating in the [100] and [111] directions. The origin
(plane 0) and positive sense of the [100], [111] directions
are chosen as defined in Fig. 5.3.2. The interplanar force
constants k_n connecting the planes determine the phonon
dispersion $\omega(k)$ along the respective propagation directions.
The relation $k_n = k_{-n}$ only holds for even $\underline{n}$ (and in
longitudinal [100] vibrations for any $\underline{n}$).

5.3.2b for [111] – and determine another set of force constants,
which is then pertinent to a different linear 'chain' of planes –
such as the one shown for [111] in Fig. 5.3.1b.

The interplanar force constants k_n or $K(\ell\kappa;\ell'\kappa')$ are
certainly well defined combinations of the interatomic ones
$\Phi_{\alpha\beta}(\ell\kappa;\ell'\kappa')$, because every probe atom "feels" displacements of
every individual atom in the displaced plane. However, we will no
longer be interested in finding the individual interatomic Φ's –
because the harmonic lattice dynamics (along the high-symmetry
directions, at least) can equally well be constructed starting from
the interplanar force constants k_n. That we eventually end up with

a large number of force constants is not a real difficulty: they are easy to handle, whatever their range is, and neither their precise determination, nor figurative physical meaning, critically depend on the smallness of the set, as in the case of the Φ's calculated in the past by numerical fitting of the data. (Anyhow, an equivalent set of Φ's, not generated by any of the well known phenomenological models, would also be rather large, if determined ab initio in direct space.)

The above picture of the "linear chain" is not new. The new ingredient is the faculty to determine the planar force constants ab initio, independently of any phenomenological model for the interactions. By switching from interatomic to interplanar force constants we have achieved two goals: 1) The ab initio evaluation of $\{k_n\}$ requires supercells which are large only in one dimension – and which thus become feasible. 2) Moreover, the $\{k_n\}$ fall off with distance faster than the corresponding $\{\Phi(\ell\kappa;\ell'\kappa')\}^n$. This property will be particularly useful in polar crystal where the $\{k_n\}$ include all electrostatic interactions.

We first illustrate the method on the example of Ge – which is sufficiently simple to demonstrate the basic steps – before addressing a more complicated system, a _polar_ crystal. We follow closely the path used in Ref. 39. In the supercells shown in Figs. 5.3.2a,b, we choose the origin of coordinates so that the atom at (000) is in the plane $\underline{0}$; positive directions [100] or [111] are such that atom at $a/4(\bar{1}11)$ is in the plane $+1$, atoms $a/4(-1,1,-1)$ and $a/4(-1,-1,1)$ in the plane -1. The symmetry of the structure implies that $k_n = k_{-n}$ only for _even_ n, for longitudinal vibrations in [100], however, $k_n = k_{-n}$ holds for any n.

The self-consistent calculations were performed on supercells with $m = 4$ to 8 (quadrupled to octupled) and displacements $|\vec{u}|$ ranging from 0.01 to 0.02 $\underline{a}$. The results from different supercells are consistent within $<0.004 \times 10^5$ dyn/cm and the force constants given in Tab. 5.1 are those obtained on the smallest of the supercells tried (which we expect to limit the roundoff errors and thus to be more reliable); they will be discussed later. Writing down the equations of motion for the linear chains shown schematically in Fig. 5.3.1 leads to 2 x 2 secular equations, which have as solutions the phonon dispersion shown in Fig. 5.3.3; before reaching this point, however, two problems require an adequate treatment: anharmonicity and spatial extent of forces. They are the topics of the next two Subsections.

5.4. Anharmonicity

We have seen already, in the context of frozen phonons, that anharmonic terms appear in the "direct" approach simultaneously

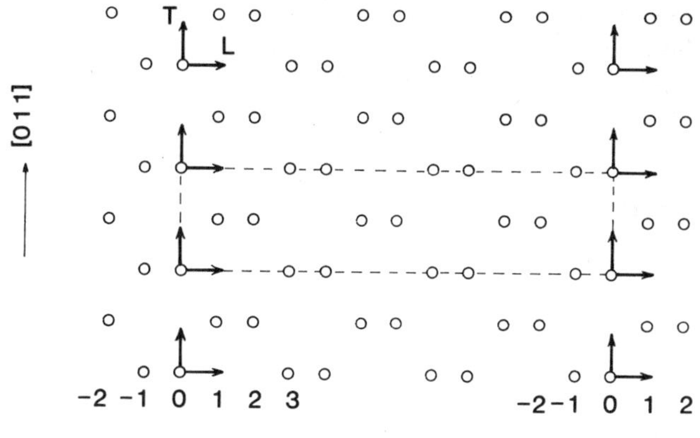

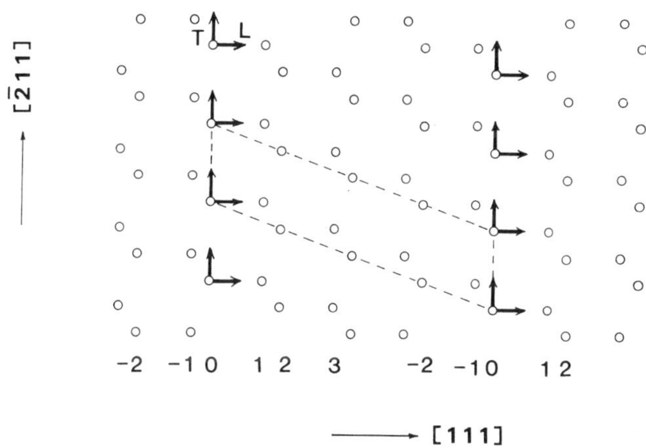

Fig. 5.3.2:   Supercells and displacement patterns used for the calculation of planar force constants in Ge which connect the atomic planes shown in Fig. 5.3.1. The origin of coordinates is chosen so that the atom (000) is in the plane 0; positive directions [100] and [111] are such that atom at a/4(111) is in the plane +1, atoms a/4(-1,1,-1) and a/4(-1,-1,1) in the plane -1.

with the purely harmonic contributions. In the force constant approach the anharmonicity can show up in two different ways: 1) The relation between the force and displacement may not be exacly linear as assumed in eq. (5.3.1); 2) The force may not have exactly the same <u>direction</u> as the displacement.

The first case[1,13] which has its analogy in previous calculations of frozen phonons[6,13], requires calculating with different values of $|u|$, so that the linear part of the variation can be extracted. In the second case, one only retains the projection of $\vec{F}$ onto the direction of $\vec{u}$ - before writing eq. (5.3.1) or before verifying its linearity. Here the anharmonic terms emerge to remind us of the limited validity of the one-dimensional representations: the symmetry-suggested decoupling into longitudinal and transverse modes is only exact within the limits of the harmonic approximation.

It turns out that the longitudinal [100] vibrations in Ge belong to the first category: one finds $F_y = F_z = 0$ at all atomic sites κ, but a perceptible cubic anharmonicity makes $F_x(u_x)$ deviate from linearity:

$$-F = k\, u + \ell\, u^2 + \ldots \qquad\qquad (5.4.1)$$

(Note that although this Section deals mainly with forces, we refer to anharmonicities as is usual in the context of energies - attempting to minimize the risk of confusion: thus cubic and quartic anharmonicities will mean $F \cong u^2$ and $F \cong u^3$ respectively.) In the general case, the contribution $\ell\, u^2$ would be eliminated by repeating calculations twice, with $u_x = +u$ and $u_x = -u$ and by averaging the two results for F_x/u_x. In the [100] longitudinal geometry, however, the symmetry simplifies the task by imposing not only the equality $k_{+n} = k_{-n}$ for the harmonic part at any n, but also by making (to all orders in $\underline{u}$) the force at the site $-n$ $|\vec{F}(\kappa=-n)|$ computed with the displacement $+u$, equal to the $|\vec{F}(\kappa=+n)|$ which is obtained with $-\underline{u}$; consequently only one calculation with a single $\underline{u}$ is sufficient and the harmonic k_n is obtained by averaging "k_n" and "k_{-n}" evaluated naively as F/u.

In a calculation with $|\vec{u}| = 0.01\ \underline{a}$ one finds the harmonic force constants listed in Tab. 5.1; to illustrate the order of magnitude of the anharmonic contributions, the differences $|\,"k_{+1}" - "k_{-1}"\,|/2$ and $|\,"k_{2}" - "k_{-2}"\,|/2$ are, at this displacement, respectively $\cong 3$ % of k_1 and ≤ 2 % of $k_{\pm 2}$.

The transverse [100] vibrations in Ge illustrate the other manifestation of anharmonicity. The displacement $\vec{u} = (0,u,u)$ produces forces with components F_y, F_z equal, but with a small nonzero component F_x which orients $\vec{F}$ slightly off the direction $\vec{u}$: at $\vec{u} = 0.007\ (0,1,1)\ \underline{a}$, the value used in Ref. 39, this "nonparallel" component of $\vec{F}$ is roughly 10 % of the parallel one at first neighbor sites, 3.5 % at 2nd neighbor sites; this contribution was found to be a consequence of cubic anharmonicity. This component is eliminated easily by projecting $\vec{F}$ onto $\vec{u}$ - which, in this case, simply means not considering F_x. After this "projection" is accomplished, there is no cubic anharmonicity left

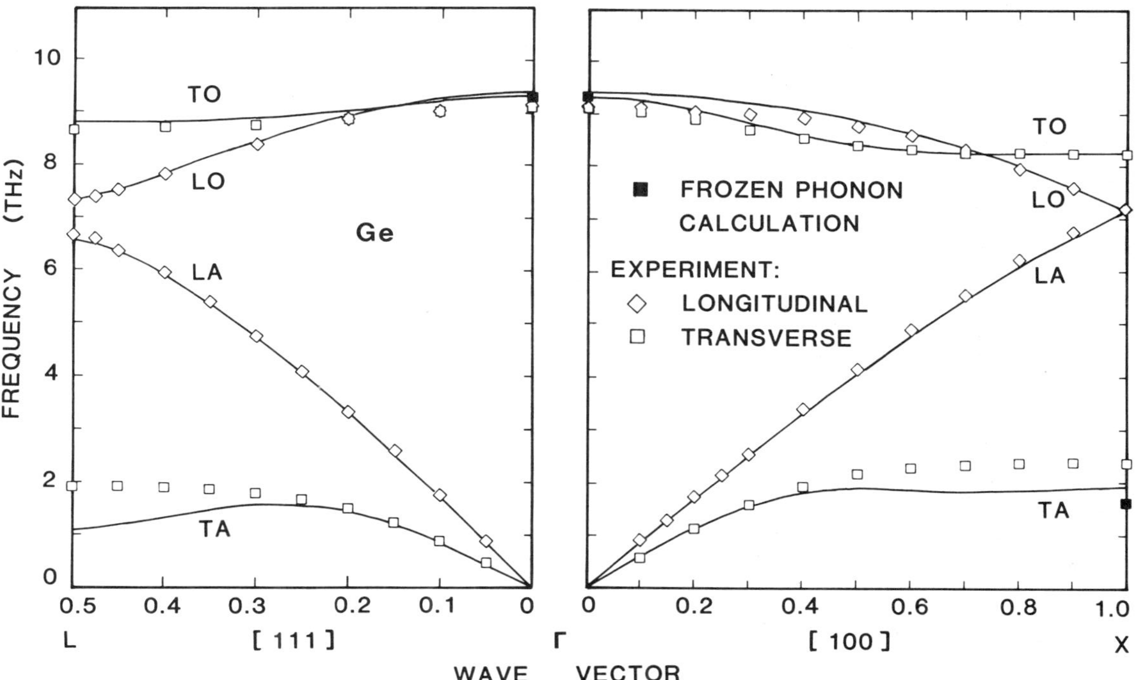

Fig. 5.3.3: Phonon dispersion $\omega(\vec{k})$ in Germanium calculated from the planar force constants determined ab initio. (From Ref. 39.)

because, as found from symmetry, every k_n evaluated as above is exactly the same, whether determined with trial displacement $+u$ or $-u$.

The anharmonicity is markedly stronger in the [111] direction, and not even approximate results could be expected from a single calculation, with only one value of the displacement. The supercell needed for evaluation of the [111] force constants was shown in Fig. 5.3.2b, the immediate environement of the displaced atoms in longitudinal modes is similar to that of the $TO(\Gamma)$ frozen phonon, Fig. 4.2.1. Based on this analogy, two calculations with displacements $\vec{u} = \pm 0.006$ (111) $\underline{a}$ had to be performed (note that the simplification of the kind used with the longitudinal [100] configuration is not possible here). The resulting forces are all parallel with the direction [111] – but rather different for the $+$ and $-$ displacements: with the above $|\vec{u}|$, the cubic term ℓu^2 in eq. (5.4.1) represents 14 % of the harmonic one for k_{+1}, 3% for k_{-1}; as in the $TO(\Gamma)$ context, it is easier to stretch the bonds than to compress them. Averaging the two results obtained with $+$ and $-\vec{u}$ eliminates the cubic anharmonicity and yields the force-constants given in Tab. 5.1; they can, at worst, contain some quartic contributions – which the analogy with the $TO(\Gamma)$ frozen phonon (see Fig. 8 in Ref. 13) suggests to be small.

The most complicated anharmonic terms are found in the transverse [111] configuration, where both "forms" mentioned above appear at the same time: forces are not parallel to the displacement, and their variation is not linear. Two calculations were performed with $\vec{u} = \pm 0.004$ (-2,1,1) $\underline{a}$ and the calculated forces were projected on the 3 perpendicular directions (-2,1,1), (111) and (0,-1,1). Whereas all projections on (0,-1,1) are zero, the "longitudinal" (111) component of the force at the first-neighbor site ($\kappa=+1$) is, with the above $|\vec{u}|$, as much as 46 % of the "transverse" (-2,1,1) one; at the sites $\kappa=-1,\pm2$, the non-parallel components are still respectively 13 % and 20 % of the parallel ones. For obtaining the harmonic force constants, only the (-2,1,1) projection of the force was retained – and (in contrast to the transverse [100] case) a noticeable anharmonic behavior was still found: $\{k_n\}$ determined with displacements $+u$ and $-u$ were averaged, in order to eliminate the cubic contributions – because they still differed considerably: for k_{-1} the cubic term ℓu^2 in (5.4.1) represents 7 % of the harmonic one. It was verified[39] by a calculation with larger $\vec{u}$ that the influence of quartic anharmonicity is negligible.

The force constants summarized in Tab. 5.1 represent the harmonic terms of eq. (5.4.1). It is clear that the anharmonic contributions, whose presence has been considered as an unpleasant perturbation to be eliminated, carry in themselves a considerable amount of valuable information, which still awaits exploration.

5.5. Spatial Extent of Forces

Evaluating $\{k_n\}$ on supercells of the type shown in Fig. 5.3.2 - $\underline{m}$-times repeated elementary cell - we have tacitly assumed that the supercell is "long" enough (i.e. $\underline{m}$ sufficiently large) so as to keep the displaced planes "decoupled": this is certainly important because, in the opposite case, we would risk the determination - as in Section 5.2, Fig. 5.2.1 - of some meaningless combinations of k_n rather than the <u>individual</u> force constants. How big does the supercell need to be in order to separate the displaced planes "sufficiently"? - The answer, obviously, depends on the spatial extent of the forces dealt with and will be different for different sets $\{k_n\}$.

The question "how far do the forces extend?" is the first question one naturally asks when considering lattice dynamics of any solid, by any method. In the context of ab initio calculation of these quantities, the answer to the question of the actual range of forces is getting a new meaning: whereas in phenomenological models the range of forces dealt with was limited "technically", by the number of parameters which the numerical fitting procedures could handle, or by the volume of experimental information that was available or that one wished to incorporate into the model, the ab initio methods allow one, at least in principle, to determine forces acting on arbitrarily distant neighbors. However, as the size of the computations increases very quickly with the size of the supercells used - i.e. with the range of interactions sought - the unrestricted amount of accessible first-hand information is still limited, viz. by the computational effort one is ready to invest. Thus awareness of the spatial extent of interactions helps to choose the supercells which are, for a given purpose, no larger than necessary. For example it is clear that it would be pointless to look for forces that are smaller than the numerical uncertainty of the calculations: it is not necessary to deal with supercells which are much larger than the minimum required by the range of interactions. On the other hand, even if the sets of meaningful force constants are obtained, the convergence properties of one or another physical quantity may be such that the last terms contribute only a little to the quantity in question and are thus beyond practical interest. Finally, it turns out that with certain cutoffs some force-series may converge better than with even larger cutoffs, and certain smaller sizes of supercell can thus be more efficient than other, even larger ones.

The <u>convergence</u> of several quantities depending on planar forces was examined in detail in Ref. 39, from which the main results are summarized in this Subsection. The notion "convergence" expresses mathematically what we intuitively feel as "range of forces" because, strictly speaking, the spatial extent of forces is infinite. Thus rather than asking "how far do the forces extend", a

Table 5.1: Interplanar force constants k_n defined by eq. (5.3.1) as obtained from the ab initio self-consistent calculations in Refs. 39 and 10; the anharmonic contributions of the lowest order are eliminated. The transverse [100] force constants were determined on sextupled supercells, all the others on the quadrupled ones; the last decimal place is not guaranteed. The labels c-c, a-a on the even-n forces correspond to cation-cation and anion-anion interactions. All force constants in 10^5 dyn/cm.

n	Ge [100]		Ge [111]		GaAs [100]			n
	longit.	transv.	longit.	transv.	longit.	transv.		
0	+2.248	+1.806	+2.161	+1.931	+2.069	+1.467	c-c	0
					+2.103	+1.755	a-a	0
+1	-1.050	-1.695	-1.142	-0.101	-0.931	-1.470		+1
-1	-1.050	-0.203	-0.900	-1.899	-0.931	-0.127		-1
±2	-0.083	+0.086	-0.054	+0.048	-0.117	+0.095	c-c	±2
					-0.081	-0.020	a-a	±2
+3	–	-0.035	-0.020	-0.044	+0.029	-0.019		+3
-3	–	-0.094	-0.029	-0.024	+0.029	-0.076		-3
±4	–	+0.029	–	–	–	+0.028	c-c	±4
					–	-0.002	a-a	±4
+5	–	-0.007	–	–	–	-0.018		+5
-5	–	-0.023	–	–	–	-0.006		-5

more rigorous formulation of the question is "what error do we incur by neglecting forces beyond the n-th neighboring plane?". Unlike most DF calculations (and unlike Section 3), the convergence of force-series discussed in this Chapter is not convergence with the number of plane-waves, but with the extent of forces in real space.

For longitudinal forces, a good estimate of their "significant" range is provided by the self-consistent electronic charge density n(r), viz. by modification of its distribution consequent to the displacement of one atomic plane. This is shown in Fig. 5.5.1 for a longitudinal [100] displacement in Ge ; so as to retain only that part of the information that is significant for $\{k_n\}$, the picture was made one-dimensional by averaging n(r) over the remaining two coordinates y, z. The perturbation in charge density produced by the displacement is strong at the first neighbor plane, weak at the second and negligible at the third.

254

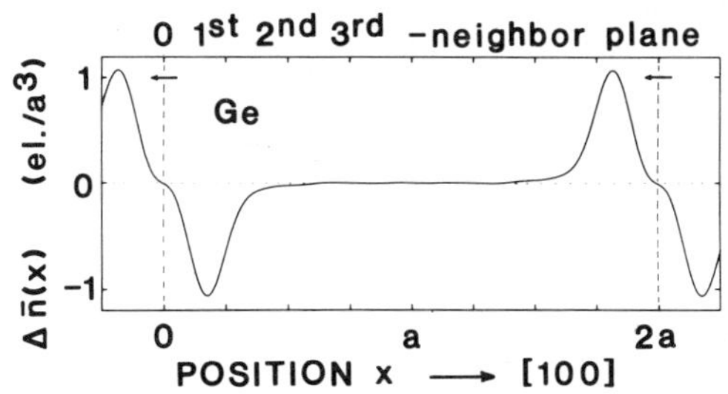

Fig. 5.5.1:   Modification in the electronic charge density of
Ge caused by a longitudinal displacement u= - 0.01 $\underline{a}$; in
electrons/a³. (From Ref. 39.)

This agrees with values of the force constants found ab initio and
given in Tab. 5.1: k_3 was found to be of the same order as the
error margin - thus only k_1 and k_2 are to be considered as
meaningful interactions and the longitudinal [100] dispersion $\omega(\vec{k})$
shown in Fig. 5.3.3 was calculated with k_1 and k_2 only.

Much more interesting is the spatial extent of the [100]
transverse forces, and interactions up to the fifth neighbors were
included for calculation of the phonon dispersion in Fig. 5.3.3. It
is argued in Ref. 39 that the range of these forces is m = 5 ± 2 -
the error margin partly resulting from the present use of less
precise local potentials. The most sensitive part of the transverse
dispersion is the TA-branch: a fairly steep slope at the origin
(elastic constant c_{44}) becomes a flat dispersion near the
Brillouin-zone edge; this behavior is characteristic for covalent
bonding. In Fig. 5.5.2a the frequency ν(TA(X)) calculated by the ab
initio force constant method is plotted versus the range of forces
$\underline{n}$: The frequency of the mode is given in terms of the transverse
force constants as

$$M\omega^2(\text{TA(X)}) = -2 \ (\ k_{-1}+2k_2+k_{+3}+k_{-5}+2k_6+k_{+7}+k_{-9}+...) \quad (5.5.1)$$

and the values ν(n) plotted represent partial sums of the series
(5.5.1), from k_{-1} through $k_{\pm n}$. The mathematically rigorous limit
of (5.5.1) for n $\to \infty$ is known: the calculated frequency of the
frozen phonon which is shown in Fig. 5.5.2a by the black dot. This
alternative approach, proceeding via energies, includes all force
interactions, up to infinite neighbors, within the same physical
approximations.

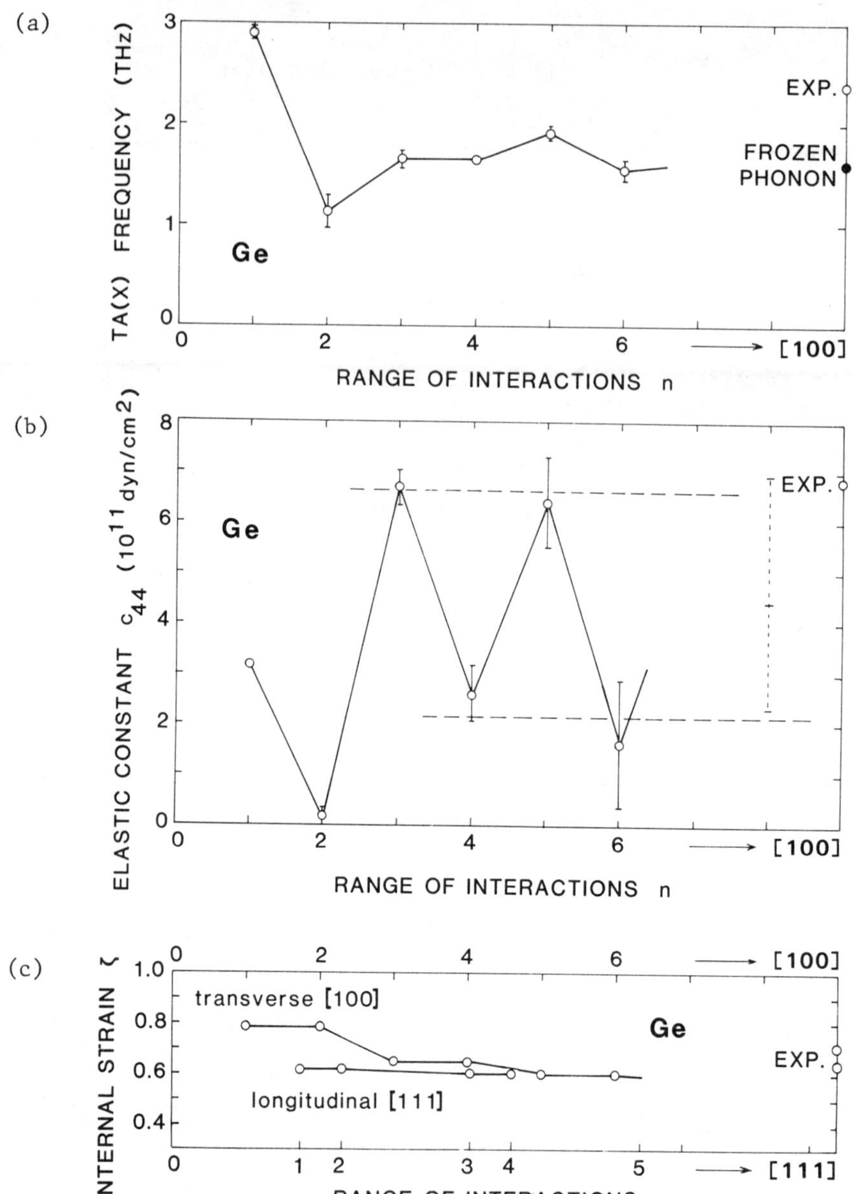

Fig. 5.5.2:   Convergence with spatial extent of forces in Ge:
different quantities are evaluated, with forces included up to
the n-th neighbors. Error bars at $\underline{n}$ correspond to ±0.004 x 10^5
dyn/cm in k_n and k_{-n}.
a) The TA(X) frequency given by eq. (5.5.1). The n → ∞ limit
of the force constant method is the frozen phonon result.

b) The elastic constant $c_{44}(n)$ evaluated from the transverse
[100] forces through eq. (5.5.2). The reason for the zig-zag
convergence is that the expression for c_{44} in terms of force
constants is an alternating series.
c) Internal strain parameter ζ calculated from eqs. (5.5.3)
and (5.6.1).(Error bars are smaller than the circles.)
(From Ref. 39.)

Inspection of Fig. 5.5.2a suggests that the limiting value is
already attained at n=3 - although as many as 6 to 7 neighbor
forces may be needed to "stabilize" its "evolution", i.e. in order
to achieve convergence. (Note that 4th- and 8th-neighbor forces do
not contribute at all to vibrations at the Brillouin-zone edge.)

In Fig. 5.5.3, the sensitive branches TA(Δ) and TA(Λ) are
plotted entirely with different cutoffs n; results corresponding to
even and odd n are plotted separately. At first sight all "partial
sums" with even n have to be eliminated; the dispersions with odd n
shown in Fig. 5.5.3 then allow us to choose between n=3,5, and 7,
values which the Fig. 5.5.2a left us with. - Judged merely by a sum
of least squares, the n=3 and n=5 curves in Fig. 5.5.3 match
experiment to about the same degree. Nevertheless the curve with
n=3 misses an important physical feature: the characteristic
flatness of the TA branch. The value n=5 is thus the smallest n
reproducing this flatness adequately. Including an even more
distant force ($k_{\pm 7}$) appears desirable in order to remove a small,

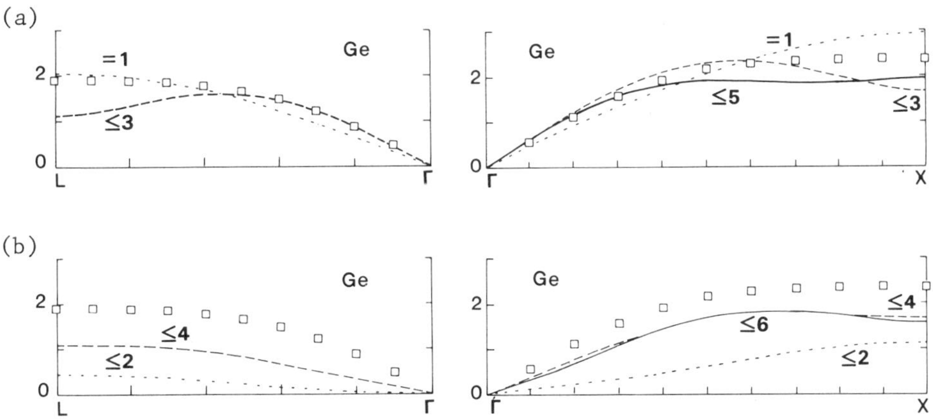

Fig. 5.5.3:   Convergence with spatial extent of forces:
    dispersion $\omega(\vec{k})$ of the most sensitive branches TA(Δ) and
    TA(Λ) in Ge calculated from the planar force constants given
    in Tab. 5.1.
    a) Odd-n cutoffs   b) Even-n cutoffs
    (From Ref. 39.)

barely visible bow on the flat part of the n=5 dispersion along[39]
[100]. Unfortunately, the values found on the octupled supercell
were less reliable and k_{+7} turned out to be of the same order as
the precision of the method. The choice of n=5 is an acceptable
compromise, we conclude that transverse forces extend <u>at least</u> to
the 5th neighbor plane. Limiting the expansion to n=3 might,
nevertheless, be justified in complex situations, if only <u>rough</u>
<u>estimates</u> of the transverse dispersion were acceptable.

A feature which might cause surprise in Fig. 5.5.3 is that
including <u>one</u> more interaction does not necessarily improve the
result: <u>two</u> have to be added. This is a remarkable property of the
slope at Γ, the origin of which can be traced back to the nature of
transverse forces in covalent crystals. The elastic constant c_{44}
giving the slope of the TA(Δ) branch is given[52], in terms of [100]
interplanar transverse force constants, as

$$4a\, c_{44} = - \sum_{n=-\infty}^{+\infty} n^2 k_n + \left[\sum_{\mathrm{odd}\ n} n\, k_n \right]^{+2} \left[\sum_{\mathrm{odd}\ n} k_n \right]^{-1}$$

$$= - \sum_{n} n^2 k_n + \zeta^2 \sum_{\mathrm{odd}\ n} k_n \tag{5.5.2}$$

where

$$\zeta = \left[\sum_{\mathrm{odd}\ n} n\, k_n \right] \left[\sum_{\mathrm{odd}\ n} k_n \right]^{-1} \tag{5.5.3}$$

is the internal strain parameter. All the summations are infinite,
and for finite cutoffs <u>n</u> the expression (5.5.2) is evaluated in
Fig. 5.5.2b; the error bars correspond to an uncertainty of 0.004 x
10^5 dyn/cm in k_n. The conspicuous zig-zag behavior of $c_{44}(n)$ is not
surprising once we notice (see Tab. 5.1) that expression (5.5.2) is
an <u>alternating series</u>.

The physical meaning of the alternation of signs of k_n will be
discussed in Section 5.8, in the meantime we notice that $c_{44}(n)$
with <u>odd</u> n's, n>1, seems to converge to a different value than the
series with <u>n</u> even. This behavior should not be surprising: as
there is no evidence that the series (5.5.2) converges <u>absolutely</u>,
we recall a well known property of alternating series, viz. that
their convergence can be accelerated by an appropriate grouping of
the terms. In spite of the increasingly large roundoff errors in
the partial sums, Fig. 5.5.2b merely suggests that the rate of
convergence of the upper dotted line is more favorable than for the
lower one. It is understood that for large <u>n</u> both of them should
reach the same limit.

The optimal "grouping" of terms in the series (5.5.2), which
is suggested by Figs. 5.5.2b and 5.5.3, is thus <u>in pairs</u>, and with

the cutoff after an <u>odd</u>-neighbor interaction. A clear physical meaning of this rule is not obvious, but it is reminiscent of assembling terms in the expression for Madelung energy in the way that makes them correspond to electrically neutral shells.

In the [111] direction, behavior of the <u>longitudinal</u> [111] dispersion was found in Ref. 39 to be similar to the longitudinal [100], except that forces at 3rd and 4th neighbors are still sizeable; the $\omega(\vec{k})$ shown in Fig. 5.3.3 was calculated with k_{+1} through k_{+3}. The <u>transverse</u> [111] dispersion is similar in all respects to the transverse [100], and the Γ-L part of Fig. 5.5.3 indicates that the same rule for optimizing the convergence will apply: cutoffs after <u>odd</u>-neighbor interactions are to be preferred. The transverse dispersion shown in Fig. 5.3.3 was calculated with k_{+1} through k_{+3}, as well; the sensitive TA branch is still far from being flat and it shows that also in the [111] direction the transverse forces extend <u>beyond</u> the 3rd neighbor plane. The conclusion of Ref. 39, that 3rd neighbor forces are not sufficient to describe adequately the flat TA dispersion, applies in the [111] direction as well; a result with 5th neighbor forces is not available.

The above results, viz. that the transverse forces extend <u>at least</u> over a distance $\cong$ 1.6 <u>a</u> (to the 5th neighbor plane) is in itself not shocking, nevertheless, it brings a certain surprise, because several Valence Force Field models[53,54] have demonstrated that excellent phonon dispersions of Ge, including the flat TA branches, can be obtained with forces effectively extending to the <u>third</u> neighbor [100] plane only, and neglecting all more distant interactions. Also the result of Herman[55] - that at least the 5th neighbor interatomic force constants are needed - is consistent, in spite of first impressions, with other phenomenological models. If we denote by $\{\Phi(n)\}$ the set of interatomic force constants between atom <u>0</u> and the shell of its n-th neighbors, then the [100] interplanar k_{+3} can be written as a linear combination of $\{\Phi(3)\},\{\Phi(5)\},\{\Phi(7)\},\ldots$ and is, according to Herman, "needed"; on the other hand, it follows from geometry that k_{+5} is only composed of $\{\Phi(7)\},\{\Phi(9)\},\ldots$ - and is thus zero when all $\{\Phi(n)\}$, n>5 are zero.

So far, different approaches have been giving clear - but different - answers to a question as simple as "how far?". We note that phenomenological results generally do not carry much weight if any disagreement with ab initio results occurs. The <u>assumption</u> at the spatial extent of forces, which every model treatment starts with, is not necessarily confirmed or disproved when the good agreement with experiment is limited to $\omega(\vec{k})$ <u>only</u>. (We remember from Section 4.3.5 that e.g. models fitted to phonon dispersion only, and matching the experimental values of frequencies excellently, can be completely unreliable in predicting quantities

other than the information fed in.) On the other hand, as the
calculations described use a rather low-quality pseudopotential,
some reservation is not out of place here either. [It can be
speculated that e.g. a "correctly" predicted (i.e. slightly higher)
frozen phonon frequency would lead to a less severe "breaking" of
the initial course of TA, which might then produce a flat branch,
without requiring forces on very distant neighbors.]

The problem of spatial extent is not yet completely settled
because, obviously, the error-margin ± 2 for the range of forces is
still rather large.

5.6. Internal Strain

We have also plotted in Fig. 5.5.2c the convergence of the
internal strain parameter ζ which, in terms of transverse [100]
force constants, is given by eq. (5.5.3), and which can be
equivalently expressed in terms of the longitudinal [111] force
constants as

$$\zeta = \left[\sum_{\text{odd } n} n\, k_n \right] \left[\sum_{\text{odd } n} k_n \right]^{-1} + 1/2 \tag{5.6.1}$$

(see Ref. 52). The calculated value $\zeta = 0.601 \pm 0.002$ – compares
well with recent experiments[57] $\zeta = 0.640 \pm 0.004$, but less well
with older ones[58] $\zeta = 0.71$. It seems to confirm the general trend
emerging recently[59]: all values of ζ calculated ab intio by
different methods[21;22,42,52,59] and on different materials lead
systematically to a lower value than experiment – often by as much
as 20 %.

5.7. Phonon Dispersion in Chains of Atomic Planes

Equations of motion for the linear chains in Fig. 5.3.1 lead,
in $\vec{k}$-space, to the dynamical matrix

$$C(\kappa\kappa'|\vec{k}) = (M_\kappa M_{\kappa'})^{-\frac{1}{2}} \sum_{\ell'} K(\ell\kappa,\ell'\kappa') \, \exp\{-i\vec{k}[\vec{x}(\ell\kappa)-\vec{x}(\ell'\kappa')]\} \tag{5.7.1}$$

which is defined in complete analogy with the general
three-dimensional case (see e.g. Ref. 51, eq. (2.1.58)) – except
that no cartesian indices α,β are needed in one dimension; instead,
different sets of force constants have to be used for different
polarizations (longitudinal, transverse) and different directions
of propagation. $K(\ell\kappa;\ell'\kappa')$ is the alternative (fully general)
notation, eq.(5.3.1b), for the interplanar force constants, they
are in most of this paper referred to as k_n (eq. (5.3.1a)).

260

The complete eigensolutions of the (5.7.1)-defined secular equation, $\omega(\vec{k}j)$ and the normalized eigenvectors $w(\kappa|\vec{k}j)$, are shown in Figs. 5.3.3 and 5.7.1. Internal consistency of the calculations is verified by the fact that four sets of force parameters, which were provided by four independent and completely different calculations, give the LO and TO branches converging to the <u>same</u> LTO(Γ) frequency, degenerate to within 1.3 %. Also, the energy of the TO(Γ)-mode agrees with our prediction using the "frozen phonon" approach (i.e. proceeding via total energies) to within 0.3 % - which illustrates the physical equivalence of both approaches. There is in general a very good agreement with experiment, except in the transverse acoustic branch (0.47 THz (-19 %) at X, 0.80 THz (-42 %) at L) which is due to the low quality of the ionic potentials used <u>(local</u> pseudopotentials) and, possibly, to the Slater Xα form for the exchange; at the L-point, part of the problem comes from the unsufficient range of forces considered. The same difficulty - and roughly the same disagreement - has already been encountered in GaAs, in the frozen phonon context (Section 4.3.9) and will be met again in the force-constant context in Section 7. Of course, this particular mode is the most difficult to reproduce by <u>any</u> ab initio method, cancellations of different contributions to its energy making this phonon energy particularly sensitive to all approximations and round-off errors. On the other hand, the imperfect agreement between the TA(X) frozen phonon calculation (1.61 THz) and the endpoint of the transverse branch (1.93 THz) suggests that the real-space convergence may not yet have been achieved by inclusion of the forces up to the 5th neighbors (see Section 5.5.)

Fig. 5.7.1 shows the second part of the eigen-solutions, the normalized (complex) eigenvectors $w(\kappa|j\vec{k})$, which are written as (real) amplitudes $|w|$ and phases ϕ; the displacement of an atom $(\ell\kappa)$ in the mode (jk) is

$$u(\ell\kappa|j\vec{k}) = M(\kappa)^{-\frac{1}{2}} |w(\kappa|j\vec{k})| \exp\{-i\omega t + i\phi(\kappa|j\vec{k}) + i\vec{k}\vec{x}(\ell\kappa)\} \tag{5.7.2}$$

The choice of phase factors in (5.7.1) and (5.7.2) is the one corresponding to the "C-type" dynamical matrix (eq. (2.1.58) of Ref. 51); the eigenvectors $\vec{w}$ (5.7.2) correspond then to those of eq. (2.1.60) in Ref. 51. We note that in Ge the amplitudes of both atoms have to be equal by symmetry; also the <u>end-points</u> of the dispersion of phases are fully determined by symmetry - but not the variation between them. Closer inspection reveals, however, that the <u>form</u> of the variation is determined (in the non-trivial cases) essentially by the first-neighbor force constants and depends little on interactions with more distant planes; this is a <u>physical</u> fact, not predictable from symmetry considerations.

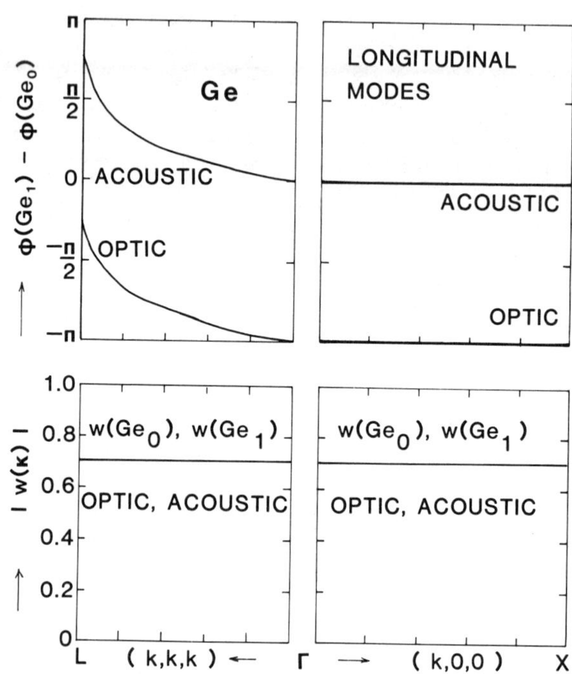

Fig. 5.7.1: Dispersion of amplitudes of eigenmodes in Ge; the
complex quantities are translated into real amplitudes and
phase factors. The actual displacements of a plane ($\ell\kappa$) in a
mode $\vec{k}j$ are then given by eq. (5.7.2). $\phi(Ge_1)$ and $\phi(Ge_0)$ are

(continued)

262

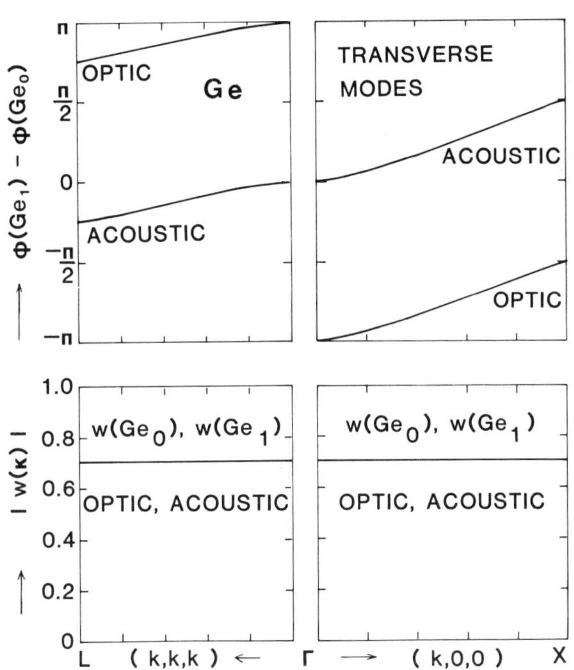

Fig. 5.7.1 (cont.):
  the $\phi(\kappa)$ of the two basis planes in eq. (5.7.2), the numbering Ge_1 and Ge_0 refers to numbering of planes in Fig. 5.3.1. Note that the factor $\vec{k} \cdot [\vec{x}(Ge_1) - \vec{x}(Ge_0)]$ in eq. (5.7.2) [which adds to $\phi(Ge_1) - \phi(Ge_0)$ given in this figure] turns out to be $\pi/2$ and $3\pi/4$ at X and L respectively. (From Ref. 39.)

263

The interplanar force constants defined by eq. (5.3.1) are summarized in Tab. 5.1. As explained in Section 5.4, care was taken to remove from them the anharmonicity of the lowest order allowed by symmetry; only the values used for calculation of phonon dispersion are quoted. Translational invariance of the supercell implies that the restoring force on the displaced plane $-k_o$ is given by

$$-k_o = \sum_{n \neq 0} k_n \qquad (5.8.1)$$

On the other hand, the sum

$$S \equiv \sum_{odd\ n} k_n \qquad (5.8.2)$$

is required to be invariant for all four columns given in Tab. 5.1, because it is proportional to the degenerate LTO(Γ) frequency, independent in Ge of both propagation and polarization directions. The first relation provides a check on the error margins and internal consistency of each column in Tab. 5.1 (respectively $+0.002$, -0.013, -0.002 and -0.001×10^5 dyn/cm.), from the second we can judge the <u>overall consistency</u> of the numerical procedures (S = -2.112, -2.057, -2.091 and -2.068 respectively (2.7 % spread)), which explains the closeness found in previous Subsection where 4 independently calculated branches converged to the same Γ to within 1.3 %. Furthermore, the value of ζ displayed in Fig. 5.5.2c shows that, for this quantity as well, two independent and completely different calculations converge to the same result within 0.5 %.

Fig. 5.8.1 displays graphically the longitudinal forces (negative of the planar force constants) resulting from the displacement of a (100) plane: the "restoring force" $-k_o$, acting on the displaced plane, aims at restoring the undistorted structure, the forces at the first and further neighbors have the same orientation as the displacement, and fall off rapidly. All longitudinal forces originate from <u>cancellation</u> between the Coulomb forces (ion-ion interactions between the unscreened $+4|e|$ cores) and the electronic forces (electron-ion interactions): k_o and k_2 have their signs determined by ion-ion interactions, while k_1 is dominated by the electronic forces. The "right-left" symmetry (k_n = k_{-n}) results from the crystal symmetry of the (distorted) structure: note that every "double bond" on the left of the displaced plane in Fig. 5.8.1 represents exactly the same two "arms" as on the right – but in a plane perpendicular to that of the figure. Consequently, in a longitudinal displacement, every bond pair at the left is stretched + bent (closing the angle) to

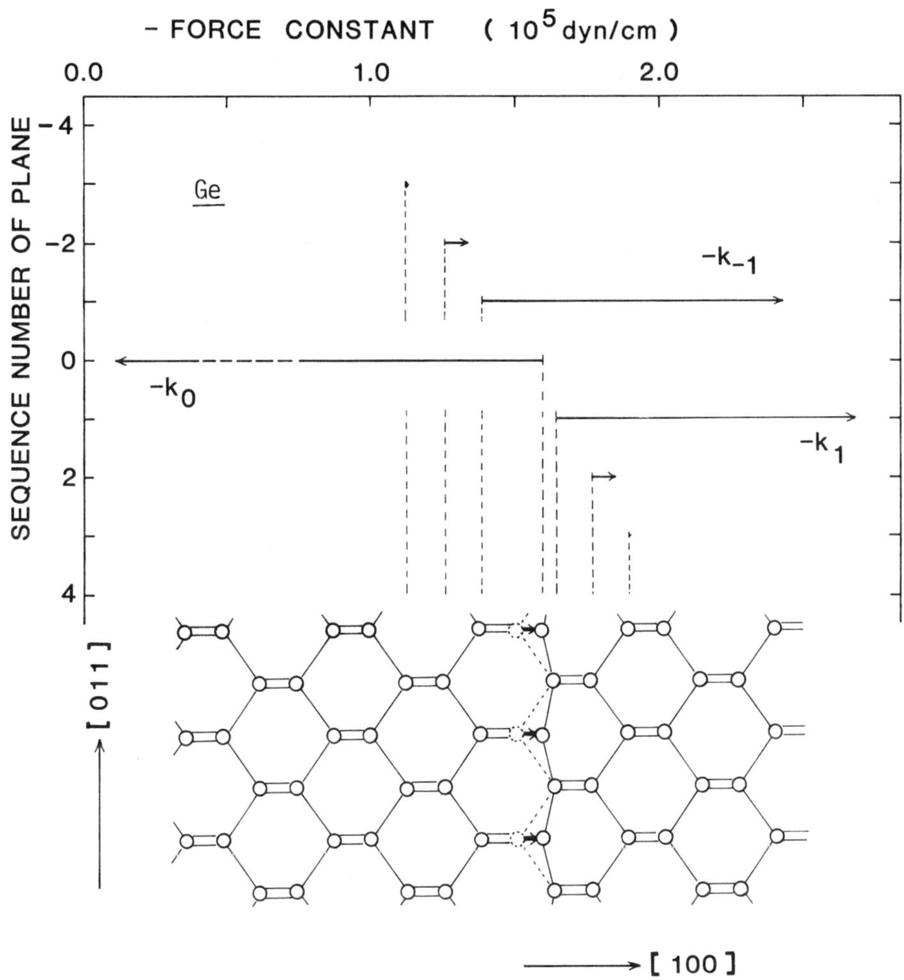

Fig. 5.8.1:   Longitudinal forces in Germanium for the [100] direction. The restoring force $-k_o$ is a sum of all other forces. The interactions are negligible at distances $\geq 3a/4$ (third neighbors). (From Ref. 39.)

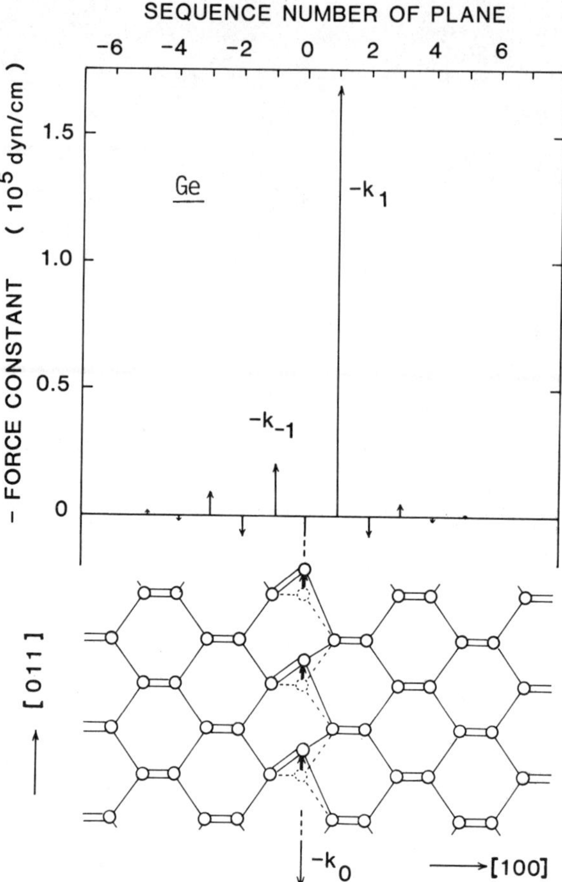

Fig. 5.8.2: Transverse forces in Germanium for the [100] direction. The restoring force $-k_0$ is a sum of all other forces. The fifth neighbor force k_{-5} is still $\cong$ 10 % of the first-neighbor one k_{-1}; the inclusion of forces extending to distant neighbors is essential for reproducing the flat TA branches of phonon dispersion, which are characteristic of covalent crystals. The force constant k_{+1} is due essentially to bond-stretching, the k_{-1} to bond-bending. The alternating signs of the force constants reflect the presence of non-central (angular) interactions (see text.) (From Ref. 39.)

266

exactly the same degree as is compressed + "unbent" (opening the angle) any pair on the right; this produces, in the harmonic approximation, the same response.

Quite a different symmetry is met in the structure with a transverse [100] displacement (Fig. 5.8.2): here the bonds at the left of the displaced plane are mainly bent, whilst those at the right are mainly stretched. The lack of symmetry between bond stretching (k_1) and bond bending (k_{-1}) force constants is the first conspicuous feature of the planar forces for <u>transverse</u> vibrations in [100] (Fig. 5.8.2). The restoring force $-k_0$ acting on the displaced plane is nearly balanced by the strong $-k_1$ (or slightly overbalanced by $-(k_1 + k_{-1})$). Although in absolute value the forces fall off with distance fairly quickly, we note that k_{-3} and k_{-5} are still respectively 46 % and 11 % of k_{-1}; these "medium range" forces are those responsible for the flat TA(X) branches - the characteristic feature of covalent compounds already mentioned; we will return to them again before the end of this Subsection.

The most interesting feature of the forces in Fig. 5.8.2 is the regular alternation of their signs. This alternation, which was responsible for the peculiar convergence properties of c_{44} in Section 5.5, can be explained[60] by <u>angular</u> interactions, which are another typical feature of covalent compounds (see e.g. Ref. 61, Fig. 5). The connection can be seen more clearly if we adopt for a while the language of phenomenological theory, viz. a description in terms of Valence Force Field potentials (see e.g. Refs. 53,54). Fig. 5.8.3 shows a chain of atoms, along the [110] direction, each atom number n is in the n-th plane (100); atom $\underline{0}$ is given a displacement u_0, as in Fig. 5.8.2. As there is a central interaction between atoms 0 and 1, the (bond-bending or bond-stretching) force F_1 follows the direction of u_0. The force on atom 2 would equally follow the direction u_0 if the 0 - 2 interaction were central, $E_A = \phi(r)$. On the contrary, if it is angular - as e.g. the one governed by a three-center potential $E_B = \frac{1}{2} k \theta_1^2$ - the force F_2 will have exactly the opposite sign: as u_0 "opens" the angle θ_1, it increases the potential energy E_B and the force F_{2B} will thus tend to "close" θ_1, in order to restore the equilibrium.

The ab initio determined force constants in Tab. 5.1 or Fig. 5.8.2 leave no doubt as to which of the two potentials mediates the 0 - 2 interaction: they clearly support the idea of <u>angular</u> forces.

The argument applies to more distant neighbors as well: for the 3rd neighbors (Fig. 5.8.3b) and 5th neighbors (not shown), both central and angular mechanisms lead to the same direction of F_3 or F_5, one which agrees with that shown in Fig. 5.8.2. For the 4th neighbors, however (Fig. 5.8.3c), the dilemma central/angular is, again, resolved by Tab. 5.1, viz. in favor of the _angular_ force F_{4B}.

The above arguments and the values of transverse force constants give a great deal of justification to the physical realism of the Valence Force Field descriptions of covalent crystals[53,54] - in spite of the disagreement on the actual range of interactions discussed in Section 5.5.

The _transverse_ [111] forces show a similar behavior to the transverse [100] ones: in particular, the alternation of signs further supports the idea of angular forces; we do not plot them because the present picture, stopping at the 4th neighbors, would necessarily be incomplete.

Finally in Fig. 5.8.4 we have represented the transverse [100] forces of Fig. 5.8.2 as a sum of the ion-ion (Coulomb interactions between the unscreened $+4|e|$ cores) and "electronic" terms (electron-ion, Hartree and exchange-correlation); in order to visualize also the barely visible distant terms, all forces are weighted by n^2. Two features are worth noting in Fig. 5.8.4: 1) The unscreened core-core interactions vanish beyond the 4th neighbors, and their contribution to k_{+4} is already rather small. 2) The force k_{+1} shows the "composition" typical of bond-stretching forces, k_{-1} that of bond-bending ones: a strong ion-ion interaction, respectively positive and negative, is opposed by a strong electronic contribution. This is the same "configuration" as found in the analogous energy diagrams (Fig. 4.4.1) which similarly reconstruct the energies of the $TO(\Gamma)$ and $TA(X)$ frozen phonons. In that context, the ion-ion energy respectively "stabilizes" and "destabilizes" the crystal structure[4], whereas the electronic contributions provide compensation, allowing a stable structure to be achieved in both cases: The crystal energy _increases_ when atoms are displaced (see Section 4.4). The force k_{+3} is similar to k_{-1}, while the other ones show various intermediate compositions.

5.9. Defects in Solids
================

We have explained in Section 5.2 why the individual interatomic force constants $\Phi_{\alpha\beta}(\ell\kappa;\ell'\kappa')$ _cannot_ be determined on supercells of reasonable size: the force constants fall off with distance rather slowly so that _several_ displaced atoms contribute to forces "detected" on other sites (Fig. 5.2.1). There is, however, a situation where decoupling of the contributions

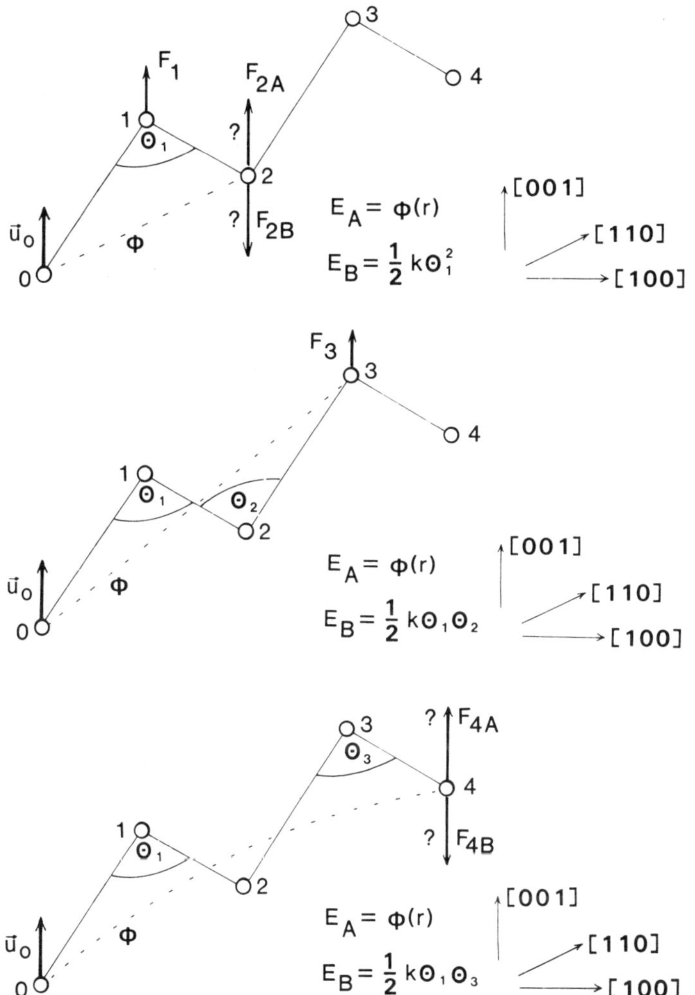

Fig. 5.8.3: Alternating signs of the transverse [100] forces (Fig. 5.8.2) reveal the presence of non-central interactions, which are typical of covalent materials; this is explained in terms of angular Valence Force Field potentials (see text). (From Ref. 39.)

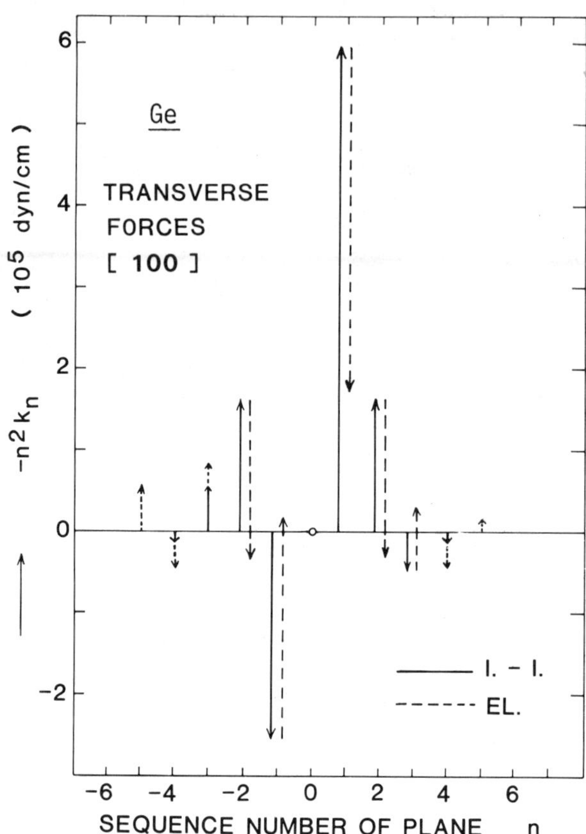

Fig. 5.8.4: Screening of ion-ion interactions in Ge by the electronic terms: the transverse [100] forces of Fig. 5.8.2 are shown as the result of competition between the Coulomb forces (ion-ion interactions between the unscreened $+4|e|$ cores), and the "electronic" forces (electron-ion, Hartree and exchange-correlation terms). The force constants k_n are weighted by the factor n^2 so as to visualize the forces on distant neighbors; their origin is purely "electronic": the electrostatic forces vanish at distances $\geq a$ (4th neighbors). Compare the "composition" of k_{+1} (bond stretching) and k_{-1} (bond bending) with that of, respectively, TO(Γ) and TA($\bar{X}$) modes in Fig. 4.4.1. (From Ref. 39.)

originating from different displaced atoms becomes possible: in the
defect problem, when the main quantity sought is the _change_ in
force constants $\Delta\Phi$ consequent to introduction of an impurity. In
case of isoelectronic defects the different $\Delta\Phi_{\alpha\beta}(\ell\kappa;\ell'\kappa')$ fall off
with distance quickly - they are supposed to, at least - and can
thus be determined on fairly small configurations. Instead of one
self-consistent calculation of force, one needs two: with the
displaced site occupied by, respectively, the defect and the
original atom. In the only case studied by this method[62], that of
substitutional Al in GaAs, the modification of forces was
undetectable at third neighbors and turned out to be fairly small
at the second nearest ones.

If the question of "decoupling" has disappeared, the defect
problem brings another difficulty, which might be an even more
serious obstacle: the _lattice relaxation_ around the imperfection.
The rearrangement of atoms around the impurity can modify the
bonding considerably, and its influence on $\Delta\Phi$ may be decisive.
Total energy calculations are very well suited to finding the new
atomic positions around the defect, by total energy minimization,
but the size of supercells required might, again, exceed the
capacity of present methods.

We have shown in Ref. 62, however, that the contribution of
lattice relaxation to $\Delta\Phi$ can be properly accounted for _without_
knowing explicitly the actual amount of the relaxation: one more
self-consistent calculation is needed, which determines the forces
acting on the impurity's neighbors. They are kept at the sites
corresponding to the perfect crystal - the only ones which are
known precisely. From the $\Delta\Phi$ determined as described above, one
merely has to _subtract_ the force due to the "non-relaxation".
However simple, this procedure is _exact_ - to the first order, i.e.
as long as the lattice relaxation is small.

6. ELECTRIC FIELDS IN AB INITIO TREATMENTS OF POLAR CRYSTALS

Compared to the homopolar substances treated in the previous
Section, lattice vibrations in polar crystals are distinguished by
the appearance of electric fields. The effective charges carried by
individual atoms or, more exactly, the electric dipoles created by
their displacements, give rise to _local_ and/or _macroscopic_ fields
that significantly influence most of the phonon-related properties.
Since their early days, the phenomenological theories of lattice
dynamics[63], aided by various dipole models, have been attempting
to grasp the effects of the local electric fields, and methods have
been developed for dealing with the divergent part of the
long-wavelength limit, with the macroscopic field.

In the self-consistent calculations of the type described in
Ref. 12 and used in these lectures, all effects of <u>local</u> electric
fields are naturally present: the "dipoles at lattice sites" of
phenomenological model-descriptions are nothing but shorthand for
the effect of screening the core-charge by valence electrons – a
process which self-consistent treatments describe in abundant
detail. Once the actual charge densities are known, all the
electrostatics of the system cores + charge-distribution is
automatically included in the self-consistent equations as the
ion-electron and electron-electron terms (potentials $V^{ion}(\vec{r})$ and
$V^{H}(\vec{r})$), and the success of the frozen phonon treatment of a <u>polar</u>
substance in Section 4 reflects this fact. Both the (local)
electrostatic and the "short-range" forces of the phenomenological
descriptions are now treated on the same footing.

There is still a limit to the all-embracing power of the
self-consistent schemes: <u>macroscopic</u> electric fields. As a
consequence of periodicity of the crystal, the relation $V(\vec{r} + n_1\vec{a}_1
+ n_2\vec{a}_2 + n_3\vec{a}_3) = V(\vec{r})$ implies that the cell average of the
corresponding $\vec{E}$ has to be zero. The self-consistent schemes of the
type described in Ref. 12 are truly <u>microscopic</u> theories, and they

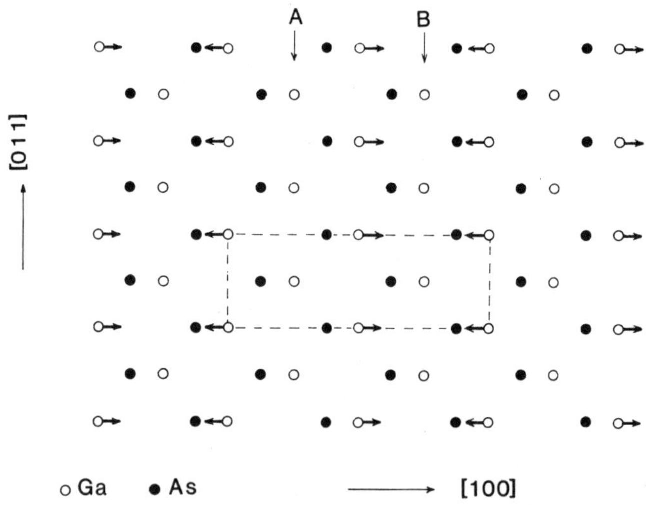

Fig. 6.1.1:   Displacement pattern used for the determination of
the effective charge e_L^*(Ga). The supercell, obtained by
repeating the elementary unit cell of GaAs along the [100]
direction, contains 4 Ga and 4 As atoms displaced in an
appropriately chosen pattern (which does not correspond to any
particular phonon eigenmode) and the effective charge is
determined from a self-consistent potential evaluated at
planes A and B, separated by distance <u>a</u>.

do not allow us to deal with macroscopic quantities, i.e.
quantities which only vary over dimensions much larger than those
of unit cells. The slowest allowed variations are, in the standard
formulation, those corresponding to wavevectors $\vec{k} = \vec{b}_{1,2,3}$ – but
not to fractions of $\vec{b}$. (We recall from Section 4 that,[1,2,3] for dealing
with frozen phonons with $\vec{k}$ at the Brillouin-zone boundary, we had
to double the unit cell, so that the $\vec{k}$ became a full reciprocal
lattice vector of a system with translational periodicity of the
supercell.) For the above reason we could not have treated in
Section 4 the LO(Γ) phonon – whose only distinction from TO(Γ) is
the presence of a constant (macroscopic) electric field.

The classical theories meet difficulties, as well, when it
comes to dealing with $\vec{k}=0$ macroscopic fields, and the usual
procedure in that case is to evaluate limits for $\vec{k} \rightarrow 0$. This
approach is not viable in the self-consistent procedures – although
attempts have been made recently[64,65] to find out how to circumvent
the problem of non-periodicity.

In this Section we merely attempt to make a contact with
phenomenological theories, and concentrate our attention on
effective charges in polar crystals; at the same time we will pay
attention to the electric fields – macroscopic or not – as they
appear in our self-consistent calculations on polar crystals with
displaced atoms.

6.1. Effective Charges

In a homopolar crystal, such as Ge, the strong electric
dipoles created in a displacement of bare core-charges $+4|e|$ are
completely cancelled by the electronic charge density, which
readily relaxes, i.e. adapts its distribution to the displacement.
In a polar crystal, however, this cancellation is not complete, and
we are interested in the resulting electric moment, which the
phenomenological theories traditionally represent as a displaced
effective charge.

Consider the displacement pattern in Fig. 6.1.1: a unit cell
of GaAs is repeated 4 times in the direction [100] and two atoms of
this supercell are displaced in opposite directions – so that
entire planes of charged sites are displaced by a small amount $\vec{u}$.
By allowing the electrons to re-distribute self-consistently and to
adapt to the "new" structure we obtain the self-consistent
potential shown in Fig. 6.1.2 (solid lines), defining the total
field felt by an electron. Compared to the potential in the
undisplaced configuration (dotted lines) the effect of displacing
the atoms consists mainly of shifting the potential upwards in one
half and downwards in the other half of the unit cell. The local
component of the field, the one rapidly varying from atom to atom,

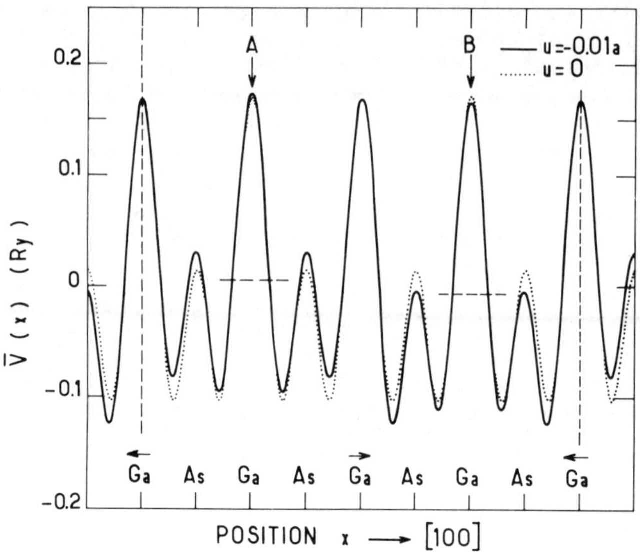

Fig. 6.1.2:   Self-consistent potential $V^{sfe}(r)$ (felt by electron) obtained from the supercell with the displacement pattern of Fig. 6.1.1 (solid line); the potential of the undistorted configuration is shown for comparison (dotted line). To make the picture one-dimensional, the potential is averaged in every plane (100) over the remaining two coordinates y, z. Both curves have identical shape in any region sufficiently far from the displaced planes but the average levels V_o are shifted when the atoms are displaced (broken horizontal lines). The magnitude of this shift reflects the electric dipoles created around the displaced planes – hence the effective charge, which is determined from ΔV_o, i.e. from the difference $\bar{V}^{sfc}(A) - \bar{V}^{sfc}(B)$. The actual calculation corresponds to the displacement u = -0.01 a shown in Fig. 6.1.1.

is essentially modified only in the immediate vicinity of the displaced atoms and the perturbation caused by the displacement is expected to be nearly completely absorbed at distance <a/2; essentially the same behavior is found for the self-consistent charge density. The upward/downward shift of the average potential V, displayed in Fig. 6.1.2 by broken horizontal lines, is due to the electric dipoles which we have created by displacing the

274

(charged) planes of atoms and which are not completely screened
out. It was to be expected that displacement of a single plane of
atoms generates a dipole moment distribution which is restricted to
a region within atomic dimensions of the displaced plane. Inside
this region there is a net electric field which causes a difference
in average electrostatic potentials between the two sides of the
plane; outside this "slab" of finite thickness, however, the
electric field and polarization are zero.

Schematically, the origin of the corresponding additional
field is shown in Fig. 6.1.3, which depicts the potential V of <u>two
capacitors</u> in an otherwise empty space. (Note that we are
systematically plotting the potential V felt by an electron, rather
than the electrostatic potential $\phi \equiv -V$.) Shifting an (infinitely
thin) plane, charged with a density σ, by a distance $\underline{u}$ amounts to
adding to the system a capacitor and thus causing the potential to
change by $\Delta V = -4 \pi \sigma \underline{u}$; another pair of "plates" displaced in the
opposite direction then brings the potential back to the original
level. This is an equivalent scheme to the situation in Fig. 6.1.2,
expressed in terms of <u>localized</u> charges. If the density σ (Fig.
6.1.3)$_2$ is the density of effective (point) charges $\sigma = e_L^*/A$, where
$A = a^2/2$ is the surface area per atom, then the ΔV is the
difference between the two average levels V_o of Fig. 6.1.2. (It is

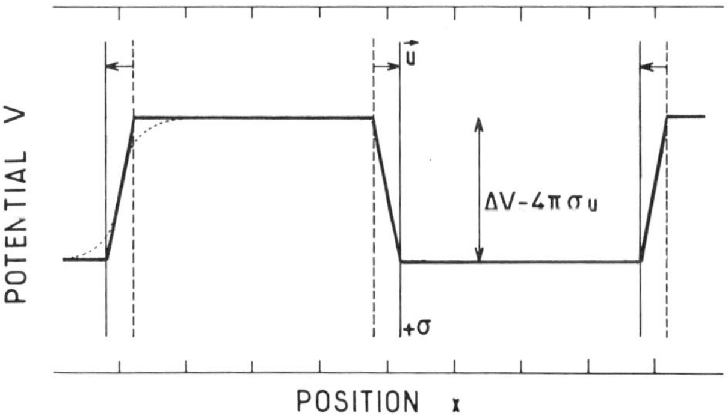

Fig. 6.1.3:   Schematic representation of the electrostatic
   potential created in an otherwise empty space by shifting the
   planes uniformly charged with density σ; the displacements
   chosen correspond to those shown in Fig. 6.1.1, and $\sigma > 0$ was
   assumed in the diagram. The potential actually plotted is the
   one effectively felt by an electron, $V \equiv -\phi(x)$; this is (a
   simplifed form of) the potential that has to be added to
   dotted lines of Fig. 6.1.2 so as to obtain the solid lines.

straightforward to show that e_L^* is a <u>longitudinal</u> charge since the
net electric fields in the dipole regions are fully screened. This
is discussed more thoroughly in Ref. 66.)

If we think of σ as the density of <u>core-charges</u> $+3|e|/A$ or
$+5|e|a$, then Fig. 6.1.3 would schematically represent the
modification of the $V^{ion}(\vec{r})$ resulting from the displacement pattern
of Fig. 6.1.1.

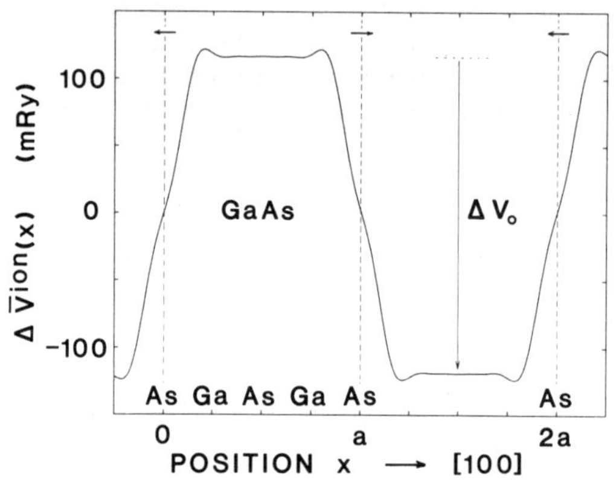

Fig. 6.1.4:   Calculated difference of ionic potentials $V^{ion}(\vec{r})$
between the configuration with the displacements of Fig. 6.1.1
and the undisplaced one. Averaged over y, z as in Fig. 6.1.2.
At distances $\geq a/4$ the potential $\Delta V^{ion}(\vec{r})$ caused by shifting
the Arsenic planes agrees with the schematic representation
given in Fig. 6.1.3: the difference ΔV_o of the two potential
levels is determined by the core charge $+5|e|$ of As. In case
of Gallium planes displaced, the ΔV_o would be reduced by
factor 3 : 5.

In the actual calculation we displaced the atoms by u=-0.01 <u>a</u>,
and Fig. 6.1.4 shows the consequent modification in the unscreened
<u>ionic</u> potential; it behaves, indeed, as in Fig. 6.1.3, except that
the width of the region with nonvanishing electric field is now
much larger than $|\vec{u}|$, approximately a/2. This only reflects the

276

fact that the charge in the plane (100) is not rigorously confined
to the "plates", as assumed in Fig. 6.1.3, and namely that the
distribution of the corresponding dipoles over the plane is not
continuous but discrete. Fig. 6.1.4 shows ΔV due to displacement of
the <u>arsenic</u> plane; the equivalent figure with gallium displaced is
analogous, except that the "difference of levels" is smaller, in
the proportion 3:5 : $\Delta V = -4\pi\sigma\underline{u}$ with $\sigma = Z^{core}/(a^2/2)$.

The <u>self-consistent</u> potential (more exactly, its modification
following the displacements) for the system in Fig. 6.1.1 will,
again, have the form shown schematically in Fig. 6.1.3, except that
σ will be the density of the <u>effective</u> charge: the dipoles causing
the "two-plateau" structure of Fig. 6.1.4 will be considerably
weakened, because of screening by the electronic charge that has
adapted its distribution to the displacements. The potential
obtained in the actual calculation (with displaced As) is shown

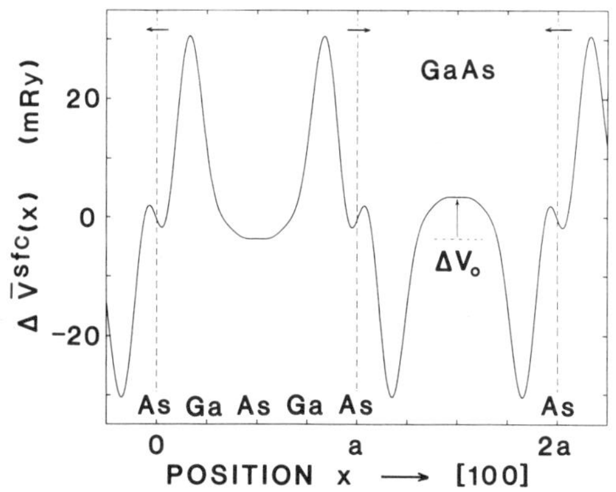

Fig. 6.1.5:   Difference of self-consistent potentials $V^{sfc}(\vec{r})$
between the configuration with displacements of Fig. 6.1.1 and
the undisplaced one; this is the potential $\Delta V^{ion}(\vec{r})$ of Fig.
6.1.4 screened. The potential difference ΔV_o between the two
potential levels has the opposite sign than in Fig. 6.1.4
because the relative shift ΔV_o^* is now governed by the
effective charge $e_L(As)$ which is negative. In the case of
displaced Gallium planes, the absolute value of ΔV_o would be
the same but its sign would be reversed. (Averaging over $\underline{y}$, $\underline{z}$
like in Fig. 6.1.2).

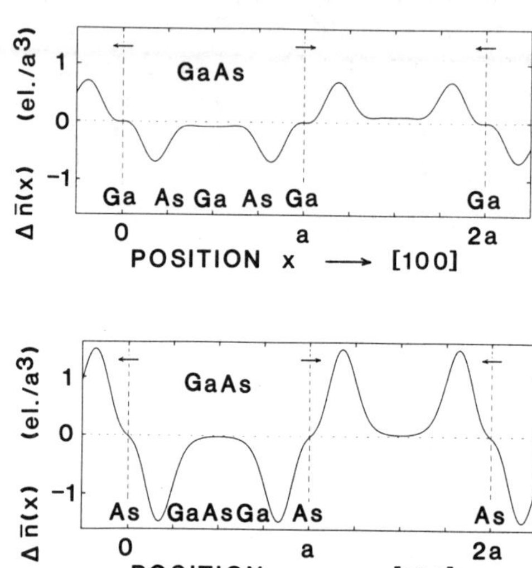

Fig. 6.1.6:   Difference $\Delta n(\vec{r})$ of electronic charge densites caused by the displacement patterns of Fig. 6.1.1, with either Gallium or Arsenic planes displaced. Averaged over y, z as in Fig. 6.1.2. In the case of Ga-displacements, the two shifted planes are not yet completely "decoupled" at distance a, and the $\Delta n(\vec{r})$ does not vanish completely in the space between the planes.

in Fig. 6.1.5. Disregarding the peaks in the vicinity of the displacement, which are of no interest for our purposes, we see that the potential difference ΔV between the two levels is not only weakened but <u>overscreened</u>: its sign is now opposite to that found in Fig. 6.1.4. The case with displaced Gallium planes is fully analogous – but the "ordering" of the two plateaux is as in Fig. 6.1.3 or 6.1.4: the planar charge density σ has the same absolute value but the opposite sign as in Fig. 6.1.5.

Writing the corresponding σ as a planar density of the (longitudinal) <u>effective</u> charge, $\sigma = e_L^*/(a^2/2)$, Fig. 6.1.5 yielded $e_L^*(As) = -0.153 \, |e|$; the analogous calculation with gallium planes displaced led to $e_L^*(Ga) = +0.161 \, |e|$. The two values have to sum to zero, and the small difference between them reflects the numerical uncertainty; one of its causes may be imperfect convergence, two others will be discussed below. The present result^{67} $e_L^* = 0.158 \pm 0.005 \, |e|$ compares with the experimental value^{67} $e_L^* \equiv e_T^*/\varepsilon_0 = 2.16/10.9 = 0.198 \, |e|$ – and we note that the calculation also provided the <u>sign</u> of the effective charge (in agreement with the established picture of <u>ionic</u> bonding), which has not been determined experimentally.

As in Section 5.5, we have to ask at which distance the two displaced planes of Fig. 6.1.1 are sufficiently decoupled, because this is clearly the basic assumption behind all reasoning of this subsection. We have plotted, in Fig. 6.1.6, the modification in electronic <u>charge density</u> consequent to the displacement used in these calculations. The variation Δn is close to zero (0.02 electrons/a^3) in the region of the "plateaux" when the As atoms are displaced, whereas it is getting a value of $\cong 0.09$ el./a^3 when the Ga planes are moved. This partly explains the discrepancy between the values found for $e_L^*(As)$ and $e_L^*(Ga)$, and it shows that even at a distance $\underline{a}$ the two planes are still not sufficiently apart; it suggests that the result for the $e_L^*(As)$ is likely to be the more reliable one.

It was pointed out in Ref. 64 that the <u>screened electrostatic potential</u> (i.e. $V^{ion} + V^H$, not including any V^{xc}) rather than the self-consistent potential should be used for determination of e_L^* – and this holds for other dielectric quantities, as well. In fact, the electromagnetic theory refers to <u>classical</u> test charges, the exchange-correlation effects enter only indirectly, through determination of the actual charge distribution. The screened potential $V^{scr} \equiv V^{ion} + V^H$ plotted in Fig. 6.1.7 for comparison is slightly different from $V^{sfc} = V^{scr} + V^{xc}$ shown in Fig. 6.1.5, and the effective charge determined from these data becomes slightly larger: $e_L^*(As) = -0.166 \, |e|$ (9 % increase) – though the numerical uncertainty of the calculation might be responsible for a part of this variation.

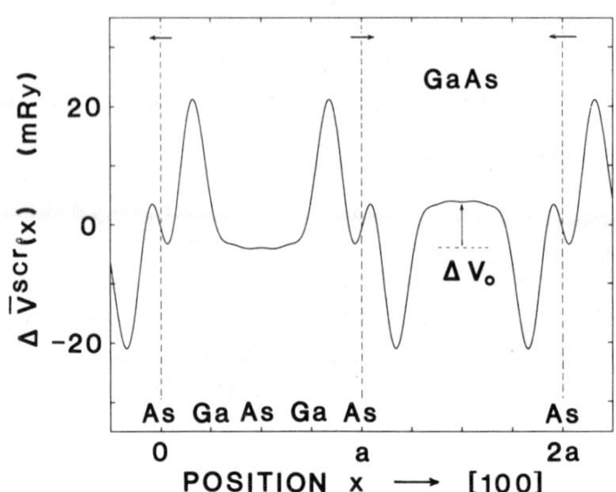

Fig. 6.1.7:   Same as Fig. 6.1.5, but the increase of the screened potential $V^{scr}(\vec{r}) \equiv V^{ion} + V^{H}$ is plotted rather than the variation of the self-consistent one $V^{sfc}(\vec{r}) \equiv V^{ion} + V^{H} + V^{xc}$.

Concluding this Subsection we note that the method described, which is discussed in full generality in Ref. 66, shows graphically the dynamic mechanisms of electronic screening that gives rise to the above effective charges; they are opposite in sign to those which would be predicted by simple macroscopic screening arguments. Although the presently available densities calculated by local pseudopotentials are far from being exact, more precise information might be extracted by using more reliable pseudopotentials[16-18,68].

6.2. Macroscopic Field

We conclude from Fig. 6.1.3 that there is no electric field in most of the volume of the supercell Fig. 6.1.1 - and even the

"potential-steps" produced by the two displaced planes compensate
each other; however, we risked not having the displaced planes
sufficiently "decoupled" at distance $\underline{a}$. A different pattern is
suggested in Fig. 6.2.1 : the supercell is of the same size - but
only $\underline{one}$ plane in each cell is displaced: the two nearest
"capacitors" are thus separated by a distance 2 $\underline{a}$. Which
potential describes the electrostatic field of this chain of
"capacitors"?

The answer is given in Fig. 6.2.2a: $\underline{inside}$ each "capacitor"
the potential rises again by $4\pi\sigma\underline{u}$; $\underline{between}$ two of them, there is a
constant $\underline{electric\ field}$ affecting several (elementary) cells. It is
clear from the diagram that the potential is periodic on the
supercell and that the supercell-averaged $\vec{E}$ is zero.

The potential in Fig. 6.2.2a is found by solving the Poisson
equation with periodic boundary conditions. It may be instructive,
however, to construct it from that of an isolated capacitor - as we
did in the previous subsection; this is illustrated in Fig. 6.2.2b.

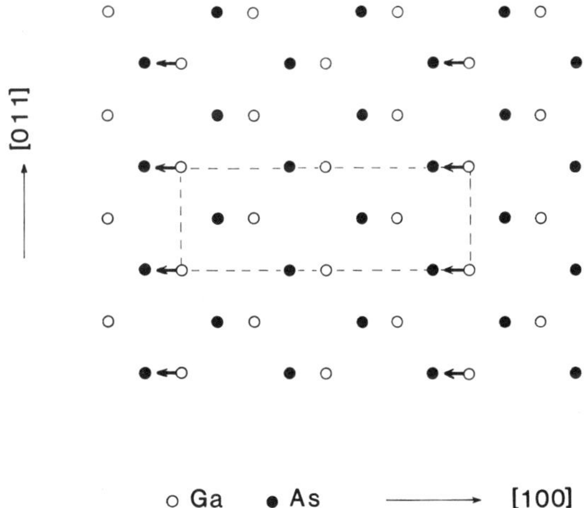

Fig. 6.2.1:   Displacement pattern in which only one plane per
    supercell is shifted; the translational unit is obtained by
    repeating the elementary unit cell of GaAs four times along
    the direction [100].

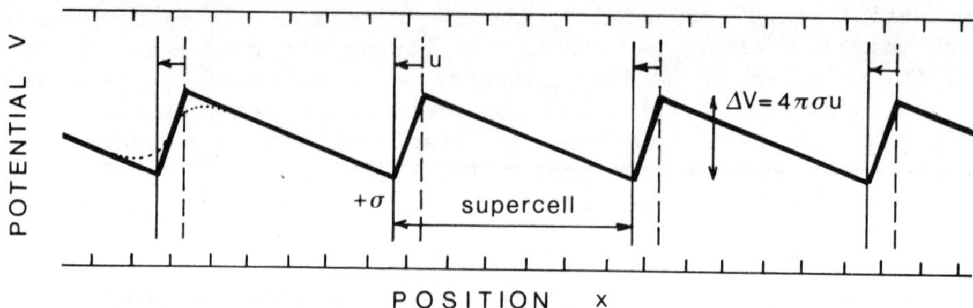

Fig. 6.2.2a: Schematic representation of the electrostatic
potential created in an otherwise empty space by slightly
displacing planes with uniform charge density σ, in a pattern
defined by Fig. 6.2.1; the potential actually felt by an
electron $V \equiv -\phi(x)$ is plotted, $\sigma > 0$ was assumed. Periodic
repetition of dipole layers has produced a constant field in a
large part of each supercell – though the (supercell–)
averaged electric field is zero, as required by the periodic
boundary conditions.

There is an electric field inside each "capacitor", but, seen from
outside, the only visible effect consists in raising the level of
the potential by $\Delta V = 4\pi\sigma u$. The next "capacitor" of the chain then
repeats this lifting, etc. We can see that this simpleminded
construction leads to a $V(\vec{r})$ which is <u>not</u> periodic on the
supercell; moreover, it builds up a <u>net potential difference</u>
between the two ends of the "chain" – thus also violating the Born
von Karman periodic boundary conditions or, if the free-end
boundaries were preferred, creating some artificial <u>surface
charges</u>. The latter, however, can be easily cancelled by imposing
from the outside an (equally artificial) <u>depolarizing field</u>, which
brings the two ends of the "staircase" to the same level, thus
cancelling the surface charges and making the supercell-averaged
field zero. The sum of the two potentials in Fig. 6.2.2b yields the
potential shown in Fig. 6.2.2a.

Compared with the previous Subsection, the new feature of the
present configuration (Fig. 6.2.1) is the appearence of a <u>constant
electric field</u> inside the supercell. Strictly speaking, we are not
encountering any $\vec{k}=0$ macroscopic field, because the potential does
vary inside each supercell. Nevertheless, from the point of view of
a few <u>elementary</u> cells, situated between two "capacitors", there is
no difference between this saw-like potential and a "real"
macroscopic field, because $\vec{E}$ is <u>constant</u> in the entire elementary
cell.

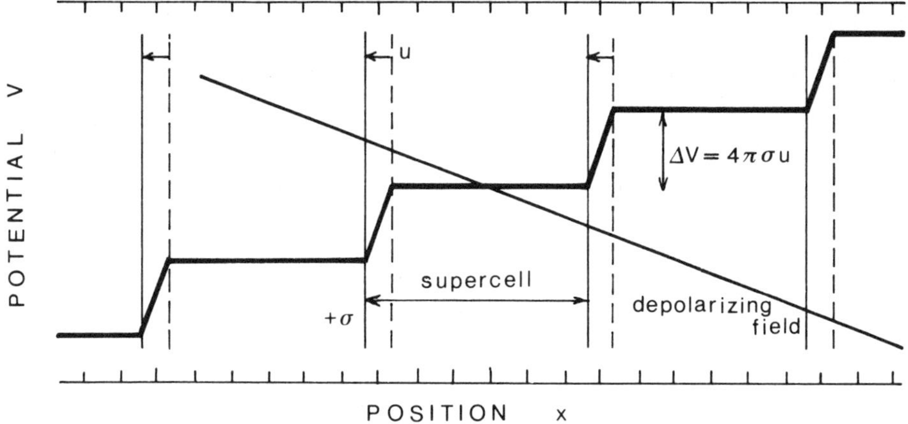

Fig. 6.2.2b: Potential of Fig. 6.2.2a can be graphically
represented as sum of two contributions: 1) Potential produced
by each dipole-layer considered separately (as e.g. in Fig.
6.1.3) - which, however, does not obey the periodic boundary
conditions. 2) Depolarizing field, constant over crystal,
which compensates the net potential difference at the two ends
of the crystal.

Potentials found in the actual calculation are very much like
the schematic representation of Fig. 6.2.2a. Modification of the
unscreened ionic potential by displacement of the As-plane is shown
in Fig. 6.2.3; an analogous picture for Ga is very similar, except
that the slope is milder in the proportion 3:5. The resemblance to
Fig. 6.2.2a (which was sketched under the assumption of <u>positive</u>
charge σ) is obvious, the only difference is in the width of the
"capacitor" which is of order $a/2$ rather than $|u|$ - a consequence
of using the <u>pseudopotentials</u>, of allowing a certain spread of the
core-charge around the "plates", and of having a discrete rather
than uniform distribution of dipoles in the plane.

As the result of screening, the actual slope of the <u>screened</u>
potential will be much weaker - when Ga planes are displaced; in
the case of As, it will have the <u>opposite sign</u> (Fig. 6.2.4): the
slopes are now governed by $e_L^*(Ga)$ and $e_L^*(As)$. Fig. 6.2.5 shows the
variation of the self-consistent charge density caused by the
displacements of Ga- and As- planes; redistribution of the charge
making up the screening (from $+3|e|\vec{u}$ to $e_L^*(Ga)\vec{u}$) is apparent. The
re-arrangement is considerably more thorough when the As-cores are
displaced - producing thus an "<u>overscreening</u>", i.e. changing the
slope corresponding to the positive $+5|e|\vec{u}$ into the one defining
the <u>negative</u> one $e_L^*(As)|\vec{u}|$.

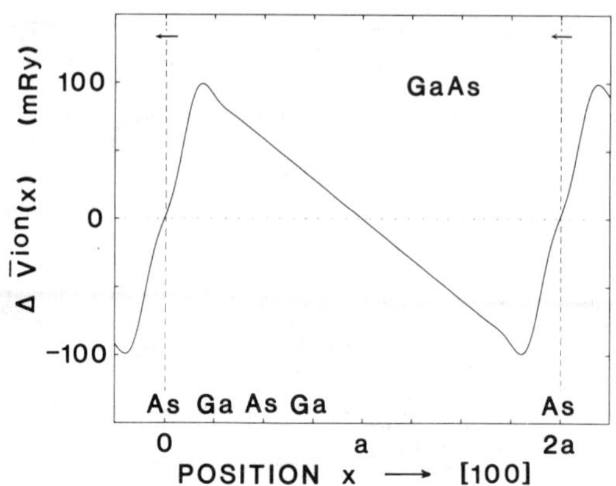

Fig. 6.2.3:   Calculated difference of ionic potentials $V^{ion}(\vec{r})$ between the configuration with the displacements of Fig. 6.2.1 and the undisplaced one. At distances $\geq$ a/4 the potential $\Delta V^{ion}(\vec{r})$ caused by shifting the Arsenic planes agrees with the schematic representation given in Fig. 6.2.2a; the slope is determined by the core charge $+5|e|$; in the case of displaced Gallium planes, the slope would be reduced by factor 3 : 5. (The potential $V^{ion}(\vec{r})$ was averaged over $\underline{y}$ and $\underline{z}$, in planes parallel with (100).)

Calculation on a 6 x GaAs supercell[10] yielded from the slopes $e_L^*(Ga) = +0.162\,|e|$ and $e_L^*(As) = -0.165\,|e|$ (i.e. $|e_L^*/e| = 0.163 \pm 0.002$) – consistently with the result 0.158 ± 0.005 of Chapter 6.1.

Having arrived at this point, it is possible to take a further step and determine the <u>force</u> acting on the central atoms in the supercell: it was amply discussed in Section 5 how this information is provided by the Hellmann–Feynman theorem. As the field in the central part of the supercell is constant (and its intensity known – one merely has to read the slope off Fig. 6.2.4), it appears as a k=0 macroscopic field to the atoms in this region; the relation[69,70]

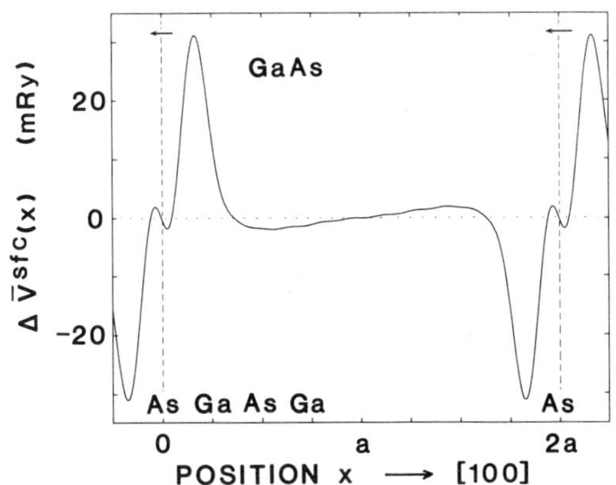

Fig. 6.2.4: Difference of self-consistent potentials $V^{sfc}(\vec{r})$ between the configuration with displacement of Fig. 6.2.1 and the undisplaced one; this is the potential $\Delta V^{ion}(\vec{r})$ of Fig. 6.2.3 screened. The slope in the center of the supercell shows the constant (macroscopic) electric field inside the cell; its intensity is governed by the effective charge $e_L^*(As)$ (negative), which explains why the slope has opposite sign to that of Fig. 6.2.3. In case of displaced Gallium planes the slope would have the same absolute value but its sign would be reversed.

$$F(\kappa) = e_T^*(\kappa)\, E^{macro} \qquad (6.2.1)$$

then makes it possible to find out the <u>transverse</u> effective charge: the value $|e_T^*/e| = 1.87 \pm 0.14$ was obtained in Ref. 10 – which compares with the experimental value ± 2.16 (Ref. 67).

Finally, using the relation

$$e_T^* = e_L^*\, \varepsilon_o \qquad (6.2.2)$$

where ε_o is the macroscopic dielectric constant, one finds[10] $\varepsilon_o = 11.4 \pm 1.6$ – compared with the experimental value 10.9.

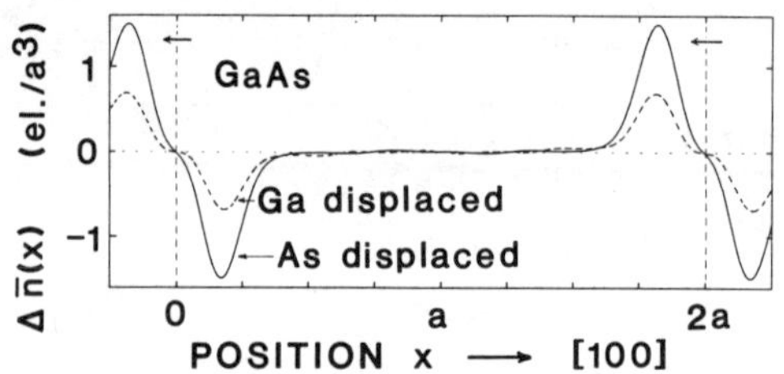

Fig. 6.2.5:   Difference $\Delta n(\vec{r})$ of electronic charge densities caused by the displacement patern of Fig. 6.2.1. This is the polarization charge that reduces the strong core charge (dipoles $Z_c(\kappa)\,\vec{u}$) into the weak effective charge (dipoles $e_L^*(\kappa)\,\vec{u}$) – even reversing the sign, in the case of the displaced Arsenic.

Disregarding the (possibly small) nonlinear effects, all the above information is obtained from a <u>single</u> self-consistent calculation, merely by understanding the electric fields in the supercell.

7. PLANAR FORCES IN POLAR CRYSTALS

Once the electric fields caused by displacement of atomic planes in polar crystals are understood, it is easy to calculate the phonon dispersion ab initio, using the method explained on Ge in Section 5. Fig. 7.0.1 shows the displacement pattern employed; it is essentially the same one as in Ge, except that <u>twice as many calculations</u> are now needed: the Ga and As atoms have to be interchanged and the calculation repeated with the As-planes displaced. The equivalent "linear chain" is shown in Fig. 7.0.2 together with the definition of force constants: to make the definition of $\{k_n\}$ clear, the displaced plane (Ga or As) is shown in Fig. 7.0.2 always at origin. The orientation of the $+[100]$ direction is the same as used in Ge (Section 5.3). With two types of atoms in each unit cell the force constants need an additional label "cation-cation" or "anion-anion"; but the labels "cation-anion" or "anion-cation" at the odd-neighbor k_n are <u>not</u> needed since k_n^{ca} is the same as k_n^{ac}, to any order in $\vec{u}-\vec{a}$ a consequence of the action-reaction principle. As two different and totally independent calculations have to lead to the same $k_{\pm 1}$ or $k_{\pm 3}$ (Fig. 7.0.2), this is an opportunity to check the numerical

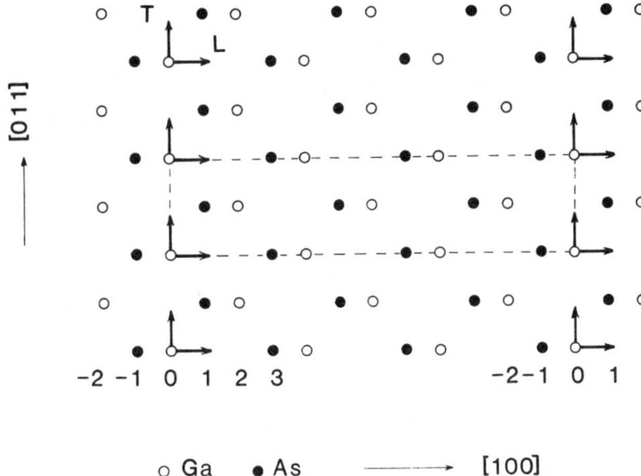

Fig. 7.0.1:   Displacement pattern used for determination of interplanar force constants in GaAs along the [100] direction. As there are two different types of atoms in the cell, twice as many calculations are needed, compared to Germanium: with Ga- and As-planes displaced. The origin and positive orientation of the [100] direction are as in Fig. 5.3.2.

accuracy by verifying the symmetry relation

$$K(\ell\kappa;\ell'\kappa') = K(\ell'\kappa';\ell\kappa) \; ; \qquad\qquad (7.0.1)$$

this is the linear-chain notation for the well-known analogous relation between the interatomic force constants (see e.g. Ref. 51, eq. (2.1.10)). The computational errors in (7.0.1) are of order 1 % for longitudinal, 4 % for the transverse k_{+1}. On the other hand, the translational invariance of the chain imposes condition (5.8.1) separately for cations and anions. For cations, this relation is numerically satisfied within 0.7 % and 0.2 % for longitudinal and transverse respectively, and for anions within 2 % and 0.3 %.

The orientation used in the <u>vibrational</u> calculations on the linear chain is the one showed in Fig. 7.0.2a: the <u>cation</u> plane is placed at the origin, and <u>anion</u> planes are at the sites ±1; plane $\kappa=0$ contains the atom at (000), the plane $\kappa=+1$ is the one containing the site a/4(111). The dynamical matrix is given by eq. (5.7.1) and treatment of anharmonic effects – which are small in the [100] direction – is as in Ge, Chapter 5.4. Even the conclusions about the spatial extent of forces turn out to be similar to Ge (Chapter 5.5), including the even/odd <u>n</u> zig-zag convergence. The essential difference is in <u>presence of a macroscopic electric field</u>.

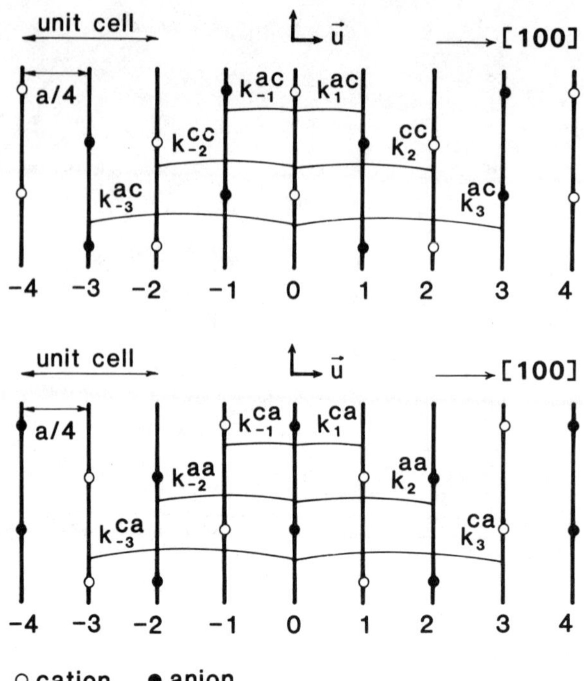

Fig. 7.0.2:   Definition of interplanar force constants for
[100] vibrations in GaAs; as two different types of planes can
be displaced (cations, anions), the force constants have to
carry additional labels ca, ac, aa, cc. (The label is omitted,
however, at the odd-neighbor k_n for which $k_n^{ac} = k_n^{ca}$ holds.)
The origin and positive orientation of the [$\bar{1}$00] direction are
chosen as in Fig. 5.3.2; in the actual (lattice dynamical)
calculation, the Gallium is assumed to be at the origin.

7.1. Longitudinal Vibrations

When we attempt, starting from the pattern of Fig. 7.0.1, to
determine the <u>longitudinal</u> force constants, a <u>constant electric
field</u> will appear along the supercell – as explained in Section 6.2
(see Figs. 6.2.2a and 6.2.4). The pattern of Fig. 7.0.1 does <u>not</u>

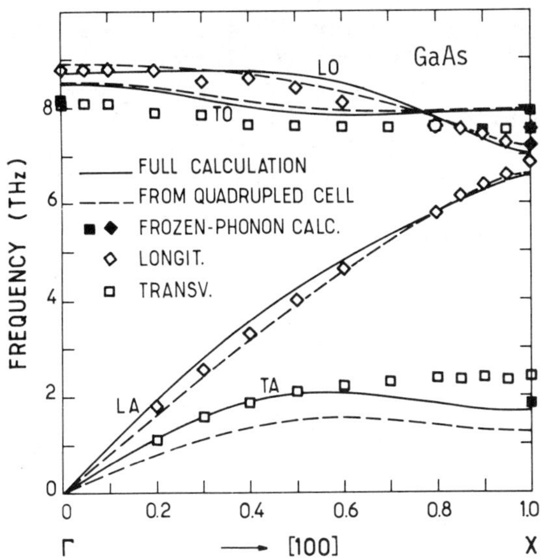

Fig. 7.1.1:   Phonon dispersion $\omega(\vec{k})$ in GaAs calculated from the planar forces (Tab. 5.1) determined ab initio. (From Ref. 10.)

correspond to any particular phonon, and the field in question has nothing to do with the dipole-dipole or other electrostatic interactions, that play such an important role in the actual vibrations of a "polar chain". The appearence of a constant field is an artifact of the displacement pattern, and it is a consequence of periodically repeating the displacement at regularly spaced intervals. (If only one plane in the entire crystal could be displaced, no field would appear, the only effect would consist in creating one "step" of potential, as in Fig. 6.2.2b). In fact, repeating the displaced plane periodically has been merely a compromise, enforced by the supercell technique employed, the $\{k_n\}$ definition itself is based on the idea of only a single plane displaced.

So far, repeating the displacement periodically has not caused any major inconvenience - except in the case where the supercell chosen was "shorter" than the range of forces; this was, however, merely a quantitative problem, adding slightly, perhaps,

to error margins. In a polar crystal, on the contrary, the
supercell-periodicity changes the picture qualitatively - because
it imposes a <u>different choice of boundary conditions</u> for the
electrostatic aspect of the problem.

Knowing that the constant field is a spurious quantity, we are
sufficiently aware of its properties from Section 6.2, to be able
to eliminate its effect on the longitudinal force constants. A
constant field produces constant forces on the atoms - to within a
variation of sign from cation to anion - which thus has to be
subtracted from all the Hellmann-Feynman supplied forces. Provided
that the supercell was chosen sufficiently large, the force on the
<u>central</u> atom in the supercell can be identified as originating
entirely from this spurious contribution. On a 4 x GaAs supercell
we found that this contribution to k_n amounted to 0.018 x 10^5
dyn/cm. On 6 x GaAs (where this value is 1.5 x weaker) we could
have verified that this force is, indeed, constant and that it only
changes its sign, when one goes from the central atom to its
neighbors.

With the effects of the spurious electric field eliminated,
the genuine longitudinal force constants are summarized in Tab.
5.1. They represent the properly screened core-core interactions
or, expressed in terms of phenomenological theories, they include
<u>both</u> short-range and Coulomb forces. This may seem somewhat
surprising, because the Coulomb contributions are known[63] to fall
off with distance slowly, thus requiring a separate treatment (such
as Ewald summation). However, here a definite advantage of working
in a one-dimensional representation appears, because using the
inter<u>planar</u> forces means that the <u>infinite sums</u> over y- and z-
coordinates <u>have already been performed</u>. (Recall from Section 5.3
or Fig. 5.2.1 that every k_n is a well defined sum of all
$\Phi_{\alpha\beta}(\ell\kappa;\ell'\kappa')$, extending as far as the inter<u>atomic</u> interactions.)
Along the remaining coordinate <u>x</u>, the longitudinal forces fall off
with distance rapidly - essentially as fast as in Ge, where any
field-generating effective charges were zero. This is easy to
understand: whereas the dipole-dipole interactions (i.e.
inter<u>atomic</u> forces) decrease as $1/r^3$, the field of a <u>plane</u> of
dipoles is confined, as in a capacitor.

The longitudinal dispersion plotted in Fig. 7.1.1 was
calculated with forces extending to 3rd neighbors only. The
corresponding force constants are displayed graphically in Fig.
7.1.2.

7.2. <u>Transverse Forces</u>

For transverse forces the polarity of the crystal does not
introduce any new complication, because the transverse
displacements of Fig. 7.0.1 do not produce any macroscopic field.

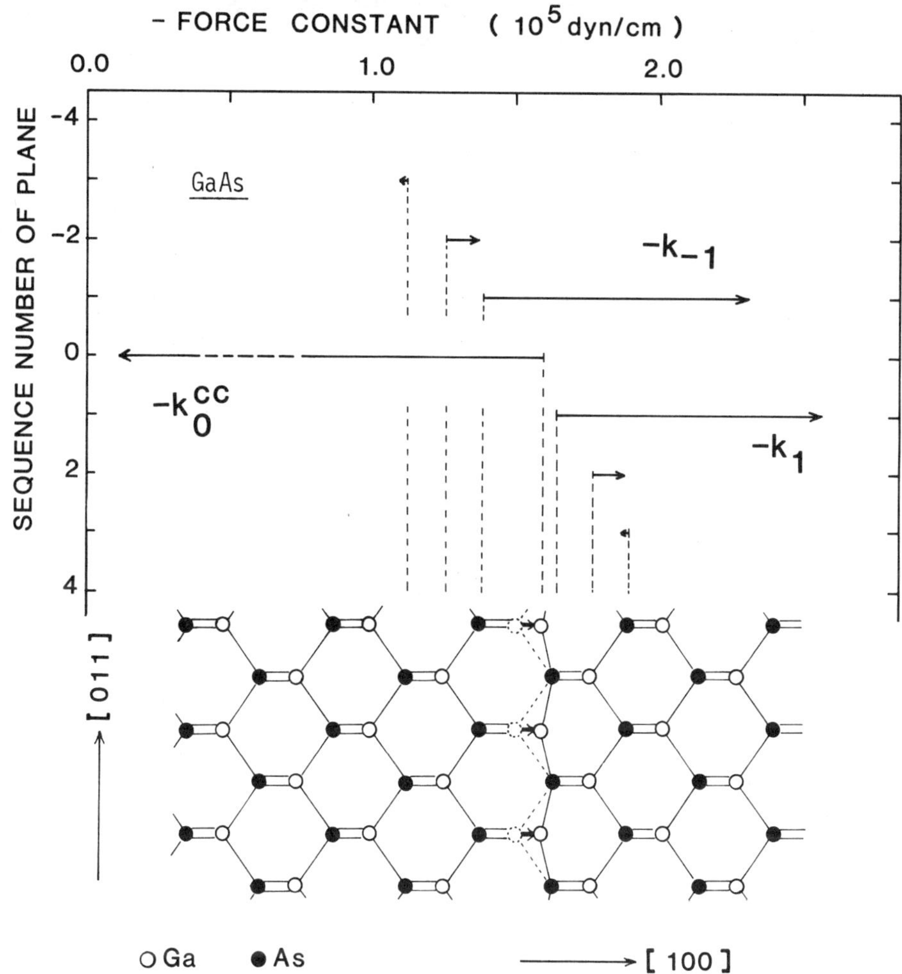

Fig. 7.1.2:   Interplanar forces determining longitudinal vibrations along [100] in GaAs. These forces include all long-range effects due to electrostatic interactions, as discussed in the text. (The contribution of the spurious macroscopic field is subtracted.) Like in Ge (Fig. 5.8.1), the interactions are negligible at distances $\geq 3a/4$. Only the forces corresponding to displaced Ga atoms are shown; if Ga $\leftrightarrow$ As planes are interchanged, the $k_{\pm 2}$ is reduced by 30 %.

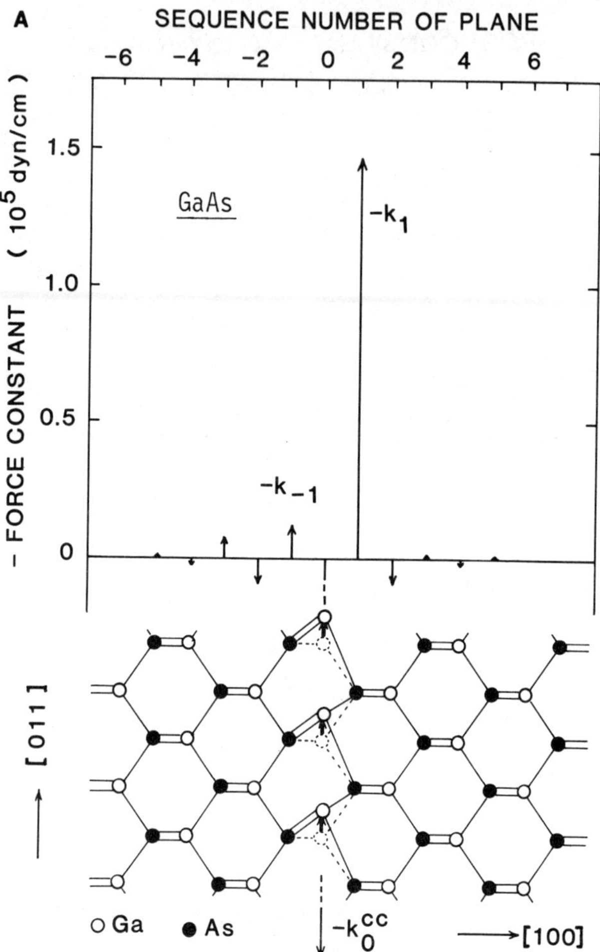

Fig. 7.2.1:   Planar forces determining transverse vibrations
along [100] in GaAs, as found from the displacement pattern T
of Fig. 7.0.1.
(a) Force constants corresponding to a displaced plane of
Ga-atoms exhibit alternating signs, as in the analogous Fig.
5.8.2 for Ge.
(b) Forces caused by the As-plane displaced, have all the same
sign and are rather weak; this is assumed to reflect a partial
cancellation of central and angular interactions (see text).

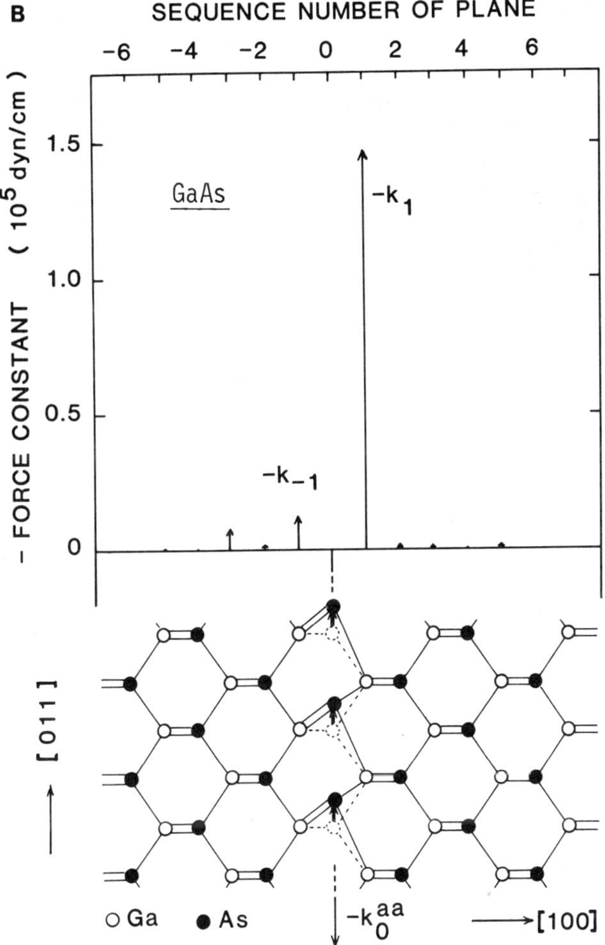

Fig 7.2.1.   Continued

With $\vec{u} = -0.007$ (011) $\underline{a}$ the procedure for the determination of $\{k_n\}$
is as straightforward as in Ge – except that $\underline{two}$ calculations are
needed instead of one (Ga-displaced, As-displaced). Forces up to
the 5th neighbor plane $\underline{at\ least}$ are required to reproduce the flat
TA branch (Fig. 7.1.1), the force constants are summarized in Tab.
5.1.

In many respects it is instructive to display the transverse
force constants in a similar manner to Ge; this is done in Fig.
7.2.1. The pattern is very similar to Fig. 5.8.2 – when the $\underline{gallium}$
plane is displaced; also the diagnostics would follow the reasoning
of Chapter 5.8: we are encountering $\underline{angular}$ interactions, typical
for covalent solids. However, in the alternative case of the
$\underline{arsenic}$ plane displaced, the behavior of the forces changes: they
will all have the same sign. Does this mean that the covalent
picture of angular forces discussed in Section 5.8 does not hold
any more when the anion is displaced and should we deduce the
presence of mainly central interactions? – It is unlikely. Although
no detailed quantitative arguments are available, we believe that
$\underline{both}$ $\underline{central}$ and $\underline{angular}$ mechanisms (shown e.g. in Fig. 5.8.3) now
play nearly equally important roles, practically cancelling each
other: note that all anion-anion force constants are nearly
vanishing, compared with the cation-cation ones.

7.3. Phonon Dispersion in GaAs

Complete dispersion curves calculated with the ab initio force
constants for Γ – X plotted in Fig. 7.1.1 are complemented by the
dispersion of eigenvectors shown in Fig. 7.3.1. As the transverse
forces caused by the displaced cations are an alternating series
(Fig. 7.2.1a), the phonon dispersion, examined as a function of the
range of forces, showed a similar zig-zag behavior as found in Ge,
Section 5.5: cutoffs after $\underline{odd}$-neighbor interactions thus have to
be preferred. The flat TA branch is reproduced when forces
extending to $\underline{fifth}$ neighbor plane are taken into account. The
agreement with frozen phonon calculations is of order 5 % – 6 %,
which sets a limit upon the influence of forces beyond the 5th
neighbor plane.

The dispersion of eigenvectors shown in Fig. 7.3.1 is similar
to that found in Ge (Fig. 5.7.1) – the effect of the lowered
crystal symmetry is apparent, namely in making the amplitudes of
displacements generally "unpredictable" from symmetry alone. The
end-points of the eigenvector dispersion are in excellent agreement
with values calculated by the frozen phonon method (or, whenever
possible, predicted by symmetry). Note that in Ge the LO(X) and
LA(X) modes are degenerate, so that different combinations of those
displayed in Fig. 5.7.1 are equivalent; the two corresponding to a

sum and difference, meaning that one sublattice of Ge vibrates
while the other is at rest, are well suited for comparison with
GaAs (Fig. 7.3.1).

8. LATEST DEVELOPMENTS

Returning to Section 6, we notice that the Density Functional
method might also be an excellent tool for dealing with, at least
some, dielectric properties: Sections 6.1 and 6.2 suggest that the
DF method can reveal static <u>dielectric response</u> of the system.
Indeed, all quantities describing the electrostatics of the
electrons and ions constituting the solid are readily available
from the solutions of the fundamental DF equations: the electronic
charge density has a well defined meaning, the Poisson equation is
inherently built into the formalism, the boundary conditions are
clearly spelled out, the theory is expressed in terms of
potentials. Describing electrostatics is, however, only a minor
part of the power of the DF method: The requirement of
self-consistency for the electronic charge density expresses
mathematically the <u>relaxation</u> of electrons. By allowing the
electrons to adapt to any new or modified situation, the DF scheme
makes it possible to extract all aspects of the "response" to the
"perturbation" in question, with all the effects of
exchange-correlation automatically included – within the limits of
the local approximation. We note that the "response" and modified
or "perturbed" system are now merely conventional translations for
what are treated, by the "direct" approach, as independent and
completely unrelated situations.

So far we have been dealing with various forms of the
"response" to <u>displacements</u> of atoms. In Section 6 also certain
electric fields have been studied, we benefited from the fact that
displacing atoms in GaAs generates dipoles and therefore electric
fields; all the reasonings of Section 6 were, however, limited to
<u>polar crystals</u> and e.g. determination of static dielectric constant
ε_o as in Section 6.2 would be impossible in Ge or other homopolar
substances. From the point of view of studying dielectric
properties, the main drawback of Section 6 was our dependence upon
the various displacement patterns: the electric fields could not be
varied at will, as an independent variable. The present Section
summarizes the <u>most recent</u> applications of the DF which tend to
fill this blank and to open the way to "direct" treatment of
<u>dielectric properties</u> of semiconductors, within the framework of
the Density Functional. They are the treatment of constant
macroscopic electric field imposed from outside (Section 8.1) and
"direct" evaluation of the individual elements of the inverse
dielectric matrix $\varepsilon^{-1}(\vec{q} + \vec{G}, \vec{q} + \vec{G}')$ (Section 8.2).

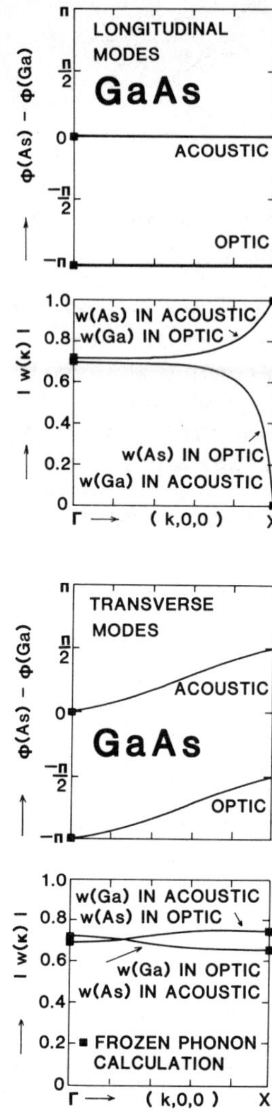

Fig. 7.3.1:   Dispersion of amplitudes for the transverse and longitudinal vibrations with $\vec{k}$ = (k,0,0) in GaAs calculated from the planar force constants given in Tab. 5.1. The values at Γ and X were also obtained by using the frozen phonon approach, Section 4.3.4 (black squares). The complex amplitudes $w(\kappa)$ are represented by real values of $|w(\kappa)|$ and phase factors $\phi(\kappa)$, the actual displacement of a plane ($\ell\kappa$) in a mode kj is given by eq. (5.7.2). The choice of phase factors relates to the basis defined in Figs. 7.0.1 and 7.0.2, with Gallium placed at the origin.

296

<u>8.1. External Macroscopic Electric Field.</u>

Physically the simplest, though certainly not the easiest, of the electrostatic problems is the response of a system to <u>constant</u> external electric field. In general, there is no principal objection to incorporating one more potential "external to the system of electrons" into the HKS equations: the ionic potential $V^{ion}(\vec{r})$ can readily be replaced by some $V^{ion}(\vec{r}) + V^{ext}(\vec{r})$, where V^{ext} is the potential of the field <u>external to the crystal</u>. The difficutly arises when V^{ext} is to represent a macroscopic field <u>constant</u> in space; it comes from the impossibility for plane-wave expansions to handle any quantity which does not have the periodicity of the lattice or superlattice. The problem was met already in Sections 6 and 4.3.10, and it is not new in solid state physics. The standard solution used in other contexts, viz. evaluating the $q \to 0$ limit, is not practical in the "direct" approach, because the smallest q's which can be handled are likely to be still too large for evaluating the $q \to 0$ limit numerically.

The solution proposed in Ref. 65 consists of a kind of "reconciliation" of periodic with aperiodic. Instead of a <u>constant</u> electric field we choose as $V^{ext}(\vec{r})$ the saw-like potential given in Fig. 8.1.1. This field is periodic on a <u>supercell</u> which repeats the elementary unit cell m-times and yet there will be a few <u>elementary</u> cells for which the external potential is a "straight line" and which will "feel" a constant electric field; it is in this part of supercell that the response to <u>constant macroscopic field</u> is to be studied. It is understood that the size of this small part of the supercell has to be equal to at least one elementary unit cell.

Imposing the saw-like potential corresponds physically to incorporating into the crystal <u>capacitor plates</u> charged + and - ; one can imagine that we are to study behavior of the dielectric in a capacitor. At this point it is practical to smooth the sharp edges (dotted lines), in order to improve the convergence of the Fourier expansions involving the saw; this corresponds to allowing the charges to spread somewhat out of the "plates".

With $V^{ion} + V^{ext}$ replacing in DF equations the usual $V^{ion}(\vec{r})$, the self-consistent solutions were found[65] on supercells repeating the elementary cells of Ge and GaAs 8 times. A typical result is shown in Fig. 8.1.2: The fast atom-to atom oscillations are uninteresting for present purposes, the effect of the external saw-like field is in the double-broken shape of the figure; in the absence of the field, the "oscillations" had been "straight". Their envelope, the underlying saw in Fig. 8.1.2, is not as steep as the saw defining $V^{ext}(\vec{r})$ – because the external field is <u>screened</u>. The ratio of the two slopes provides, by definition, the static dielectric constant ε_o.

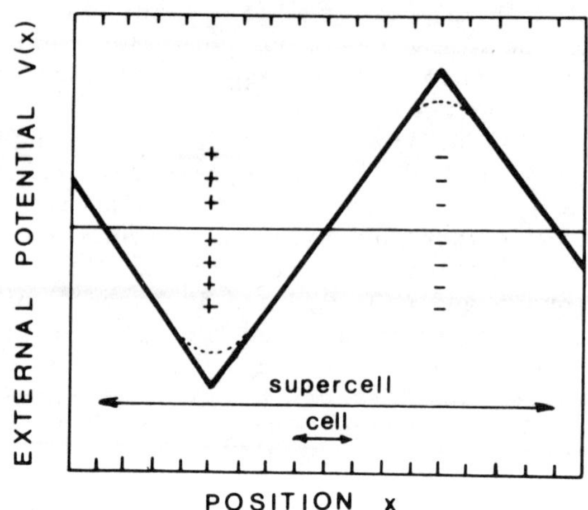

Fig. 8.1.1:   Saw-like potential $V^{ext}(x)$, periodic on supercell, used for producing the constant electric field in at least one elementary cell. The potential can be imagined by incorporating into the system a sequence of capacitor plates, charged alternately + and −. By allowing the charge to spread out of the plates slightly, the sharp edges get rounded and Fourier expansions shortened.

In order to better visualize the screened electric field and to determine the <u>slope</u> of the underlying saw in Fig. 8.1.2, we have subtracted from Fig. 8.1.2 the self-consistent potential corresponding to the "unperturbed" situation, i.e. to the one without external field; the difference $\Delta V(x)$ shown in Fig. 8.1.3 represents the <u>screened</u> saw-like external potential of the Fig. 8.1. (As mentioned already in Section 6.1, the screened potential $V^{scr} = V^{ion} + V^{ext} + V^{H}$, rather than the self-consistent one $V^{sfc} = V^{ion} + V^{ext} + V^{H} + V^{XC}$, has to be considered in dealing with all dielectric quantities.) Comparison of the slopes in Figs. 8.1.1 and 8.1.3 yielded the ratio $\varepsilon_o = 19.08$ for Ge and 13.60 in GaAs, which in both cases is 20 % above the experimental values.

298

A closer inspection of Fig. 8.1.3 reveals that the screened potential is not represented by perfectly straight lines; the bumps, barely visible in Ge but clearly apparent in GaAs, are <u>not</u> computational noise and they reflect a well defined physical effect: <u>microscopic variation of local fields</u>. By treating in the same way $n(\vec{r})$ instead of $V^{scr}(\vec{r})$, the corresponding local variations of electronic charge density were obtained as well[65]. All local variations have their origin in the microscopic inhomogeneity of solid; they are traditionally studied by the linear response theory, which relates them to the off-diagonal elements of the inverse dielectric matrix; this is explained in the articles by A. Baldereschi and R. Resta in the present volume. A quantitative comparison of the local variations in charge density with the predictions[71,72] using the RPA dielectric matrix showed excellent agreement between both approaches, both in magnitude and detailed shape.

The dielectric constant ε_0 can equivalently be determined from the <u>polarization charge</u> accumulated at the "capacitor plates", a quantity which is also available from the $\Delta n(\vec{r})$; in the case of a polar crystal, also the effective charges can be determined from the same self-consistent solutions, by evaluating the Hellmann–Feynman forces. For more details the reader is referred to Ref. 65.

8.2. Inverse Dielectric Matrix.

The shape of Fig. 8.1.1, with the sharp edges rounded, is not dissimilar to a simple sine wave and it is natural to ask how the self-consistent solutions of the DF equations would change in presence of an external potential having a form of a <u>plane wave</u>. Obviously, we are returning to reciprocal space and one can expect the answer to be simple enough when formulated in terms of Fourier components: we are about to find the individual elements of the response matrix $\varepsilon^{-1}(\vec{q}+\vec{G},\vec{q}+\vec{G}')$.

The linear response of a crystal to a small external perturbation V^{ext} is given by the inverse dielectric matrix ε^{-1} as

$$V^{scr}(\vec{q}+\vec{G}) = \sum_{\vec{G}'} \varepsilon^{-1}(\vec{q}+\vec{G},\vec{q}+\vec{G}')\, V^{ext}(\vec{q}+\vec{G}') \qquad (8.2.1)$$

where V^{scr} is the electrostatic potential actually felt by a classical test charge; $\vec{G}$, $\vec{G}'$ are reciprocal lattice vectors, and $\vec{q}$ lies inside the first Brillouin zone. General properties of the response matrix ε^{-1} are thoroughly discussed elsewhere in this Volume[73,74].

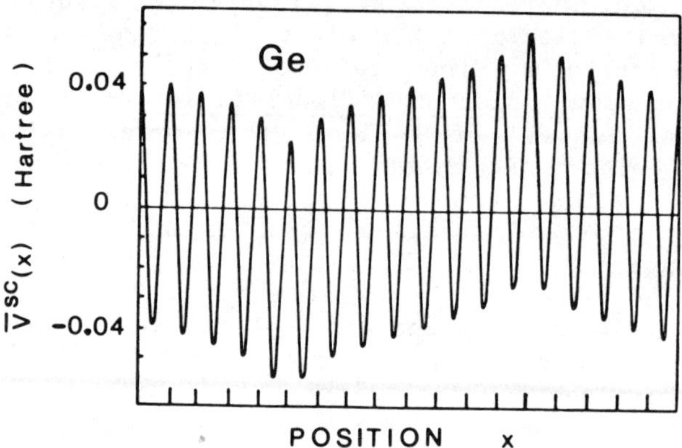

Fig. 8.1.2:   The actual self-consistent potential $V^{sfc}(\vec{r})$ calculated for Ge in the external field shown in Fig. 8.1.1. Averaging over $\underline{y}$ and $\underline{z}$ coordinates like in Fig. 6.1.2. (From Ref. 65.)

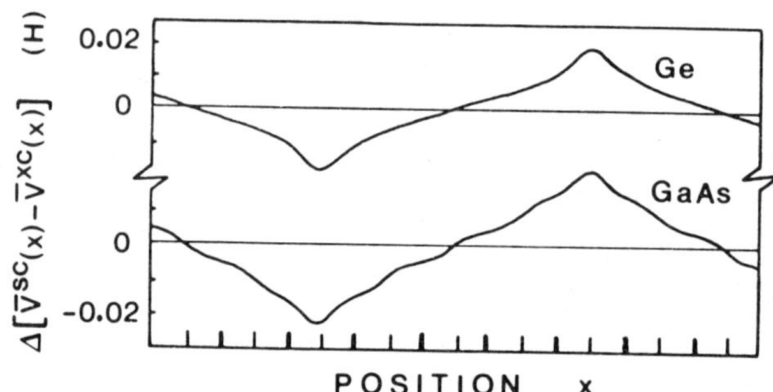

Fig. 8.1.3:   Screened saw-like potential of Fig. 8.1.1 is obtained when the fast atom-to-atom oscillations are removed from Fig. 8.1.2 by subtracting the unperturbed self-consistent potential. $\Delta V^{scr}(\vec{r}) \equiv \Delta V^{ion} + \Delta V^{H}$ is plotted. The ratio of slopes between Fig. 8.1.1 and 8.1.3 determines the dielectric constant of the material. Small deviations from straight-line shape reflect the microscopic variations of local fields. (From Ref. 65.)

300

As in Section 8.1, we can replace the $V^{ion}(\vec{r})$ by $V^{ion}(\vec{r})$ + $V^{ext}(\vec{r})$ and re-do the self-consistency; expressed in terms of the dielectric response theory, the V^{ext} is a "perturbation" of the system and we are looking for the response. The screened potential $V^{scr}(\vec{r}) \equiv V^{ion}(\vec{r}) + V^{H}(\vec{r})$ becomes, in the presence of the external field, $V^{scr}(\vec{r}) + \Delta V^{scr}(\vec{r}) \equiv V^{ion}(\vec{r}) + V^{ext}(\vec{r}) + V^{H}(\vec{r}) + \Delta V^{H}(\vec{r})$ – so that the sought <u>response</u> to $V^{ext}(\vec{r})$ is

$$\Delta V^{scr}(\vec{r}) = V^{ext}(\vec{r}) + \Delta V^{H}(\vec{r}) \quad ; \tag{8.2.2}$$

the ΔV^{scr} and V^{ext} are related by eq. (8.2.1).

Choosing as external potential a <u>single plane wave</u>

$$V^{ext}(\vec{r}) = V_o \exp(i\vec{g}'\vec{r}) \tag{8.2.3}$$

the resulting modification $\Delta V^{scr}(\vec{r})$ will <u>not</u> be limited to a single component and will read

$$\Delta V^{scr}(\vec{r}) = \sum_{\vec{g}} \Delta V^{scr}(\vec{g}) \exp(i\vec{g}\vec{r}). \tag{8.2.4}$$

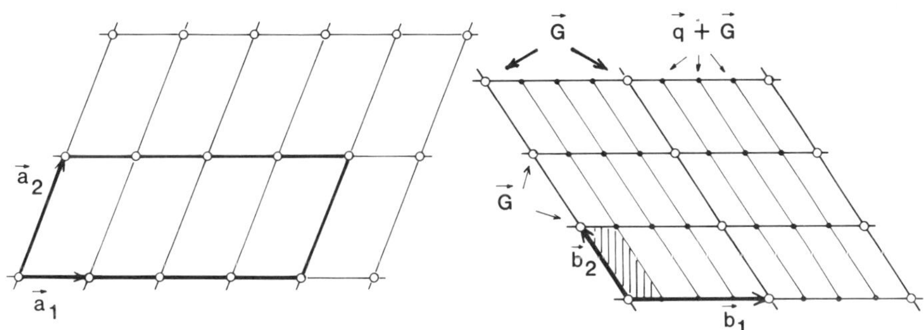

Fig. 8.2.1: Direct and reciprocal lattice vectors of a supercell. An elementary unit cell defined by a_1, a_2, a_3 determines b_1, b_2, b_3, hence the reciprocal lattice vectors $\vec{G}$. For a supercell defined by the translations $4a_1$, a_2, a_3, the reciprocal space is spanned by vectors $1/4\, b_1$, b_2, b_3 (the shaded area); their combinations (black dots) can be written as $\vec{q} + \vec{G}$, where $\vec{q}$ lies inside the 1st Brillouin zone (of the elementary cell). By using different supercells, the individual elements of the inverse dielectric matrix $\varepsilon^{-1}(\vec{q}+\vec{G},\vec{q}+\vec{G}')$ can be determined "directly" for different values of $q \neq 0$.

As the coefficients of the expansion (8.2.4) are related to the "perturbation" (8.2.3) by the eq. (8.2.1)

$$V(\vec{g}) = \varepsilon^{-1}(\vec{g},\vec{g}')\, V_o,$$ (8.2.5)

the ΔV^{scr} yields, from a <u>single</u> self-consistent calculation, an <u>entire column</u> of the ε^{-1} matrix.

An explanation of how one can deal with the <u>fractional</u> reciprocal lattice vectors, the $\vec{q}$'s in eq. (8.2.1) is needed, because so far the procedure might give the impression of evaluating ε^{-1} at $\vec{q} = 0$ only. Indeed, starting with self-consistent solutions for <u>elementary</u> unit cell, and choosing for the periodicity g' of the "perturbing" potential (8.2.3) one of the reciprocal lattice vectors $\vec{G}$, then also the "response" (8.2.4) is given by a set of $V(\vec{G})$'s, without any $\vec{q} \neq 0$. However, starting from a conveniently chosen <u>supercell</u>, the reciprocal lattice vectors $\vec{g}$ defined by the supercell will necessarily be of the form $\vec{q}+\vec{G}$, with $\vec{G}$ corresponding to the elementary cell and $\vec{q}$ lying in the 1st Brillouin zone. This is illustrated in Fig. 8.2.1, where a very simple supercell is chosen, one which repeats $\vec{a}_1$ four times: The unit cell of the corresponding reciprocal space becomes 4 times smaller and the $\vec{g}$-vectors are thus multiples of $\vec{b}_1/4$, and combinations.

A few typical values obtained by the above method on Ge are shown in Tab. 8.1, more results for GaAs (at both $q = 0$ and $\vec{q} \neq 0$) are given in Ref. 75. Very little comment is possible at present, because results from different sources are based on different pseudopotentials (i.e. different band structures) and do not use the same approximations (RPA, DF). Nevertheless, the inclusion in the present treatment of the <u>exchange</u> seems to account for large part of the differences: the off-diagonal elements in Tab. 8.1 are systematically larger (in absolute value) than in RPA, the diagonal ones are about 20 % lower.

So far the standard method for determination of ε^{-1} has consisted of inverting a large matrix $\varepsilon^{+1}_{76}(\vec{q}+\vec{G},\vec{q}+\vec{G}')$ with elements computed through the Adler-Wiser formula[76]; all the RPA-results in Tab. 8.1 were computed in this way. This approach requires, for every individual matrix element, the evaluation of extensive sums over all valence <u>and</u> conduction bands, and is limited to the RPA approximation. (The exchange-correlation effects can then be incorporated as corrections.) The method is rather cumbersome and meets convergence problems concerning both number of conduction bands to be included and dimension of the ε^{+1} matrix to be inverted.

The main advantage of the present method is in evaluating the elements of ε^{-1} "directly" (without needing to invert the ε^{+1}

Table 8.1: Some elements of the symmetrized inverse dielectric matrix $|\vec{G}|/|\vec{G}'|\ \varepsilon^{-1}(\vec{G},\vec{G}')$ for Ge, calculated by the present "direct" method[75] and through the Adler-Wiser formula[76] (RPA).

$\vec{G}$	$\vec{G}'$	$\varepsilon^{-1}(\vec{G},\vec{G}')$ [a]	RPA[b]	RPA + Xα[c]
000	000	......	+0.063	+0.062
111	111	+0.517	+0.610	+0.552
200	200	+0.570	+0.669	+0.615
200	111	−0.048	−0.045	−0.043
222	111	+0.049	+0.043	+0.049
022	111	−0.048	−0.043	−0.045
1-1-1	−111	+0.077	+0.048	+0.075
200	−111	−0.014	−0.007	−0.016
311	200	+0.050	+0.041	+0.047
020	200	+0.023	+0.013	+0.017
−200	200	+0.015	+0.011	+0.022

(a) Ref. 75.   (b) Ref. 79.   (c) Ref. 80,81.

matrix) and particularly in requiring knowledge of the _valence bands only_: we are _explicitly_ using the fact that ε^{-1} is a property of the ground state. These factors make the "direct" method by 1 − 2 orders of magnitude more efficient numerically; moreover, the effects of exchange and correlation are included automatically − within the limits of the DF approximation. On the other hand, the "direct" method cannot handle the elements $\varepsilon^{-1}(\vec{q}+\vec{G},\vec{q}+\vec{G}_1')$ with simultaneously $\vec{q}=0$ and $\vec{G}=0$ (or $\vec{G}'=0$): in particular $\varepsilon^{-1}(0,\vec{G}')$, $\varepsilon^{-1}(\vec{G},0)$ and $\varepsilon^{-1}(0,0)$ are not accessible. Whilst the first two are zero by symmetry, the $\varepsilon^{-1}(0,0)$ corresponds to macroscopic screening. We notice, however, that the limit $\varepsilon^{-1}(q,q)$ for $q \to 0$ can be evaluated by another "direct" method, viz. the one discussed in Section 8.1: the limit is the inverse of the static dielectric constant.

8.3. Discussion

A _numerical_ evaluation of the above $q \to 0$ limit is the idea behind an alternative method for ab initio determination of static dielectric constant, presented recently by McKitterick[64]. By essentially the same method as described in Section 8.2, McKitterick has calculated $\varepsilon^{-1}(\vec{q},\vec{q})$ for GaAs at $\vec{q} = 2\pi/a(q,q,q)$; four self-consistent calculations on doubled, quadrupled, sextupled and octupled supercells yielded values of $\varepsilon^{-1}(q,q)$ at q= 0.5, 0.25, 0.167 and 0.125 which, in turn, made possible to extrapolate to zero by a polynomial of the type $A + Bq^2 + Cq^4$. A similar idea of numerical evaluation of limits is applied in Ref. 64 also to determination of <u>effective charges</u> and of the <u>piezoelectric constant</u>.

The most recent of the "direct" methods concerns the evaluation of the <u>elements of the ε^{+1} dielectric matrix</u>[84]; it is based on the fact that the modification in the electronic <u>charge density</u> $\Delta n(\vec{r})$ is related to the modification in <u>total potential</u> $V^{src}(\vec{r})$ by the <u>polarizability matrix</u> $\chi(\vec{q}+\vec{G},\vec{q}+\vec{G}')$. The method does not require achieving self-consistency and provides the elements of $\varepsilon^{+1}_{+1}(q+G,q+G')$ <u>within the RPA</u> approximation. After inversion of the ε^{+1} the combination of the two direct methods (for ε^{-1} and ε^{+1}) seems to offer a possibility of switching on and off, at will, the exchange-correlation; its effects on various physical properties could then be studied in detail. Besides the original work Ref. 77, the reader is referred to the articles[73,74] of R. Resta and A. Baldereschi in this Volume.

9. CONCLUSION

The lecture notes explained two "direct" approaches for dealing with dynamical properties of solids from first principles: the <u>"frozen phonon"</u> method applies to isolated phonons corresponding to high-symmetry points of reciprocal space; the interplanar <u>force constant</u> method treats entire phonon dispersion along a selected propagation direction - still required to be one of high symmetry. The macroscopic <u>electric field</u> encountered when applying the above methods to polar crystals could have been handled within the same scheme, and its understanding suggested that the <u>dielectric response</u> in semiconductors could also be treated from first principles.

The work summarized in these lecture notes is based on extensive use of the <u>Density Functional</u> method, and a local expression selected for the exchange-correlation operator is the only essential approximation. The DF theory, originally thought as method for evaluation of total <u>energy</u>, turned out to be equally efficient for determination of <u>forces</u> acting on atoms, macroscopic <u>stresses</u> and, most recently, of details of the microscopic <u>dielectric response</u>.

304

A number of new results obtained with the "direct" ab initio
methods include phonon frequencies, anharmonicities, predictions of
displacement patterns, soft-mode phase transitions, effective
charges, dielectric constant, local field variations, elements of
inverse dielectric matrix, etc.; they were all obtained from the
same fundamental equations. The Density Functional method opens a
way to unified description of ground state properties of solids:
static, dynamic and dielectric ones. Though all the partial results
above are interesting by themselves, they are even more important
by providing further tests of and support for the validity of the
Density Functional theory.

Acknowledgments

I would like to acknowledge the invaluable advice of R. M.
Martin on several controversial issues encountered while this text
was being written; I have also benefited from discussions with H.
Bilz, V. Heine, R. Resta, O. H. Nielsen, C. and N. Churcher. I am
grateful for the hospitality extended to me by the Cavendish
Laboratory, Cambridge. This work was partly supported by a NATO
grant and the computer resources were provided by the Scientific
Committee of CCVR (Centre de Calcul Vectoriel pour la Recherche,
France).

APPENDIX 1: PSEUDOATOMS Ga AND As

Calculations using pseudopotentials instead of dealing with
all-electron problem rely on the validity of what is called "Frozen
Core Approximation"; as for the behavior of the charge densities of
the valence electrons, it is expected to resemble the "real"
(all-electron) charge distribution as closely as possible outside
the core-region.

Whereas the latter requirement is trivially satisfied with the
norm-conserving non-local potentials[16-18], which are generated
starting from atomic wave-functions, the realism of any local
potential can be questioned, and its "efficiency" in mimicking the
real charge densities is worth testing. In this Appendix we
summarize results of self-consistent atomic calculations for Ga and
As, based on the same Density Functional theory as used throughout
this work, on the same assumption of the Slater $X\alpha = 0.8$ exchange,
but the full nuclear potential $-Z/r$ is replaced by the local
pseudopotential used in the solid. The Hohenberg-Kohn-Sham

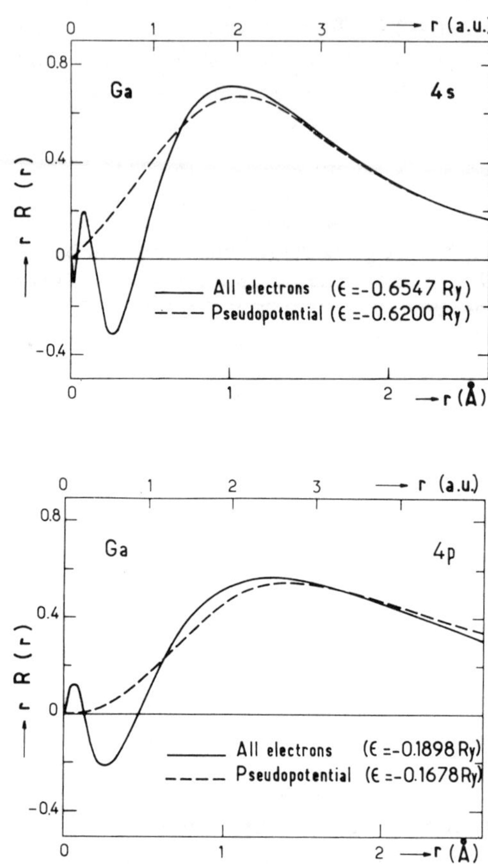

Fig. A1.2: Ga and As atoms and pseudo-atoms: radial parts of the
wavefunctions (solid lines) ane pseudo-wavefunctions (broken
lines), as calculated for valence electrons from the
self-consistent potentials and pseudopotentials of Fig. A1.1.

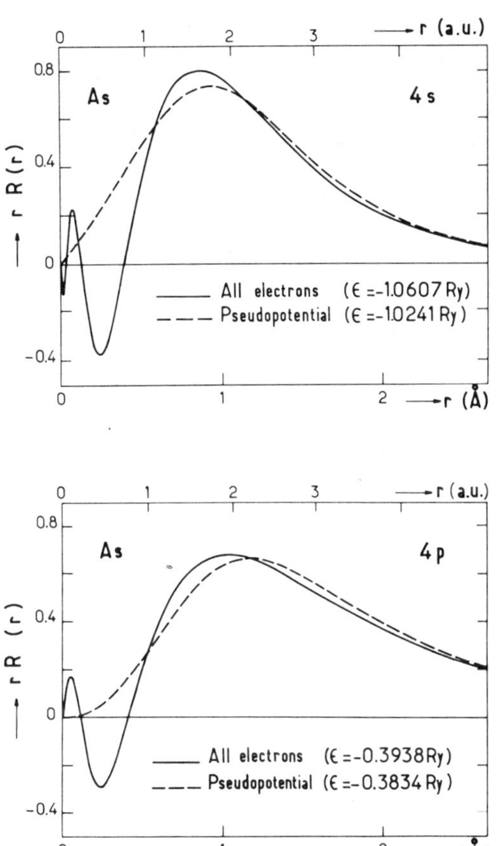

The pseudo-wavefunctions approximate the genuine all-electron wavefunctions outside the core, but do not show any oscillations within the core region. The corresponding eigenvalues ε_i agree to within <36 mRy over a range of $\cong$ 1 Ry. All functions $R_n(r) \, Y_{\ell m}(\theta,\phi)$ normalized to unity.

equations are integrated in real space, a procedure based on the standard Herman-Skillman scheme[78] is used; the one-electron wavefunctions are assumed to be of the form

$$\psi(\vec{r}) = R_n(r)\, Y_{\ell m}(\theta,\phi) \quad , \tag{A1.1}$$

and the charge-distributions of the incompletely filled shells are spherically averaged. The results are compared to all-electron calculations in Fig. A1.1 to A1.3. The deep all-electron potential is seen to be replaced by a shallow self-consistent pseudopotential (Fig. A1.1), the pseudo-wavefunctions do not show any oscillations inside the core region (Fig. A1.2) and the pseudo-charge densities agree, approximately, with the authentic charge distribution of valence electrons. The eigenvalues show errors of <36 mRy over a range of $\cong$ 900 mRy.

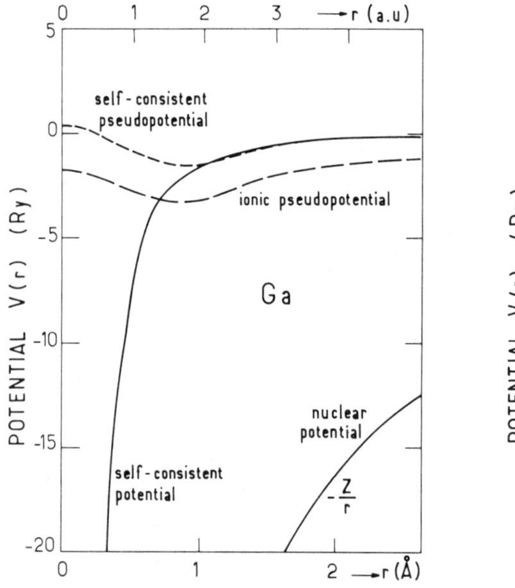

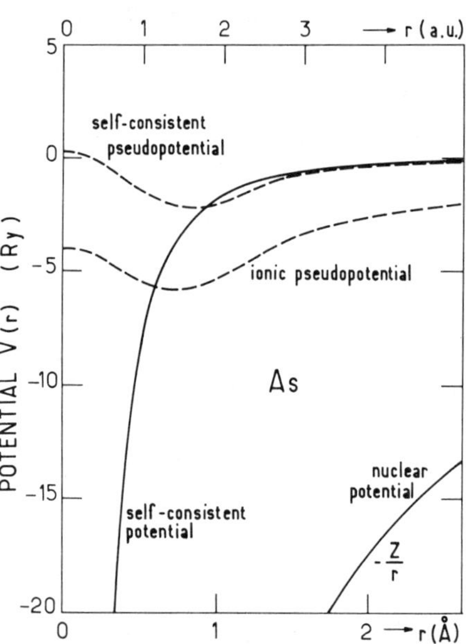

Fig. A1.1: Ionic pseudopotentials of Fig. 2.1 plotted in direct space (broken line) and compared with the full nuclear potential $-Z/r$ (solid line); when used for self-consistent determination of the (pseudo-)atomic charge densities, the screening converts them into, respectively, self-consistent pseudopotential (broken line) and self-consistent potential (solid line). The exchange factor $\alpha = 0.8$ and spherical averaging of the charge density of incomplete shells were used in solving self-consistently the equations of DF, as described in Ref. 78.

308

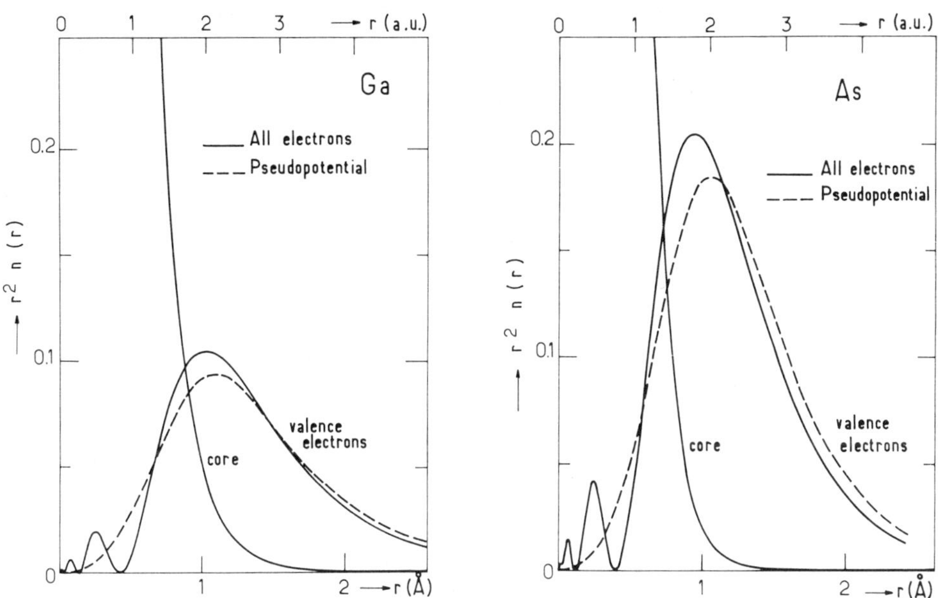

Fig. A1.3: Ga and As atoms and pseudoatoms: Charge densities n(⃗r) calculated from the self-consistent potentials and pseudopotentials shown in Fig. A1.1. Solid lines: all-electron problem; broken lines: pseudized problem. The main effect of replacing the full nuclear potential by a pseudopotential, and of reducing the all-electron problem to that of valence electrons alone, consists in suppressing the oscillations in the core region and approximating the charge density outside the core. The densities corresponding to valence electrons are obtained from the wavefunctions and pseudo-wavefunctions of Fig. A1.2 by averaging over the angular degrees of freedom.

REFERENCES

1. V. Heine, D. Weaire in: "Solid State Physics", Vol. 24, H. Ehrenreich et al., eds., Academic Press (1970).

2. P. Hohenberg, W. Kohn, Phys. Rev. 136, B864 (1964); W. Kohn, L. J. Sham, ibid. 140, A1133 (1965); L. J. Sham, W. Kohn, ibid. 145, B561 (1966).

3. D. J. Chadi, R. M. Martin, Solid State Commun. 19, 643 (1976).

4. H. Wendel, R. M. Martin, Phys. Rev. Lett. 40, 950 (1978); Phys. Rev. B19, 5251 (1979).

5. M. T. Yin, M. L. Cohen, Phys. Rev. Lett. 45, 1004 (1980).

6. K. Kunc, R. M. Martin, Phys. Rev. B24, 2311 (1981).

7. B. N. Harmon, W. Weber, D. R. Hamann, Phys. Rev. B 25, 1109 (1982).

8. V. L. Moruzzi, J. F. Janak, A. R. Williams, "Calculated Electronic Properties of Metals", Pergamon Press, New York (1978).

9. K. Kunc, R.M. Martin, J. Phys (Paris) 42 - Suppl. C6, 649 (1981).

10. K. Kunc, R.M. Martin, Phys. Rev. Letters 48, 406 (1982).

11. M. T.Yin, M. L. Cohen, Phys. Rev. B 25, 4317 (1982).

12. R. M. Martin, this Volume.

13. K. Kunc., R.M. Martin, in: "Ab initio Calculation of Phonon Spectra", J. T. Devreese et al., eds., Plenum Press, New York (1983), pp. 65-99.

14. J. R. Chelikowsky, M. L. Cohen, Phys. Rev. B13, 826 (1976).

15. W. E. Pickett, S. G. Louie, M. L. Cohen, Phys. Rev. B17, 815 (1978).

16. D. R. Hamann, M. Schluter, C. Chiang, Phys. Rev. Lett. 43, 1494 (1979).

17. G. B. Bachelet, D. R. Hamann, M. Schluter, Phys. Rev. B 26, 4199 (1982).

18. G. P. Kerker, J. Phys. C 13, L189 (1980).

19. P. O. Loewdin, J. Chem. Phys. 19, 1396 (1951); D. Brust, Phys. Rev. 134, A1137 (1964).

20. H. J. Monkhorst, J. D. Pack, Phys. Rev. B13, 5188 (1976).

21. O.H. Nielsen, R.M. Martin, Phys. Rev. Letters 50, 697 (1983).

22. O. H. Nielsen, R. M. Martin, this Volume.

23. F. D. Murnaghan, Proc. Nat. Acad. Sci. USA 30, 244 (1944).

24. see e.g. M. T. Yin, M. L. Cohen, Phys. Rev. B 26, 5668 (1982).

25. E. Holzschuh, Phys. Rev. B 28, 7346 (1983).

26. K. J. Chang, M. L. Cohen, Solid State Commun. 50, 487 (1984).

27. M. T. Yin, M. L. Cohen, Solid State Commun. 38, 625 (1981).

28. S. Froyen, M. L. Cohen, Solid State Commun. 43, 447 (1982).

29. S. Froyen, M. L. Cohen, Phys. Rev. B 28, 3258 (1983).

30. S. Froyen, M. L. Cohen, Phys. Rev. B 29, 3770 (1984).

31. A. Mooradian, G. B. Wright, Solid State Commun. 4, 431 (1966).

32. G. Dolling, J.L.T.Waugh, in: "Lattice Dynamics", R.F.Wallis, ed., Pergamon Press, London (1965), p.19.

33. R. Trommer, H. Muller, M. Cardona, P. Vogl, Phys. Rev. $\underline{B21}$, 4869 (1980).

34. K. Kunc, H. Bilz, <u>in</u>: "Proceedings of the International Conference on Neutron Scattering, Gatlingburg 1976", R. M. Moon, ed., ORNL, Tennessee (1976), p. 195.

35. "Landolt-Börnstein", New Series, Group III, Vol. 17, O. Madelung, ed., Springer (1982), p. 236.

36. G. Nilsson, G. Nelin, Phys. Rev. $\underline{B3}$, 364 (1971).

37. M. T. Yin, M. L. Cohen, Solid State Commun. $\underline{43}$, 391 (1982).

38. M. T. Yin, M. L. Cohen, Phys. Rev. B $\underline{26}$, 3259 (1982).

39. K. Kunc, P. Gomes Dacosta, to be published.

40. Pui K. Lam, M. L. Cohen, Phys. Rev. B $\underline{25}$, 6139 (1982).

41. J. Ihm, M. T. Yin, M. L. Cohen, Solid State Commun. $\underline{37}$, 491 (1981).

42. K.-M. Ho, C.-L. Fu, B. N. Harmon, W. Weber, D. R. Hamann, Phys. Rev. Letters $\underline{49}$, 673 (1982).

43. K.-M. Ho, C.-L. Fu, B. N. Harmon, Phys. Rev. B $\underline{28}$, 6687 (1983).

44. W. Weber, Habilitationsschrift; will appear <u>in</u>: "Electronic Structure of Complex Systems", P. Phariseau and W. Temmerman, eds., Plenum Press (1984).

45. H. Goldstein, "Classical Mechanics", Addison-Wesley Inc. (1956)

46. L. D. Landau, E. M. Lifshitz, "Mechanics", Pergamon Press (1960).

47. J. A. Reissland, "The Physics of Phonons", J. Wiley & Sons, (1973).

48. R. P. Feynman, Phys. Rev. $\underline{56}$, 340 (1939).

49. J. C. Slater, J. Chem. Phys. $\underline{57}$, 2389 (1972).

50. J. Ihm, A. Zunger, M. L. Cohen, J. Phys. $\underline{C12}$, 4409 (1979); J. Ihm, M. L. Cohen, Solid State Commun. $\underline{29}$, 711 (1979).

51. A. A. Maradudin, E. W. Montroll, G. H. Weiss, I. P. Ipatova, "Solid State Physics", Suppl. 3, H. Ehrenreich et al., eds., Academic Press, (1971).

52. M. Cardona, K. Kunc, R.M. Martin, Solid State Commun. $\underline{44}$, 1205 (1982).

53. R. Tubino, L. Piseri, G. Zerbi, J. Chem. Phys. $\underline{56}$, 1022 (1972)

54. M. J. P. Musgrave, J. A. Pople, Proc. Roy. Soc. (London) A $\underline{268}$, 474 (1962); H. L. McMurry, A. W. Solbrig, Jr., J. K. Boyter, C. Noble, J. Phys. Chem. Solids $\underline{28}$, 2359 (1967); B. D. Singh, B. Dayal, phys. stat. solidi $\underline{38}$, 141 (1970).

55. F. Herman, J. Phys. Chem. Solids $\underline{8}$, 405 (1959).

56. W. Weber, Phys. Rev. Letters $\underline{33}$, 371 (1974); Phys. Rev. B $\underline{15}$, 4789 (1977).

57. A. Segmuller, H.R. Neyer, Phys. Kond. Materie $\underline{4}$, 63 (1965).

58. C.S.G.Cousins, private communication.

59. N. E. Christensen, Solid State Commun. $\underline{50}$, 177 (1984).

60. R. M. Martin, private communication.

61. R.M. Martin, Solid State Commun. $\underline{8}$, 799 (1970); Phys. Rev. B $\underline{1}$, 4005 (1970).

62. K. Kunc, Physica $\underline{116}$B, 52 (1983).
63. M. Born, K. Huang, "Dynamical Theory of Crystal Lattices", Oxford University Press (1954).
64. J. B. Mc Kitterick, Phys. Rev. B $\underline{28}$, 7384 (1983).
65. K. Kunc, R. Resta, Phys. Rev. Letters $\underline{51}$, 686 (1983).
66. R. M. Martin, K. Kunc, Phys. Rev. $\underline{B24}$, 2081 (1981).
67. G. Lucovsky, R. M. Martin, E. Burstein, Phys. Rev. $\underline{B4}$, 1367 (1971).
68. A. Zunger, M. L. Cohen, Phys. Rev. $\underline{B18}$, 5449 (1978).
69. L. J. Sham, Phys. Rev. $\underline{188}$, 1431 (1969).
70. R. M. Pick, M. H. Cohen, R. M. Martin, Phys. Rev. B $\underline{1}$, 910 (1970).
71. A. Baldereschi, R. Car, E. Tosatti, Solid State Commun. $\underline{32}$, 757 (1979).
72. R. Resta, A. Baldereschi, Phys. Rev. B $\underline{23}$, 6615 (1981).
73. A. Baldereschi, R. Resta, this Volume.
74. R. Resta, A. Baldereschi, this Volume.
75. K. Kunc, E. Tosatti, Phys. Rev. B $\underline{29}$, 7045 (1984).
76. S. L. Adler, Phys. Rev. $\underline{126}$, 413 (1962); N. Wiser, Phys. Rev. $\underline{129}$, 62 (1963).
77. A. Fleszar, R. Resta, Phys. Rev. B, to be published.
78. F. Herman, S. Skillman, "Atomic Structure Calculations", Prentice Hall Inc. (1963).
79. A. Baldereschi, E. Tosatti, Phys. Rev. B $\underline{17}$, 4710 (1978).
80. P. E. Van Camp, V. E. Van Doren, J. T. Devreese, Phys. stat. solidi (b) $\underline{110}$, K133 (1982); Phys. Rev. $\underline{B24}$, 1096 (1981).
81. P. E. Van Camp, V. E. Van Doren, J. T. Devreese, private communication.

STRESS: CONCEPTS AND APPLICATIONS

Ole H. Nielsen

NORDITA, Blegdamsvej 17, DK-2100 Copenhagen, Denmark

and

Richard M. Martin

Xerox Palo Alto Research Center, 3333 Coyote Hill Road
Palo Alto, CA 94304

The stress theorem determines the stress from the electronic
ground state of any quantum system with arbitrary strains and
atomic displacements. We derive this theorem in reciprocal space,
within the local-density-functional approximation. The evaluation
of stress, force and total energy permits, among other things, the
determination of complete stress-strain relations including all
microscopic internal strains. We describe results of *ab-initio*
calculations for Si, Ge, and GaAs, giving the equilibrium lattice
constant, all linear elastic constants c_{ij} and the internal strain
parameter ζ.

1. Introduction

Total energy calculations of the quantum mechanical ground
state have advanced significantly in recent years, and have been
applied to an ever-increasing number of different systems and
physical properties. The core of most of this work is the
density-functional theory of Hohenberg, Kohn and Sham, and in
particular the local-density approximation (see, e.g., Lundqvist
and March, 1983). This theory is based on the variational
principle and applies to the ground-state of a quantum system. It
is, however, not restricted to the state of globally lowest

energy, but is equally applicable to systems with external constraints such as volume and crystal symmetry.

This is precisely the concept that leads to very many applications: Often a system may be characterized by a geometric arrangement of its nuclei that is stable over long periods of time, compared to the time it takes the electron system to find its ground state (this is the adiabatic approximation). In this case the nuclei are thought of as classical particles at fixed positions in space, whose Coulomb fields impose constraints upon the electrons. By varying the nuclear positions in some desired way and calculating the electronic ground state, the total energy of the entire system can be evaluated as function of nuclear position, and knowledge is gained of the physical behaviour of the system.

Notwithstanding the role of total energy as the fundamental quantity, many advantages are obtained when considering directly the derivatives of total energy with respect to structural parameters. Such quantities are defined by considering the change in total energy due to an infinitesimal change in some parameter λ which constrains the system (e.g., the position of a nucleus). The derivative $\partial E_{Total}/\partial\lambda$ is the *conjugate force* corresponding to λ. Even though a change in λ disturbs the wavefunction to first order in λ, the total energy changes only to *second order* in the wavefunction because of the variational principle. Thus the conjugate force only contains terms that are *explicit* derivatives with respect to λ of the expression for total energy.

This general result is well known as the "Hellmann-Feynman" theorem when λ represents the position **x** of a nucleus. The force **F** that the system exerts on the nucleus is the expectation value of minus the gradient of $V(\mathbf{x})$, where V is the potential that acts on the nucleus. This theorem was originally derived by Ehrenfest (1927), and was used in Hellmann's (1937) treatise to establish the forces in a molecule. Feynman (1939) independently derived the result for molecules. We will refer to the result simply as the "force theorem".

A different type of structural parameter was considered recently by the present authors, namely λ representing a homogeneous macroscopic *strain* defined as the linear scaling of all particle positions as $\mathbf{x}\rightarrow(1+\epsilon)\mathbf{x}$. The ϵ is a constant 3x3 strain tensor, and $\epsilon=0$ corresponds to some reference configuration. The conjugate force is in this case defined as the macroscopic *stress* σ, and an explicit general expression denoted the "stress theorem" is derived by Nielsen and Martin (1983). The result is a generalization of the quantum virial theorem (Born et al., 1926), and uses the idea of scaling introduced in the elegant paper by Fock (1930). McLellan (1974) has discussed a similar result.

The combined force and stress theorems are necessary and sufficient to describe the general equation-of-state of a quantum system, i.e. the relations of force and stress to displacements and strain, and thus constitute a powerful tool in the study of structural and dynamic properties of matter. For example, phonon properties can be studied in great detail by imposing regular nuclear displacements and calculating the restoring forces. The reader is referred to the papers by Martin and by Kunc (this volume).

One application of the stress theorem is the study of elastic properties of solids, which becomes straightforward when a suitable finite macroscopic strain is applied to the solid. When the wavefunctions of the distorted solid are known, the stress tensor is evaluated with the stress theorem. In the harmonic approximation elastic constants are defined as the ratio of stress to strain, and it is furthermore possible to go to large strains to obtain all nonlinear elastic properties. In general it is necessary to be concerned with internal strains that may appear microscopically owing to the lower symmetry of the strained solid. In section 6 we show in detail how this problem is solved by combining the stress and force theorems.

The present paper is organized as follows: Section 2 deals with the pseudopotential technique for solving the Schrödinger equation, section 3 derives the stress theorem expressed in reciprocal space within the local-density approximation. Section 4 comments on a number of technical but nevertheless important points in *ab-initio* calculations. Section 5 deals with calculations on the semiconductors Si, Ge, and GaAs, whose elastic properties is the topic of section 6.

2. Ab-initio pseudopotential calculations

The local-density approximation describes the electronic many-body system in terms of a single-particle-like Schrödinger equation for each occupied state in the system. The electrons are, besides external nuclear Coulomb potentials, subjected to Coulomb repulsion from the other electrons (Hartree potential) and to exchange and correlation (x-c) potentials. The latter describe the interaction of each electron with its own surrounding x-c hole (see, e.g., von Barth and Williams, 1983).

Having formulated the Schrödinger equation to be solved by iteration until selfconsistency, it is necessary to select a practical method for performing actual calculations. Among the available *ab-initio* methods, the recent developments in pseudopotential theory makes this technique advantageous in studies of a wide range of systems. The present paper is based upon results obtained with this method.

A crucial development in pseudopotential theory is the formulation of normconserving pseudopotentials (Hamann et al., 1979; Kerker, 1980). From a local-density calculation of the all-electron free atom, the relatively weak pseudopotentials which bind only the valence electrons are constructed. The valence pseudo-wavefunctions do not contain the oscillations necessary to orthogonalize to the core, but are instead smooth functions which are much easier to handle in calculations on real solids. The features of such potentials are discussed in detail by, e.g., Bachelet et al. (1982), who also present pseudopotentials for all atoms from H to Pu.

The normconserving pseudopotentials have proven to work well for many systems, notably semiconductors (see, e.g., Kunc, this volume) and their surfaces (see, e.g., Northrup and Cohen, 1982), ionic compounds (Froyen and Cohen, 1984), and simple metals (see, e.g., Lam and Cohen, 1981). Applications to transition metals also exist (see, e.g., Greenside and Schlüter, 1983). The pseudopotential approximation becomes less satisfactory when valence and core electrons begin to have large overlap, both because of the pseudo-wavefunctions lacking nodes, and because the x-c potential in the core region should also account for the presence of the core electrons. The latter problem can in many cases be treated well by "nonlinear" pseudopotentials (Louie et al., 1982).

Given the valence electron pseudopotentials of the free atoms, we next wish to carry out selfconsistent calculations for a particular solid. The spectrum of methods that apply to all-electron solids can in principle equally well be applied with a pseudopotential instead of the Coulomb potential. However, the weak strength of the pseudopotential and the resulting slow variation of the wavefunctions makes it possible to apply basis sets that would be intractable for the all-electron problem, and which present very attractive features. This work, and a large body of other works, expand wavefunctions in terms of plane waves. From the early days of solid-state physics the usefulness of this basis set has been realized (see, e.g., Herring, 1940), and it has been discussed pedagogically by Cohen, Heine and Phillips (1982). The main advantages are that the basis set is "unbiased" by the structure to which it is being applied, and that systematic improvements are obtained simply by increasing the number of Fourier components. The drawback is that basis sets sometimes must be chosen very large, setting a limit on the systems that can be treated. However, this limit moves steadily with the improvements in modern computing facilities.

With plane-wave representation of wavefuntions the method for calculation of total energy has been described by Wendel and Martin (1978, 1979). Explicit expressions covering also nonlocal (angular momentum dependent) pseudopotentials and forces were given by Ihm, Zunger and Cohen (1979). The method is described by Martin (this volume).

3. Stress calculations

The stress theorem relies upon the variational principle applied together with a strain-scaling of the quantum system, as discussed in detail by the present authors elsewhere (Nielsen and Martin, to be published). The strain scales particle positions as $\mathbf{x} \to (1+\varepsilon)\mathbf{x}$, and by definition the macroscopic stress $\sigma_{\alpha\beta}$ per volume Ω (α and β denote cartesian coordinates) is derived from the total energy E_{Tot} by

$$\sigma_{\alpha\beta} = \Omega^{-1}\, \partial E_{Tot}/\partial \varepsilon_{\alpha\beta} \tag{1}$$

In the case of plane-wave basis sets the scaling proceeds on the reciprocal-space vectors as $\mathbf{G} \to (1+\varepsilon)^{-1}\mathbf{G}$, which is seen by the definition $\mathbf{a}_i \cdot \mathbf{b}_j = \delta_{ij}$, where $\mathbf{a}_i$ and $\mathbf{b}_j$ are real- and reciprocal-lattice primitive translation vectors, respectively. Thus one finds the derivative of reciprocal-space vectors given by

$$\partial G_\gamma / \partial \varepsilon_{\alpha\beta} = -\, \delta_{\alpha\gamma} G_\beta \tag{2}$$

It is noted that structure factors, $\exp(i\mathbf{G}\cdot\mathbf{x})$, are unchanged since $\mathbf{a}_i \cdot \mathbf{b}_j = \delta_{ij}$. Volume Ω times charge density $\rho(\mathbf{G})$ (or Ω times wavefunction products $\psi\psi^*$) are unchanged by construction of the scaling argument: Wavefunctions are "stretched" but their normalization preserved. The terms in the stress are derived as follows, by using the expressions of Ihm et al. (1979) for total energy (in Hartree atomic units):

i) The kinetic energy results in a stress

$$-\sum_{\mathbf{k},\mathbf{G},i} \left| \psi_i\left(\mathbf{k}+\mathbf{G}\right) \right|^2 \left(\mathbf{k}+\mathbf{G}\right)_\alpha \left(\mathbf{k}+\mathbf{G}\right)_\beta \tag{3}$$

where $\mathbf{k}$ runs over the first Brillouin zone, $\mathbf{G}$ over the reciprocal lattice, and i over occupied states. This term is the momentum flux in the α-direction averaged over all planes with normal β.

ii) The potential energy due to local, spherically symmetric potentials V_τ^L gives

$$- \sum_{\mathbf{G},\tau}{}' S_\tau(\mathbf{G}) \left[\frac{\partial V^L_\tau(G)}{\partial(G^2)} 2\mathbf{G}_\alpha \mathbf{G}_\beta + V^L_\tau(G)\delta_{\alpha\beta} \right] \rho(\mathbf{G})^* \qquad (4)$$

where τ labels the atoms, S_τ denotes the structure factor, and Σ' denotes a summation excluding $\mathbf{G}{=}\mathbf{0}$. This term is similar to the "virial" of the ion-electron interaction encountered in the virial theorem.

iii) Nonlocal potentials $\Delta V_{1\tau}{}^{NL}$ that are superposed on $V_\tau{}^L$ to describe the dependence of the pseudopotential with angular momentum 1 (1 is often ≤ 2) give

$$\sum_{\mathbf{k},\mathbf{G},\mathbf{G}',i,l,\tau} S_\tau(\mathbf{G}-\mathbf{G}') \frac{\partial[\Delta V^{NL}_{l,\tau}(\mathbf{k}+\mathbf{G},\mathbf{k}+\mathbf{G}')]}{\partial \varepsilon_{\alpha\beta}} \Psi_i\!\left(\mathbf{k}+\mathbf{G}\right) \Psi_i\!\left(\mathbf{k}+\mathbf{G}'\right)^* \qquad (5)$$

where the derivative of $\Delta V_{1\tau}{}^{NL}$ is straightforwardly performed using Eq. (2), but since the expression is quite lengthy we defer it to a forthcoming paper (Nielsen and Martin, to be published).

iv) The $\mathbf{G}{=}\mathbf{0}$ component of the Coulomb interaction formally diverges, but is cancelled due to charge neutrality as shown by Wendel and Martin (1979). Since the pseudopotential differs from the Coulomb potential at short distances, the average non-Coulombic part (denoted α_τ) must be accounted for. This is the term denoted "αZ" by (Ihm et al., 1979) which gives an isotropic stress

$$- \delta_{\alpha\beta} \left(\Sigma_\tau \alpha_\tau\right) \left(\Omega^{-1} \Sigma_\tau Z_\tau\right) \qquad (6)$$

with Z_τ denoting the ionic charge.

v) The Hartree electron-electron repulsion contributes

$$\frac{1}{2}\cdot 4\pi \sum_{\mathbf{G}}{}' \frac{\left|\rho(\mathbf{G})\right|^2}{G^2} \left(\frac{2\mathbf{G}_\alpha \mathbf{G}_\beta}{G^2} - \delta_{\alpha\beta} \right) \qquad (7)$$

in analogy with Eq. (4).

vi) The exchange and correlation energy is in real space

$$E_{xc} = \int \varepsilon_{xc}(\rho(\mathbf{r})) \, \rho(\mathbf{r}) \, d^3\mathbf{r} \qquad (8)$$

Since $\rho(\mathbf{r})d^3\mathbf{r}$ is scaling-invariant we obtain the stress

$$\Omega^{-1} \int (\partial\varepsilon_{xc}/\partial\rho) \, (\partial\rho/\partial\varepsilon_{\alpha\beta}) \, \rho(\mathbf{r}) \, d^3\mathbf{r} =$$

318

$$\Omega^{-1} \, \delta_{\alpha\beta} \int [\epsilon_{xc}(\rho) - \mu_{xc}(\rho)] \, \rho(\mathbf{r}) \, d^3\mathbf{r} = \tag{9}$$

$$\delta_{\alpha\beta} \, \Sigma_G [\epsilon_{xc}(G) - \mu_{xc}(G)] \, \rho(G)^*$$

where $\mu_{xc}(\rho) = d(\rho\epsilon_{xc})/d\rho$ is the x-c potential. The Eq. (9) is a diagonal stress tensor, which implies that the x-c stress is just an isotropic pressure and does not contain any shears. This is a property of the local-density approximation, where an electron interacts with its own *spherical* x-c hole (von Barth and Williams, 1983). More general density functionals will include anisotropic stress.

vii) The ion-ion interaction in a constant neutralizing background gives an energy γ_{Ewald} calculated by the Ewald transformation (Fuchs, 1935; Ihm et al., 1979). The stress becomes

$$\frac{\partial \gamma_{Ewald}}{\partial \epsilon_{\alpha\beta}} = \frac{\pi}{2\Omega\epsilon} \sum_{G \neq 0} \frac{e^{-G^2/4\epsilon}}{G^2/4\epsilon} \left| \sum_\tau Z_\tau e^{i\mathbf{G}\cdot\mathbf{x}_\tau} \right|^2 \left[\frac{2G_\alpha G_\beta}{G^2} (G^2/4\epsilon + 1) - \delta_{\alpha\beta} \right] +$$

$$\tag{10}$$

$$\tfrac{1}{2}\epsilon^{\frac{1}{2}} \sum_{\tau\tau'T} Z_\tau Z_{\tau'} H'(\epsilon^{\frac{1}{2}}D) \frac{D_\alpha D_\beta}{D^2} \bigg|_{(\mathbf{D}=\mathbf{x}_{\tau'}-\mathbf{x}_\tau+\mathbf{T} \neq 0)} + \frac{\pi}{2\Omega\epsilon} \left(\sum_\tau Z_\tau \right)^2 \delta_{\alpha\beta}$$

Here ϵ denotes a convergence parameter (and not the strain $\epsilon_{\alpha\beta}$) which may be chosen for computational performance. $\mathbf{T}$ denotes the real-lattice translation vectors, and $\mathbf{x}_\tau$ the atomic positions in the unit cell. The function $H'(x)$ is

$$H'(x) = \partial(\mathrm{erfc}(x))/\partial x - x^{-1} \, \mathrm{erfc}(x)$$

with $\mathrm{erfc}(x)$ denoting the complementary error function.

The terms i) through vii) add up to the total stress $\sigma_{\alpha\beta}$ per unit volume, and are calculated in exactly the same way as total energy and forces(Wendel and Martin, 1979; Ihm et al., 1979).

4. Method of calculation

The Schrödinger equation consists of kinetic energy, ionic pseudopotentials, and the Hartree and exchange-correlation screening potentials. When wavefunctions and energy eigenvalues have been found by numerical solution, they are substituted into the expressions for total energy and force (Ihm et al., 1979) and for stress (See section 3). However, carrying out selfconsistent calculations using normconserving pseudopotentials requires

attention to a number of details. However technical these may
seem, they form the basis for much progress in recent years. In
this section we discuss briefly a set of problems which have to be
dealt with in most calculational schemes, and refer to the
literature for more complete treatments.

Firstly, the pseudopotentials are constructed by a chosen
scheme, for example as in (Hamann et al., 1979), (Kerker, 1980),
or (Bylander and Kleinman, 1984). A particular form for the
local-density functional must also be chosen. The quality of the
pseudopotential may be tested by comparing its excited-states
eigenvalues and total-energies to those of the all-electron atom,
which should reflect the accuracy to be expected in the solid
calculation. The potential for each value of l (angular momentum)
is Fourier transformed. One of these potentials is chosen to be
the local potential which is independent of l. Usually the
highest l considered in the free atom is chosen. The difference
between the remaining potentials and the local one is then defined
as the nonlocal correction potential Δv^{NL}, whose matrix elements
in a plane wave basis can be evaluated as in (Heine and Weaire,
1970). A useful set of pseudopotentials covering H to Pu is given
by Bachelet et al. (1982) in terms of Gaussian parameters. The
present work utilizes such potentials.

Secondly, there exists several techniques for integrating
approximately over the **k**-points of the first Brillouin zone (BZ).
For materials with fully occupied bands (e.g., semiconductors) the
special points method is by far the most efficient (Chadi and Cohen,
1974; Monkhorst and Pack, 1976). The method appeals to the tight-
binding picture of atomic interactions, integrating a definite
number of interactions *exactly* with a suitably chosen set of **k**-
points. For metallic systems it is necessary to exhaust the
irreducible BZ with a fine mesh, and to choose a method of
assigning occupation numbers to the electron states. Several
methods prevail, and we refer to Fu and Ho (1983) for a detailed
comparison of two schemes.

The symmetry of the crystal must be taken into account to
determine which special points are equivalent, and to find the
relative phase-factors between these points. For an automatic
symmetry analysis giving all group operations of general lattices
we refer to a suitable subset of the routines published by Warren
and Worlton (1972; 1974), which have been extremely useful in the
present work.

Thirdly, a cutoff must be introduced on the reciprocal-
lattice **G**-vectors to be included in the basis set. For a given **k**-
point in the irreducible BZ the plane-wave kinetic energy $\frac{1}{2}(\mathbf{k}+\mathbf{G})^2$
should be less than some upper bound, typically in the range of 5-
50 Ry. There exists an ambiguity when structures are calculated

at different volumes: The plane-wave kinetic energy scales as $(\text{volume})^{-2/3}$, so that a fixed *energy cutoff* leads to a varying number of plane waves. The argument is that the wavefunctions should be Fourier-analyzed with a definite spatial resolution. The other possibility is to maintain a *fixed number* of plane waves even when the volume is varied, leading to much smoother total-energy curves. We have chosen the former of the two procedures.

There remains the question of how large a cutoff is sufficient. One may calculate the total energy or differences in energy between structures as a function of cutoff to determine the convergence of these quantities. One may also calculate energy, stress and forces as a function of a single structural parameter at a given cutoff, and check whether the numerical derivative of total energy agrees with the stress and forces.

Fourth, matrix diagonalization becomes an important issue with the large Hamiltonian matrices that are often encountered. The EISPACK package (Boyle et al., 1977)) is much used and is quite susceptible to computer vectorization. New techniques for large matrices are appearing, based on iterative determination of eigenvalues and -vectors (See, e.g., Wood and Zunger, 1984).

Fifth, a starting guess of the screening potential (Hartree plus x-c) must be chosen. One common practice is to use some suitable dielectric function for linear screening of the ionic pseudopotential. Another method takes the overlapping charge densities of the free atoms and uses them to calculate the screening potential. Achieving selfconsistency from this initial guess is a *non-linear* problem which may proceed by different routes. The n-th cycle in selfconsistency receives an input screening potential $V_{in}^{(n)}$ and from the resulting wavefunctions is produced the output charge density and potential $V_{out}^{(n)}$. In general it is necessary to mix $V_{in}^{(n)}$ and $V_{out}^{(n)}$ to obtain an input for the (n+1)-th cycle, in order to damp the response of the electron system. The simplest scheme employs a mixing coefficient α which should be smaller than but close to unity in unstable situations, i.e.,

$$V_{in}^{(n+1)} = \alpha\, V_{in}^{(n)} + (1-\alpha)\, V_{out}^{(n)} \tag{11}$$

A test of Eq. (11) is shown by Martin and Kunc (1983). A variant of Eq. (11) uses an α that varies with G^2, e.g. $(1-\alpha)=1/\varepsilon(G^2)$, where ε is the dielectric function. Another scheme constructs and utilizes the dielectric *matrix* of the electron system, see Ho et al. (1982).

A very efficient scheme uses the quasi-Newton method to locate the zero-point of a general function (see review by Dennis and Moré, 1977). The method was applied to selfconsistent iterations by Bendt and Zunger (1982). Considering $V_{in}^{(n)}$ and $V_{out}^{(n)}$ as N-dimensional vectors (N is the number of G-vector stars, or a few of the shortest stars), a "coordinate" $x^{(n)}$ and a "response function" $F(x^{(n)})$ is defined as

$$x^{(n)} = V_{in}^{(n)} \tag{12a}$$

$$F(x^{(n)}) = V_{out}^{(n)} - V_{in}^{(n)} \tag{12b}$$

We wish to find an x_0 such that $F(x_0)=0$, which is the condition of selfconsistency. If $H^{(n)}$ is the inverse Jacobian matrix, linear extrapolation gives the next estimate of x_0 as

$$x^{(n+1)} = x^{(n)} - H^{(n)} F(x^{(n)}) \tag{13}$$

The central idea is to guess a $H^{(1)}$ (e.g., $H_{ij}^{(1)}=(1-\alpha)\delta_{ij}$) and update it as values of $x^{(n)}$ and $F(x^{(n)})$ are calculated. Memory is thus maintained of all previous cycles. Dennis and Moré (1977) review several schemes: The one due to Broyden gives

$$H^{(n+1)} = H^{(n)} + (s^{(n)}-H^{(n)}y^{(n)})(s^{(n)\dagger}H^{(n)})/(s^{(n)\dagger}H^{(n)}y^{(n)}) \tag{14}$$

where the numerator is a dyadic, the denominator an inner product, and $\dagger$ denotes Hermitean conjugation. The vectors $y^{(n)}$ and $s^{(n)}$ are

$$y^{(n)} = F(x^{(n+1)}) - F(x^{(n)}) \tag{15a}$$

$$s^{(n)} = x^{(n+1)} - x^{(n)} \tag{15b}$$

Thus selfconsistency proceeds by doing cycle n=1 and n=2 with some simple scheme, after which Eqs. (13)-(15) are used to update $x^{(n+1)}$ and $H^{(n+1)}$. The convergence is usually very rapid, as illustrated by Bendt and Zunger (1982), and below in section 6.

Sixth, when evaluating total energy during the selfconsistent cycles, it is important that every term in the total-energy expression is constructed with the *output* quantities of the given cycle. If the eigenvalue sum is used for total energy, it is realized that it corresponds to the *input* potential of that cycle. The error introduced is easily corrected by noting that the sum equals kinetic plus potential energy. Since kinetic energy is calculated with the output wavefunctions, there remains to use the output potential in the potential energy. One should therefore add the term

$$\sum_{G \neq 0} \left[V(G)_{output} - V(G)_{input} \right] \rho(G)^* \tag{16}$$

322

to the sum of eigenvalues, where V denotes the screening potentials, i.e. Hartree plus x-c potentials. The $G=0$ term is absent from Eq. (16) since it originates solely from μ_{xc} and can be included as a constant eigenvalue shift in the total energy. The term Eq. (16) is furthermore an indicator of the degree of selfconsistency achieved.

Seventh, calculation of the exchange and correlation terms, Eqs. (8) and (9), are readily done in real space when the charge density $\rho(G)$ is Fourier transformed and consequently $\mu_{xc}(\rho(r))$ and $\varepsilon_{xc}(\rho(r))$ are evaluated. Another Fourier transform yields $\mu_{xc}(G)$.

5. Ab-initio calculations on Si, Ge, and GaAs

We have performed calculations of stress, force and total energy with the methods described above, applying them to the semiconductors Si, Ge, and GaAs. For Si we use the form derived by Bachelet et al. (1981) employing Wigner correlation. The recently developed potentials covering H to Pu (Bachelet et al., 1982) employing Ceperley-Alder correlation are used for Ge and GaAs. In the latter cases we use the $l=2$ potentials as local potentials, since this is the most consistent procedure for high angular momenta, and it furthermore makes the local as well as non-local potentials significantly softer than with the choice suggested in (Bachelet et al., 1982).

A large number of plane waves (≈ 550) are used in all calculations, corresponding to $\hbar^2(k+G)^2/2m \leq 24$ Ry kinetic energy, in order to eliminate computational uncertainties due to cutoffs. This large cutoff is necessary with the present potentials, which are relatively hard-core, in order to achieve the high accuracy needed for reliable values of elastic constants, etc., which are the second derivatives of total energy. The plane waves of high energy (>12 Ry for Si, >16 Ry for Ge and GaAs) are treated by second-order Löwdin perturbation theory (Löwdin, 1951; Brust, 1964; Wendel and Martin, 1978). The Löwdin-wave cutoffs were chosen by a test starting from no Löwdin waves, lowering the cutoff to the minimum required for reproducing accurately the all-exact-waves calculation. In the distorted lattices suffering large strains we include the same number of plane waves as in the undistorted crystal, i.e., we apply an ellipsoidal cutoff instead of a spherical one.

The Brillouin zone (BZ) k-integration is performed by the special points method, suitably generalized for distorted lattices: The set of undistorted k-points (Monkhorst and Pack, 1976) is strained ($k \rightarrow (1+\varepsilon)^{-1}k$) and the symmetry of the strained lattice is used to identify equivalent **k**-points. This procedure assures a smooth transition from high- to low-symmetry lattices. For Si we use the sets of 2 and 10 special **k**-points in the irreducible fcc BZ, as noted. From these calculations we estimate the error introduced by using the smaller set of special **k**-points to be <1% for the lattice constant and ≈5% for the elastic constants. Tests show that going beyond 10 special points has neglegible effect on the calculated quantities. For Ge and GaAs we have used only 2 fcc special **k**-points, since their pseudopotentials require a larger numerical effort than the one used for Si. When the lattice is distorted, the number of inequivalent **k**-points increases, e.g. from 2 to 3 or 5.

The equilibrium structure is diamond for Si and Ge, and zincblende for GaAs, as was verified in recent theoretical work (Yin and Cohen, 1982; Froyen and Cohen, 1982). With the given structure as the only input we have calculated the lattice constant, a, using the stress theorem. A first calculation of pressure P at a guessed lattice constant permits a very good final estimate of a from, say, experimental B and $\partial B/\partial P$. Two calculations of pressure near this value gives by linear interpolation the final lattice constant where P=0, as well as the bulk modulus from the slope of pressure. The results are given in Table 1, showing good agreement with experiments. The deviations are state-of-the-art accuracy, and are believed to be due mainly to the local-density approximation, since our results agree well with all-electron LMTO-ASA calculations (Glötzel et al., 1980). The present results also agree well with other pseudopotential calculations (Yin and Cohen, 1982; Ihm and Joannopoulos, 1981; Froyen and Cohen, 1982).

Table 1. Lattice constants and elastic properties. Brackets denote deviations from experimental values.

		Si	Ge	GaAs
a	(Å)	5.40 (−1%)	5.59 (−1%)	5.55 (−2%)
B	(Mbar)	0.93 (−6%)	0.72 (−6%)	0.73 (−7%)
c_{11}	(Mbar)	1.59 (−5%)	1.30 (−1%)	1.23 (−1%)
c_{12}	(Mbar)	0.61 (−6%)	0.45 (−9%)	0.53 (−7%)
c_{44}	(Mbar)	0.85 (+6%)	0.63 (−8%)	0.62 (+3%)
ζ		0.53 (−27%)	0.44 (−39%)	0.48 (−37%)
$\omega_{TO(\Gamma)}$		15.64 (−1%)	9.05 (−1%)	8.09 (−1%)

6. Elastic properties of Si, Ge, and GaAs

Calculations of the full stress tensor is a method ideally suited to the derivation of elastic constants, since it contains up to six independent pieces of information that otherwise would require extensive calculations of total energy. The c_{11} and c_{12} elastic constants can be found from the stress-strain relation with the application of an ε_1-strain. (The Voigt notation is used, see e.g. (Nye, 1957), i.e. $11 \rightarrow 1$, $22 \rightarrow 2$, $33 \rightarrow 3$, $23 \rightarrow 4$, $13 \rightarrow 5$, $12 \rightarrow 6$; thus $\varepsilon_{11} = \varepsilon_1$, $\varepsilon_{23} = \frac{1}{2}\varepsilon_4$, $\sigma_{11} = \sigma_1$ and $\sigma_{23} = \sigma_4$). This strain scales the x-dimension by $(1 + \varepsilon_1)$ while maintaining the y- and z-dimensions. By symmetry there are no internal displacements in the present lattices for any ε_1. For small strains the harmonic approximation defines the relations $c_{11} = \sigma_1/\varepsilon_1$, $c_{12} = \sigma_2/\varepsilon_1$, and $c_{44} = \sigma_4/\varepsilon_4$, with the strains and stresses depicted in Fig. 1.

With a strain of $\varepsilon_1 = -0.004$ we obtain the c_{11} and c_{12} given in Table 1. The differences from experiment are -6% for Si, and up to -9% for Ge and GaAs, where where only two special **k**-points are used, resulting in a lower accuracy as estimated above. These numbers agree well with the independently calculated bulk moduli B

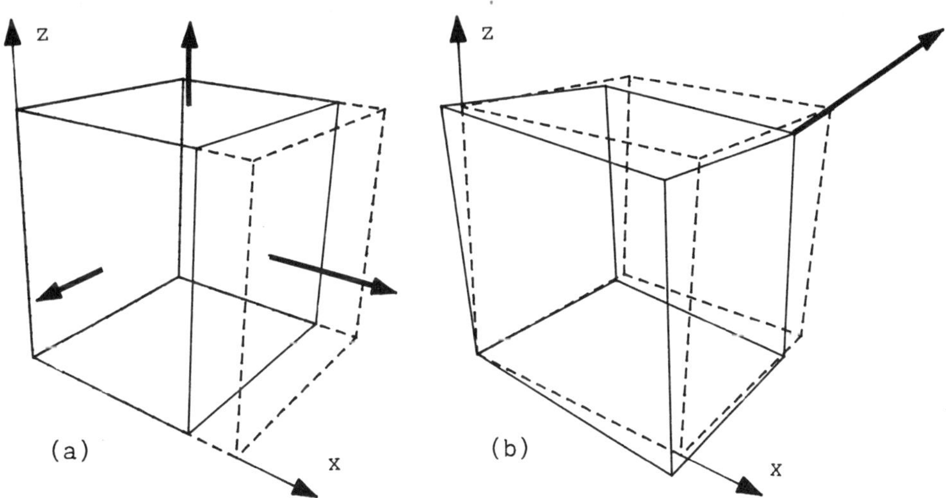

Figure 1. Perspective view of cubes (dashed lines) deformed by strains to take new shapes (full lines). Thick arrows indicate the resulting directions of stress exerted by the solid. (a) A strain $\varepsilon_1 < 0$ along (100) resulting in σ_1 and $\sigma_2 = \sigma_3$ stresses. (b) A strain $\varepsilon_4 = \varepsilon_5 = \varepsilon_6 < 0$ along (111) resulting in a stress in the same direction.

$= (c_{11}+2c_{12})/3$. The shear modulus $\frac{1}{2}(c_{11}-c_{12})$ was given previously for Si by Wendel and Martin (1979) and for Si and Ge by Yin and Cohen (1982), by calculating the total energy for a number of large volume-conserving strains. Fitting the energy curve with an assumed equation-of-state yielded the harmonic shear modulus, in reasonable agreement with the present results.

The calculation of the elastic constant c_{44} is inherently more complicated than for c_{11} and c_{12}. A strain $\varepsilon_4=\varepsilon_5=\varepsilon_6$ along the (111) direction of a zincblende lattice makes the ($\bar{1}$11) atomic bond inequivalent to the (-1,-1,1), (-1,1,-1) and (1,-1,-1) bonds. The atomic positions in the unit cell are no longer completely determined by symmetry and a static stretching of the (111)-bond similar to a TO(Γ) phonon is allowed. Kleinman (1962) formulated this in terms of an internal strain parameter ζ. The value $\zeta=0$ corresponds to a perfect strain of atomic positions: $\mathbf{r}\rightarrow(1+\varepsilon)\mathbf{r}$. An actual change of the (111) bond length of magnitude $\zeta\varepsilon_4 a\sqrt{3}/4$ defines ζ so that $\zeta=1$ corresponds to a perfectly rigid (111)-bond of length $a\sqrt{3}/4$. In practice ζ should fall in the range $0<\zeta<1$.

In order to obtain values for ζ, c_{44} and $\omega_{TO(\Gamma)}$, the total energy of the distorted crystal can be analyzed within the adiabatic and harmonic approximations, where the energy increase per unit cell due to displacements u_τ of the atoms τ in the unit cell, and a macroscopic strain ε is given by:

$$\Delta E_{tot} = \tfrac{1}{2}\Sigma_{\tau\tau'}\; \mathbf{u}_\tau\; \Phi(\tau,\tau')\; \mathbf{u}_{\tau'} + \Sigma_\tau\; \mathbf{u}_\tau\; T(\tau)\; \varepsilon\; +\; \tfrac{1}{2}\varepsilon\, \mathbf{c}\, \varepsilon \tag{17}$$

The force constant matrix for atomic displacements is denoted $\Phi(\tau,\tau')$, $T(\tau)$ is the third-rank internal strain tensor, $\mathbf{c}$ is a fourth rank elastic constant tensor. Tensor contraction is understood throughout. Thus the restoring force on an atom is

$$F(\tau) = -\Sigma_{\tau'}\, \Phi(\tau,\tau')\mathbf{u}_{\tau'} - T(\tau)\; \varepsilon \tag{18}$$

and the macroscopic stress is

$$\sigma = \Sigma_\tau\; \mathbf{u}_\tau\, T(\tau)\; + \mathbf{c}\, \varepsilon \tag{19}$$

The form of tensors in a cubic crystal is given by, e.g., Nye (1957), and we find for a relative atomic displacement $\mathbf{u} = u(1,1,1)$ along the (111)-bond and a strain $\varepsilon = \frac{1}{2}\varepsilon_4(1-\delta_{\alpha\beta})$ the force $\pm F(1,1,1)$ on the two atoms, where

$$F = \Phi(\zeta a/4\; \varepsilon_4\; -\; u) \tag{20}$$

The force constant Φ equals $\frac{1}{2}\mu\omega^2_{TO(\Gamma)}$, where μ is the reduced mass of the two atoms, and $\omega_{TO(\Gamma)}$ is the frequency of the transverse-optic phonon at the Γ-point. The stress is similarly given by $\sigma_{\alpha\beta} = \sigma_4(1-\delta_{\alpha\beta})$ with

$$\sigma_4 = c_{44}{}^{(0)} \, \varepsilon_4 - \Omega^{-1} \, \Phi \, \zeta \, \text{au}/4 \qquad (21)$$

where $c_{44}{}^{(0)}$ denotes the elastic constant that would appear in the absence of internal displacements. For a given value of ε_4, the actual physical displacement u is defined by F=0, so that the internal displacement u equals $\zeta a/4 \, \varepsilon_4$. Inserted into Eq. (20) this gives the stress-strain relation

$$\sigma_4 = [c_{44}{}^{(0)} - \Omega^{-1} \, \Phi(\zeta a/4)^2] \, \varepsilon_4 = c_{44} \, \varepsilon_4 \qquad (22)$$

which defines the physically measured elastic constant c_{44}.

Two independent calculations now suffice to determine $\omega_{TO(\Gamma)}$, c_{44} and ζ. (1) With $\varepsilon_4 = 0$ and a small value of $u = u^{(1)}$, the force $F^{(1)}$ determines $\omega_{TO(\Gamma)}$, and the stress $\sigma_4{}^{(1)}$ determines $\zeta = 4 \, \sigma_4{}^{(1)} \, \Omega/F^{(1)}a$. (2) With a small $\varepsilon_4{}^{(2)}$ and $u = 0$, the stress $\sigma_4{}^{(2)}$ determines $c_{44}{}^{(0)} = \sigma_4{}^{(2)}/\varepsilon_4{}^{(2)}$, and the force $F^{(2)}$ provides an independent calculation of ζ as $(4u^{(1)}F^{(2)})/(a\varepsilon_4{}^{(2)}F^{(1)})$.

Thus the two force calculations determine ζ without necessarily calculating the stress. The strains and displacements are illustrated in Fig. 2.

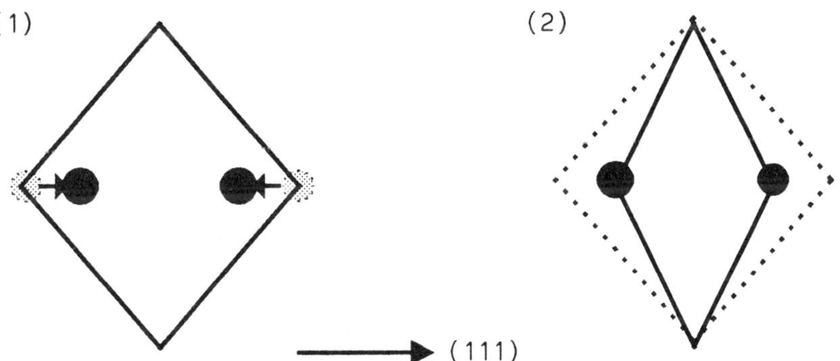

Figure 2. Schematic illustration of the two distortions used to calculate ζ, $\omega_{TO(\Gamma)}$ and c_{44}. (1) An internal displacement along (111). (2) A strain along (111) and no displacement. Rhombes indicate the unit cell, full circles the atoms. Original atomic positions and unit cell are shown as dotted.

The calculations are performed with $u^{(1)}/a = 0.002$ and $\varepsilon_4^{(2)} = -0.004$, respectively, and the results are given in Table 1. The phonon frequencies are within 1% of experimental values, and are in good agreement with previous calculations (Yin and Cohen, 1982; Froyen and Cohen, 1982). The deviations from experimental c_{44} are of similar magnitude as for B, c_{11} and c_{12}. The values of c_{44} are approximately 80% of $c_{44}^{(0)}$, showing that accurate calculations of the internal strain ζ are crucial to the determination of c_{44}, since ζ enters Eq.(10) as ζ^2. The independently calculated values of ζ are consistent within < 1% for Si and < 5% for Ge and GaAs, and the average values are given in Table 1.

The present theoretical values for ζ show interesting differences of 30-40% from two recent x-ray diffraction experiments (D'Amour et al., 1982; Cousins et al., 1982) which we believe are much larger than the calculational uncertainty in view of the accuracy obtained for all other quantities, including $\omega_{TO(\Gamma)}$ and c_{44}. Furthermore, the deformation potentials considered in (Nielsen and Martin, to be published; Christensen, 1984) lends independent experimental and theoretical support to the present values. The puzzle noted in (Nielsen and Martin, 1983) persists at present, however.

It is interesting to consider the rearrangements of electronic charge as the crystal is strained, in order to investigate the physics of elastic deformation. A large strain ($\varepsilon_4 = -0.03$) applied along the (111) axis in GaAs gives a charge density as shown in Fig. 3. The figure shows a relative increase of the bond charge on the bond oriented along (111) and a decrease of the bond charge along (1,1,-1) (and its equivalents). We find that bond charges adjust to screen approaching ions, a picture that holds also for a TO(Γ) phonon displacement which has the same symmetry as the lattice strained along (111).

For achieving self-consistency, the question may be raised about the rate of convergence of stress, forces, and other quantities as compared to that of total energy. A common argument is that total energy will converge faster than other quantities since it is a variational quantity which depends on errors in wavefunctions only to the second order. A test of convergence has been performed for the case of a Si crystal strained along the (111) direction, but with no internal strain ($\zeta = 0$). In Fig. 4 we show the convergence of total energy, force, stress, potential, charge density, and the convergence monitor given by Eq. (16). Although total energy has an extremely small error relative to its absolute value, the difference relative to the corresponding undistorted Si crystal converges no faster than stress or force. The figure shows in addition the effectiveness of the Broyden scheme (see section 4) for achieving selfconsistency relative to

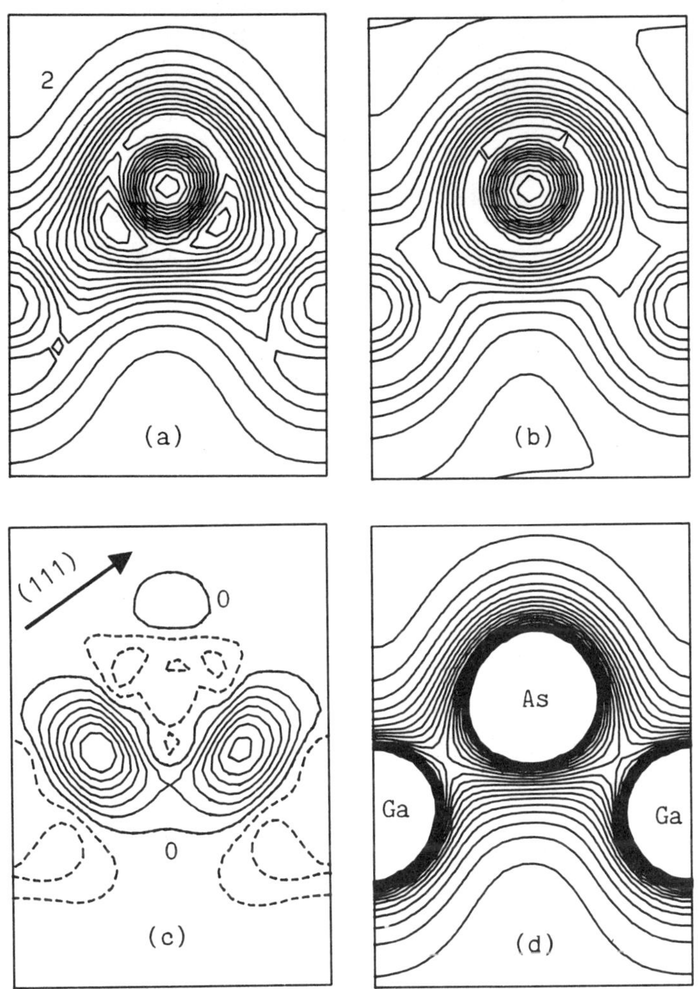

Figure 3. Charge density of GaAs subjected to a strain along (111) (ε_4=-0.03) including internal strain. The figure shows the plane of the zig-zag chain along (110). (a) Valence pseudo-density. (b) Density of overlapping free pseudo-atoms. (c) Shows (a) minus (b), the "deformation density". Outermost solid contours indicate zero, negative contours are dashed. (d) Adding the density of free all-electron atoms to (c), giving the "true" solid charge density. Contour intervals are 2 electrons per unit cell volume in (a), (b), and (d), and 1 electron per unit cell volume in (c).

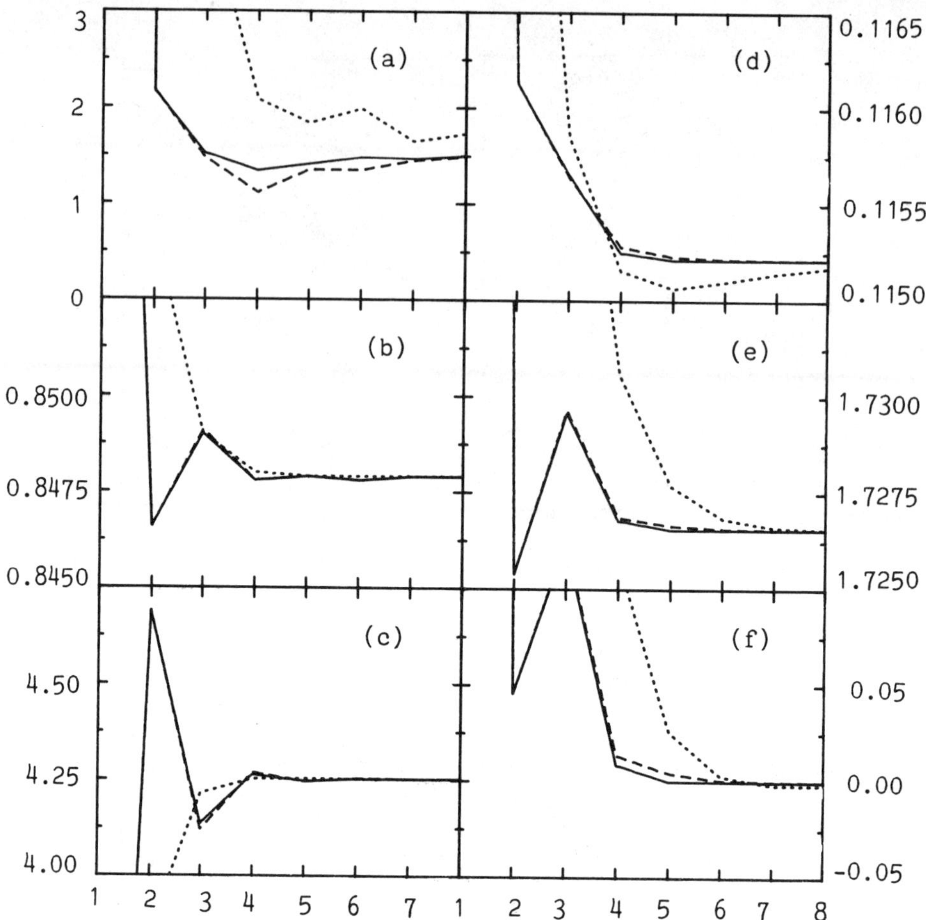

Figure 4. Plots of various quantities as function of cycle number (1 to 8) in selfconsistency for Si with a (111)-strain of $\varepsilon_4 = -0.004$ and no internal strain ($\zeta = 0$). Two fcc special **k**-points (i.e. 5 **k**-points in total) and a plane-wave cutoff of 12 Ry are used. The initial guessed potentials were linearly screened ionic pseudopotential from (Bachelet et al., 1981). (a) Total energy difference relative to the undistorted lattice, in units of 10^{-3} eV. (b) One component of restoring force in units of millidynes. (c) Stress σ_4 in units of kbar. (d) The **G**=(111) Fourier component of the input potential in units of Ry/cell. (e) The **G**=(111) Fourier component of pseudo-charge density in units of electrons/cell. (f) The eigenvalue correction measuring the degree of selfconsistency (Eq. (16)) in units of eV. Full lines give results using Broyden's update (Eqs. (13)-(15)), dashed lines refer to mixing with a dielectric function, and dotted lines refer to simple mixing with an $\alpha = 0.5$.

330

the simpler mixing schemes: The simple mixing with $\alpha=0.5$ converges rather slowly, whereas the mixing with a dielectric function (which has $\alpha\approx0$ for large $\mathbf{G}$-vectors) converges as fast as the Broyden scheme in the first few cycles (the first *two* cycles are identical). However, the numbers plotted in Fig. 4 reveal that final convergence is obtained in cycle 5 with Broyden's scheme, and only in cycle 7 or 8 with the dielectric scheme.

7. Conclusion

We have discussed total-energy calculations within the local-density-functional approximation, employing *ab-initio* pseudo-potentials. The stress theorem (Nielsen and Martin, 1983) has been derived for calculations in reciprocal space. A number of remarks on the technical, but nevertheless important, aspects of carrying out calculations have been given. These include **k**-point sampling and symmetry, and efficient methods for achieving selfconsistency. We show detailed results on the lattice constants and all linear elastic properties of the semiconductors Si, Ge, and GaAs. Results for the convergence of a number of quantities during selfconsistency iteration are given. It is shown that stress and forces converge as fast as the corresponding total-energy differences. The Broyden scheme is found to give the fastest convergence.

The combined calculation of stress, forces and total energy have thus been shown to constitute a powerful and complete method for the study of structural, elastic and dynamic properties of solids. Using the stress theorem permits rapid determination of lattice parameters, for cubic as well as non-cubic crystals. This leads to large savings in computational effort, compared to direct calculations of the total energy with subsequent numerical differentiation.

Acknowledgements

We have benefitted from enlightening discussions with W. C. Herring. The present calculations would not have been possible without the solution of the density-functional equations that was implemented by K. Kunc. This work was supported in part by ONR contract N00014-82-C-0244.

References

d'Amour, H., Denner, W., Schulz, H., and Cardona, M.,
    J. Appl. Crystallogr. **15**, 148 (1982).

Bachelet, G. B., Greenside, H. S., Baraff, G. A., and
     Schlüter, M., Phys. Rev. B $\underline{24}$, 4745 (1981).
Bachelet, G. B., Hamann, D. R., and Schlüter, M.,
     Phys. Rev. B $\underline{26}$, 4199 (1982).
von Barth, U., and Williams, A., in "Theory of the Inhomogeneous
     Electron Gas", (Plenum, New York, 1983).
Bendt, P., and Zunger, A., Phys. Rev. B $\underline{26}$, 3114 (1982).
Born, M., Heisenberg, W., and Jordan, P., Z. Phys. $\underline{35}$, 557 (1926).
Boyle et al., EISPACK guide (Springer, Berlin, 1977).
Brust, D., Phys. Rev. $\underline{134}$, A1337 (1964).
Bylander, D. M., and Kleinman, L.,
     Phys. Rev. B $\underline{29}$, 1534 and 2274 (1984).
Chadi, D. J., and Cohen, M. L., Phys. Rev. B $\underline{8}$, 5747 (1974).
Christensen, N. E., Solid State Commun. $\underline{50}$, 177 (1984).
Cohen, M. L., Heine, V., and Phillips, J. C.,
     Sci. Am. $\underline{246}$, No. 6, p. 66 (1982).
Cousins, C. S. G., Gerward, L., Staun Olsen, J., Selsmark, B.,
     and Sheldon, B. J., J. Appl. Crystallogr. $\underline{15}$, 154 (1982).
Dennis, J. E., Jr., and Moré, J. J., SIAM Review $\underline{19}$, 46 (1977).
Ehrenfest, P., Z. Phys. $\underline{45}$, 455 (1927).
Feynman, R. P., Phys. Rev. $\underline{56}$, 340 (1939);
     Undergraduate thesis (MIT, 1939, unpublished).
Fock, V., Z. Phys. $\underline{63}$, 855 (1930).
Froyen, S., and Cohen, M. L., Phys. Rev. B $\underline{28}$, 3258 (1983).
Froyen, S., and Cohen, M. L., Phys. Rev. B $\underline{29}$, 3770 (1984).
Fu, C.-L., and Ho, K.-M., Phys. Rev. B $\underline{28}$, 5480 (1983).
Fuchs, K., Proc. Roy. Soc. London A $\underline{151}$, 585 (1935).
Glötzel, D., Segall, B., and Andersen, O. K.,
     Solid State Commun. $\underline{36}$, 403 (1980).
Greenside, H. S., and Schlüter, M., Phys Rev. B $\underline{27}$, 3111 (1983).
Hamann, D. R., Schlüter, M., and Chiang, C.,
     Phys. Rev. Lett. $\underline{43}$, 1494 (1979).
Heine, V., and Weaire, D., Solid State Physics $\underline{24}$, 320 (1970).
Hellmann, H., "Einführung in die Quantenchemie",
     (Deuticke, Leipzig, 1937), pp. 61 and 285.
Herring, W. C., Phys. Rev. $\underline{57}$, 1169 (1940).
Ho, K.-M., Ihm, J., and Joannopoulos, J. D.,
     Phys. Rev. B $\underline{25}$, 4260 (1982).
Ihm, J., Zunger, A., and Cohen, M. L., J. Phys. C $\underline{12}$, 4409 (1979).
Ihm, J., and Joannopoulos, J. D., Phys. Rev. B $\underline{24}$, 4191 (1981).
Kerker, G., J. Phys. C $\underline{13}$, L189 (1980).
Kleinman, L., Phys. Rev. $\underline{128}$, 2614 (1962).
Lam, P. K., and Cohen, M. L., Phys. Rev. B $\underline{24}$, 4224 (1981).
Louie, S. G., Froyen, S., and Cohen, M. L.,
     Phys. Rev. B $\underline{26}$, 1738 (1982).
Löwdin, P. O., J. Chem. Phys. $\underline{19}$, 1396 (1951).
Lundqvist, S., and March, N. H., eds., "Theory of the
     Inhomogeneous Electron Gas", (Plenum, New York, 1983).

Martin, R. M., and Kunc, K., in "Ab-initio Calculations of Phonon
    Spectra", eds. Devreese, J. T., Van Doren, V. E., and
    Van Camp, P. E. (Plenum, New York, 1983), p. 49.
McLellan, A. G., Amer. J. Phys. **42**, 239 (1974).
Monkhorst, H. J., and Pack, J. D., Phys. Rev. B **13**, 5188 (1976)
Nielsen, O. H., and Martin, R. M.,
    Phys. Rev. Lett. **50**, 697 (1983).
Nielsen, O. H., and Martin, R. M., to be published.
Northrup, J. E., and Cohen, M. L.,
    J. Vac. Sci. Technol. **21**, 333 (1982).
Nye, J. F., "Physical Properties of Crystals" (Oxford, 1957).
Warren, J. L., and Worlton, T. G.,
    Comput. Phys. Commun. **8**, 71 (1974); *ibid* **3**, 88 (1972).
Wendel, H., and Martin, R. M., Phys. Rev. Lett. **40**, 950 (1978);
    Phys. Rev. B **19**, 5251 (1979).
Wood, D. M., and Zunger, A., Preprint.
Yin, M. T., and Cohen, M. L., Phys. Rev. B **26**,
    3259 and 5668 (1983).

PSEUDOPOTENTIALS AND TOTAL ENERGY CALCULATIONS: APPLICATIONS TO
CRYSTAL STABILITY, VIBRATIONAL PROPERTIES, PHASE TRANSFORMATIONS,
AND SURFACE STRUCTURES

Steven G. Louie

Department of Physics, University of California, and
Lawrence Berkeley Laboratory, Berkeley, CA   94720

ABSTRACT

   A review of the applications of the pseudopotential method
and total energy techniques to the electronic and structural properties of solids is presented. With this approach, it has recently
become possible to determine with accuracy crystal structures,
lattice constants, bulk moduli, shear moduli, cohesive energies,
phonon spectra, solid-solid phase transformations, and other static
and dynamical properties of solids. The only inputs to these
calculations, which are performed either with plane wave or LCAO
bases, are the atomic numbers and masses of the constituent atoms.
Calculations have also been carried out to study the atomic and
electronic structure of surfaces, chemisorption systems, and interfaces. Results for several selected systems including the covalent
semiconductors and insulators and the transition metals are discussed.
The review is not exhaustive but focuses on specific prototype
systems to illustrate recent progress.

I.   INTRODUCTION

   In the past decade, it has become possible to compute with
good accuracy a number of electronic and structural properties
for simple solids and their surfaces without resorting to empirical
means. Among the quantities obtainable from these calculations
are crystal structures, lattice constants, bulk moduli, shear
moduli, cohesive energies, phonon spectra, electron-phonon and
phonon-phonon interactions, solid-solid phase transformations,
and other static and dynamical properties. This development has
opened up many exciting possibilites for the study of condensed

matter since one is now in a position to predict properties of systems
which were formerly inaccessible to theory or experiment. Possible
applications are numerous--prediction of new materials; determination
of the structure of surfaces, interfaces, and defects in solids;
study of phase stability at high pressure; study of alloy stability;
and so forth.

Several factors have contributed to the present success of
ab initio calculations for real materials systems. These include
the development of approximations to the density functional formalism,
refinements in band structure calculational techniques, invention
of the ab initio pseudopotentials, and development of techniques
for calculating total energies. Equally important, of course,
is the availability of modern high speed computers.

This set of lectures reviews some of the many recent theoretical
accomplishments in this area.[1] Experimental results are discussed
only in so far as they bear on a theoretical result. In addition,
in order to limit the scope of this review, emphasis will be placed
on a few theoretical techniques--primarily the pseudopotential
approach. Specific prototype systems are considered to illustrate
the accomplishments of the theory for semiconductors, insulators,
and transition metals. Some details of the calculations and results
will be given, but the reader should go to the original papers
for more specifics.

The organization of the lectures is as follows. A brief
review of the theoretical techniques is given in Sec. II. This
includes a discussion on the density functional formalism, generation
of ab initio pseudopotentials, and techniques for band structure
calculations. The bulk systems are discussed in Sec. III. The
static structural properties are presented in Sec. IIIA. These
results establish the accuracy of the calculations. Examples
will be given for semiconductors, insulators, and transition metals.
The vibrational properties are discussed in Sec. IIIB. Phonon
frequencies are calculated using the frozen phonon technique.
Complete dispersion curves along symmetry directions in the Brillouin
zone are obtained from calculated force constants. Calculations
of anharmonic terms and phonon-phonon interaction matrix elements
are also presented. In Sec. IIIC, results for solid-solid phase
transitions are presented. The stability of group IV covalent
materials under pressure is discussed. Also presented is a calculation on the temperature- and pressure-induced crystal phase transitions in Be. In Sec. IV, we discuss the application of pseudopotential calculations to surface studies. Silicon and diamond surfaces
will be used as the prototypes for the covalent semiconductor
and insulator cases while surfaces of niobium and palladium will
serve as representatives of the transition metal cases. In Sec. V,
the validity of the local density approximation is examined. The
results of a nonlocal density functional calculation for Si and

336

Ge are presented. Finally, Sec. VI presents a summary and conclusions.

II. THEORETICAL TECHNIQUES

In this section, we briefly summarize the major theoretical techniques employed in contemporary pseudopotential calculations.[2] The approach involves the use of _ab initio_ pseudopotentials[3-7] to compute the electron-core interactions and local density functionals[8-14] for evaluating electron-electron interactions. Several different basis sets[2,15,16] can be used to solve the electron wave equation. The particular choice is determined by the system of interest. Finally, the total energies and forces are calculated using a momentum space scheme.[17]

A. Density Functional Formalism

One major difficulty in _ab initio_ calculations of the properties of electronic systems is an adequate treatment of the many body electron-electron interaction. The most commonly used approach for condensed matter systems is that of the density functional formalism.[8,9] Hohenberg and Kohn[8] established that the electronic energy of a system of interacting electrons in an external potential V_{ext} is a functional of the electron density. This is usually written in the form

$$E[n] = T_0[n] + \int d\vec{r}\; n(\vec{r})\, V_{ext}(\vec{r}) + E_H[n] + E_{xc}[n] \tag{1}$$

The first term is the kinetic energy of noninteracting electrons of the same density; the second term gives the energy of interaction with the external potential; the third term is the electrostatic or Hartree energy; and the last term contains the rest, the exchange-correlation energy which is an universal functional of n. Further, the total energy is minimized for the correct ground state density, i.e.

$$\frac{\delta E}{\delta n} = 0 \tag{2}$$

at the physical $n(\vec{r})$. Given the energy functional, the problem of finding the ground state energy is reduced to solving a set of effective one electron equations with a local potential, the Kohn-Sham equations[9] (in atomic units):

$$\left\{ -\frac{1}{2} \nabla^2 + V_{ext}(\vec{r}) + V_H(\vec{r}) + \mu_{xc}[n] \right\} \psi_i(\vec{r}) = \varepsilon_i \psi_i(\vec{r}) \tag{3}$$

The exchange-correlation part of the effective potential is given by $\mu_{xc} = \delta E_{xc}/\delta n$ and the density n is obtained from the one-particle wavefunctions

$$n(\vec{r}) = \sum_{i=1}^{N} |\psi_i(\vec{r})|^2 \tag{4}$$

where N is the number of electrons in the system.

This formalism, therefore, reduces the many-body problem to an effective single-particle problem which in principle gives the exact ground state energy. The central difficulty is specification of E_{xc}. The most widely used approach is the local density approximation (LDA)[9]:

$$E_{xc}^{LDA} = \int d\vec{r} \; n(\vec{r}) \; \varepsilon_{xc}^{hom}(n(\vec{r})) \tag{5}$$

where $\varepsilon_{xc}^{hom}(n)$ is the exchange-correlation energy density of the homogeneous electron gas of density n. Several parameterizations of electron gas data are in common use.[10-14,18] The procedure in a typical calculation is: (1) Determine V_{ext} from the constituent elements and an assumed geometry. (2) Calculate the electron energies and wavefunctions and determine $n(\vec{r})$. (3) Evaluate the total energy E[n].

B. Ab Initio Pseudopotentials

In the pseudopotential approach, V_{ext} is replaced by an ionic pseudopotential which combines the nucleus and the core electrons into one inert entity, and the self-consistent field equations (Eqs. 3 and 4) are carried out only for the valence electrons. The ionic pseudopotentials, unlike the all-electron potentials, are nonsingular by construction near the nuclei leading to smooth valence electron wavefunctions which greatly facilitate the calculations. Moreover, since the core electrons which do not influence the properties of the solid phase are removed from the problem, much higher numerical precisions can be achieved. Systems involving heavy atoms are not much more complicated than those with light ones. Table I depicts the precision requirements for calculating various cohesive and structural quantities. The only disadvantage of the pseudopotential approach is that the potential is nonlocal, i.e., angular momentum dependent. This, however, only imposes a small inconvenience in the calculations.

Several simple schemes[3-7] have been formulated to extract ab initio ionic pseudopotentials from atomic calculations. The basic procedure is to generate a potential by inversion of the Kohn-Sham equation. Angular-momentum-dependent screened atomic pseudopotentials, V^{ℓ}, are first constructed with the constraints that: (1) the valence eigenvalues from the all-electron calculation and those from the pseudopotential calculation agree for a chosen prototype configuration and (2) the all-electron wavefunctions

338

Table I. Precision requirement for various structural
          quantities.

Precision requirement
  cohesive energy   0.01 Ry/atom
  lattice constant, bulk modulus   0.001 Ry/atom
  phonon frequencies   0.0001 Ry/atom

Total crystal energy (referred to dissociated entities)
  all-electron calculation   ~-1000 Ry/atom
  pseudopotential calculation   ~-10 Ry/atom

---

and the pseudowavefunctions agree beyond a chosen core radius,
r_c. With these constraints, it can be shown that the potentials
have two centrally desirable properties. The electrostatic potential
produced outside r_c is identical for the all-electron and the
pseudocharge distribution; the scattering properties of the all-
electron atoms are reproduced with minimum error as the electronic
eigenvalues move away from the prototype atomic levels. These
two properties ensure a good transferability of the pseudopotentials.
The final bare-ion pseudopotentials, V_{ion}^{ℓ}, are extracted from the
neutral potentials by subtracting from each neutral V^{ℓ} the Coulomb
and exchange and correlation potentials due to the pseudovalence
charge density.

In Fig. 1, the ionic pseudopotential for Si generated using the
Hamann-Schlüter-Chiang scheme[3] is depicted. It is evident that
much of the strong attractive Coulomb point ion potential near the core
is reduced in the pseudopotential formalism. A comparison of
the all-electron wavefunction with the pseudowavefunction is illustrated in Fig. 2. It should be remarked that the exact form of
the pseudopotential in the core region is nonunique since the
potential is only required to reproduce the scattering properties
of the ion core outside of a certain radius. The different generation
schemes produce very different looking potentials. This feature
can, in fact, be well utilized in tailoring the form of the ionic
pseudopotential to suit the particular techniques used in a band
structure calculation. Figure 3 shows an Si ionic pseudopotential
generated using the Kerker scheme.[4] Identical values were obtained
from the two potentials for quantities of interest.

The _ab initio_ pseudopotentials together with the local
density approximation have proven to yield excellent results as
will be illustrated by the examples in this review. The basic
assumptions in their construction are the frozen-core approximation
and a decoupling of the core charge in the determination of the
exchange-correlation potential seen by the valence electrons.

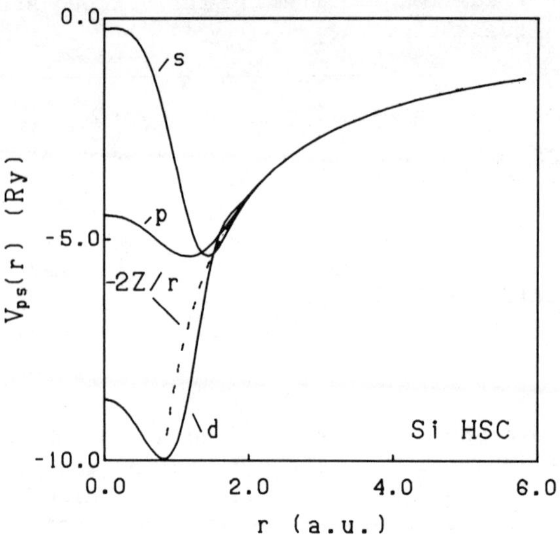

Fig. 1. Nonlocal ionic pseudopotential for Si generated using the Hamann-Schlüter-Chiang scheme Ref. 3). The potentials for angular momentum 1 = 0, 1, and 2 are shown. The dash line represents the Coulomb point ion potential.

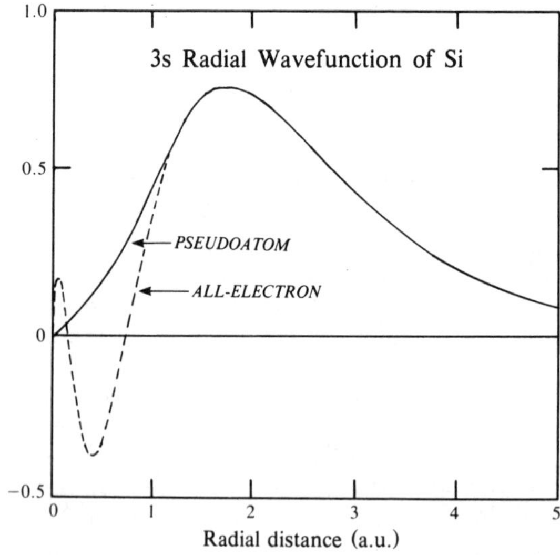

Fig. 2. Comparison of the 3S pseudowavefunctions and all-electron wavefunction for Si.

340

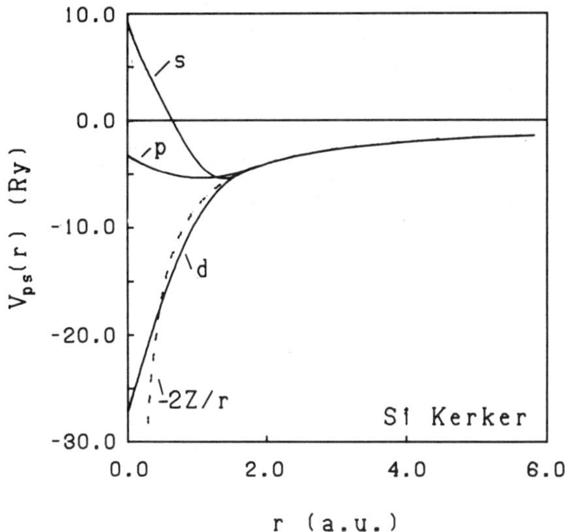

Fig. 3. Nonlocal ionic pseudopotential for Si generated using the
Kerker scheme (Ref. 4).

Since the local density exchange-correlation functionals[10-14] are
nonlinear in the charge density, the second assumption is not
strictly valid when there is a large overlap between the valence
and core charge densities or when spin-density-functional calculations
are performed. However, this problem can easily be eliminated
by the use of nonlinear ionic pseudopotentials[7] where the core
charge density is explicitly carried along in a calculation for
the purpose of evaluating the exchange-correlation potentials
and energy densities.

C. Basis Sets

To solve the Kohn-Sham equations with pseudopotentials, the
standard approach is to expand the electron wavefunctions by a
plane wave set in reciprocal space lattice vectors. The electron
structure is obtained by diagonalization of the Hamiltonian matrix.
This basis set has been mostly employed for semiconductor studies
because of the relatively smooth pseudopotentials and delocalized
electron wavefunctions of these systems. There are several advantages
for using plane waves. The Hamiltonian matrix elements are simple
to evaluate. Test of convergence in the basis expansion can be
done by simply increasing the number of plane waves used. Moreover,
the calculation of Hellmann-Feynman forces is the less involved
in a plane wave basis.

For systems with highly localized electrons such as the transi-

tion metals and large-gap insulators, a plane wave basis set would
not be suitable. Other more judicious choices of basis functions
have to be used. One approach is to use a mixed-basis set.[15] The
electron wavefunction is expanded in a combined set of plane waves
and Bloch sums of atom-centered Gaussian orbitals. The mixed-
basis set is most suitable for efficient description of systems
with both highly localized (atomic-like) electrons and delocalized
(plane wave-like) electrons.

Another approach is to use a linear combination of atomic-
like orbitals (LCAO) basis.[16] This is probably the most efficient
basis for calculating the total energy of complex systems. For
some elements, with suitable choices of Gaussian orbitals centered
on each atom, the number of basis functions per atom needed to
obtain an accurate charge density can be reduced by one or more
orders of magnitude as compared with a plane wave expansion. For
example, instead of requiring $N \sim 300$ plane waves to describe
the carbon pseudowavefunction in diamond, only 12 Gaussian orbitals
are needed for each atom. Since, in general, the computer memory
required in a calculation is proportional to N^2 and the time required
to N^3, this represents orders of magnitude of savings. Moreover,
with the LCAO approach, results can be more easily interpreted
in terms of the chemical bonds. Other factors also contribute
to the numerical efficiency of this method. With the potential
also expanded in Gaussians, all the matrix elements consist of
integrals of Gaussians and polynomials. These integrals can be
evaluated in closed form. The essential matrix elements are wavevector
independent. Thus, once these elements have been determined,
they may be stored and retrieved as necessary for the wavevector
in question. Furthermore, with the LCAO basis, there are several
simplified levels of self-consistency[19,20] that may be carried
out for those complex systems for which full point by point self-
consistency would not be practical.

D. <u>Total Energy Expressions</u>

Once the single particle wave equation has been solved, the
total energy of the system may be evaluated. It is usually cast
in the form

$$E_{total} = \sum_{i=1}^{N} \varepsilon_i - \frac{1}{2} \int V_H(\vec{r}) n(\vec{r}) d\vec{r} - \int \mu_{xc}[n] n(\vec{r}) d\vec{r}$$

$$+ \int \varepsilon_{xc}[n] n(\vec{r}) d\vec{r} + E_{ion-ion} \tag{6}$$

where the sum is over all occupied states, ε_i are the one-electron
eigenvalues, n is the self-consistent charge density, ε_{xc} is the
exchange-correlation energy density, and μ_{xc} is the exchange correla-

tion potential. $E_{ion-ion}$ is the electrostatic interaction energy among the bare ions.

The expression in Eq. 6, however, does not have the variational property given by Eq. 2 unless n is the exact fully self-consistent charge density. This is because the sum of electron eigenvalues, the first term on the right of Eq. 6, contains information on both the input charge and the output charge of a particular iterative cycle in the calculation. Thus, E_{total} is not a functional of a single density. For many applications, an alternate expression which makes explicit use of the density-functional variational principle is desirable. Consider a situation where an initial charge density n_{in} (e.g., from the superposition of atomic charge densities) is used and the single particle equation is solved to obtain a new charge density $\tilde{n}$. One can show[16] that the total energy can be written as

$$E_{total} = \sum_{i=1}^{N} \varepsilon_i - \int V_H[n_{in}]\tilde{n}(\vec{r})\,d\vec{r} + \frac{1}{2}\int V_H[\tilde{n}]\tilde{n}(\vec{r})\,d\vec{r}$$

$$- \int \mu_{xc}[n_{in}]\tilde{n}(\vec{r})\,d\vec{r} + \int \varepsilon_{xc}[\tilde{n}]\tilde{n}(\vec{r})\,d\vec{r} + E_{ion-ion} \qquad (7)$$

which is accurate to second order in $\Delta n = n - \tilde{n}$. In Eq. 7, all contributions from n_{in} (and, hence, from the input screening potential) are explicitly removed from the energy expression. The sole function of n_{in} is to provide a way for obtaining a good approximation to n. Often, it is possible to obtain accurate ground-state properties by using the superposition of atomic charge for n_{in} and only solve the Kohn-Sham equation once to obtain $\tilde{n}$. In this fashion, many self-consistent cycles may be eliminated in obtaining an accurate solution for the total energy.

To evaluate the total energy terms in Eq. 7 for periodic or quasiperiodic systems, it is most convenient to use the formalism of Ihm et al.[17] This formalism expresses the energy in momentum space and involves the Fourier transform of the potential and charge density. The energy per atom is written as (for the case of an elemental solid with one atom per unit cell)

$$E_{total} = \frac{1}{N}\sum_{n,\vec{k}} \varepsilon_{n,\vec{k}} - \Omega_a \sum_{\vec{G}\neq 0} [V_H^{in}(\vec{G}) + \mu_{xc}^{in}(\vec{G})]\tilde{n}(\vec{G})$$

$$+ \frac{1}{2}\Omega_a \sum_{\vec{G}\neq 0} \tilde{V}_H(\vec{G})\tilde{n}(\vec{G}) + \Omega_a \sum_{\vec{G}} \tilde{\varepsilon}_{xc}[\vec{G}]\tilde{n}(\vec{G})$$

$$+ \gamma_{Ewald} + \alpha_1 Z \qquad (8)$$

where

$$\gamma_{Ewald} = \frac{1}{2} \sum_{\vec{R} \neq 0} \frac{Z^2}{|\vec{R}|} - \lim_{\vec{G} \to \vec{0}} \frac{4\pi Z^2}{\Omega_a |\vec{G}|^2}$$

$$\alpha_1 Z = \frac{Z}{\Omega_a} \int \left(V_{ion} + \frac{Z}{r} \right) d^3 r.$$

Here Ω_a is the atomic volume in the crystal, N is the number of
atoms in the crystal, and the sum of the first term is over all
occupied bands. We use atomic units, and $\vec{G}$ denotes the reciprocal
lattice vectors.

III. BULK PROPERTIES

In this section, we describe the applications of the density
functional pseudopotential scheme to the bulk properties of solids.
Since this is a very active area, only specific prototypical calculations are featured to illustrate major subareas. The results
on semiconductors were calculated using the plane wave method
whereas results on transition metals and insulators were obtained
either using the mixed basis approach or the LCAO approach.

The total energy must be computed very precisely if it is
to be used to calculate structural and other ground-state properties.
An estimate of the precision requirements for the total energy
for calculation of the cohesive energy, the lattice constant or
bulk moduli, and phonon frequencies are given in Table I. A major
advantage of the pseudopotential method as mentioned in the previous
section is that the energies of the core electrons which are of
the order of -10^3 to -10^4 Ry per atom are removed from the total
energies of both the isolated atoms and those of the solid state.
This leads to a major enhancement in the precision of a calculation.

A. Static Structural Properties

Calculations have been carried out for a number of solids.
The first applications were to the semiconductors.[21,22] However,
for the purpose of illustration here, we first discuss the results
for diamond.[16] The calculation was carried out using the LCAO
basis with three Gaussian exponents for each of the s, p_x, p_y,
and p_z orbitals totaling 12 basis functions per carbon atom. The
calculated total energy as a function of volume E(V) is present
in Fig. 4. The points are the computed values, and the curve
is a fit of the results to the Murnaghan[23] equation of state.
The minimum and curvature of E(V) near the minimum determine the
lattice constant and bulk modulus. The cohesive energy can be
evaluated by comparing the energy for the solid including a zero-point
motion contribution and the isolated pseudoatom ground-state energy.

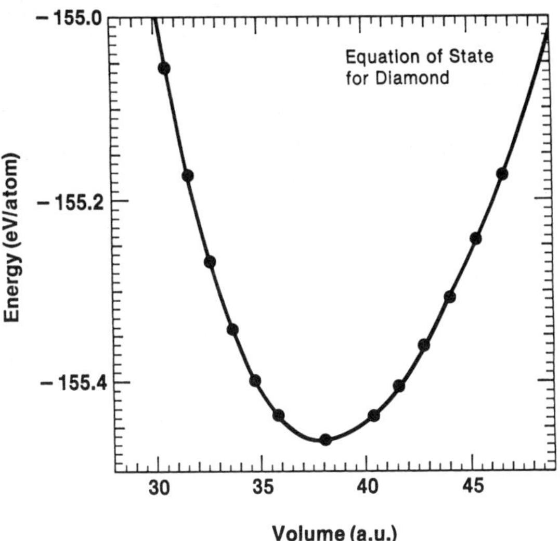

Fig. 4. Total energy vs. volume for carbon in the diamond structure.
The continuous curve is the Murnaghan equation of state fit
to the calculated points. (from Ref. 16)

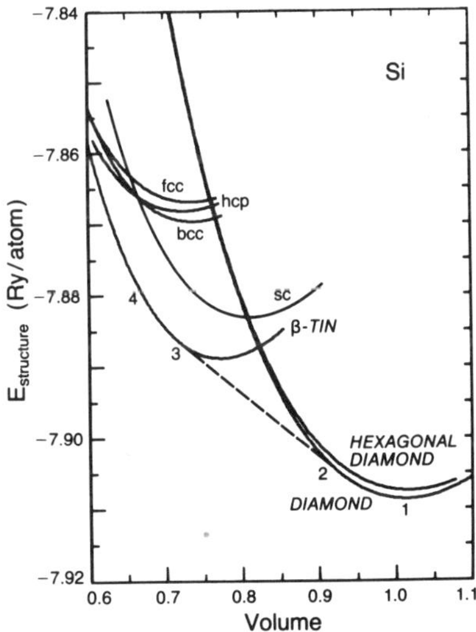

Fig. 5. Total energy curves for various assumed crystal structures
of Si as a function of volume normalized to the observed
volume. The dashed line is the common tangent between the
diamond and white tin phases. (from Ref. 21)

Table II. Ground-state properties of diamond. (after Ref. 16)

Ground state	Experiment	Theory	
		LCAO	Plane waves
Cohesive energy (in eV)	7.37	7.84	8.10
Lattice constant (in Å)	3.567	3.560	3.60
Bulk modulus (in Mbar)	4.42	4.37	4.33
Pressure derivative of bulk modulus	4	3.54	

Further, by fitting the calculated points to an equation of state such as the Murnaghan form:

$$E(V) = \frac{B_0 V}{B_0'(B_0'-1)} \left[B_0'\left(1 - \frac{V_0}{V}\right) + \left(\frac{V_0}{V}\right)^{B_0'} - 1 \right] + E(V_0) \qquad (9)$$

we obtain both the equilibrium bulk modulus B_0 and its pressure derivative B_0'. The results[16] are compiled in Table II together with the experimental values and results from a self-consistent plane wave calculation.[24] There is excellent agreement between theory and experiment and between the two theoretical calculations. The LCAO total energy was evaluated using the variational expression (Eq. 7) with $\tilde{n}$ obtained from a potential generated using superposition of carbon sp^3 atomic charges. This and the fact that only 12 basis functions per atom were used account for the slight difference in the cohesive energy between the two calculations. We note that because of the localized nature of the carbon bond in diamond approximately 250 plane waves per atom were used for the plane wave basis.

In Fig. 5, the total energy of Si is given as a function of volume for seven different crystal structures.[21] As expected and consistent with experiment, the diamond structure has the lowest energy. Since these curves were generated with the atomic number and several assumed crystal structures as the only input, the method can be used to predict crystal structures and to study solid-solid structural phase transformations. The calculated structural results[21] for Si and Ge in the diamond structure are

Table III. Comparison of calculated and measured static properties of Si and Ge. (from Ref. 21)

	Lattice constant ($\overset{\circ}{A}$)	Cohesive energy (eV/atom)	Bulk modulus (Mbar)
Si			
Calculation	5.451	4.84	0.98
Experiment	5.429	4.63	0.99
Ge			
Calculation	5.655	4.26	0.73
Experiment	5.652	3.85	0.77

Table IV. Static properties for some III-V semiconductors. (from Ref. 22)

	GaAs	GaP	AlAs	AlP
$a(\overset{\circ}{A})$				
th	5.570	5.340	5.641	5.420
exp	5.653	5.451	5.662	5.451
B_o ($10^{10} Nm^{-2}$)				
th	7.25	8.97	7.41	8.65
exp	7.48	8.87	7.70	8.60

summarized in Table III. Agreement with experiment to within 1% is found for the lattice constants; the cohesive energies and bulk moduli are given to around 10%.

Structural calculations have not been limited to the group IV elements. The plane wave calculations have been extended to the III-V semiconductors[22] and the simple metals[7,25,26] like Be, Na, and Al. Some results for these crystals are listed in Tables IV and V. In the case of Be, it was possible to calculate the Poisson ratio and the c/a ratio. Using the LCAO approach, these calculations have also been extended to study the transition metals. Some representative results[27] are presented in Table VI for the cases of Mo and W. The transition metals are much more difficult to

Table V. Static structural properties of Al and Be. (after Refs. 25 and 26)

Aluminum

	$a(\text{Å})$	B_o ($\times 10^{12}$ dyne cm^{-2})
Theory	4.01	0.715
Experiment	4.02	0.722

Beryllium

	$a(\text{Å})$	$c(\text{Å})$	c/a	B_o ($\times 10^{12}$ dyne cm^{-2})
Theory	2.25	3.57	1.58	1.368
Experiment	2.2858	3.5842	1.567	1.14 – 1.27

Poisson ratio

Theory	−0.05
Experiment	−0.01 – −0.05

deal with because of their much stronger potentials and their very localized electrons. The LCAO method, nevertheless, yields results as accurate as those of the semiconductors which are considered state-of-the-art. Similar calculations have also been carried out using the mixed basis method.[28]

The accuracy of the results for all the materials mentioned are, therefore, comparable and, in general, in very good agreement with experiment. For special cases like Na,[7] which has a small cohesive energy and low electronic density so that correlation effects are important, the results are sensitive to the choice of the LDA exchange-correlation potential. For most systems, the use of different exchange-correlation potentials only results in a few percent change in the computed values. In addition to the structural properties, the LDA pseudopotential calculations provide very accurate valence charge densities $n(\vec{r})$. This is illustrated by Fig. 6 in which the calculated charge density of graphite[29] is compared with the experimental density obtained from x-ray measurements.[30]

B. Vibrational Properties

Given the capacity to evaluate total energy for arbitrary crystal structures, it is clear that we should be able to probe energy changes with atomic positions and, therefore, be able to calculate lattice vibrational properties.

Table VI. Comparison of calculated and measured
           static properties of Mo and W. (from
           Ref. 27)

	Lattice constant (Å)	Cohesive energy (eV/atom)	Bulk modulus (Mbar)
Mo			
Calculation	3.09	7.16	2.86
Experiment	3.15	6.82	2.73
W			
Calculation	3.12	8.56	3.43
Experiment	3.16	8.9	3.23

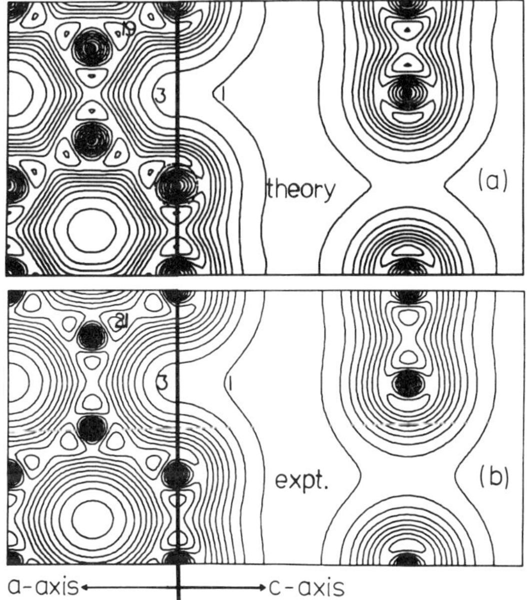

Fig. 6. Contour plots of valence-electronic-charge density for gra-
         phite: (A) Present results and (b) results from analysis
         of x-ray data. Contour values are given in units of 0.1
         e/A^3. Atomic positions are denoted by filled circles. Two
         planes are shown--one containing an a axis and the other con-
         taining the c axis and both intersecting at 90° along a C-C
         bond. In (a), the dashed circle denotes the pseudopotential
         radius. (from Ref. 29)

1. <u>Phonon Dispersions</u>. The frequency and eigenvector of
an individual phonon mode can be obtained using the frozen phonon
technique.[16,25,28,31-35] In this approach, the usual Born-Oppenheimer
approximation[36] is made. The electrons are assumed to be in the
ground state defined by the instantaneous ionic configuration. In
the calculation, the crystal is distorted with atomic displacements
$\vec{u}_{\vec{k}}$ corresponding to a particular phonon mode (For simplicity, we
specialize to the case of one atom per unit cell), for example,

$$\vec{u}_{\vec{k}}^{i} = u_o \, \hat{e}_{\vec{k}} \cos(\vec{k} \cdot \vec{R}_i + \delta_{\vec{k}}) \tag{10}$$

where $\vec{R}_i$ is the equilibrium position of the ith atom, u_o is the
amplitude of the distorted wave, and $\delta_{\vec{k}}$ is the phase factor. If the
wave vector $\vec{k}$ is commensurate with the bulk reciprocal lattice,
the resultant deformed lattice is just another crystal with reduced
symmetry. The energy difference per atom between the two crystals
can be calculated as a function of the displacement amplitude u_o
and fit to the expansion

$$\Delta E = \frac{1}{2!} K_2 u_o^2 + \frac{1}{3!} K_3 u_o^3 + \frac{1}{4!} K_4 u_o^4 + \dots \tag{11}$$

K_2 as shown below is related to the phonon frequency

$$M\omega_{\vec{k}}^2 = K_2 \tag{12}$$

for a zone center or zone boundary phonon and

$$\frac{1}{2} M\omega_{\vec{k}}^2 = K_2 \tag{13}$$

for a phonon mode of arbitrary $\vec{k}$. The higher order terms K_3 and
K_4 are the anharmonic terms which give rise to the phonon-phonon
interactions.

Thus, from Eqs. 12 and 13, the phonon frequency can be evaluated
from the curvature of the calculated energy vs. displacement curve
for small displacements. These results can be extended to
the case of compounds and to general wave vectors where the lack
of symmetry requires the calculation and diagonalization of the
dynamical matrix to obtain the phonon frequencies and polarization
vectors. Moreover, this approach allows a detailed investigation
of the role of core-core, electron-core, electron kinetic, and
electron-electron energies to determine the vibrational frequencies
of the solids examined. This kind of information has been valuable
in analyzing and understanding phonon anomalies in semiconductors
and transition metals.

Figure 7 illustrates the changes in the crystal total energy
as a function of the amplitude for a $\vec{k} = 0$ optical phonon frozen

350

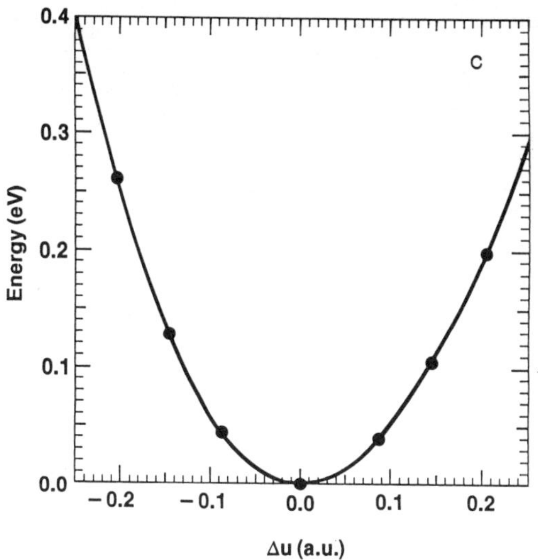

Fig. 7. Frozen-phonon energy vs. bond displacement. (from Ref. 16)

into the diamond crystal.[16] The atomic motions for this phonon
mode are particularly simple. The motion may be taken to be a
constant volume uniaxial distortion of the diamond crystal along
the ⟨111⟩ direction. From the quadratic term of the curve in Fig. 7,
the zone center optical phonon frequency was obtained to within
1% of the experimental value. Compared to the structural properties
calculations, the only additional input is the ionic masses. The
calculated phonon frequencies for some selected phonon modes are
presented in Table VII and Table VIII for diamond[35] and Si[31] respectively. As seen in Table VIII, phonon mode Grüneisen parameters
have also been obtained. It is probably fair to say that phonon calculations using this ab initio approach give results which are more
accurate than most empirical fits. In addition to obtaining accurate
frequencies, the calculations are of great value because they allow
detailed analysis of the electronic response to lattice distortions
and, hence, the mechanisms causing phonon anomalies.[28] The calculations also provide microscopic information not obtainable otherwise
and provide benchmarks for the validity of phenomenological models.

The frozen phonon technique has been applied with equal success
to the metals.[25,28] The only added complication in these calculations
is that a large number of $\vec{k}$-points in the Brillouin zone is needed
to sample over the Fermi surface for convergent results because
of the small energies involved.

It is also possible to obtain whole phonon dispersion curves
along some direction in $\vec{k}$-space from the ab initio calculations.[35,37,38] This can be done in two equivalent ways. One way is

Table VII. Frozen phonon calculations of phonon frequencies for selected modes of diamond. (from Ref. 35)

Mode	ω_{theory} (cm^{-1})	$\omega_{expt.}$ (cm^{-1})	$\Delta\omega$(cm^{-1})
LTO(Γ)	1346.5	1332	15
LO($\vec{k}$ = 1/3 to X)	1353.3	--	
LO($\vec{k}$ = 2/3 to X)	1328.2	--	
LOA(X)	1219.5	1185	34
TO(X)	1173.9	1069	104
TA(X)	772.1	807	-35

Table VIII. Total energy calculations of phonon energies and Grüneisen parameters for a few phonon frequencies of Si. (from Ref. 31)

	LTO(Γ)	TA(X)	TO(X)	LOA(X)
Phonon frequencies (THz)				
$F_{calc.}$	15.16	4.45	13.48	12.16
$F_{expt.}$	15.53	4.49	13.90	12.32
Deviation	-2.4%	-0.9%	-3.0%	-1.3%
Grüneisen parameter				
$\gamma_{calc.}$	0.92	-1.50	1.34	0.92
$\gamma_{expt.}$	0.98	-1.40	1.50	~0.90

by computing the Hellmann-Feynman forces on atoms resulting from displacing a plane of atoms. This is done by considering a supercell of N layers and, thus, obtaining the first N nearest neighbor force constants. Results[38] for Si in the [100] direction of the Brillouin zone calculated this way are given in Fig. 8. Another equivalent way which does not involve the Hellmann-Feynman forces directly is to perform several frozen phonon calculations along a $\vec{k}$-direction for a given branch. Given the calculated frequencies, one can invert the dynamical matrix to obtain the planar force constants up to several nearest-neighbor planes until convergence is achieved. Results[35] for the longitudinal branches of diamond in the [100]

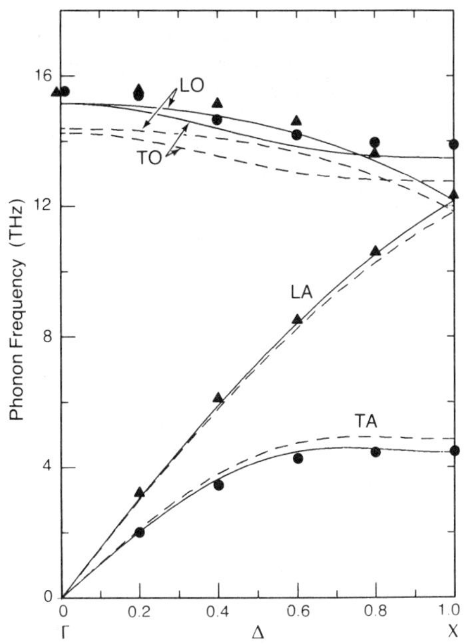

Fig. 8.   Calculation of phonon dispersions for Si in the [100] direction using Hellmann-Feynman forces. The dashed line is the result when plane waves up to an energy of 6 Ry are used; the solid line is for 10 Ry. The triangles and dots represent measured points. (from Ref. 38)

direction obtained using this scheme are given in Fig. 9.

   2. _Anharmonic terms and phonon-phonon interactions_. As seen from Eq. 11, in addition to the phonon frequencies, the frozen phonon calculations yield higher order anharmonic terms and, hence, information on phonon-phonon interactions. They are extremely valuable information since these terms are often not directly measurable and cannot be reliably gotten from phenomenological models.

To illustrate the kind of useful information that can be obtained, we consider here in some detail one example--calculation of the optical phonon-phonon interactions[35] in diamond. This will serve both to illustrate the power of the method and to shed some light into the phenomenon of two-phonon Raman anomaly in diamond.[39] It was observed that the two-phonon Raman spectrum of diamond has an anomalous sharp peak (not seen for Si and Ge) at 2667 cm^{-1} which is at an energy 3 cm^{-1} higher than twice the optical phonon frequencies at Γ. Despite a number of theoretical works,[40,41] the nature and origin of this peak is still a mystery. One particularly intriguing explanation was the two-phonon bound state theory by Cohen and Ruvalds.[40] They proposed that a two-phonon bound state is formed

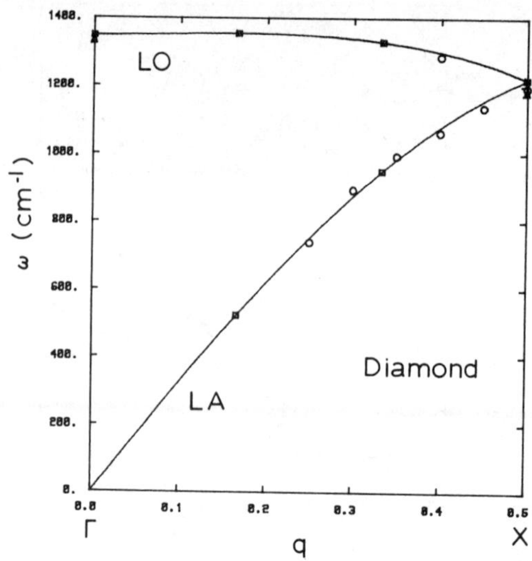

Fig. 9.   Calculation of phonon dispersions for diamond in the [100] direction by extracting force constants from frozen-phonon results. Only the longitudinal modes are plotted. The squares are frozen-phonon results. The triangles and dots represent measured points. (from Ref. 35)

giving rise to a split-off level at the anomalous peak position if the optical phonons near the Brillouin zone center interact through a <u>positive</u> quartic anharmonic interaction Φ

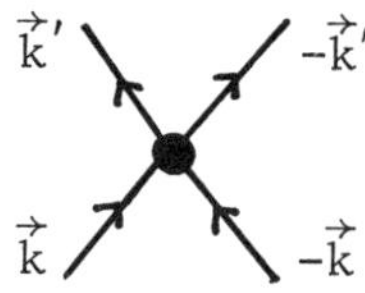

which is greater than certain critical value. For 15 years, this theory was neither confirmed nor disproved because of the lack of information on the optical phonon anharmonic terms.

In the following, we summarize a calculation on Φ for the optical phonons at $\vec{k} = 0$. For simplicity, we again specialize our discussion to one basis atom per unit cell and later generalize to the case of the diamond structure. The total energy of the crystal is expanded in a series in the atomic displacements $\vec{\xi}$ from the equilibrium positions.

$$E_{tot} = E_0 + (T + E_2') + E_3 + E_4 + \ldots \tag{14}$$

where

T = kinetic energy

E_o = constant

$$E_2' = \frac{1}{2!} \sum_{\substack{\vec{m}i \\ \vec{n}j}} \Phi_{ij}^{\vec{m}\vec{n}} \, \xi_i^{\vec{m}} \, \xi_j^{\vec{n}}$$

$$E_3 = \frac{1}{3!} \sum_{\substack{\vec{m}i \\ \vec{n}j \\ \vec{p}k}} \Phi_{ijk}^{\vec{m}\vec{n}\vec{p}} \, \xi_i^{\vec{m}} \, \xi_j^{\vec{n}} \, \xi_k^{\vec{p}}$$

$$E_4 = \frac{1}{4!} \sum_{\substack{\vec{m}i \\ \vec{n}j \\ \vec{p}k \\ \vec{q}l}} \Phi_{ijkl}^{\vec{m}\vec{n}\vec{p}\vec{q}} \, \xi_i^{\vec{m}} \, \xi_j^{\vec{n}} \, \xi_k^{\vec{p}} \, \xi_l^{\vec{q}}$$

with $i,j,\ldots$ = cartesian indices and $\vec{m},\vec{n},\ldots$ = real space lattice vectors. If we denote the eigenvectors of the harmonic part, $T + E_2'$, by

$$Q_{i\lambda}^{\vec{m}\vec{k}} = \frac{1}{\sqrt{N}} \, \hat{e}_{i\lambda}^{\vec{k}} \, e^{i\vec{k}\cdot\vec{m}} \tag{15}$$

where λ is the polarization index, then the normal mode coordinates a's are given by

$$\xi_i^{\vec{m}} = \sum_{\vec{k}\lambda} Q_{i\lambda}^{\vec{m}\vec{k}} \, a_\lambda^{\vec{k}} \tag{16}$$

The energy terms in Eq. 14 may be written as

$$E_2' = \frac{1}{2!} \sum_{\substack{\vec{k}\vec{k}' \\ \lambda\lambda'}} \Phi_{\lambda\lambda'}^{\vec{k}\vec{k}'} \, a_\lambda^{\vec{k}} \, a_{\lambda'}^{\vec{k}'} \tag{17a}$$

$$E_3 = \frac{1}{3!} \frac{1}{\sqrt{N}} \sum_{\substack{\vec{k}\vec{k}'\vec{k}'' \\ \lambda\lambda'\lambda''}} \Phi_{\lambda\lambda'\lambda''}^{\vec{k}\vec{k}'\vec{k}''} \, a_\lambda^{\vec{k}} \, a_{\lambda'}^{\vec{k}'} \, a_{\lambda''}^{\vec{k}''} \tag{17b}$$

and

$$E_4 = \frac{1}{4!}\,\frac{1}{N} \sum_{\substack{\vec{k}\vec{k}'\,\vec{k}''\vec{k}''' \\ \lambda\lambda'\,\lambda''\lambda'''}} \Phi^{\vec{k}\vec{k}'\,\vec{k}''\vec{k}'''}_{\lambda\lambda'\,\lambda''\lambda'''}\, a^{\vec{k}}_{\lambda}\, a^{\vec{k}'}_{\lambda'}\, a^{\vec{k}''}_{\lambda''}\, a^{\vec{k}'''}_{\lambda'''} \qquad (17c)$$

The Φ's are then the generalized elastic constants. The third and fourth order Φ's are directly proportional to the matrix elements for the three- and four-phonon processes respectively.

For example, since the a's are the normal mode coordinates,

$$\Phi^{\vec{k}\vec{k}'}_{\lambda\lambda'} = \delta_{\vec{k},-\vec{k}'}\, \delta_{\lambda,\lambda'}\, \kappa(\vec{k},\lambda) \quad . \qquad (18)$$

Therefore,

$$E_2' = \frac{1}{2} \sum_{\vec{k}\lambda} \kappa(\vec{k},\lambda)\, a^{\vec{k}}_{\lambda}\, a^{*\vec{k}}_{\lambda} \qquad (19)$$

Also, the kinetic energy is given by

$$T = \frac{M}{2} \sum_{\vec{k}\lambda} \dot{a}^{\vec{k}}_{\lambda}\, \dot{a}^{*\vec{k}}_{\lambda} = \frac{M}{2} \sum_{\vec{k}\lambda} \omega^2(\vec{k},\lambda)\, a^{\vec{k}}_{\lambda}\, a^{*\vec{k}}_{\lambda} \qquad (20)$$

Equating $E_2 = T$ mode by mode, one obtains

$$\omega(\vec{k},\lambda) = \sqrt{\frac{\kappa(\vec{k},\lambda)}{M}} \quad . \qquad (21)$$

For the purpose of understanding the two-phonon bound state problem, we are interested in the $k \approx 0$ phonons and their interactions. We, therefore, need to calculate $\Phi^{000}_{\lambda\lambda'\lambda''}$ and $\Phi^{0000}_{\lambda\lambda'\lambda''\lambda'''}$ where λ denotes the various optical modes. For example, $\Phi^{000}_{\lambda\lambda'\lambda''}$ gives the amplitude for the process

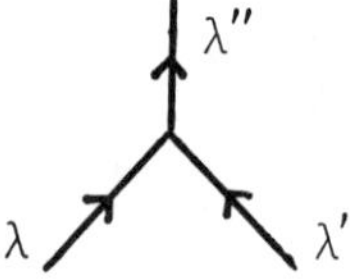

and $\Phi^{0000}_{\lambda\lambda'\lambda''\lambda'''}$ the amplitude for

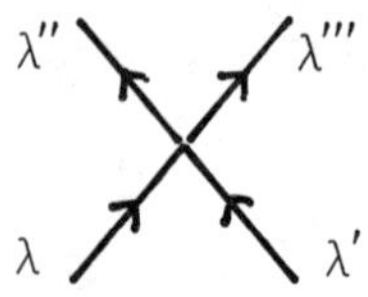

Here, we will focus on calculating only these $\vec{k} = 0$ terms. Generalization to other cases involving finite $\vec{k}$'s can be done straightforwardly.

In a frozen-phonon calculation, instead of traveling waves

$$\xi_i^{\vec{m}} = \frac{1}{\sqrt{N}} \sum_{\vec{k}\lambda} a_\lambda^{\vec{k}} \, \hat{e}_{i\lambda}^{\vec{k}} \, e^{i\vec{k}\cdot\vec{m}} \tag{22}$$

standing waves of the form

$$\xi_i^{\vec{m}} = \sum_{\vec{k}\lambda} u_\lambda^{\vec{k}} \, \hat{e}_{i\lambda}^{\vec{k}} \, \cos(\vec{k}\cdot\vec{m}) \tag{23}$$

are considered. For the case of diamond and $\vec{k} = 0$, a standing wave consisting of a linear combination of the optical modes is used. We may choose the three polarization vectors to be $\hat{x}$, $\hat{y}$, and $\hat{z}$. Then, the atomic displacements for the two carbon atoms in the unit cell are of the form

$$\vec{\xi}^{\vec{m}}(1) = \vec{u} \, \cos(\vec{k}\cdot\vec{m})$$

$$\vec{\xi}^{\vec{m}}(2) = -\vec{u} \, \cos(\vec{k}\cdot\vec{m}) \tag{24}$$

where $\vec{u} = (u_x, u_y, u_z)$. The frozen wave is equivalent to having

$$a_\lambda^{\vec{k}} = \begin{cases} \sqrt{2N} \, u_\lambda & \text{for } \vec{k} = 0, \ \lambda = x,y,z \\[2mm] 0 & \text{otherwise} \end{cases}$$

and the various energy terms (per atom) in Eq. 17 become

$$E_2' = \frac{1}{2}\left(\frac{1}{2N}\right) \Phi_{\lambda\lambda'}^{00} \, a_\lambda^o \, a_{\lambda'}^o = \frac{1}{2} \Phi_{xx}^{00} (u_x^2 + u_y^2 + u_z^2)$$

$$E_3 = \frac{1}{3!} \frac{1}{(2N)^{3/2}} \sum_{\lambda\lambda'\lambda''} \Phi_{\lambda\lambda'\lambda''}^{000} \, a_\lambda^o \, a_{\lambda'}^o \, a_{\lambda''}^o$$

$$= \frac{1}{3!} \left\{ \Phi_{xyz}^{000} u_x u_y u_z + \Phi_{xxx}^{000} u_x u_x u_x + \ldots \right\}$$

$$E_4 = \frac{1}{4!} \frac{1}{(2N)^2} \sum_{\lambda\lambda'\lambda''\lambda'''} \Phi_{\lambda\lambda'\lambda''\lambda'''}^{0000} \, a_\lambda^o \, a_{\lambda'}^o \, a_{\lambda''}^o \, a_{\lambda'''}^o$$

$$= \frac{1}{4!} \left\{ \Phi_{xxxx}^{0000} u_x u_x u_x u_x + \ldots \right\} \tag{25}$$

Comparing the above expression with a general Taylor's series expansion for E_{total}, we see that the Φ's are simply the various order derivatives of E_{total} with respect to the u_i's, for example,

$$\Phi^{oo}_{xx} = \frac{\partial^2 E}{\partial x^2} = E_{xx}$$

$$\Phi^{ooo}_{xyz} = \frac{\partial^3 E}{\partial x \partial y \partial z} = E_{xyz} \quad , \text{ etc.}$$

The symmetry of the diamond crystal structure dictates that many of the Φ's are zero. One can easily show that the only terms up to fourth order not required by symmetry to vanish are the following:

$$E_{xx} = E_{yy} = E_{zz} \equiv \kappa$$

$$E_{xyz} = \gamma$$

$$E_{xxxx} = E_{yyyy} = E_{zzzz} \equiv \alpha$$

$$E_{xxzz} = E_{yyzz} = E_{xxyy} \equiv \beta \tag{26}$$

All other terms that are not related to the above terms by permutation of the indices are zero. As noted before, κ gives the optical phonon frequency; γ gives the amplitudes for three-phonon processes; and α and β give the amplitudes for the bare four-phonon processes.

To obtain κ, γ, α, and β, several $\vec{u}$'s of different magnitudes and directions are used to calculate the total energy of distorted crystals. Figure 10 illustrates the calculation of κ, α, and β. Plotted are curves for $\Delta E/u^2$ vs. u^2 for $\hat{u}$ along the (001), (110), and (111) directions. (For the (111) direction, the plotted values are those averaged over positive and negative $\vec{u}$'s. This is to remove the third-order component in the energy. There is no third-order contribution for the other two directions because of symmetry.) The intercept at $u = 0$ is the value for κ. The slopes of the curves correspond to α, $\alpha/2 + 3\beta/2$, and $\alpha/3 + 2\beta$ for the (100), (110), and (111) directions respectively. Also, by plotting $\Delta E/u^2$ vs. u for the (111) direction, one extracts the value for the third-order term γ. The calculated results are summarized in Table IX.

The theoretical value for α is negative whereas for β it is positive and small compared to the magnitude of α. This implies that the phonon-phonon interaction of the form

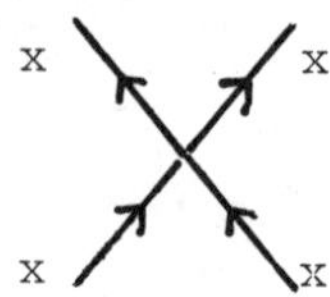

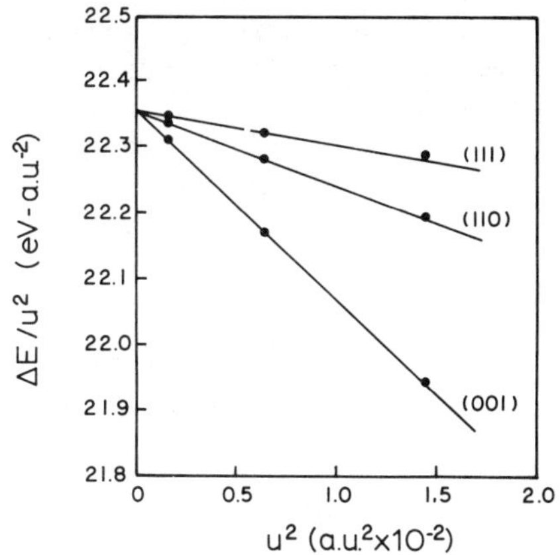

Fig. 10. Plots of $\Delta E/u^2$ vs. u^2 for $\vec{k} = 0$ frozen-optical phonons in diamond (see text). (ΔE is energy/cell.) (from Ref. 35)

Table IX. Bare harmonic and anharmonic parameters for $\vec{k} = 0$ optical phonons in diamond (see text). (from Ref. 35)

$\dfrac{2\kappa}{2!}$	$\dfrac{2\gamma}{3!}$	$\dfrac{2\alpha}{4!}$	$\dfrac{2\beta}{4!}$
22.36	-24.49	-28.98	2.08

is attractive, but interaction of the form

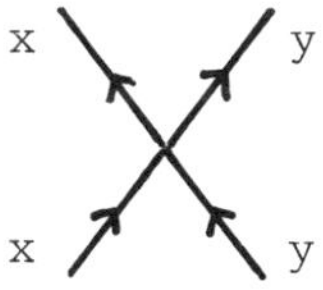

is weakly repulsive.

The parameters α and β, however, only describe the direct four-phonon terms. To address the two-phonon bound state question, we have to consider other possible four-phonon diagrams of comparable

strength. They may be obtained using perturbation theory. Diagrams
up to order γ^2 are given in Fig. 11. These are phonon-phonon interactions via an exchange of a third phonon. It turned out that because
of symmetry the diagrams in Fig. 11 do not effect processes of
the type xx → xx. Therefore, α remains unchanged. However, processes
of the type xy → xy or xx → yy will have other contributions in
addition to the direct term. Thus, the parameter is renormalized.
For scattering of the type xy → xy, four additional diagrams contribute (c-f in Fig. 11) leading to a renormalized value

$$\beta' = \beta - \frac{2\gamma^2}{3\kappa} \tag{27}$$

and for scattering of the type xx → yy, also four additional diagrams
contribute (a-d in Fig. 11) leading to a renormalized value

$$\beta'' = \beta - \frac{2\gamma^2}{\kappa} \ . \tag{28}$$

There are now three effective four-phonon interactions of the types

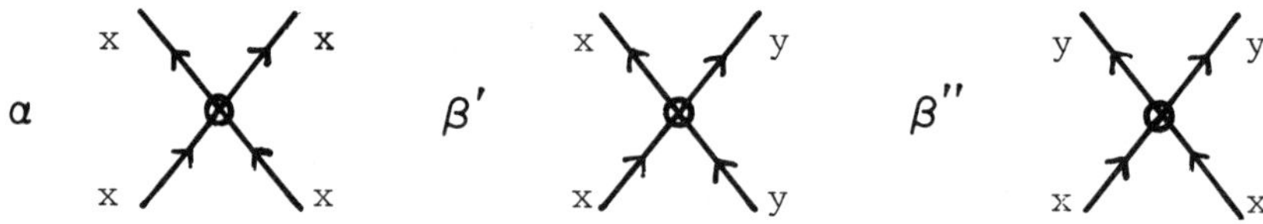

The values for α, β', β'' are tabulated in Table X. Note that all
parameters are now negative.

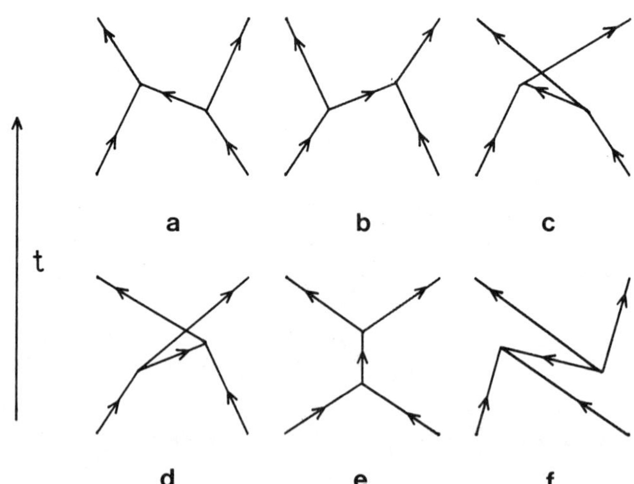

Fig. 11. Diagrams for phonon-phonon scattering via exchange of
          another phonon.

Table X. Renormalized fourth-
order anharmonic
parameters for $\vec{k} = 0$
optical phonons in
diamond (see text).

$\dfrac{2\alpha}{4!}$	$\dfrac{2\beta'}{4!}$	$\dfrac{2\beta''}{4!}$
-28.98	-11.33	-38.16

This work contributed in two major ways. First, the calculation
shows that it is now possible for the first time to calculate from
first principles phonon-phonon interaction parameters that are
inaccessible from experiment. Second, since all the effective
four-phonon terms (direct plus mediated processes up to γ^2) are
attractive for the $\vec{k} = 0$ optical phonons in diamond, the formation
of the two-phonon bound state is unlikely in this system.

C. Structural Phase Transformations

In this subsection, we describe several examples of applications
of the total energy pseudopotential method to structural phase
transformations induced either by pressure or temperature.

By calculating the total energy of an element or compound at
various crystal structures, one can determine the relative stability
among the structures considered. For example, in Fig. 5, the total
energy for Si as a function of volume for seven common crystal
structures is displayed.[21] The diamond structure curve has the
lowest energy minimum. Hence, it is the stable structure for Si
as found in experiment. At small volumes which can be achieved
under pressure, the diamond structure energy is no longer the lowest.
A solid-solid structural phase transition to a lower energy structure
should occur.

Near the experimental equilibrium volume, the theoretical
curves (Fig. 5) show that Si in the hexagonal diamond (homopolar
wurtzite) structure has energy closest to the cubic diamond curve.
However, since the hexagonal diamond curve lies higher in energy
than cubic diamond over the whole range of volumes, no pressure-
induced transition is predicted. This is again consistent with
experimental observations. On the other hand, at volume smaller
than 0.82 of normal volume, the β-Sn structure has lower energy
than the cubic diamond structure. Therefore, a structural transformation should occur as a function of pressure.

At zero or low temperatures, the critical pressure at which Si transforms from the diamond structure into the β-Sn structure can be calculated by examing the enthalpy

$$H = E + PV \tag{29}$$

for the two structures. The transition occurs at a pressure when the enthalpy of the two phases is equal, that is, when

$$P = -\left(\frac{E_2 - E_1}{V_2 - V_1}\right) . \tag{30}$$

From the defintion of P, we see that the diamond-β-Sn transition occurs at the volume where the first common tangent (Gibbs line) can be drawn between the diamond curve and the β-Sn curve. In Fig. 5, the Gibbs line is given by the dashed line. Thus, the points 2 and 3 label the transition volumes between the structures and the slope of the Gibbs line provides the transition pressure. A calculation of the energetics of Ge in various crystal structures showed that Si and Ge behave very similarly under pressure. Both materials transform to a metallic β-Sn phase at a pressure around 100 kbars. The theoretical results[21] (Table XI) are in excellent agreement with experiment particularly for the transition volumes. These results are remarkable considering that the only input to

Table XI. Comparison of the calculated and measured transition volumes ($V_t^{d,\beta}$) of the diamond and β phases, their ratios (V_t^{β}/V_t^{d}), and the transition pressures (P_t) for Si and Ge. Volumes are normalized to the measured zero-pressure volumes. (from Ref. 21)

	V_t^{d}	V_t^{β}	V_t^{β}/V_t^{d}	P_t (kbar)
Si				
Calculation	0.928	0.718	0.774	99
Experiment	0.918	0.710	0.773	125
Deviation	1.1%	1.1%	0.1%	−20%
Ge				
Calculation	0.895	0.728	0.813	96
Experiment	0.875	0.694	0.793	100
Deviation	2.3%	4.9%	2.5%	−4%

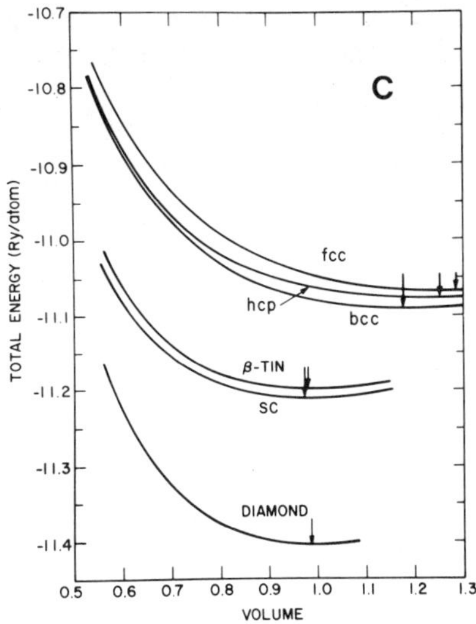

Fig. 12.   Total energy versus volume (normalized to the experimental
           volume) for six structures of carbon.   Vertical arrows
           denote the minimum energy in each structure.   (from Ref. 24)

the calculations are the atomic numbers and an assumed set of crystal
structures.

Carbon[24,42-44], on the other hand, behaves differently under
pressure among these three group IV covalent elements. Fig. 12
shows that carbon remains in the diamond structure over a very
large volume (and pressure) range. No transition to the β-Sn
structure or to another structure was predicted. This is consistent
with observations that diamond remains unchanged up to megabars
pressure range. More recent calculations[42-44], however, showed
that diamond will transform to a BC-8 structure, a structure of
distorted tetrahedrons with eights atoms per unit cell, at about
12 Mbars. Hence, carbon is predicted to be stable in the open
diamond structure up to extremely high pressure at least for the
structures considered. Comparing Si and Ge with diamond, the extra-
ordinary stability of carbon in the tetrahedral structure is probably
attributable to the large bond-bending restoring forces of the
sp^3 carbon bond which may be traced back to the lack of d states
nearby the sp complex. The diamond structure is, of course, only
a metastable state of carbon whose lowest energy phase is that
of graphite. The energy difference between these two phases is
extremely small, and at present, it is not completely clear how
the graphite-diamond transition occurs. Calculations of the kind
discussed here should help in understanding such transitions.

Several calculations along the same line have also been performed

Table XII. Transition pressures and structures for III-V
semiconductors. (after Ref. 22)

	Transition Pressure (kbar)		Structure	
	Theory	Exp.	Theory	Exp.
GaAs	160	160-190	β-Sn, rocksalt, NiAs	orthorhombic?
GaP	217	200-240	β-Sn, rocksalt	β-Sn
AlP	93	140-170	rocksalt, NiAs	fcc?
AlAs	76	---	rocksalt, NiAs	--

for the III-V semiconductors[22] and metals[25,45,46] under pressure.
For the metals, high pressure phase transitions were predicted
by comparing the total energies for the fcc, bcc, and hcp phases,
and the results are consistent with experiment and with those using
other techniques.[47] For the III-V compounds, several high pressure
phases were calculated to have approximately the same energy. Hence,
at present, for some of the III-V compounds, it was not possible
to decide the appropriate high-pressure phase, but the pressure
of the transition was predicted (Table XII), and agreement with
experiment was found in cases where data is available.

Another important class of studies is to investigate temperature-
induced structural transitions. This has been performed for the
metal Be.[48] At ambient pressure, Be transforms from a low temperature
hexagonal close-packed (hcp) structure to a high temperature body-
centered cubic (bcc) structure at approximately 1530 K before it
melts at about 1560 K.[49] This transition is interesting in several
respects: there is a large entropy difference between the two
phases at the transition, and the bcc phase exists only for a very
small range of temperature. The structural transition temperature
decreases with increasing pressure. Furthermore, although the
face-centered cubic (fcc) phase is calculated to have a lower energy
than the bcc phase at zero temperature and zero pressure, the transition is from hcp to bcc.

In order to study strucural phase transitions at finite

temperature and pressure, the Gibbs free energy G(P,T) must be considered.[48] The most stable crystal phase at a given T and P is the one with the lowest Gibbs free energy[50]

$$G(P,T) = H - TS = E + PV - TS \tag{31}$$

where E is the internal energy and S is the entropy and both are dependent on T. Within the adiabatic approximation, G(P,T) of a crystal can be separated into two contributions, a static lattice part G_{st} and a vibrational part G_{ph}. G_{st} includes the electronic energy and the static Coulomb interaction between the ions fixed at their equilibrium positions. G_{ph} includes both the zero-point motion and the phonon contribution.

In the work of Lam et al.,[48] the Gibbs free energies for the hcp, fcc, and bcc structures were calculated. Several approximations were used to make the calculation tractable. First, the temperature dependence of G_{st} is neglected. The justification is that electronic excitations are negligible compared with phonon excitations at temperatures which are small compared to the Fermi temperature of the metal. Hence,

$$G_{st}(P) = E_{st}(T = 0) + P_{st}V \tag{32}$$

which is just the zero temperature enthalpy and is calculated using

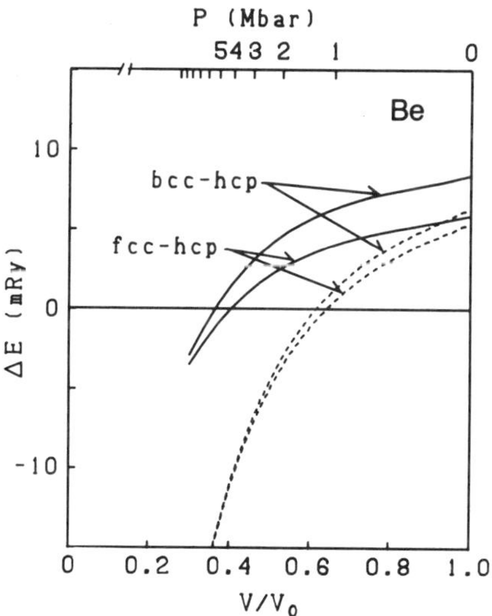

Fig. 13. Static lattice energy differences, fcc-hcp (A) and bcc-hcp (B), as functions of normalized volume (full curves) and energy differences with zero-point energy included (broken curves). (from Ref. 48)

the _ab initio_ total energy method as discussed (Fig. 13). Second,
the contributions to the pressure from the phonons is recognized
to be very small compared with that from the static lattice energy
and is, therefore, neglected in the calculation of G_{ph}. Thus,

$$G_{ph}(T) = E_{ph}^{O} + E_{ph}(T) - TS_{ph} \tag{33}$$

where E_{ph}^{O} is the zero-point energy. This is just the expression
for the Helmholtz free energy which can be evaluated within the
harmonic approximation using[51]

$$G_{ph}(T) = \sum_{i} \frac{\hbar\omega_i}{2} + k_B T \sum_{i} \ln\left[1 - \exp\left(-\frac{\hbar\omega_i}{k_B T}\right)\right] \tag{34}$$

where ω_i is the phonon frequency. Finally, since a completely
ab initio calculation of the full phonon spectrum was not practical,
the above expression was evaluated in the following way. An approximate phonon spectrum was obtained by representing the phonon frequency
with a finite Fourier series. The coefficients of the series were
determined to give the correct sound velocities calculated from
the elastic constants. For the hcp phase, experimentally determined
elastic constants were available and were used. The elastic constants
for the fcc and bcc phases were computed by calculating the static
lattice energy for the appropriate distortions of the lattice (Table
XIII).

The calculated differences in Gibbs free energy for the three
structures at P = 0 are presented in Fig. 14. At zero temperature
and zero pressure, the most stable phase is the hcp phase as observed
experimentally. As the temperature increases, the Gibbs free energy
for the bcc phase decreases relative to the hcp value and becomes
the lowest one at temperature approximately equal to 1500 K. The
observed transition is at about 1530 K. The high temperature bcc
phase is, thus, stabilized by the large entropy term, $-TS_{ph}$. The
origin of this large entropy is associated with the presence of
low-energy (soft) phonon modes. These soft phonon modes exist
in the bcc phase because the $C_{11}-C_{12}$ elastic constant is anomalously
small in this phase (see Table XIII). Also, the calculation showed
that although the fcc Gibbs free energy is lower than that of the
bcc phase at T = 0 and P = 0, its phonon free energy is considerably
larger than that of the bcc phase. Therefore, the fcc phase is
never stabilized by temperature.

These total energy methods should be applicable to the study
of other solid-solid structural transitions such as the ω-phase
of Zr. A goal in the future in this area would be to attempt to
predict crystal structures and to suggest experimental studies which
might give rise to new materials with desirable properties. For
example, as mentioned above, although graphite has a lower energy,
diamond does exist. One aim would be to understand transitions of
this kind and attempt to predict others.

Table XIII. Elastic constants
of Be (10^{12} dyne
cm^{-2}). (after
Ref. 48)

	HCP[†]	FCC[‡]	BCC[‡]
C_{11}	2.95	2.20	1.20
C_{12}	0.26	0.63	1.10
C_{44}	1.71	1.86	1.70

[†]Experimentally determined elastic constants from Silversmith and Averbach (1970).
[‡]Calculated elastic constants.

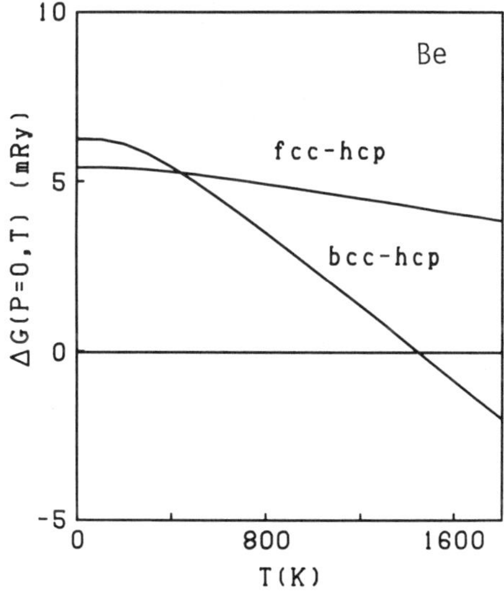

Fig. 14. Differences in the Gibbs free energy (A: fcc-hcp; B: bcc-hcp) as a function of temperature at P = 0. (from Ref. 48)

IV. SURFACES AND CHEMISORPTION SYSTEMS

Research in the properties of surfaces and related systems represents a major area in condensed matter science. The present

interests in these systems are generated both by their fascinating
fundamental properties and by their practical importance in fields
such as catalysis and device applications. However, despite a
great deal of activity in the past decades, an understanding of
the electronic and structural properties of surfaces of real materials
remains a challenge for both theory and experiment. In this section,
we discuss several representative examples of surface studies using
pseudopotential methods.

Among the goals for theoretical work in this area are: determination of the geometric structures, understanding of the nature
of surface electronic states and chemisorption bonds, theory for
mechanisms for surface atomic rearrangements (relaxations and reconstructions), prediction of chemisorption energetics, and analysis
of spectroscopic and other experimental data. Most of the past
studies have been centered on the search for surface states and
characterization of their properties. However, with the advance
of the total energy methods such as those described in the previous
sections, recent work begins to address the questions of surface
geometries and energetics.

One major constraint in theoretical surface calculations is
the lack of translational symmetry because of the surface. This
can be overcome in two ways. One approach[52] is to match extended
electronic wavefunctions from the bulk crystal to decay states
representing the surface. Another[53] is to use a thin slab of the
crystal to model two surfaces. Further, if a supercell geometry
is employed with the slabs, standard band structure techniques
may be used. The slab approach is most common and is the method
employed for the studies discussed in this section. Another constraint in surface studies is the requirement of somewhat detailed
self-consistency in the calculations because of the asymmetrical
rearrangement of charge near the surface. Quantities such as work
functions, atomic rearrangements, and surface state energies are
sensitive to such charge rearrrangements.

A. Surfaces of Semiconductors and Insulators

Theoretical research in this area has been focused mainly
on the group IV and the III-V materials. In this subsection, for
illustration purposes, we describe in some detail two surfaces,
the (111) surfaces of silicon and diamond, and briefly mention others.

The (111) surfaces of the tetrahedral elements C, Si, and
Ge are found to undergo a remarkable variety of surface reconstructions.[54] A possible common denominator is the apparent occurrence
of a 2×1 reconstruction on all three surfaces. The similarity
of the angle-resolved photoemission results[55-57] suggests that
a common structure may be responsible.

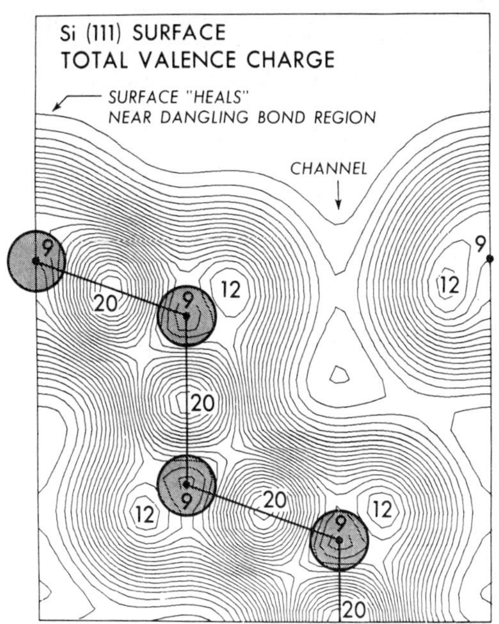

Fig. 15. Total valence electron charge density for the Si(111)-(1×1)
surface. A (110) plane is shown with the top of the diagram representing the semiconductor surface. The cores are
shown as shaded discs, and heavy lines represent bonds.
Charge density contours are normalized to e/Ω_c where Ω_c is
the bulk unit cell. (from Ref. 53)

The Si(111) surface is probably the most studied semiconductor
surface. Yet, the details of the atomic and electronic structure
are still considered open subjects. Experimental interest remains
high because it is possible to cleave Si in vacuum and produce
clean surfaces which can be studied with a host of techniques.
Theoretically, this surface is considered to be the prototype semiconductor surface.

In the 1970's, the ideal 1×1 Si(111) surface was studied using
the self-consistent pseudopotential method.[53] The charge density
was found to smooth out and "heal" the cut bonds (Fig. 15). A
dangling-bond surface state band with wavefunctions highly localized
at the cut bonds was found to lie in the semiconductor energy gap.
Since there is one broken bond per surface unit cell and each has
one electron, the surface dangling-bond band is half full leading
to a metallic surface. Experimentally, however, Si(111) is semiconducting. The nonmetallic nature is generally believed to be associated with reconstructions or movements of the atoms at the surface.
Low energy electron diffraction (LEED) measurements show that Si(111)
reconstructs into a metastable 2×1 pattern when cleaved and into

a stable 7×7 pattern upon annealing. Laser-annealed surfaces, on
the other hand, show a 1×1 pattern; however, the surface may not
be well ordered and the exact interpretation is controversial.

Because of the complexity of the 7×7 geometry, most theoretical
research has been directed to the 2×1 reconstruction. Until recently,
the commonly accepted model for this surface was a buckling model[54]
in which alternate rows of surface atoms are raised and lowered.
This model has two inequivalent atoms per surface unit cell and,
hence, should give rise to a semiconducting surface if the displacement
were large enough. Calculations[53] using the buckling model as
input gave results consistent with this picture. The correctness
of the buckling model, however, has been challenged. Angle-resolved
photoemission measurements[55-57] give an energy dispersion curve,
$E(\vec{k})$, for the surface states which is in disagreement with theory
constrained to a buckled 2×1 reconstruction. Recently, total energy
and force calculations[58,59] also show that the buckling model
is unfavorable.

In the <u>ab initio</u> surface calculations, as in the bulk cases
discussed in Sec. III, the exact geometry is no longer a required
input. The surface structure is determined by minimizing the total
energy with respect to the coordinates of atoms in the first several
layers for a given topology. An equivalent approach is to calculate
the Hellmann-Feynman forces on each atom and move the surface atoms
until all forces are zero. In either approach, the calculation
must be done iteratively since when atoms are moved, new forces

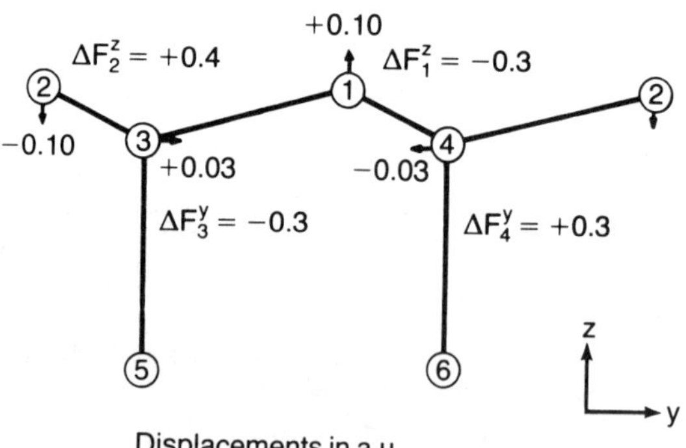

Fig. 16. Side view of a 2×1 buckling distortion used to test the
stability of the ideal 1×1 Si(111) surface. (from Ref. 58)

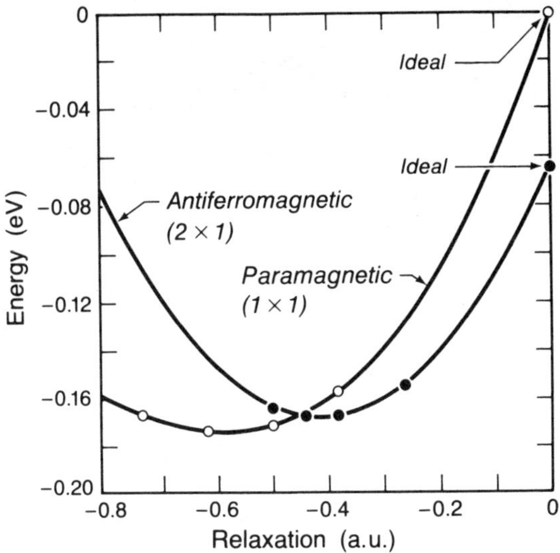

Fig. 17.   Energy as a function of relaxation of the surface layer,
the antiferromagnetic 2×1 and the paramagnetic 1×1 states
of Si(111).   The zero of energy is the ideal paramagnetic
surface.   (from Ref. 63)

develop on their neighbors.   Several cycles are usually needed
to achieve a minimum energy, zero force structure.   When this approach
was used for Si(111),[58] the surface was found to resist buckling.
Forces developed on a buckled surface to restore the atoms back
to a relaxed 1×1 configuration (Fig. 16).

A possible state in which the surface behaves electronically
semiconducting and yet remains in the 1×1 geometry is one with a gap
arising from electron-electron interaction.   An antiferromagnetic
phase for the 1×1 Si(111) surface has been predicted[58,60-62] but not
observed.   The most recent calculation[63] placed the antiferromagnetic
phase at a lower energy than the paramagnetic phase (Fig. 17)
for the ideal geometry, but with relaxation, it is difficult to
determine whether the occurrence of this state is feasible.   As
seen in Fig. 17, the difference in total energy between the paramagnetic and the antiferromagnetic phases is very small.   Also, it is
sensitive to the approximations used in the calculations.[63]

Since the 1×1 geometry is predicted to be stable with respect
to buckling, this suggests that the observed 2×1 structure may
result from a distortion of a different kind.   Motivated by the
angle-resolved photoemission data, Pandey[64] suggested that the
2×1 structure is a π-bonded chain geometry.   A schematic ball and
stick model of the ideal and the π-bonded chain structures are
shown in Fig. 18.   The characteristic six-fold ring geometry of

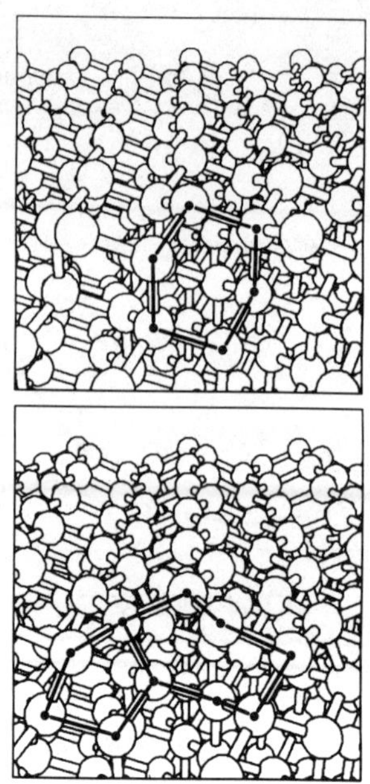

Fig. 18.   Geometry of ideal 1×1 Si(111) surface with six-fold rings
           (upper).   The π-bonded chain geometry with five-fold and
           seven-fold rings (lower).

the ideal 1×1 structure and its decomposition into five-fold and
seven-fold rings for the 2×1 π-bonded chain surface are depicted.
It is evident that the two structural models have very different
topologies.   Thus far, the π-bonded chain geometry gives the lowest
calculated total energy among the various models proposed in the
literature.[59]   This geometry is stabilized by the dangling bonds
moving into near-neighbor positions where they can participate
in π bonding.   It has also been shown that at least one path existed
going from the ideal geometry to the π-bonded chain with a barrier
of only ~0.01 eV.[59]   The cleaving process could easily supply enough
energy for this transition to occur.   The resulting chain geometry
is lower in energy by ~0.2 eV than the relaxed ideal (1×1) geometry.

One can also compare the calculated surface state dispersion
with experimental results for confirmation once a minimum energy
structure is determined.   The theoretical surface state bands[59,64]
were found to be in good agreement with angle-resolved photoemission
data (Fig. 19).   The agreement of the $\vec{k}$-dependent of the surface
state energies with experiment is impressive considering that only
the atomic number and some geometical restrictions went into the
theory.   The absolute position of the experimental points in Fig. 19
has been shifted up rigidly by 0.3 eV for ease of comparison.   This
kind of discrepancy arises from using local density functional

372

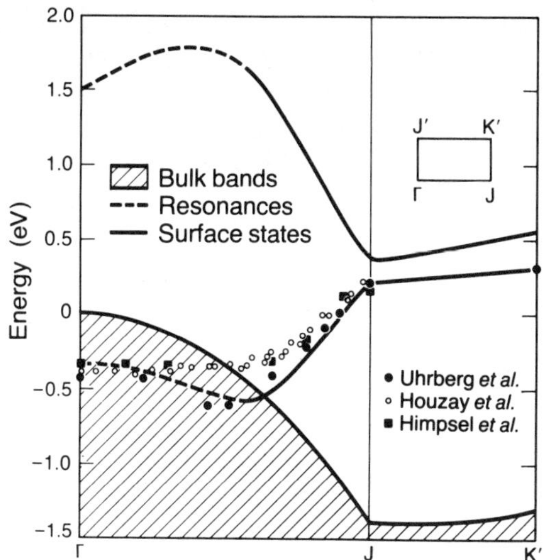

Fig. 19. Calculated electron energy dispersion curve for Si(111)
          in the energy optimized π-bonded chain structure. The
          experimental points have been shifted upwards by 0.3 eV
          so that the theory and experiment are aligned at the
          J point. (after Ref. 59)

theory which is well-known to give excellent ground-state properties
but too small excitation energies. Calculations for the 2×1
Ge(111) surface[65] also showed that the π-bonded chain geometry
has the lowest energy and a surface band dispersion similar to
Si. We should note that agreements between theory and experiment
for $E(\vec{k})$ for a low energy structure are necessary conditions for
determining surface reconstructions but not sufficient conditions.
Other geometries may give similar $E(\vec{k})$ curves, and at present,
there are no certain tests to determine that a given structure
corresponds to the absolute minimum and not a local minimum in
energy.

     Total energy calculations have also been carried out to
study a variety of proposed reconstructions on the diamond (111)
surface.[19,66] For this case, the LCAO basis was used because
of the localized nature of the carbon electron wavefunctions.
Both a 1×1 phase obtained by polishing and a 2×2/2×1 phase obtained
by annealing to above ~1000 C were observed by LEED. (LEED cannot
distinguish between a true 2×2 or disordered domains of 2×1 for
this surface; the similarity of the angle-resolved photoemission
to that of Si and Ge suggests the latter.) No surface states
were observed in the gap for the 1×1 diamond surface. However,
recent experiments have shown that this is a hydrogen-terminated
surface. The 2×2/2×1 surface, on the other hand, is believed

Table XIV. Calculated total energies of C(111) 1×1
           and 2×1 surface reconstruction models.
           (from Ref. 66)

Surface model	Energy eV/(surface atom)
Ideal 1×1	0.00
Relaxed 1×1	−0.37
Buckled ($\Delta z = \pm 0.26$ Å)	0.35
Chadi π-bonded molecule	0.28
Seiwatz single chain	1.30
Ideal Pandey π-bonded chain	−0.05
Relaxed Pandey π-bonded chain	−0.47
Same with ±2% dimerization	−0.46
Same with ±4% dimerization	−0.43
Same with ±6% dimerization	−0.38
Fully relaxed Pandey chain	−0.68

to be hydrogen free and clean, and unlike Si and Ge whose 2×1
phases are metastable, the diamond 2×2/2×1 structure appears
to be thermodynamically stable.

Similar to the case of the Si(111) surface, many models[67-70]
have been proposed for the 2×1 diamond surface. Results[66]
of the total energy calculation for these models are summarized
in Table XIV. The energy per surface atom for the ideal 1×1
model is used as a zero of energy. Relaxing the first two surface
bonds (Fig. 20(a)) lowers the energy by 0.37 eV. Four topologically
distinct 2×1 models have been tested: the buckling model,[54]
the Pandey π-bonded chain model,[67] the Chadi molecule model,[71]
and the Seiwatz single-chain model.[72] As in Si, buckling of
the 1×1 surface is found to raise the energy. Of the other three
models, the Pandey chain model[73] clearly has the lowest energy.
The Chadi molecule model, in second place, has not been relaxed
further because the calculated surface state dispersion is inconsis-
tent with angle-resolved photoemission data.[55]

The energy of the Pandey chain model shown in Fig. 18 was
minimized by adjusting the four surface-most bond lengths to
give the "relaxed" structure of Fig. 20(b) lowering the energy
to −0.47 eV. A surprising feature of the resulting geometry
was the 8% lengthening of the subsurface interlayer bond. The
surface chain bond, on the other hand, was only contracted by
4% to a value approximately midway between that of graphite and
diamond. Contrary to some speculations,[67] dimerization of the

374

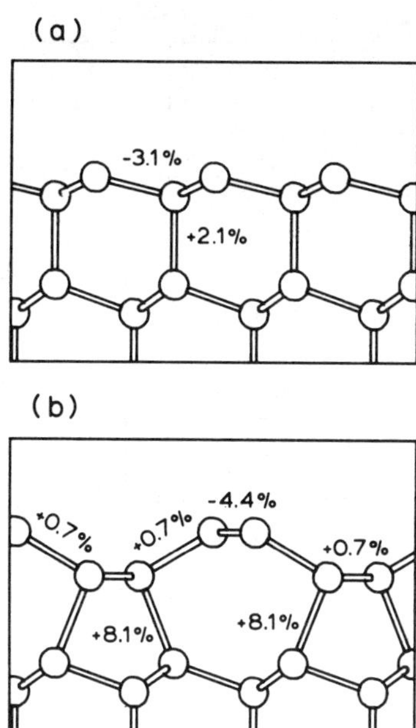

Fig. 20. Illustration of bond length changes (with respect to
bulk) which occur upon relaxation of (a) 1×1 and (b)
2×1 Pandey chain models for the diamond (111) surface.
(from Ref. 66)

chain does not lower the surface energy. However, the structure
can be further relaxed by allowing the atoms below the first
two layers to move. This further relaxation relieves some of
the bond angle strains on the third layer atoms. The final geometry
has an energy lowered by an additional 0.21 eV per surface atom.

It is instructive to compare the diamond results with those
of Si where the ideal and relaxed Pandey chain models have energies
(per surface atom) of −0.22 and −0.36 eV respectively compared
to the ideal 1×1 surface. In diamond, the ideal Pandey model
is less favorable, but relaxations are more important. This
can be attributed to the highly directional nature of the carbon
sp^3 bonds which implies that bond angle variations are more costly.
(The ratio of the bond−bending force constant to the bond−stretching
force constant is twice as large in diamond than that in Si.)
Evidently, the large bond angle distortions in the third layer
are more costly in diamond and the relaxations which relieve
bond angle strains more important. This is consistent with the
8% expansion of the diamond subsurface interlayer bond which
can be ascribed to bond weakening resulting from bond angle
distortion.

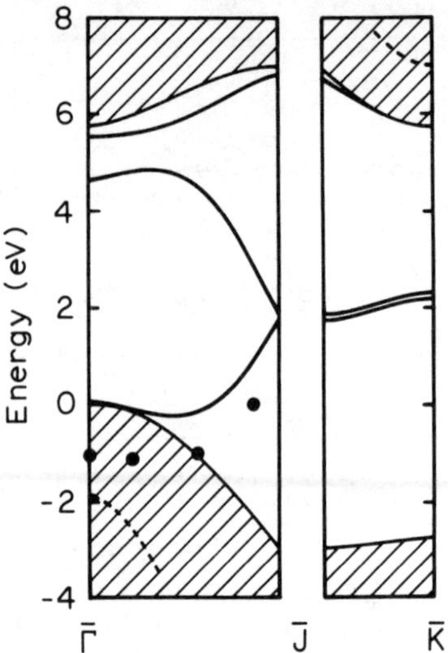

Fig. 21.   Calculated surface bands (solid lines) and resonances
(dashed lines) for the 2×1 diamond (111) surface for
fully relaxed Pandey chain model. The bulk projected
band structure (shaded) and the experimental data of
Ref. 55 (black dots) are shown for comparison. (from
Ref. 66)

Figure 21 shows the calculated surface band structure for
the fully relaxed Pandey chain model. Experimental angle-resolved
photoemission data are shown for comparison. The dispersion
of the calculated surface band is in good agreement with experiment.
However, the calculated band is too high by a rigid shift of ~1 eV.
As mentioned before, such a shift is also observed for the surface
bands of Si and Ge (by 0.3 and 0.8), and a correlation effect
may perhaps be invoked to explain this discrepancy.[59,65]

From these calculations, it is, therefore, fair to conclude
that, taken both total energies and spectroscopic data into consideration, a Pandey π-bonded chain topology is most likely to be the
correct structure for the 2×1 phase of the (111) surfaces of
diamond, Si, and Ge.

The methods described here have been applied to the other
faces of the group IV elements such as the Si(100) surface[74,75]
and to other semiconductor surfaces.[76] One important application
which we have not discussed is the study of chemisorption, both
on the geometry and energetics of adsorbates on surfaces. Figure 22 illustrates the result from a study of Al on the GaAs(110)
surface.[76] Energy surfaces for Al atoms adsorbed on the surface
were determined and comparisons made between various possible

376

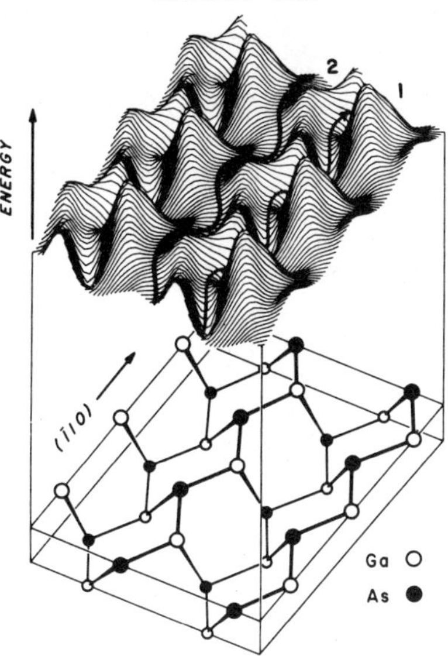

Fig. 22.   Three-dimensional plot of the total energy for an Al
atom adsorbed on the GaAs(110) surface. Two favorable
paths for the surface migration of Al atoms are indicated
as 1 and 2. It is clear in the figure that channel
1 follows the valleys and channel 2 traces plateaus.
Corresponding atomic positions for the first two surface
layers are also illustrated. (from Ref. 76)

adsorbate sites. Another important application is to use the
total energy methods to calculate microscopic interaction parameters
among the various structural units on a surface. These parameters
can then be used in the study of statistical, temperature-dependent
properties of the surface employing techniques such as the renormalization group method. This kind of analysis has recently been
done for the structural properties and phase diagram of the Si(100)
surface.[75] Yet, another area which we have not discussed is defects
on surfaces. It is likely that total energy studies will be
applied in the near future to determine the lowest energy reconstructions near defects such as steps.

B.   Surfaces of Transition Metals

In recent years, research in metal surfaces has been focused
primarily on the transition metals because of their richness
in phenomena and their technological importance. Compared to
the semiconductor surfaces, transition metals surfaces are much
more difficult to treat theoretically owing to the coexistence

of the localized d-electrons and the delocalized sp electrons.
As a consequence, the theoretical study of these surfaces has
lagged somewhat behind that of semiconductor surfaces. Virtually
all work has been done only on the electronic properties. _Ab
initio_ determination of surface structures via total energy calculations similar to those discussed in the previous subsection is
yet to be done. From the electronic structure calculations,
one can, nevertheless, still gain much insight into the nature
of these surfaces and, in many cases, provide detailed explanations
of experimental observations. Our discussion here will, thus,
be limited to surface electronic structures.

Several methods[77-83] have been developed to calculate the
surface electronic structure self-consistently for transition
metal systems. All of these involve modeling the surfaces by
thin slabs (or by repeated slabs in the case of the supercell
approach) and expanding the electron wavefunctions in some basis
sets. In conjunction with pseudopotentials, the mixed basis
or the LCAO basis are most commonly employed. With basically
the surface geometry as input, these calculations yield the
work function, surface states, adsorbate states, surface charge
densities, densities of states, and often information on preferred
sites of adsorption. Surface states are shown to be important
in the interpretation of spectroscopic measurements, and chemisorption
studies give valuable information concerning the nature of the
surface chemical bond.

We discuss below the Nb(001) and Mo(001) surfaces as examples
of bcc transition metal surfaces and the Pd(111) surface as an
example of an fcc transition metal surface. As an illustration
for chemisorption, we consider the case of hydrogen on the Pd(111)
surface. Except for the case of the Nb(001) surface which was
done with a plane wave basis, the calculations were performed
using the mixed basis method in a repeated slab geometry with
slab size usually 7 to 11 atomic layers thick.

The calculated surface energy band structure for the Nb(001)
surface[77] is presented in Fig. 23 along high symmetry directions
in the two-dimensional Brillouin zone. The vertical and horizontal
crosshatching shows the allowed bulk states (the projected band
structure) of various symmetries. The dash curves are the surface
bands (either _bona fide_ or strong resonance). These results
demonstrated, in a fully self-consistent calculation, that transition
metal surfaces support a variety of surface states. The surface
states can have different angular momentum character and exist
over a wide range of energies and over different portions of
the two-dimensional Brillouin zone.

The characters of the surface states shown in Fig. 23 are
mainly atomic d-like and are highly localized in the surface

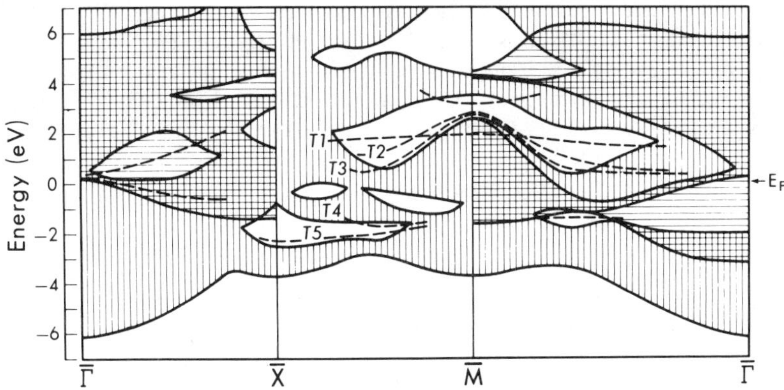

Fig. 23. Surface bands (dashed curves) and the projected band structure for the Nb(001) surface. (from Ref. 77)

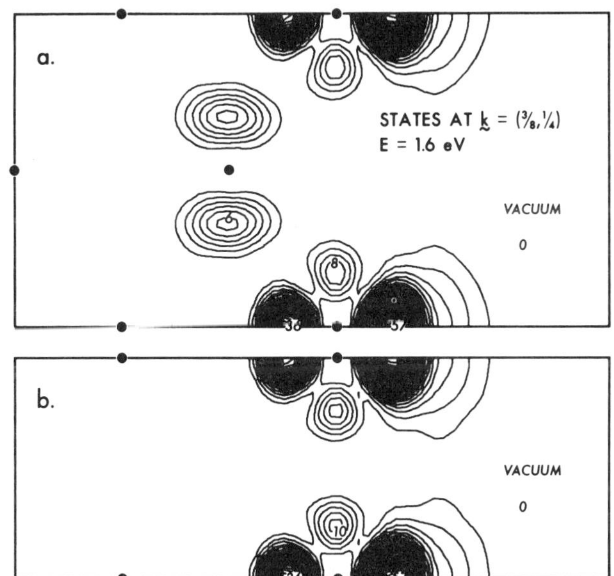

Fig. 24. Charge density distribution of a T1 (see Fig. 23) surface state at $\vec{k} = (3/8,1/4)2\pi/a_c$ (with a_c the bulk crystal lattice constant) plotted on (a) the (110) plane and (b) the (100) plane. Charge density is given in relative units. (from Ref. 77)

region. The charge density (square of the wavefunction) of one
of the prominent surface states is presented in Fig. 24. This
state is associated with the band labeled Tl in Fig. 23 which
extends over a large region in the two-dimensional Brillouin
zone and has very flat dispersion. As seen in Fig. 24, states
in the Tl surface band are $d_{3z^2-r^2}$-like and are highly localized
on the surface layer with practically no overlap between neighboring
surface atom consistent with the weak $\vec{k}$-dependent of this band.

Calculation for the Mo(001) surface[84] yielded a surface
electronic structure qualitatively similar to that of the Nb
surface. Surface states of a variety of character are found
to exist in roughly the same energy ranges and positions in $\vec{k}$-space.
It is found that the existence of some of these states are sensitive
to the details of the self-consistent potential. The calculated
surface states have been successfully used to interpret many
of the surface sensitive spectral features observed in angle-resolved
photoemission experiments.

One important prototypical study in this area is that on
the Pd(111) surface and the interaction of hydrogen with this
surface.[78,86,87] This system is of great interest because of
its fundamental importance in chemisorption theory and in technological areas such as catalysis, hydrogen storage, and hydrogen
embrittlement.

The calculated surface state bands for the clean Pd(111)
surface are shown in Fig. 25 together with that of a monolayer
of hydrogen adsorbed on the surface three-fold sites.[78,86] The
computed work function of 5.8 eV agrees well with the measured
value of 5.6 eV. There is a significant surface charge rearrangement
in going to self-consistency so that the ad hoc constraint of
atom-by-atom neutrality used in many non-self-consistent calculations
is likely too rigid. Figure 25 shows that, as in the case of
bcc metals, many surface states exist with most of them d-like
and highly localized. An interesting exception is the surface
state at ~2 eV above the Fermi level which is sp-like and penetrates
some distance into the bulk.

The calculation also yielded valuable information on the
symmetry and origin of the individual surface states. Knowing
the symmetry of the surface state wavefunctions would greatly
facilitate the identification of these states in experiment by
use of selection rules. High resolution angle-resolved photoemission
measurements[87,88] have recently been performed to search for
the surface states predicted in Fig. 25. The agreement between
experiment and theory is very good for both the energy positions
and the wavefunction symmetry of the states (Table XV). Furthermore,
it is shown[89,90] that one can understand the surface states on
other (111) surfaces for the nearby elements in the Periodic

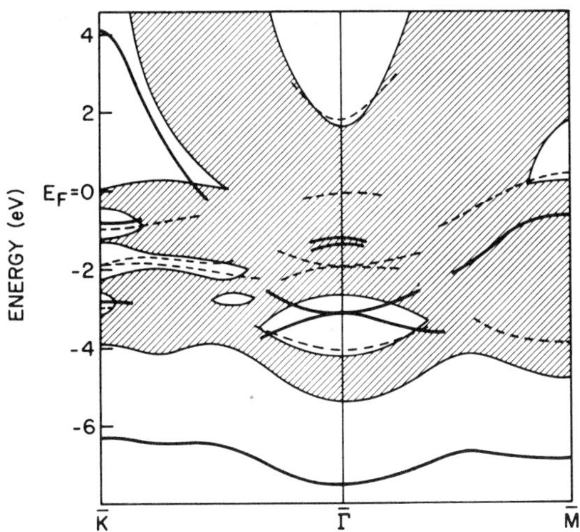

Fig. 25.   Localized states at the Pd(111) surface.   Dash curves
are for states of the clean surface, and solid curves
are for states of the H—covered surface with H in the
three—fold centered C-site.   (after Refs. 78 and 86)

Table in terms of the Pd(111) results using a rigid-band interpretation.

The surface states strongly affect the spectroscopic properties
of the surface.   Figure 26 shows the calculated local density
of states (LDOS) for the fourth layer from the surface, the second
layer from the surface, the surface layer, and a region one layer
thick beyond the surface layer for the Pd(111) surface.   The
LDOS gives the energy spectrum for electrons in a particular
region Ω_i and is defined

$$N_i(E) = \sum_{\vec{k}_\parallel,n} \int_{\Omega_i} |\psi_{\vec{k}_\parallel,n}(\vec{r})|^2 d^3r \; \delta(E - E_n(k_\parallel)) \qquad (35)$$

where $k_\parallel$ is the wavevector parallel to the surface, n is the
band index, and ψ is the electron wavefunction.   As seen from
the figure, the fourth layer LDOS is virtually identical to the
bulk Pd DOS, but the surface layer LDOS is very different.   It
is enhanced in the region from 0 to -2 eV below the Fermi level
E_F and is noticeably narrower.   This enhancement arises from
the existence of the many surface states near E_F (Fig. 25).

Table XV. Localized states on the Pd(111) surface. (Energies are in eV measured relative to E_F.) Experimental data from Refs. 87 and 88. (after Refs. 78 and 86)

	Clean		H-adsorbed	
	Theory	Expt.	Theory	Expt.
$\bar{\Gamma}$	1.9	1.7	-1.3 / -1.4	-1.2
	-0.2	-0.3	-3.2	-3.1
	-2.0	-2.2	-7.5	-7.9
	-4.1	---		
$\bar{K}$	-1.0	-0.3	-0.8	-1.0
	-1.9	-2.1	-2.8	-2.8
	-3.0	---	-6.3	-5.9
$\bar{M}$	---	-1.0	-0.6	---
	-3.8	---	---	-4.1
			-6.8	-6.4

Most importantly, these results explain a characteristic adsorbate-independent reduction of photoemission signal from 0-2 eV below E_F observed when different adsorbate species are chemisorbed on the Pd(111) surface.[91,92] (Similar effects are seen in Ni and Pt.) It is interpreted[78] that this reduction is largely due to the removal (shifting to lower energies) of surface states and resonances by the adsorbates since angle-integrated measurements basically probe the surface LDOS. Figure 27(a) shows the difference between the layer 4 LDOS and that of the surface layer. The negative parts of the curve correspond to regions where there are excess state density at the surface. This is to be compared with the curves in Fig. 27(b) which are the observed differences in photoemission intensity between the clean and the adsorbate-covered surfaces. The deficit can, therefore, clearly be associated with the removal of surface states. The efffect is adsorbate-independent because the states involved are characteristic of the substrate. This picture has since been confirmed in detail by a direct calculation of H chemisorbed on the Pd(111) surface.[86]

Experimentally, hydrogen dissociates as the molecule adsorbs on the Pd(111) surface and forms a (1×1) monolayer at low tempera-

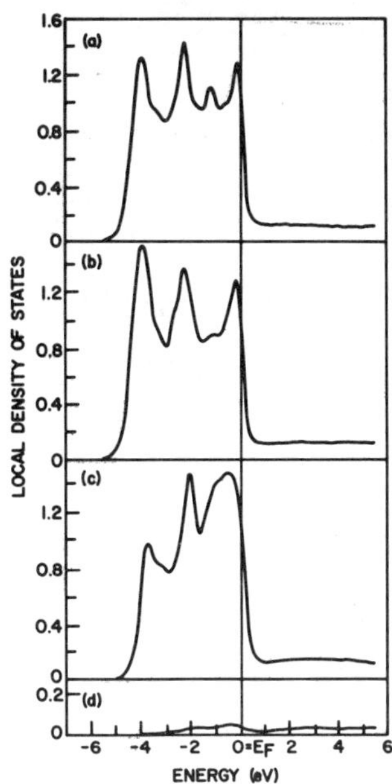

Fig. 26. Calculated local density of states for (a) layer 4, (b)
layer 2, (c) layer 1 (the surface layer), and (d) a region
one layer thick beyond the surface layer for the Pd(111)
surface. (from Ref. 78)

ture.[91,92] The exact structure, however, is not determined.
In the calculations,[86] three possible structural models were
considered for monolayer coverage: (A) H on top of every surface
Pd atom; (B) H on a three-fold site over a hole in the second
layer; and (C) H on a three-fold site over an atom in the second
layer. The H–Pd bond length was fixed at the sum of the Pd metallic
and the H covalent radii.

The calculated densities of states were compared with angle-
integrated photoemission data to distinguish the three possibilites.
The first possibility, site A, was ruled out as the preferred
site. The two three-fold sites B and C produce very similar
spectra which agree well with the experimental data and are essentially indistinguishable from one another.

The calculated surface band structure for H on site C is
shown in Fig. 25. The H adatoms induce extensive changes in
the surface electronic structure of the clean Pd(111) indicating
a strong surface chemical bond. The two most striking H–induced
features are the narrow H–Pd bonding adsorbate band which appears
about 2 eV below the Pd bulk d bands, and the 4 eV wide anti–bonding
H–Pd band just above E_F in a gap in the projected band structure

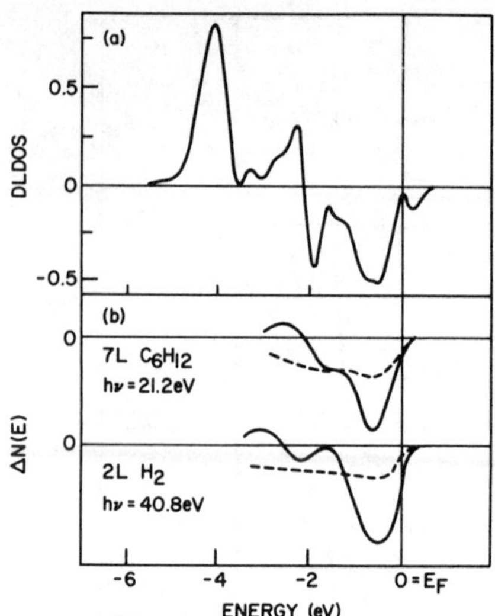

Fig. 27. (a) Calculated difference in LDOS between layer 4 and
the surface layer for the Pd(111) surface. (b) Adsorbate-
induced differences in photoemission intensities from
Ref. 92. The dash curves indicate the estimated attenua-
tion if uniform attentuation of the d-band were to
occur. (from Ref. 78)

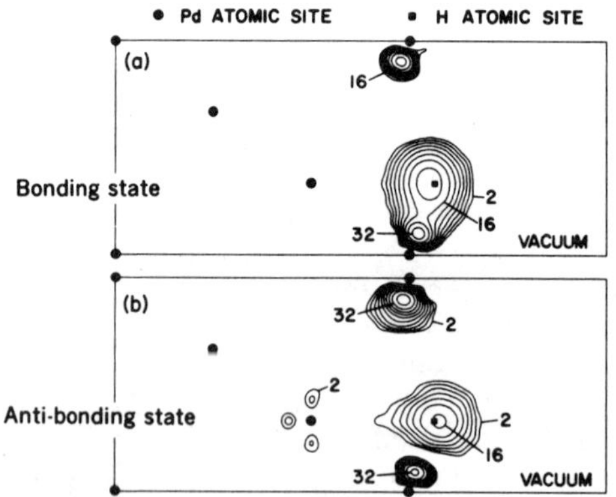

Fig. 28. Charge density contour plots for adsorbate states at
K̄ at (a) -6.3 eV and (b) 4.1 eV [H/Pd(111) at C sites].
The charge densities are given in relative units and
plotted for a (1̄10) plane. (after Ref. 86)

384

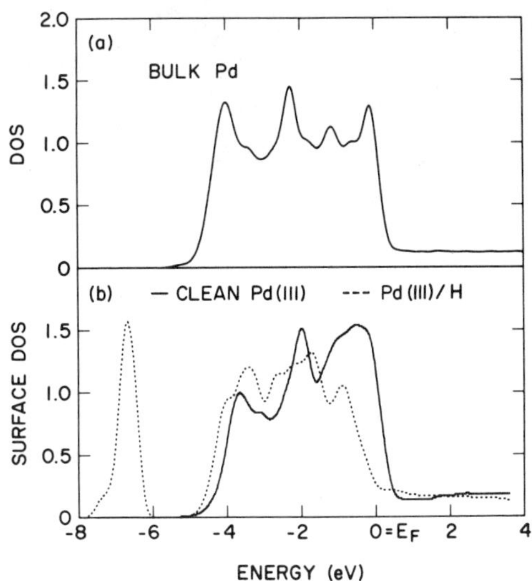

Fig. 29.   Calculated LDOS for (a) bulk Pd and (b) Pd(111) surface
           with and without a monolayer of H in the C site.   (from
           Ref. 86)

near $\bar{K}$. The intrinsic surface states of clean Pd(111) are strongly
affected by H adsorption--some disappear and contribute to the
H-Pd adsorbate states, some have their character intact but move
to lower energies, and some change their character.

An examination of the charge distribution of the H-Pd adsorbate
states shows that the wavefunctions of these states are almost
completely localized on the H atoms and the first Pd layer (Fig. 28).
The strong bonding at the surface is predominantly from the interaction between the Pd 4d and the H 1s orbitals. The highly localized
nature of the H-substrate interaction explains the virtually
identical surface band structures for the sites B and C.

The calculated surface density of states is presented in
Fig. 29 together with the bulk density of states. The bonding
Pd-H band appears as a distinct peak at -6.5 eV. The reduction
of the amplitude of the density of states near E_F is primarily
due to removal of intrinsic surface states from this energy region.
This can also be seen in Fig. 25 which shows that near E_F all
intrinsic surface states are shifted to lower energies confirming
the interpretation given above on the universal reduction of
photoemission intensity in this region. Another important result
from the calculation is the absence of a well-defined peak corresponding to anti-bonding H-Pd states as would be predicted by simple
chemisorption models.[93,94] This is a band structure effect.
Because of the symmetry of the surface, the anti-bonding states

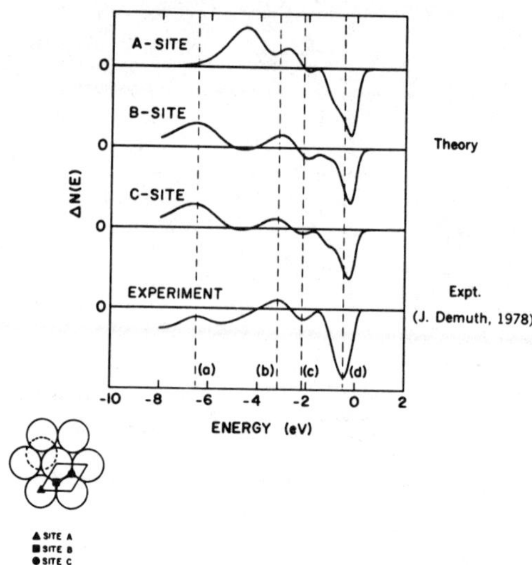

Fig. 30.   Comparison of the calculated H-induced photoemission
difference spectra with the experimental difference
curve.   (from Ref. 86)

can only exist in a limited portion of the two-dimensional Brillouin
zone. The large band width of these states also contributed to their
invisibility in the spectrum.

The theoretical difference spectra ΔN for the three chemisorption geometries are presented in Fig. 30 together with the experimental curve. The theoretical ΔN's are evaluated from the LDOS
with final-state and matrix element effects ignored.[86] As seen
in the figure, the most prominent feature (structure (d)) in
the experimental curve is well reproduced in all three theoretical
curves. This structure as discussed above mainly results from
the removal of surface states/resonances from this particular
energy region; it is, hence, insensitive to the nature of the
adsorbate or to its position. The comparison clearly rules out
the case with H on each Pd atom. The spectra for sites B and
C, on the other hand, are very similar and reproduce remarkably
well the experimental spectrum. These results, therefore, show
that hydrogen prefers a three-fold site and interacts strongly
with the Pd d-states.

In addition to the low-temperature angle-integrated photoemission data,[92] the surface state dispersions for H on Pd(111) have been
measured recently by high resolution angle-resolved photoemission

386

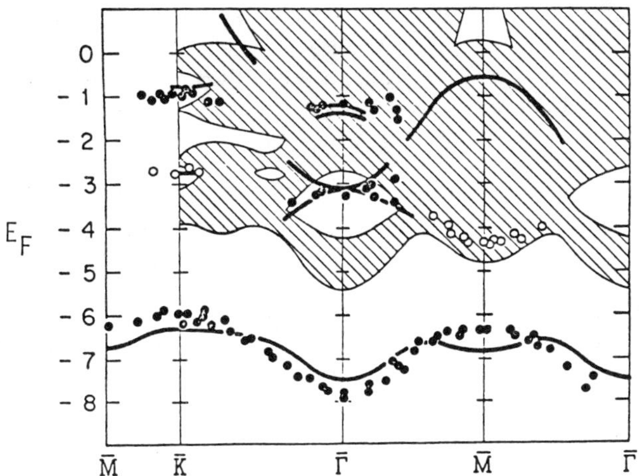

Fig. 31.   Calculated and measured surface states for the H(1×1)
Pd(111) surface at low temperature phase. The shaded
regions are the calculated projection of the bulk bands.
The heavy lines are calculated surface states or resonances; the circles are data. The open circles indicate
experimental peaks with some uncertainty. (after Ref. 87)

techniques at temperatures ranging from nitrogen to above room
temperature.[87] The predicted surface band structure is shown
to be in excellent agreement with the low temperature data.
A comparison between theory and experiment is shown in Fig. 31.
(Also, see Table XV.) The only possible discrepancy is for states
near $\bar{\mathrm{M}}$ at which experiment also has a larger uncertainty. These
calculations, which have treated both the clean and the adsorbate-
covered surfaces in equal footing, give, as a consequence, a
complete and very accurate description of the chemisorption process
at low temperature.

The situation at higher temperature is, however, more complex.
Experiments indicate that there appear to be two H/Pd(111) phases
with the more stable high temperature phase showing no sign of
any H-induced features, at least in near normal emission spectra.[87]
Although the exact H coverage is not certain, these results suggest
a structural transformation with the H moving from the low-temperature three-fold sites into some other positions as the temperature
rises. The same phenomenon is also observed for H on the (111)
surfaces of Ni and Pt. At present, there is no satisfactory
model for the high temperature phase. Total energy calculations
on these systems would help in clarifying the situation.

V. BEYOND LOCAL DENSITY APPROXIMATION

In the previous sections, we have seen that pseudopotential local density calculations give impressive results for a variety of ground-state properties. There are, however, some systematic discrepancies. The cohesive energy is quite generally overestimated, a result usually attributed to the failure of the local density approximation (LDA). Also, when the eigenvalues from the Kohn-Sham equations are interpreted as quasiparticle energies, the minimum gaps for semiconductors and insulators are consistently underestimated by 30-50%.[95-97]

The problem of the cohesive energy, in principle, can be corrected with a better exchange-correlation functional. The band gap problem is more difficult because density functional formalism does not explicitly provide information on the excitation energies. The quasiparticle energies should, in general, be obtained from the one-particle Green's function.[98,99] However, recently Sham and Schlüter[100] have shown that there is a formal relationship between the minimum gap E_g of an insulator and the Kohn-Sham gap ε_g obtained from a difference in eigenvalues:

$$E_g = \varepsilon_g + \Delta \tag{36}$$

where Δ is the discontinuity in the functional derivative of the exchange-correlation functional across a gap. Since there is not yet a calculation for Δ for a real material, there remains the important question of whether the discrepancy for the LDA results arises largely from inadequacies of the LDA or the neglect of Δ.

It is, therefore, important from both practical and fundamental points of view to carry out the calculation beyond the LDA. In this section, we discuss one such calculation[101] on Si and Ge using a generalized weighted density functional scheme.

The usual LDA is based on an intrinsically metallic system, the limit of a nearly homogeneous electron density. A semiconductor differs from this limit in two respects--the charge density is localized in bonds and the electron spectrum has a gap which changes the screening properties qualitatively from those of the metals. Several schemes going beyond the LDA have been proposed for overcoming the first: gradient corrections,[8,102] self-interaction corrections,[103] and the weighted density approximation (WDA).[101] [104-106] With the exception of the WDA calculations reported in Refs. 101 and 106, none of these schemes have been carried out fully for semiconductors. In particular, structural properties have only been investigated beyond the LDA in one study.[101] The semiconductor screening aspect of the problem has been addressed partially by calculating the exchange-correlation energy of a

model electron gas which reproduces the characteristic screening
of a semiconductor arising from the gap,[18] but the resulting
LDA exchange-correlation functional only gave small improvements
for the spectrum of Si.

In a complete formulation, both the inhomogeneity and the
effect of the gap should be treated on an equal footing.[101]
Table XVI summarizes the band energies at selected points for
Si and Ge calculated using the WDA both with metallic screening
and with semiconductor screening in the exchange-correlation
functional. In this approach, the exchange-correlation energy
of an interacting electron system is expressed in terms of the
exchange-correlation hole charge n_{xc} (in Ry units)[105]:

$$E_{xc}[n] = \int\int d\vec{r}\,d\vec{r}' \; \frac{n(\vec{r})n_{xc}(\vec{r},\vec{r}')}{|\vec{r} - \vec{r}'|} \tag{37}$$

This hole arises from the exchange and dynamical Coulomb interactions
between electrons which lead to a local depletion of electron
density around a give electron. The exchange-correlation hole
is related to a pair correlation function

$$n_{xc}(\vec{r},\vec{r}') = n(\vec{r}')G_n(\vec{r},\vec{r}') \tag{38}$$

and the exchange-correlation hole must contain precisely one
unit of charge

$$-1 = \int d\vec{r}'\,n_{xc}(\vec{r},\vec{r}') \tag{39}$$

In the LDA, E_{xc} is approximated by replacing the exact G_n
by the homogeneous electron gas result evaluated for the local
density. The argument of the density prefactor in Eq. 38 is
further changed to the local point $\vec{r}$:

$$n_{xc}^{LDA}(\vec{r},\vec{r}') = n(\vec{r})G^{hom}\left(|\vec{r} - \vec{r}'|;n(\vec{r})\right) \tag{40}$$

The LDA exchange-correlation hole is, therefore, spherically
symmetric and centered on the electron by construction. The
LDA is successful for ground-state properties despite these limita-
tions because E_{xc} depends only on the spherical average of the
hole charge and because the sum rule (Eq. 39) is satisfied leading
to some systematic cancellation of errors.[107]

In the WDA,[105] a better description of the exchange-correlation
hole is attempted. The proper density prefactor is retained,
and G^{hom} is evaluated for a density averaged essentially over
the size of the hole:

$$n_{xc}^{WDA}(\vec{r},\vec{r}\,') = n(\vec{r}\,')G^{hom}\left(|\vec{r} - \vec{r}\,'|;\bar{n}(\vec{r})\right) \tag{41}$$

The parameter $\bar{n}$ is determined at each point by requiring the
sum rule (Eq. 39) be satisfied. The exchange-correlation hole
needs no longer remain centered on the electron and depends nonlocally
on the charge density which may be highly inhomogeneous.

Because simple formulations of G^{hom} are not available, the
calculations[101] were carried out with the analytic ansatz proposed
by Gunnarsson and Jones[108]:

$$G(x;\bar{n}) = C(\bar{n})\left[1 - \exp(-[\lambda(\bar{n})/x]^5)\right] \tag{42}$$

The parameters C and λ are determined by demanding that G in
Eq. 42 reproduces the homogeneous limit. Effectively G is constructed
to have the same zeroth and inverse r moment as some G^{hom}. In
Table XVI, three sets of calculations are presented together
with the experimental values.[109-111] The first column shows
the results of LDA calculations with the Ceperley-Alder (CA)
exchange-correlation data[14] which agree well with previous calculations. The second column (labeled WDA(CA)) shows the results
for calculations where $C(\bar{n})$ and $\lambda(\bar{n})$ are determined from usual
electron gas data (CA). Results including semiconductor screening
using the Levine-Louie (LL) model[18] are given in the third column
(labeled WDA(LL)). For both Si and Ge, the discrepancy with
experiment for the minimum gap is significantly reduced. The
conduction band energies at X and L are consistently improved.
However, the direct gaps are not substantially improved. As
seen from Table XVI, the improved description of charge density
inhomogeneity and that of electron screening both appear to play
equally important roles.

There is a clear qualitative difference between the LDA
and the WDA. Figure 32 shows the exchange-correlation hole (Eq.
38) for an electron at the antibonding site in Si. In the LDA,
as discussed earlier, the hole is spherically symmetric and centered
on the electron (Fig. 32(a)). In the WDA, the hole is shifted
towards the bond charge by the proper density prefactor in Eq.
41 (Fig. 32(b)). The exchange-correlation potential and, hence,
the spectrum are affected. The effect on the structural properties
is shown in Table XVII. The results for the WDA(LL) are, indeed,
as good as the LDA results and compare favorably to experiment
and results discussed in Sec. III. The importance of treating
inhomogeneity and the effects of the gap on screening on an equal
footing is again emphasized as the WDA(CA) does not give results
as well as either the LDA or WDA(LL). We also note that the
cohesive energy in the WDA(CA) is underbound by ~30%. The overestima-
tion of cohesive energies found generally in LDA calculations
is often attributed to underbinding in the atomic calculation.

Table XVI. Comparison of the calculated band energies
at selected points to experiment for Si
and Ge. Energies are reported in eV and
relative to the valence band maximum.
(from Ref. 101)

	LDA	WDA(CA)	WDA(LL)	EXP.[a]
Si				
E_g	0.56	0.71	0.90	1.17
$\Gamma_{25',v}$	0.00	0.00	0.00	0.00
$\Gamma_{15,c}$	2.57	2.69	2.81	3.40
$\Gamma_{2',c}$	3.29	3.27	3.34	4.2
$X_{4,v}$	-2.86	-2.80	-2.74	-2.9
$X_{1,c}$	0.71	0.87	1.06	1.30[c]
$L_{3',v}$	-1.21	-1.18	-1.16	-1.2±.2
$L_{1,c}$	1.55	1.62	1.73	2.1[d]
$L_{3,c}$	3.40	3.54	3.68	3.9±.1[b]
Ge				
E_g	0.52	0.60	0.69	0.74
$\Gamma_{25',v}$	0.00	0.00	0.00	0.0
$\Gamma_{2',c}$	0.74	0.74	0.78	0.89
$\Gamma_{15,c}$	2.58	2.71	2.79	3.25±.1
$X_{4,v}$	-3.03	-2.97	-2.92	-3.15±.2
$X_{1,c}$	0.80	0.95	1.11	1.3±.2
$L_{3',v}$	-1.38	-1.35	-1.33	-1.4±.3
$L_{1,c}$	0.52	0.60	0.69	0.74
$L_{3,c}$	3.74	3.88	3.99	4.3±.2

[a] Ref. 109 except where noted.

[b] Ref. 110

[c] Estimated from conduction band minimum and longitudinal effective mass.

[d] Ref. 111

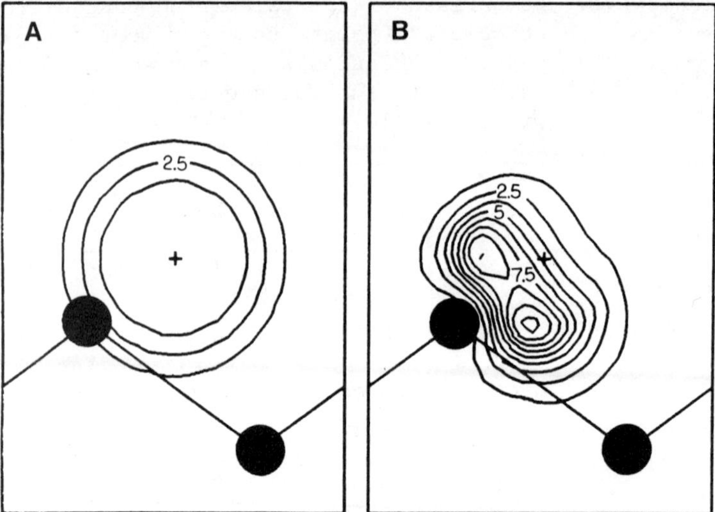

Fig. 32.   Contour plot of the exchange-correlation hole charge
for an electron at the anti-bonding site in the (1$\bar{1}$0)
plane of Si (electron location denoted by +) for the
LDA (a) and for the WDA(LL) (b).   The contour interval
is 1.25 electrons per unit cell.   (from Ref. 101)

In the case of the WDA(CA), most of the change in the cohesive
energy is due to an enhanced binding in the atom.   There is a
relative small change in the total crystalline energy as one
goes from the LDA to the WDA(CA).   The LL model is not directly
applicable to isolated atoms, so the cohesive energy was not
obtained.

The above calculation, thus, demonstrates that by going
beyond the LDA, a minimum gap in Si and Ge much closer to experiment
can be obtained without degrading the description of structural
properties in the solid state.   However, for accurate quantitative
cohesive energies and excitation energies, much improvement is
still needed in the theory.

VI.   SUMMARY AND CONCLUSIONS

The aim of this article was to review some of the recent
progress in calculating the electronic and structural properties
of condensed matter using the _ab initio_ pseudopotential density
functional approach.   Specific examples have been given for a
variety of properties and systems.   These include the static
structural properties, the vibrational properties, phonon-phonon
interactions, solid-solid structural phase transitions, surface

Table XVII. Comparison of the calculated equilibrium struc-
             structural properties to experiment for Si and
             Ge (after Ref. 101)

	Lattice constant $\overset{\circ}{A}$	Bulk modulus (kbar)	Cohesive energy (eV)
Si			
LDA	5.40	940	5.28
Present work			
WDA(CA)	5.48	850	3.19
WDA(LL)	5.39	940	--
Exp.	5.43	990	4.63
Ge			
LDA	5.60	730	4.67
Present work			
WDA(CA)	5.68	620	2.64
WDA(LL)	5.61	700	--
Exp.	5.65	770	3.85

electronic and geometric structures, and so forth. The method
is shown to be equally applicable to semiconductors, insulators,
simple metals, and transition metals. With basically the atomic
number and atomic mass of the constituent elements as input,
many of the above properties have been calculated to within a
few percent of experiment.

Currently, there is considerable activity in applying and
extending these methods to study other systems and properties.
Only a subset of the applications have been discussed here.
Several important areas have been omitted. One area is defects
in solids. Calculations in impurities, vacancies, interstitials,
line defects, and plane defects have been performed using similar
methods. Another area is interfaces. Applications are being
made to the metal-semiconductor interfaces (Schottky barriers),
semiconductor-semiconductor interfaces (heterojunctions), and
superlattices. The study of molecules and clusters is a third
area. Also, since the method can be used to calculate electron-phonon
couplings, phenomena such as resistivity, superconductivity,
ultrasonic attentuation, and so forth, can now be studied using
these ab initio methods. Results from research in all these
areas have been impressive and very encouraging.

The use of the density functional formalism, however, restricts
the applications to properties and phenomena related to the ground
state (or differences in ground states). Excited states or quasi-
particle energies, at present, cannot be justifiably extracted
from these calculations. As discussed in Sec. V, the calculated
energy band gaps for most semiconductors and insulators are too
small by 30-50% compared to observed values when a local density
approximation is used. Many workers are currently investigating
modifications to the LDA or exploring other new avenues to extend
these calculations to excited states.

New ideas and refinements to present methods will surely
continue to develop for this important field. However, even
if one limits the studies to ground-state properties, future
applications are numerous and exciting. With the availability
of modern supercomputers, this general approach should have the
power to predict the existence of new materials, novel properties
and phenomena, and to extend the study of materials to physical
situations (such as those at extreme pressures or at complex
interfaces) which are difficult or impossible to examine experimentally.

ACKNOWLEDGMENTS

This work was supported by National Science Foundation Grant
No. DMR83-19024 and by a program development fund from the Director
of the Lawrence Berkeley Laboratory.

REFERENCES

1.   For related topics, see articles by K. Kunc and by R. M.
          Martin in this volume.
2.   Introductory reviews on background materials are available
          in M. L. Cohen, Phys. Today 32:40 (1979), and M. L.
          Cohen, Phys. Scripta T1:5(1982).
3.   D. R. Hamann, M. Schlüter, and C. Chiang, Phys. Rev. Lett.
          43:1494 (1979).
4.   G. Kerker, J. Phys. C 13:L189 (1980).
5.   A. Zunger and M. L. Cohen, Phys. Rev. B 18:5449 (1978);
          20:4082 (1979).
6.   T. Starkoff and J. D. Joannopoulos, Phys. Rev. B 16:5212
          (1977).
7.   S. G. Louie, S. Froyen, and M. L. Cohen, Phys. Rev. B 26:1738
          (1982).
8.   P. Hohenberg and W. Kohn, Phys. Rev. 136:B864 (1964).
9.   W. Kohn and L. J. Sham, Phys. Rev. 140:A1133 (1965).
10.   E. Wigner, Phys. Rev. 46:1002 (1934).
11.   L. Hedin and B. I. Lundqvist, J. Phys. C 4:2064 (1971).
12.   U. von Barth and L. Hedin, J. Phys. C 5:1629 (1972).

13. O. Gunnarsson, B. I. Lundqvist, and J. W. Wilkens, Phys. Rev. B 10:1319 (1974).

14. D. M. Ceperley, Phys. Rev. B 18:3126 (1978); D. M. Ceperley and B. I. Alder, Phys. Rev. Lett. 45:566 (1980). These results are parameterized in J. P. Perdew and A. Zunger, Phys. Rev. B 23:5048 (1981).

15. S. G. Louie, K. M. Ho, and M. L. Cohen, Phys. Rev. B 19:1774 (1979).

16. J. R. Chelikowsky and S. G. Louie, Phys. Rev. B 29:3470 (1984).

17. J. Ihm, A. Zunger, and M. L. Cohen, J. Phys. C 12:4401 (1979).

18. Z. Levine and S. G. Louie, Phys. Rev. B 25:6310 (1982).

19. D. Vanderbilt and S. G. Louie, Phys. Rev. B (in press).

20. J. R. Chelikowsky, S. G. Louie, D. Vanderbilt, and C. T. Chan, Int. J. of Quant. Chem. (in press).

21. M. T. Yin and M. L. Cohen, Phys. Rev. B 26:5668 (1982).

22. S. Froyen and M. L. Cohen, Phys. Rev. B 28:3258 (1983).

23. F. D. Murnaghan, Proc. Nat. Acad. Sci. U.S.A. 3:244 (1944).

24. M. T. Yin and M. L. Cohen, Phys. Rev. B 24:6121 (1981).

25. P. K. Lam and M. L. Cohen, Phys. Rev. B 24:4224 (1981); B 25:6139 (1982).

26. M. Y. Chou, P. K. Lam, and M. L. Cohen, Solid State Comm. 42:861 (1982).

27. C. T. Chan, J. R. Chelikowsky, and S. G. Louie (to be published).

28. K. M. Ho, C. L. Fu, B. N. Harmon, W. Weber, and D. R. Hamann, Phys. Rev. Lett. 49:673 (1982); Phys. Rev. B 29:1575 (1984).

29. N. A. W. Holzwarth, S. G. Louie, and S. Rabii, Phys. Rev. B 26:5382 (1982).

30. R. Chen, P. Trucano, and R. F. Stewart, Acta Crystallogr. Sec. A 33:823 (1977).

31. M. T. Yin and M. L. Cohen, Phys. Rev. Lett. 45:1004 (1980); Phys. Rev. B 26:3259 (1982).

32. K. Kunc and R. M. Martin, Phys. Rev. Lett. 48:406 (1982).

33. B. N. Harmon, W. Weber, and D. R. Hamann, Phys. Rev. B 25:1109 (1982).

34. O. H. Nielsen and R. M. Martin, Phys. Rev. Lett. 50:697 (1983).

35. D. Vanderbilt, S. G. Louie, and M. L. Cohen (to be published).

36. M. Born and R. Oppenheimer, Ann. Physk 84:457 (1927).

37. K. Kunc and R. M. Martin, J. de Physique, Colloque C 6:649 (1981).

38. M. T. Yin and M. L. Cohen, Phys. Rev. B 25:4317 (1982).

39. S. A. Solin and A. K. Ramdas, Phys. Rev. B 1:1687 (1970).

40. M. H. Cohen and J. Ruvalds, Phys. Rev. Lett. 23:1378 (1969).

41. R. Tubino and J. L. Birman, Phys. Rev. B 15:5843 (1977), and references therein.

42. R. Biswas and R. M. Martin (to be published).

43. S. Fahy and S. G. Louie (to be published).

44. M. T. Yin , Phys. Rev. B (in press).

45. P. K. Lam and M. L. Cohen, Phys. Rev. B 27:5986 (1983).
46. W. Maysenholder, S. G. Louie, and M. L. Cohen (to be published).
47. J. A. Moriarty and A. K. McMahan, Phys. Rev. Lett. 48:809 (1982).
48. P. K. Lam, M. Y. Chou, and M. L. Cohen, J. Phys. C 17:2065 (1984).
49. F. Aldinger and G. Petzow, Beryllium Science and Technology, ed. D. Webster and G. J. London, Plenum, New York, Vol. 1 (1979), p. 235.
50. F. Reif, Fundamentals of Statistical and Thermal Physics, McGraw-Hill, New York (1965), p. 294.
51. R. P. Feynman, Statistical Mechanics, Benjamin, Reading (1972), p. 9.
52. J. A. Appelbaum and D. R. Hamann, Rev. Mod. Phys. 48:3 (1974).
53. M. Schluter, J. R. Chelikowsky, S. G. Louie, and M. L. Cohen, Phys. Rev. B 12:4200 (1975).
54. For a recent review, see D. Haneman, Adv. Phys. 31:165 (1982).
55. F. J. Himpsel, D. E. Eastman, P. Heimann, and J. F. van der Veen, Phys. Rev. B 24:7270 (1981).
56. F. Houzay, G. Guichar, R. Pinchaux, G. Jezequel, F. Solal, A. Barsky, P. Steiner, and Y. Petroff, Surf. Sci. 132:40 (1983), and references therein.
57. J. M. Nicholls, G. V. Hansson, R. I. G. Uhrberg, and S. A. Flodstrom, Phys. Rev. B 27:2594 (1983).
58. J. E. Northrup. J. Ihm, and M. L. Cohen, Phys. Rev. Lett. 47:1910 (1981).
59. J. E. Northrup and M. L. Cohen, Phys. Rev. Lett. 49:1349 (1982).
60. G. Allan and M. Lannoo, Surf. Science 63:11 (1977).
61. C. B. Duke and W. K. Ford, Surf. Sci. 111:L685 (1981).
62. R. Del Sole and D. J. Chadi, Phys. Rev. B 24:7430 (1981).
63. J. E. Northrup and M. L. Cohen, Phys. Rev. B 29:5944 (1984).
64. K. C. Pandey, Phys. Rev. Lett. 47:1913 (1981).
65. J. E. Northrup and M. L. Cohen, Phys. Rev. B 27:6553 (1983).
66. D. Vanderbilt and S. G. Louie, Phys. Rev. B 29 (in press).
67. K. C. Pandey, Phys. Rev. B 25:4338 (1982).
68. W. S. Yang and F. Jona, Bull. Am. Phys. Soc. 28:415 (1983).
69. S. V. Pepper, Surf. Sci. 12:47 (1982).
70. D. J. Chadi, Proceedings of the IXth Vacuum Congress/Vth International Conference on Solid Surfaces, Madrid, Spain, 1983; and J. Vac. Sci. Technol. (in press).
71. D. J. Chadi, Phys. Rev. B 26:4762 (1982).
72. R. Seiwatz, Surf. Sci. 2:473 (1964).
73. This unrelaxed Pandey model has all bulk bond lengths except for the graphite length for surface chain bonds.
74. J. Ihm, M. L. Cohen, and D. J. Chadi, Phys. Rev. B 21:4592 (1980); M. T. Yin and M. L. Cohen, ibid. 24:2303 (1981).
75. J. Ihm, D. H. Lee, J. D. Joannopoulos, and J. J. Xiong, Phys. Rev. Lett. 51:1872 (1983).

76. J. Ihm and J. D. Joannopoulos, Phys. Rev. B 26:4429 (1982).
77. S. G. Louie, K. M. Ho, J. R. Chelikowsky, and M. L. Cohen, Phys. Rev. Lett. 37:1289 (1976); Phys. Rev. B 15:5627 (1977).
78. S. G. Louie, Phys. Rev. Lett. 40:1525 (1978).
79. J. A. Appelbaum and D. R. Hamann, Solid State Comm. 27:881 (1978).
80. C. S. Wang and A. J. Freeman, Phys. Rev. B 11:793 (1979).
81. J. R. Smith, J. G. Gay, and F. J. Arlinghaus, Phys. Rev. B 21:2201 (1980).
82. H. Krakauer, M. Posternak, and A. J. Freeman, Phys. Rev. B 19:1706 (1979).
83. O. Jepsen, J. Madsen, and O. K. Andersen, Phys. Rev. B 18:605 (1978).
84. G. P. Kerker, K. M. Ho, and M. L. Cohen, Phys. Rev. Lett. 40:1593 (1978).
85. S. L. Wang, T. Gustafsson, and E. W. Plummer, Phys. Rev. Lett. 39:822 (1977).
86. S. G. Louie, Phys. Rev. Lett. 42:476 (1979).
87. W. Eberhardt, S. G. Louie, and E. W. Plummer, Phys. Rev. B 28:465 (1983).
88. P. D. Johnson and N. V. Smith, Phys. Rev. Lett. 49:290 (1982).
89. S. G. Louie, P. Thiry, R. Pinchaux, Y. Petroff, D. Chandesris, J. LeCante, Phys. Rev. Lett. 44:549 (1980).
90. N. A. W. Holzwarth and J. R. Chelikowsky (to be published).
91. H. Conrad, G. Ertl, J. Kuppers, and E. E. Latta, Surf. Sci. 58:578 (1976).
92. J. E. Demuth, Surf. Sci. 65:369 (1977), and unpublished.
93. D. M. Newns, Phys. Rev. 178:1123 (1959).
94. K. Schonhammer, Solid State Comm. 22:51 (1977).
95. For semiconductors, refer, for example, to J. Ihm and J. Joannopoulos, Phys. Rev. B 24:4191 (1981); M. T. Yin and M. L. Cohen, Phys. Rev. B 26:5668 (1982); S. Froyen and M. L. Cohen, Solid State Comm. 43:447 (1982).
96. D. R. Hamann, Phys. Rev. Lett. 42:662 (1979).
97. For the case of wide gap insulators, see, for example, R. A. Heaton and C. C. Lin, Phys. Rev. B 22:3629 (1980); S. B. Trickey, F. R. Green, Jr., and F. W. Averill, Phys. Rev. B 8:4822 (1973).
98. G. Strinati, H. J. Mattausch, and W. Hanke, Phys. Rev. B 25:2867 (1982), and references therein.
99. C. S. Wang and W. E. Pickett, Phys. Rev. Lett. 51:597 (1983).
100. L. J. Sham and M. Schlüter, Phys. Rev. Lett. 51:1888 (1983)
101. M. S. Hybertsen and S. G. Louie, Solid State Comm. (in press).
102. D. C. Langreth and M. J. Mehl, Phys. Rev. B 28:1809 (1983), and references therein.
103. J. P. Perdew and A. Zunger, Phys. Rev. B 23:5048 (1981).
104. J. A. Alonso and L. A. Girifalco, Solid State Comm. 24:135 (1977); Phys. Rev. B 17:3735 (1978).

105. O. Gunnarsson, M. Jonson, and B. I. Lundqvist, Solid State Comm. 24:765 (1977); Phys. Rev. B 20:3136 (1979).

106. F. Manghi, G. Rieglar, C. M. Bertoni, C. Calandra, and G. B. Bachelet, Phys. Rev. B 28:6157 (1983).

107. O. Gunnarsson and B. I. Lundqvist, Phys. Rev. B 13:4274 (1976).

108. O. Gunnarsson and R. O. Jones, Phys. Scr. 21:394 (1980).

109. Landolt-Bornstein, Zahlenwerte und Funktioned aus Naturwissenschaften und Technik, N.S. Vol. III17a, Springer-Verlag, New York (1982).

110. W. E. Spicer and R. C. Eden, Proceedings of the Ninth International Conference of the Physics of Semiconductors, Moscow, 1968, Nauka, Leningrad, Vol. 1 (1968), p. 61.

111. R. Hulthen and N. G. Nilsson, Solid State Comm. 18:1341 (1976).

II. EXPERIMENTAL ANALYSIS

SURFACE PHONON CALCULATIONS IN METALS AND

COMPARISON WITH EXPERIMENTAL TECHNIQUES

V. Bortolani, A. Franchini and G. Santoro

Dipartimento di Fisica, Universita' di Modena and
Gruppo Nazionale Struttura della Materia
41100 Modena, Italy

ABSTRACT

There has been considerable recent progresses in the experimental
techniques to detect surface phonons over the entire Brillouin zone
for noble and transition metals with the high resolution electron
energy loss spectroscopy and with the inelastic atomic scattering.
However the cross sections are in general not strictly proportional
to the density of states so that in the interpretation of the
experimental spectra it is necessary a knowledge of the scattering
mechanisms.

In these lectures we will present the theory of atomic
scattering focusing in particular on the atom surface potential.
This potential is separated in an attractive part of the Van der
Waals type and in a repulsive part related to the surface charge
which is approximate as a superposition of atomic charges. The
lateral Fourier trasform of this potential, which enters in the
cross sections, has a gaussian form which is essential in order to
explain the falling off of the Rayleigh peaks at the zone boundary.

The bulk phonons are evaluated within a microscopic approach
based on a force constants parametrization. We include central and
angular forces in order to simulate the anisotropy of the electron
gas produced by the presence of d levels. The surface phonons are
evaluated, with these force constants, for a sufficiently thick
slab in order to avoid interference effects between the modes of
the two surfaces.

We also show that is necessary to modify the surface force
constants in order to explain the atom scattering data. We will
outline in a perturbative pseudopotential approach that the effect
of the surface on the electron gas can reduce the surface force
constants.

1. INTRODUCTION

The problem of lattice vibrations, which is one of the major
task of solid state physics, has recently become of interest in
order to obtain information on the dynamics and the electronic
properties of solid surfaces. The rapid development of
experimental techniques makes it possible to obtain information on
surface phonons, surface structure and atom-surface interaction
potentials.

The surface Brillouin scattering technique allows one to
detect long wavelength acoustic phonons for clean[1] and coated[2]
surfaces. For a long time it has been recognized[3] that the atomic
scattering technique has a remarkable potential for obtaining new
information on surface lattice dynamics over the entire Brillouin
zone. However, technical difficulties, connected with the
possibility of obtaining a monochromatic atomic beam, have held up
progress until recently[4,5]. Similarly for the high resolution
electron energy loss spectroscopy[6] it was not possible to detect
surface phonons until it was realized that by raising the electron
incident energy to a few hundred eV, the cross section contains
peaks associated with the excitation of surface phonons and
superimposed on a broad background.

These techniques require incident monochromatic beams with
energy resolution of a few meV. In the case of bulk phonons a high
resolution incident beam was obtained in the early 1960's[7] by using
thermal neutrons and by determining their energy and wavelength for
the angles at which they are Bragg reflected from a single crystal.
Since neutrons are very penetrating (1 cm), the probability that a
neutron will be reflected from a surface is so low that they cannot
be used in order to detect surface properties. The use of a
neutron beam for the detection of surface excitations would require
a very intense source in order to perform grazing incidence
experiments. This kind of very intense source is not available up
to now. Nevertheless, the use of neutrons is the appropriate tool
to detect phonon excitations. In fact the unknown short range
nucleon interaction between the neutron and the target nucleus can
be replaced by a delta function interaction whose strength is
adjusted to give the experimentally known scattering length[8]. With
this simple interaction the cross section can be evaluated
analitically in the Born approximation and it results strictly
proportional to the delta-functions conserving energy and

402

momentum[9]. In the case of photons, neutral atoms and electrons, the interaction potential with the surface cannot be approximated by this simple delta function interaction and the determination of the cross section depends on the chosen type of interaction. For this reason the analysis of the data is more complicated. This explain the difficulties encountered in the determination of surface phonon dispersion relations.

In these lectures we will focus on the determination of surface phonons over the entire Brillouin zone and on the study of the atom-surface interaction potential. We will compare our results for noble and transition metals with the experimental data obtained recently with inelastic neutral atom scattering techniques[5]. We will present the theory of bulk phonons [0] within the framework of a force constant parametrization. In section 2 we will consider a central two body and in Section 3 a three body potential[11] in order to give a good description of the measured bulk phonons and to avoid the Cauchy relations. Surface phonons are treated with the slab method and the theory is presented in Section 4. We take a slab formed by a large number of atomic planes in such a way that interference effects due to the excitation on the two surfaces can be neglected. In particular we will discuss the symmetry properties of slabs formed by a regular stacking of ideal (100) FCC planes. This analysis is of great utility for the study of the relevant modes in a very thick slab. The knowledge of the symmetries of the slab allows one to reduce drastically the size of the dynamical matrix to make the problem more manegeable. The atom-surface cross section is discussed in Section 5. We will see that the cross section is not proportional to the surface phonon correlation function but it is strongly modulated by the matrix elements of the interaction potential. To understand the role played by these matrix elements we will present in Section 6 the theory of the atom-surface potential. We will prove the existence of a lateral potential cut-off which, in momentum space, reduce drastically the events with large momentum transfer. The final Section 7 is devoted to a comparison with the experimental time-of-flight spectra for Ag (111) and parallel momentum along the $[11\bar{2}]$ and $[1\bar{1}0]$ symmetry directions. We will show that in order to explain the huge peak observed in the continuum part of the spectrum one has to lower the surface force constants in order to introduce phenomenologically in the model the effect due to the termination of the electron gas in the surface region.

2. BULK PHONONS

The problem of bulk lattice dynamics of atoms interacting through central forces is well known[10]. We will treat it only for introducing the notation. For simplicity we will consider a

Bravais lattice. We write the potential energy of the crystal as a
sum of pairwise interactions between the atom i and the atom j
located at $\vec{R}_i$ and $\vec{R}_j$ respectively.

$$V_c = \frac{1}{2} \sum_{i,j} v(R_{ij}) \tag{2.1}$$

with $R_{ij} = |\vec{R}_i - \vec{R}_j|$. $\vec{R}_i = \vec{R}_i + \vec{u}_i$ represents the actual position of the
atom i. $\vec{R}_i$ is the static equilibrium position and $\vec{u}_i$ is the
dynamical displacement. In the harmonic approximation the
potential can be written as:

$$V_c = V_c^0 + \sum_{1,\alpha} \Phi_\alpha(\vec{R}_1) u_{1\alpha} + \frac{1}{2} \sum_{\substack{1,\alpha \\ 1',\beta}} \Phi_{\alpha\beta}(\vec{R}_1,\vec{R}_{1'}) u_{1\alpha} u_{1'\beta} \tag{2.2}$$

The parameter Φ_α evaluated in the static equilibrium configuration
(indicated by the subscript O) is given by:

$$\Phi_\alpha(\vec{R}_1) = \frac{1}{2} \sum_{i,j} \frac{\partial v}{\partial R_{ij}} \frac{\partial R_{ij}}{\partial R_{1\alpha}} \bigg|_0 \tag{2.3}$$

We define the first order tangential force constant α_{ij} by:

$$\alpha_{ij} = \frac{1}{R_{ij}} \frac{\partial v}{\partial R_{ij}} \bigg|_0 \tag{2.4}$$

Bulk equilibrium implies that V_c is minimum for a uniform variation
of the lattice parameter a_0. For a uniform expansion of the
crystal the new atomic position becomes:

$$R_i = R_i \frac{a}{a_0} \tag{2.5}$$

The total potential can be written as:

$$V_c = V_c^0 + \frac{1}{2} \frac{a-a_0}{a_0} \sum_{i,j} \alpha_{ij} R_{ij}^2 + 0\left[\left(\frac{a-a_0}{a_0}\right)^2\right] \tag{2.6}$$

To fulfill the bulk equilibrium condition we should impose the
condition:

$$\sum_{i,j} \alpha_{ij} R_{ij}^2 = 0 \qquad (2.7)$$

This equation imposes a constraint among the various first order tangential force constants. The parameters $\Phi_{\alpha\beta}$, the second order derivatives of the potential, are defined by:

$$\Phi_{\alpha\beta}(\vec{R}_1,\vec{R}_{1'}) = \left. \frac{\partial^2 V_c}{\partial R_{1\alpha} \partial R_{1'\beta}} \right|_0 \qquad (2.8)$$

Translational symmetry requires that Eq. (2.9) depends only on the relative ion positions $\vec{R}_1 - \vec{R}_{1'}$. By taking into account the dependence of Eq. (2.1) on the distance R_{ij} we have:

$$\Phi_{\alpha\beta}(\vec{R}_1 - \vec{R}_{1'}) = \frac{1}{2} \sum_{i,j} \left\{ \frac{\partial^2 v}{\partial R_{ij}^2} \frac{\partial R_{ij}}{\partial R_{1\alpha}} \frac{\partial R_{ij}}{\partial R_{1'\beta}} + \frac{\partial v}{\partial R_{ij}} \frac{\partial R_{ij}}{\partial R_{1\alpha} \partial R_{1'\beta}} \right\} \qquad (2.9)$$

We define the second order radial force constant β_{ij} as:

$$\beta_{ij} = \left. \frac{\partial^2 v}{\partial R_{ij}^2} \right|_0 \qquad (2.10)$$

β_{ij} is related to the stretching forces between atom i and atom j, while α_{ij} is connected to bending forces. The central force $\vec{F}_n^{(c)}$ acting on the n-th atom in the crystal is given by:

$$\vec{F}_n^{(c)} = -\nabla_{\vec{u}_n} V_c = -\sum_j \left\{ \alpha_{nj} \vec{u}_{nj} + (\beta_{nj} - \alpha_{nj}) \frac{\vec{R}_{nj} \cdot \vec{u}_{nj}}{R_{nj}} \frac{\vec{R}_{nj}}{R_{nj}} \right\} \qquad (2.11)$$

where $\vec{u}_{nj}$ is the relative displacement of two atoms ($\vec{u}_{nj} = \vec{u}_n - \vec{u}_j$). The sum over j includes also the n-th term which gives the interaction of the n-th atom with itself. Generally one should take away this self interaction. However this force is already zero by virtue of the expression for the force given by Eq. (2.11). The equation of motion of the lattice for atoms having mass M has the form:

$$M \ddot{\vec{u}}_n = \vec{F}_n^{(c)} \qquad (2.12)$$

Lattice translational invariance allows one to expand the

displacements in a Bloch type series of the form:

$$\vec{u}_n = \frac{1}{\sqrt{M}} \sum_{\vec{q}} e^{i\vec{q}\cdot\vec{R}_n} e^{i\omega t} \vec{e}(\vec{q})$$ (2.13)

The sum over $\vec{q}$ can be eliminated if one considers a single monochromatic wave propagating through the crystal. The $\vec{e}(\vec{q})$ is the polarization vector of the mode having frequency ω and momentum $\vec{q}$. Substituing Eq. (2.13) in Eq. (2.14) we can cast the equation of motion in matrix form:

$$\sum_{\beta} \{D_{\alpha\beta}(\vec{q}) - \omega^2 \delta_{\alpha\beta}\} e_{\beta}(\vec{q}) = 0$$ (2.14)

where the dynamical matrix is given by:

$$D_{\alpha\beta}(\vec{q}) = \frac{1}{M} \sum_{1} \Phi_{\alpha\beta}(\vec{R}_1) e^{i\vec{q}\cdot\vec{R}_1}$$ (2.15)

By using Eq. (2.11) we find that the dynamical matrix $D_{\alpha\beta}$ for central pairwise forces can be written as:

$$D_{\alpha\beta}(\vec{q}) = \frac{1}{M} \sum_{1} \left\{ \alpha_1 \delta_{\alpha\beta} + (\beta_1 - \alpha_1) \frac{R_{1\alpha} R_{1\beta}}{R_1^2} (1 - e^{i\vec{q}\cdot\vec{R}_1}) \right\}$$ (2.16)

With the use of normal coordinates Eq. (2.13) we reduce the problem of the solution of the infinite set of equations (2.12) to the problem of solving a set of three linear equations in the three unknowns e_{α} for any value of $\vec{q}$.

3. ANGULAR FORCES

In the case of FCC transition and noble metals one has to introduce non central interactions. In fact for central forces should hold the Cauchy relation between the elastic constants c_{12} and c_{44} i.e. $c_{12} = c_{44}$ which is strongly violated for these metals. In order to account for the non central nature of the ion-ion interaction we assume a potential energy of the type:

$$V_A = \frac{1}{3!} \sum_{i,j,k} w(\cos\theta_{ijk}) \tag{3.1}$$

where θ_{ijk} is the angle formed by the vectors $(\vec{R}_i - \vec{R}_j)$ and $(\vec{R}_i - \vec{R}_k)$. As for central force we have to expand (3.1) up to second order in the displacements u_i. We introduce a first and second order angular force constants by :

$$\gamma_{ijk} = \frac{\partial w}{\partial \cos\theta_{ijk}}$$

$$\eta_{ijk} = \frac{\partial^2 w}{\partial \cos\theta^2_{ijk}} \tag{3.2}$$

The first order contribution to Eq. (3.1) can be written in the form:

$$V_A^{(1)} = \frac{1}{3} \sum_{i,j,k} \gamma_{ijk} \frac{\vec{\phi}(ijk) \cdot \vec{u}_{ij}}{R_{ij}} \tag{3.3}$$

where:

$$\vec{\phi}(ijk) = \hat{r}_{ik} - \cos\theta_{ijk} \hat{r}_{ij} \tag{3.4}$$

and $\hat{r}_{ij} = \vec{R}_{ij}/R_{ij}$ is the unity vector.
The equilibrium condition requires that the force acting on the n-th atom arising from this part of the potential must vanish i.e.

$$\sum_{j,k} \left(\gamma_{njk} \frac{\vec{\phi}(njk)}{R_{nj}} - \gamma_{jnk} \frac{\vec{\phi}(jnk)}{R_{jn}} \right) = 0 \tag{3.5}$$

This relation implies that the γ_{njk} cannot be chosen arbitraly. The second order contribution is:

$$V_A^{(2)} = \frac{1}{12} \left[- \sum_{i,j,k} \gamma_{ijk} \frac{\vec{A}(ijk) \cdot (\hat{r}_{ij} \times \vec{u}_{ij})}{R_{ij}} + (j \rightleftarrows k) \right.$$

$$\left. + \sum_{i,j,k} \eta_{ijk} \frac{\vec{\phi}(ijk) \cdot \vec{u}_{ij}}{R_{ij}} + (j \rightleftarrows k) \right] \tag{3.6}$$

where:

$$\vec{A}(ijk) = \frac{1}{R_{ij}} \{2(\hat{r}_{ij}\cdot\vec{u}_{ij})\hat{r}_{ij}x\hat{r}_{ik} + \cos\theta_{ijk}\hat{r}_{ij}x\vec{u}_{ij}\} +$$

$$\frac{1}{R_{ik}} \{\hat{r}_{ij}x\vec{u}_{ik}-(\hat{r}_{ik}\cdot\vec{u}_{ik})\hat{r}_{ij}x\hat{r}_{ik}\} \tag{3.7}$$

The force due to $V_A^{(2)}$ can be split in a part depending on the first order force constant $\vec{F}_n^{(A,2)}(\gamma)$ and in a part depending on the second order force constant $\vec{F}_n^{(A,2)}(\eta)$. We obtain:

$$\vec{F}_n^{(A,2)}(\gamma)= -\frac{1}{3}\sum_{j,k}\left[\gamma_{njk}\left[-\frac{1}{R_{nj}^2}\{(\hat{r}_{nk}\cdot\vec{u}_{nj})\hat{r}_{nj}+(\hat{r}_{nj}\cdot\vec{u}_{nj})\hat{r}_{nk}+\right.\right.$$

$$\cos\theta_{njk}\vec{u}_{nj}-3\cos\theta_{njk}(\hat{r}_{nj}\cdot\vec{u}_{nj})\hat{r}_{nj}\} +$$

$$\frac{1}{R_{nj}R_{nk}}\{\vec{u}_{nk}-(\hat{r}_{nk}\cdot\vec{u}_{nk})\hat{r}_{nk}-(\hat{r}_{nj}\cdot\vec{u}_{nk})\hat{r}_{nj} +$$

$$\left.\cos\theta_{njk}(\hat{r}_{nk}\cdot\vec{u}_{nk})\hat{r}_{nj}\}\right] -$$

$$\gamma_{jnk}\left[-\frac{1}{R_{jn}^2}\{(\hat{r}_{jk}\cdot\vec{u}_{jn})\hat{r}_{jn}+(\hat{r}_{jn}\cdot\vec{u}_{jn})\hat{r}_{jk} +\right.$$

$$\cos\theta_{jnk}\vec{u}_{jn}-3\cos\theta_{jnk}(\hat{r}_{jn}\cdot\vec{u}_{jn})\hat{r}_{jn}\} +$$

$$\frac{1}{R_{jn}R_{jk}}\{\vec{u}_{jk}-(\hat{r}_{jk}\cdot\vec{u}_{jk})\hat{r}_{jk}-(\hat{r}_{jn}\cdot\vec{u}_{jk})\hat{r}_{jn} +$$

$$\left.\left.\cos\theta_{jnk}(\hat{r}_{jk}\cdot\vec{u}_{jk})\hat{r}_{jn}\}\right]\right] \tag{3.8}$$

and:

$$\vec{F}_n^{(A,2)}(\eta) = -\frac{1}{3}\sum_{j,k}\{\frac{\eta_{njk}}{R_{nj}}\psi(njk)\vec{\phi}(njk) - (n\rightleftarrows j)\} \tag{3.9}$$

where :

$$\psi(njk) = \frac{\vec{\phi}(njk)\cdot\vec{u}_{nj}}{R_{nj}} + \frac{\vec{\phi}(nkj)\cdot\vec{u}_{nk}}{R_{nk}} \tag{3.10}$$

As one can see both contribution to the force depend on the relative displacements of two atoms so that an uniform translation does not produce any force on the crystal. Similarly a rotation of the crystal does not introduce any force in the crystal. For this reason the dynamical matrix which can be obtained from Eqs. (3.8) and (3.9) is invariant under rotation and translation of the crystal. The general expression of the dynamical matrix including both central and angular interaction is very cumberstome. For this reason we specialize the results to FCC crystals for which experimental results are available at present. We will consider interactions extending up to second nearest neighbors for the angular interaction $V_A^{(2)}$ and to fourth neighbors for the central forces. These interactions are sketched in Fig. 1.

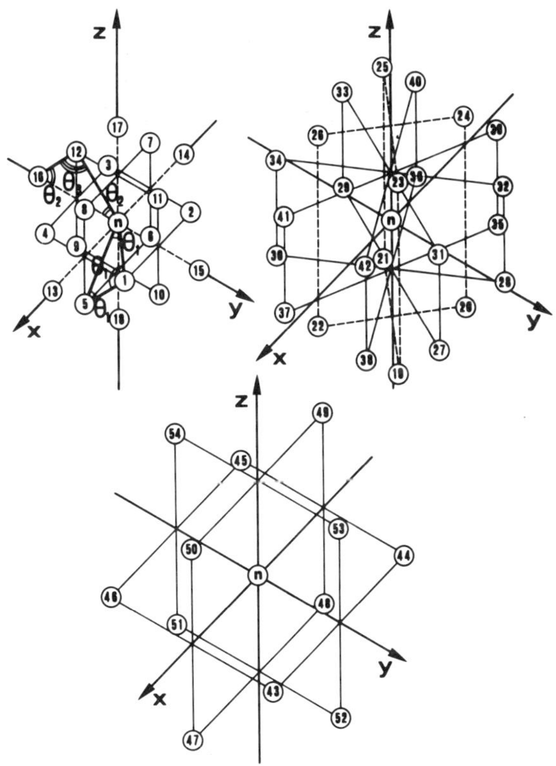

Fig. 1 Nearest neighbors for an FCC crystal. Atoms labelled from 1 to 12 are the first n.n.; from 13 to 18 second n.n.; from 19 to 42 third n.n. and from 43 to 54 fourth n.n. $\theta_1 = \pi/6$, $\theta_2 = \pi/4$ and $\theta_3 = \pi/2$ are the angles considered in the interactions.

In this case the dynamical matrix can be cast in the form:

$$D_{11}=\Lambda_1 c_y c_z+\Lambda_2 c_x(c_y+c_z)+\Lambda_3(1-c_{2x})+\Lambda_4(2-c_{2y}-c_{2z})+$$
$$\Lambda_6 c_{2x} c_y c_z+\Lambda_7 c_x(c_y c_{2z}+c_{2y}c_z)+\Lambda_8 c_{2y}c_{2z}+\Lambda_9 c_{2x}(c_{2y}+c_{2z})$$

$$D_{12}=D_{21}=\Lambda_5 s_x s_y+\Lambda_{10}\{s_x s_y c_{2z}+2c_z(s_x s_{2y}+s_{2x}s_y)\}+\Lambda_{11}s_{2x}s_{2y}$$

$$D_{13}=D_{31}=\Lambda_5 s_x s_z+\Lambda_{10}\{s_x s_z c_{2y}+2c_y(s_x s_{2z}+s_z s_{2x})\}+\Lambda_{11}s_{2x}s_{2z}$$

$$D_{22}=\Lambda_1 c_x c_z+\Lambda_2 c_y(c_x+c_z)+\Lambda_3(1-c_{2y})+\Lambda_4(2-c_{2x}-c_{2z})+$$
$$\Lambda_6 c_{2y}c_x c_z+\Lambda_7 c_y(c_x c_{2z}+c_z c_{2x})+\Lambda_8 c_{2x}c_{2z}+\Lambda_9 c_{2y}(c_{2x}+c_{2z})$$

$$D_{23}=D_{32}=\Lambda_5 s_y s_z+\Lambda_{10}\{s_y s_z c_{2x}+2c_x(s_y s_{2z}+s_z s_{2y})\}+\Lambda_{11}s_{2y}s_{2z}$$

$$D_{33}=\Lambda_1 c_x c_y+\Lambda_2 c_z(c_x+c_y)+\Lambda_3(1-c_{2z})+\Lambda_4(2-c_{2x}-c_{2y})+$$
$$\Lambda_6 c_{2z}c_x c_y+\Lambda_7 c_z(c_x c_{2y}+c_y c_{2x})+\Lambda_8 c_{2x}c_{2y}+\Lambda_9 c_{2z}(c_{2x}+c_{2y}) \quad (3.11)$$

The parameters involving central and angular interactions up to second neighbors are:

$$\Lambda_1 = -4(\alpha_1^B+3\delta_1^B)$$

$$\Lambda_2 = -2(\beta_1^B+\alpha_1^B+6\delta_1^B+8\delta_3^B)$$

$$\Lambda_3 = 2(\beta_2^B+4\delta_3^B)$$

$$\Lambda_4 = 2(\alpha_2^B-2\delta_3^B)$$

$$\Lambda_5 = 2(\beta_1^B-\alpha_1^B+3\delta_1^B-8\delta_3^B) \qquad (3.12)$$

Notice that for this type of interactions there are only five linearly independent combinations of the six force constants α_1^B, β_1^B, $\delta_1^B=\eta_1^B/3a_0^2$, α_2^B, β_2^B, $\delta_2^B=\eta_2^B/3a_0^2$ and $\delta_3^B=\eta_3^B/3a_0^2$. The central interaction relative to third and fourth neighbors give rise to the parameters

$$\Lambda_6 = -\frac{8}{3}(2\beta_3^B + \alpha_3^B)$$

$$\Lambda_7 = -\frac{4}{3}(\beta_3^B + 5\alpha_3^B)$$

$$\Lambda_8 = -4\alpha_4^B$$

$$\Lambda_9 = -2(\beta_4^B + \alpha_4^B)$$

$$\Lambda_{10} = \frac{4}{3}(\beta_3^B - \alpha_3^B)$$

$$\Lambda_{11} = 2(\beta_4^B - \alpha_4^B) \tag{3.13}$$

In this case the number of force constants is less than the number of parameters involved in the dynamical matrix. We have defined:

$$c_i = \cos(a_0 q_i/2) \qquad s_i = \sin(a_0 q_i/2)$$

$$c_{2i} = \cos(a_0 q_i) \qquad s_{2i} = \sin(a_0 q_i) \tag{3.14}$$

To solve the secular problem one should have at disposal the values of the force constants. These could be evaluated in principle from the knowledge of the total potential. Since the available total potential do not produce results which compare accurately with the measured bulk phonon frequencies we prefer to use the force constants as free parameters to be fitted to experimental data.

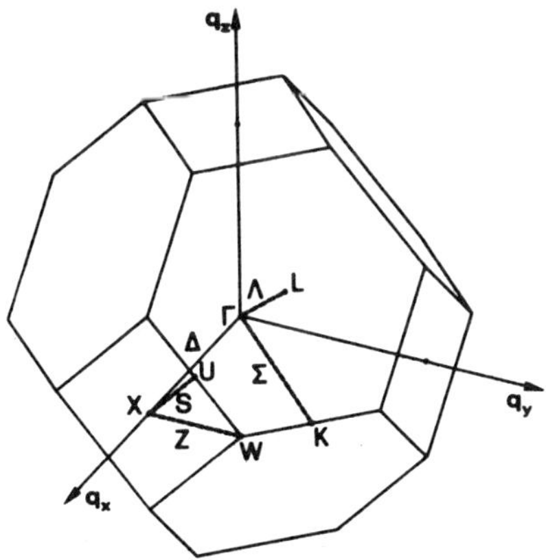

Fig. 2 Brillouin zone for FCC. Symmetry points and symmetry lines are indicated.

We use a weighted least square fitting procedure:

$$\delta \sum_{n,j} \frac{1}{\sigma^2_{nj}} \{\Omega^2_{nj} - \omega^2_{nj}\}^2 = 0 \tag{3.15}$$

The Ω_{nj} are the experimental values of the frequency of the mode with momentum $\vec{q}_n$ and branch index j and σ_{nj} are standard deviations. The variation must be performed with respect to the force constants. The Ω_{nj} are taken from the available neutron scattering experiments data with $\vec{q}$ along the symmetry directions. In Fig. 2 is drawn the Brillouin zone of FCC crystals to specify the notation used. We also include in our fitting procedure the experimental elastic constants. In this manner our results reproduce very well in the long wavelenght limit the velocities of sound in the various directions. The expression for the ω^2_{nj} are derived by solving the secular problem Eq. (2.14) for $\vec{q}$ along the symmetry directions. The solution of Eq.(3.15) gives the parameters Λ_i appearing in Eqs. (3.12) and (3.13). As one can see from Eq. (3.12) from the knowledge of the five Λ_i is only possible to obtain the value of five force constants. The sixth force constant should be linearly dependent on the other five ones. For this reason we have chosen $\delta^B_3 = 0$.

To illustrate the quality of the results that can be obtained with this procedure we will consider Pt. We choose Pt because in the experimental phonon frequencies it is well evident a Kohn anomaly in the Σ direction. We will consider the following parametrisations for the force constants:

central first nearest neighbors   (1C)

central forces up to second n.n.   (2C)

central forces up to second n.n.
and angular forces up to second n.n.   (2CA)

central forces up to fourth n.n.   (4C)

central forces up to fourth n.n.
and angular forces up to second n.n.   (4CA)

We include in the least square fitting procedure the equations for the elastic constants that we assume to be known within 2% error. In Figs. 3a, 3b, 3c, 3d and 3e are drawn the results of the fitting procedure. Model (1C) reproduce very poorly all the transverse branches. Even model (2CA) understimate the transverse branches particularly in the Z direction. Long range forces are needed to explain the last direction, model (4C).

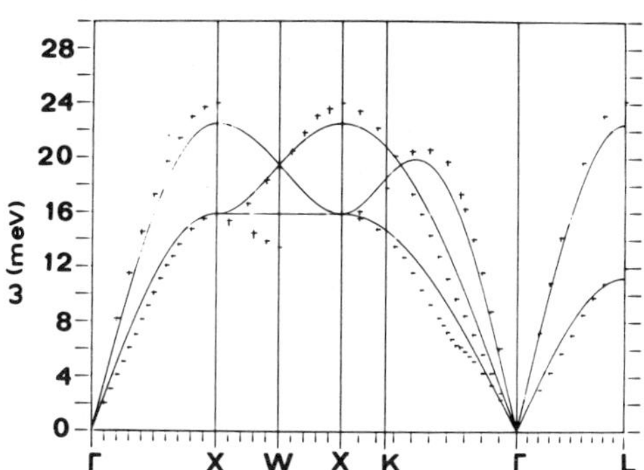

Fig. 3a Bulk phonon frequencies of Pt obtained with the (1C) model. The experimental points are from Ref. 12.

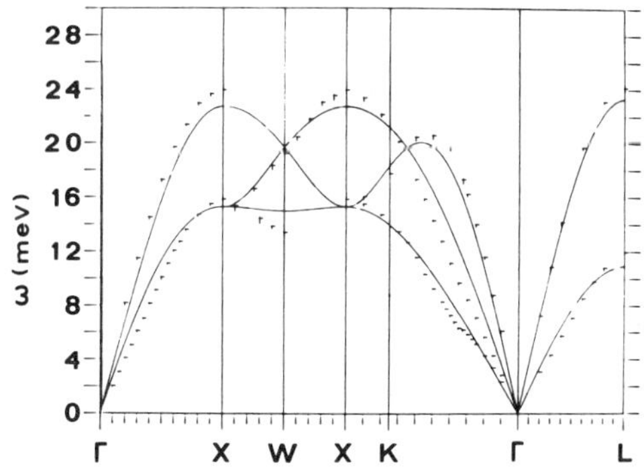

Fig. 3b Bulk phonon frequencies of Pt obtained with the (2C) model.

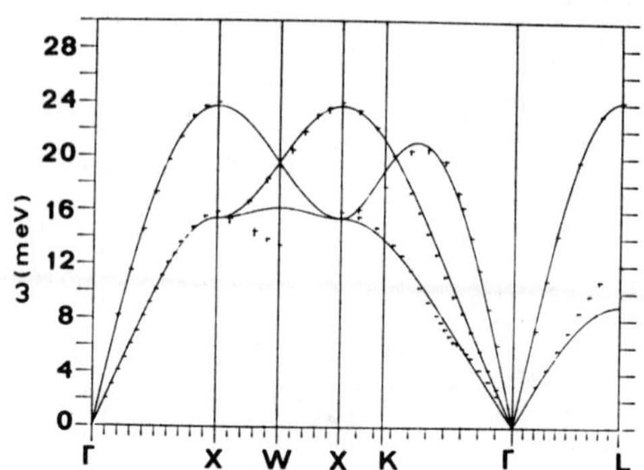

Fig. 3c Bulk phonon frequencies of Pt obtained with the (2CA) model.

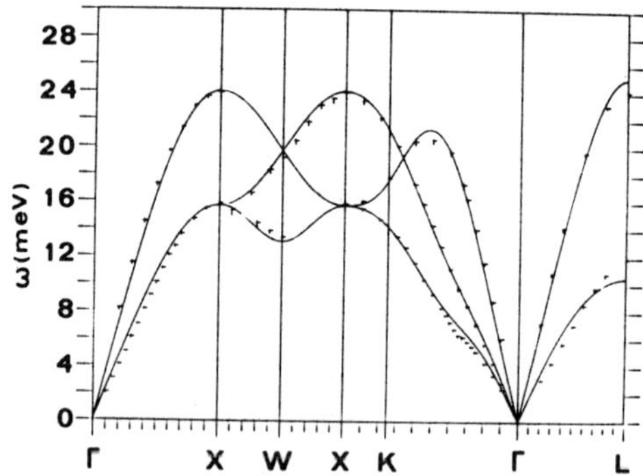

Fig. 3d Bulk phonon frequencies of Pt obtained with the (4C) model.

414

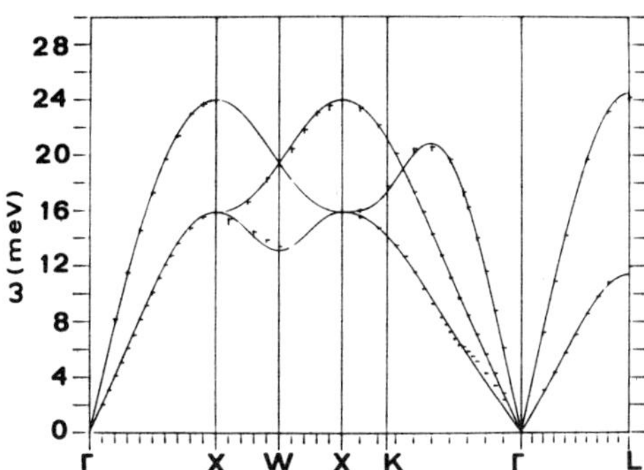

Fig. 3e Bulk phonon frequencies of Pt obtained with the (4CA) model.

The inclusion of angular forces, model (4CA), gives a very good overall fit and also reproduce the Kohn anomaly. In Table 1 are reported the force constants for the five different models. As one can see the values of the leading terms β_1^B and β_2^B are

Table 1. Central and angular force constants (in THz^2) of Pt fitted to the experimental bulk phonon frequencies

model	α_1^B	β_1^B	δ_1^B	α_2^B	β_2^B	δ_2^B	α_3^B	β_3^B	α_4^B	β_4^B
1C	0	3.69	–	–	–	–	–	–	–	–
2C	-0.18	3.95	–	0.18	-0.14	–	–	–	–	–
4C	-0.53	4.64	–	-0.03	0.29	–	0	0.17	0.14	-0.61
2CA	0.25	5.70	-0.19	-0.25	0.35	-0.18	–	–	–	–
4CA	0.13	5.62	-0.24	-0.05	0.33	-0.05	-0.02	0.27	0.01	-0.43

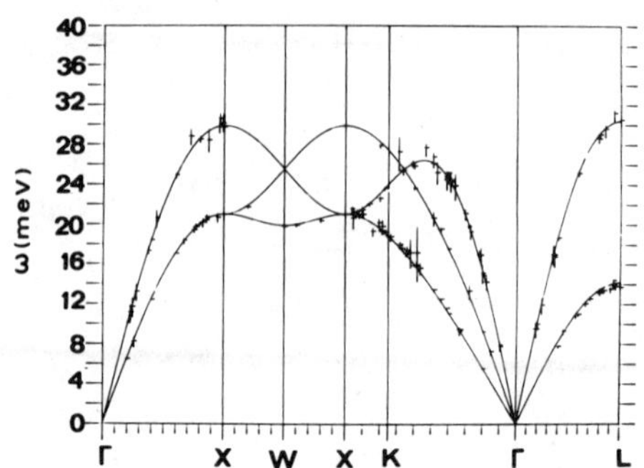

Fig.   4   Bulk phonon frequencies for Cu evaluated   with   the   (4CA)
         model.   Experimental data from Ref.   13.

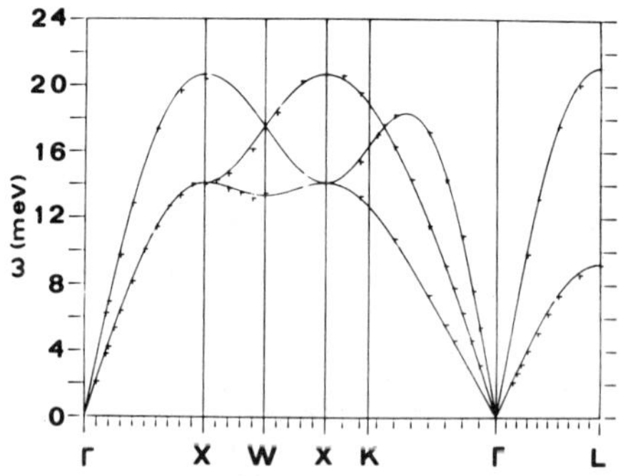

Fig.   5   Bulk phonon frequencies for Ag evaluated   with   the   (4CA)
         model.   Experimental data from Ref.   14.

416

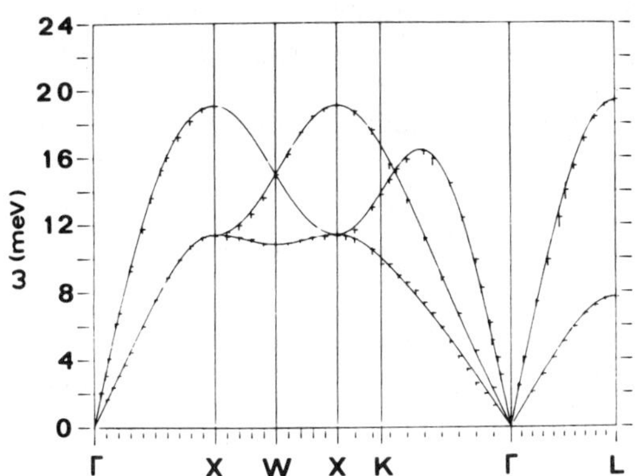

Fig. 6 Bulk phonon frequencies for Au evaluated with the (4CA) model. Experimental data from Ref. 15.

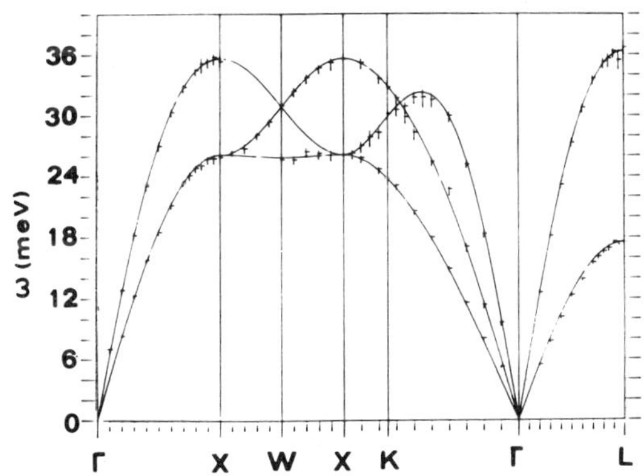

Fig. 7 Bulk phonon frequencies for Ni evaluated with the (4CA) model. Experimental data from Ref. 16.

practically converging to a definite value. Notice that β_1^B evaluated with the Johnson potential turn out to be close to the values reported in Table 1 for the (4CA) model. This seems to indicate that the leading force constant evaluated with a suitable fitting procedure can be directely related to the potential. The same conclusions can be also drawn for the noble metals and for Ni. In Figs. 4, 5, 6 and 7 are reported the calculated phonon frequencies for these metals for the (4CA) model.
We want to stress that for this model the speed of sound in all the directions is reproduced to within 5% of accuracy.

4. SURFACE PHONONS

In the previous sections we have presented the theory to determine the force constants from neutron scattering experiments. To simulate the surface we will now consider the lattice dynamics of a slab. In principle one should deal with a semiinfinite crystal in order to avoid the interference effects between the two surfaces. However, the splitting in energy due to the second surface is only of importance for the longwavelegth acoustic phonons which are very penetrating. Apart this small q region, we found that a slab of 45 atomic planes is sufficient to eliminate the splitting in phonon frequencies.

For a slab formed by N planes with one atom per unit surface cell of area A, the dynamical problem can be written in the form:

$$\sum_{1',\beta} \{D_{\alpha\beta}^{11'}(\vec{Q}) - \omega_j^2(\vec{Q})\delta_{\alpha\beta}\delta_{11'}\}\, e_{1'\beta}(\vec{Q},j) = 0 \qquad (4.1)$$

where 1 and 1' label the planes of the slab and the $\vec{e}_1(\vec{Q},j)$ are the eigenvectors corresponding to the eigenvalues $\omega_j(\vec{Q})$.

The elements of the dynamical matrix can be determined by taking the two dimensional Fourier transform of the forces Eqs. (2.11), (3.8) and (3.9) acting on the various planes. The detailed expressions for these matrix elements in the case of the (4CA) model are too lengthly. For reason of space we give in the Appendix the expressions for the forces relative to the (100) ideal surface from which one can determine the dynamical matrix. In these expressions the force constants in the surface region are considered to be different from those of the bulk.

As one can see from Eq. (4.1) the size of the dynamical matrix is 3Nx3N. For a slab formed by 45 atomic planes one has to diagonalize a 135x135 matrix. From the numerical point of view this is a rather complicate problem. To reduce the size of the dynamical matrix it is convenient to analyze the symmetry of the slab.

418

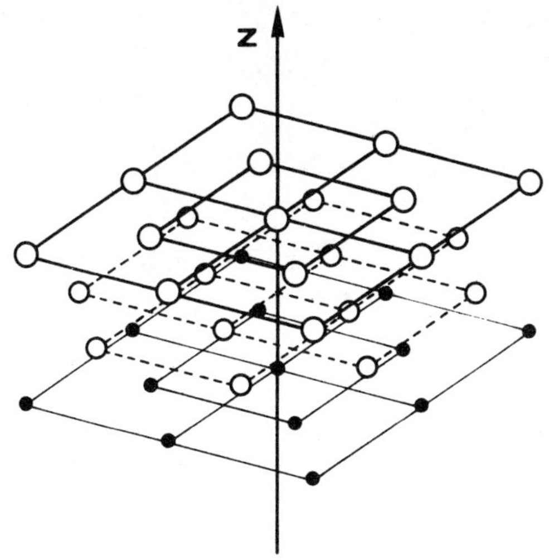

Fig. 8   Geometry of the (100) surface of an FCC crystal. Large
circles represent the surface atoms. Small circles refer
to atoms on the first layer below the surface. Full
circles represent atoms on second layer.

We will firstly consider the (100) surface. In this case we
take an odd number of ideal planes in order to have two equivalent
surfaces. The geometry of (100) surface is drawn in Fig. 8.   For
a general point $\vec{Q}$ inside the two dimensional Brillouin zone (2DBZ)
depicted in Fig. 9 the point group $G(\vec{Q})$ is formed by the identity
$\{E|0\}$ and by the inversion operation $\{IC_{2z}|0\}$ with respect to the z
axis. By chosing the central plane as the x-y plane of reference
the last operation change z in -z whitout modifying x and y. $G(\vec{Q})$
is isomorphic to the point group C_s. This group has only two
one-dimensional representations $\Gamma^{(1)}$ and $\Gamma^{(2)}$. The character table
of this group has therefore the form:

	$\{E|0\}$	$\{IC_{2z}|0\}$
$\Gamma^{(1)}$	1	1
$\Gamma^{(2)}$	1	-1

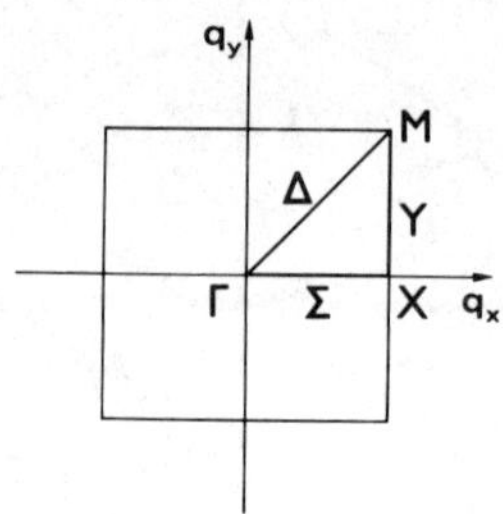

Fig. 9 Two dimensional Brillouin zone for the (100) surface. The
symmetry points and lines are indicated.

According to group theory, the eigenvectors of the dynamical matrix
transform according to the irreducible representations $\Gamma^{(1)}$ and
$\Gamma^{(2)}$. By using the projectors method [17] we can write:

$$\vec{u}_1^{(i)} = A_1 \sum_R \chi^{(i)}(R)P(R)\vec{u}_1 \qquad (4.2)$$

where R corresponds to the symmetry operations $\{E|0\}$ and $\{IC_{2z}|0\}$.
$P(R)$ is the projection operator on the subspace R and $\chi^{(i)}(R)$ is
the character of R in the i-th irreducible representation $\Gamma^{(i)}$. A_1
is a normalization factor. Explicitly we have:

$$P(\{E|0\})\vec{u}_1 = \vec{u}_1 = (u_1,v_1,w_1) \qquad (4.3)$$

and:

$$P(\{IC_{2z}|0\})\vec{u}_1 = (u_{-1},v_{-1},-w_{-1}) \qquad (4.4)$$

Notice that in the last equation, z change of sign and the
displacement $\vec{u}_1$ is rotated according to IC_{2z} and is transferred to
the -1 plane. By taking advantage of these symmetry properties we
can write the symmetrized displacements as:

$$u_{1\alpha}^{(\pm)} = A_1(u_{1\alpha} \pm p_\alpha u_{-1\alpha}) \qquad (4.5)$$

with $\vec{p}=(1,1,-1)$, $A_1=1/\sqrt{2}$ for $l\neq0$ and $A_0=1/2$. The $\vec{u}_1^{(+)}$ transform according to the identity representation $\Gamma^{(1)}$. These modes have a node for z component on the central plane. The $\vec{u}_1^{(-)}$ transform according to $\Gamma^{(2)}$ and have a node for the in-plane component on the central plane. The dynamical matrix must be invariant under the symmetry operations of $G(\vec{Q})$ which implies:

$$D_{\alpha\beta}^{-1-1'} = p_\alpha p_\beta D_{\alpha\beta}^{11'} \tag{4.6}$$

By using Eqs. (4.5), (4.6) and (4.1) we can block-diagonalize the dynamical matrix. The first block refer to the $u_1^{(+)}$ modes. The matrix elements of this block are related to those of the original matrix by:

$$D_{\alpha\beta}^{(+)11'} = D_{\alpha\beta}^{11'} + p_\beta D_{\alpha\beta}^{1-1'} \qquad 1,1'=1,(N-1)/2 \quad \alpha,\beta=1,2,3$$

$$D_{\alpha\beta}^{(+)10} = \sqrt{2}D_{\alpha\beta}^{10} \qquad 1=1,(N-1)/2 \quad \alpha=1,2,3 \quad \beta=1,2$$

$$D_{\alpha\beta}^{(+)01'} = \sqrt{2}D_{\alpha\beta}^{01'} \qquad 1'=1,(N-1)/2 \quad \alpha=1,2 \quad \beta=1,2,3$$

$$D_{\alpha\beta}^{(+)00} = D_{\alpha\beta}^{00} \qquad \alpha,\beta=1,2 \tag{4.7}$$

The second block refers to $u_1^{(-)}$ and is given by:

$$D_{\alpha\beta}^{(-)11'} = D_{\alpha\beta}^{11'} - p_\beta D_{\alpha\beta}^{1-1'} \qquad 1,1'=1,(N-1)/2 \quad \alpha,\beta=1,2,3$$

$$D_{\alpha\beta}^{(-)10} = \sqrt{2}D_{\alpha\beta}^{10} \qquad 1=1,(N-1)/2 \quad \alpha=1,2,3 \quad \beta=3$$

$$D_{\alpha\beta}^{(-)01'} = \sqrt{2}D_{\alpha\beta}^{01'} \qquad 1'=1,(N-1)/2 \quad \alpha=3 \quad \beta=1,2,3$$

$$D_{\alpha\beta}^{(-)00} = D_{\alpha\beta}^{00} \qquad \alpha,\beta=3 \tag{4.8}$$

This separation allows one to reduce the size of the matrices that one has to diagonalize. This results in a noticeable reduction of computing time.

We have performed calculations for Pt (100) with $N=19$ within the (4CA) model by using Eqs. (4.7) and (4.8). For such a number of planes, as one can see from Figs. 10 and 11, the spectra are nearly the same. Notice that in Fig. 11 around the Γ point, the Rayleigh wave has a parabolic behavior since it corresponds to flexural modes of the slab. Apart this small region the

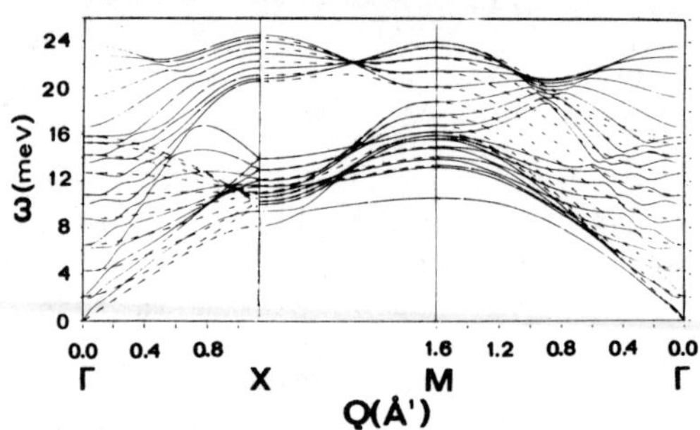

Fig. 10   Surface phonon frequencies of Pt determined   for   the $\vec{u}^{(+)}$ modes Eq. (4.7). In the calculation we have used a slab formed by 19 planes.

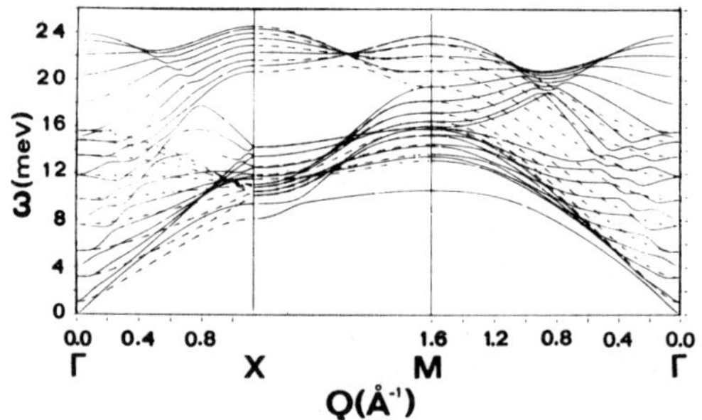

Fig. 11   Surface phonon frequencies of Pt determined   for   the $\vec{u}^{(-)}$ modes Eq. (4.8). In the calculation we have used a slab formed by 19 planes.

frequencies are almost identical so that in the subsequent
calculations we will use only Eq. (4.7).

We pass now to investigate the effects of the bulk
parametrization. In Fig. 12 are presented the surface phonons for
Pt by using the (1C) model for a slab formed by 65 atomic planes.
The full lines in the Δ and Σ directions refer to modes polarized
in the sagittal plane defined by the momentum $\vec{Q}$ and the normal to
the surface. In the Y direction the full lines refer to modes
that, on the surface, are polarized in a plane with Miller indices
(1,-1,0). The dashed lines in the Δ and Σ directions are relative
to Shear Horizontal modes. In Fig. 13 are reported the
calculations for the (4CA) model. As one can clearly see the use
of the (1C) model deeply modify the surface phonon spectrum with
respect to the (4CA) model. The band width of all the branches and
the relative gaps are, in the (1C) case, too narrow with respect to
the experimental values determined from neutron data. Furthermore
the shape of the branches is very crytical to the value of the
force constant β_1^B . In this case the results are strongly
model-dependent. We want to stress that the modification in the
branches by the use of the (1C) model with respect to the use of
the (4CA) model are more pronounced than in the bulk case. This
indicate that it is necessary to have a very good parametrization
of the bulk properties in order to study surface dynamics. In
Figs. 14, 15, 16 are drawn the surface phonons for noble metals.

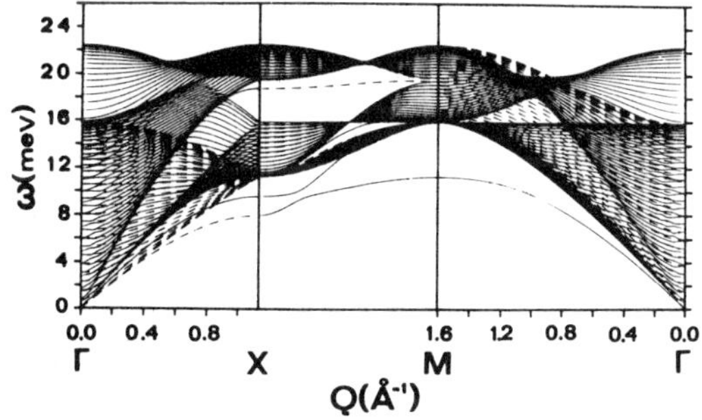

Fig. 12 Surface phonons for Pt. Results for the (1C) model with
N=65.

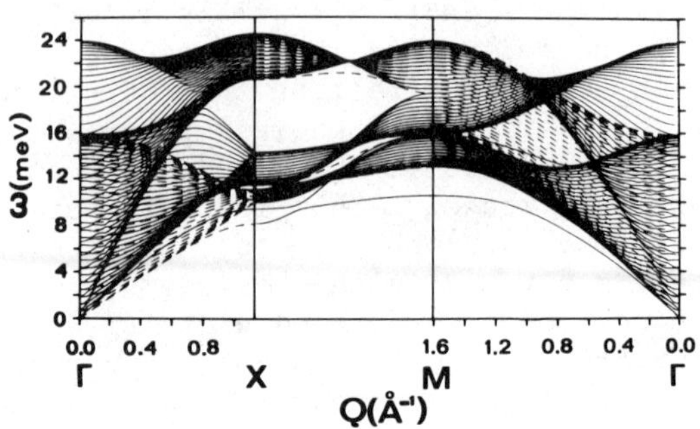

Fig. 13 Surface phonons for Pt. Results for the (4CA) model with
          N=65.

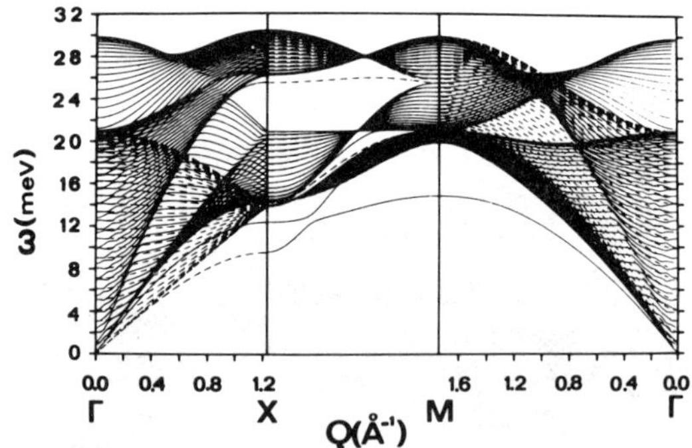

Fig. 14 Surface phonons for Cu. Results for the (4CA) model with
          N=65.

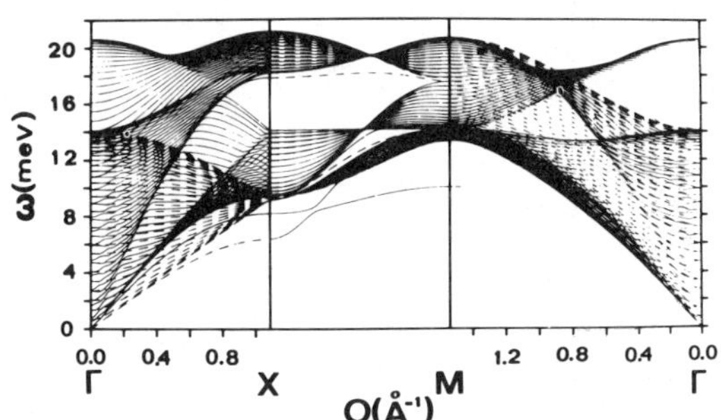

Fig. 15   Surface phonons for Ag.   Results for the (4CA) model with N=65.

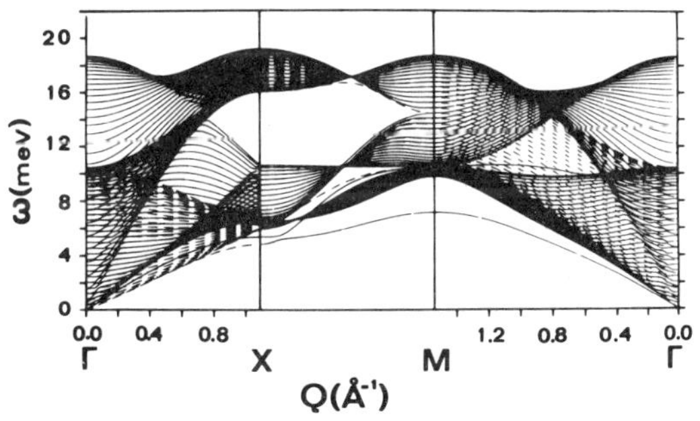

Fig. 16   Surface phonons for Au.   Results for the (4CA) model with N=65.

In the Δ direction there is evidence of a single localized mode polarized in the sagittal plane. The penetration depth decreases by increasing $\vec{Q}$. At the M point it becomes a Shear Vertical mode almost localized on the surface. In the upper lens is present a localized mode that becomes resonant for Au. In the Y direction there are different localized modes and resonances. There are two surface modes which hybridize mostly in passing from Cu to Au. The upper mode becomes resonant in the continuum for Cu and Ag, while for Au emerges as a localized state in the first lens. Also in the upper lens there is a localized mode which moves toward the upper branch in passing from Cu to Au.

To show the effect of the modification of the surface force constants on the surface modes we illustrate in Fig. 17 the calculation for Ag (100) with $\beta_1^S = 0.48\,\beta_1^B$. As one can see there is evidence of a new localized sagittal mode near to the Rayleigh wave in the Σ direction. The available experimental data[18] are plotted in the same figure. Unfortunately these data have not enough resolution to prove the existence of this new branch even if they do not disagree with the theoretical results. In Fig. 18 are drawn the phonons for Ni. In the Σ direction are also indicated the EELS experimental results[6]. Apart the large $\vec{Q}$ region the agreement is satisfactory.

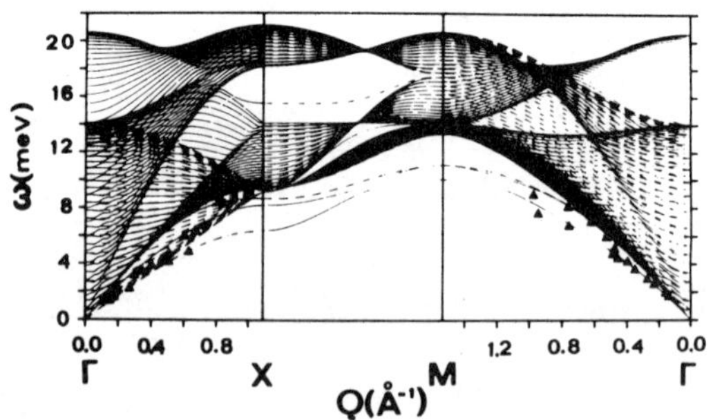

Fig. 17 Surface phonons for Ag. Results for the (2CA) model with N=65. The surface force constant β_1^S is taken to be $0.48\,\beta_1^B$.

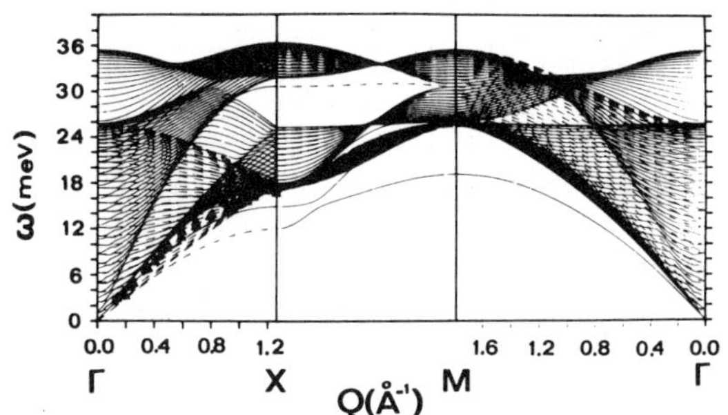

Fig. 18 Surface phonons for Ni. Results for the (4CA) model with N=65. The experimental data (triangles) are from Ref. 5.

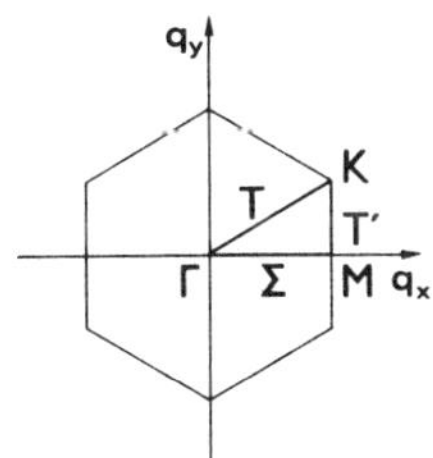

Fig. 19 2DBZ for the (111) surface. Symmetry points and lines are indicated.

Recently Rahman et al.[6] have explained the disagreement in the large Q region in terms of the geometrical inward relaxation of the surface which in turn produces an enhancement of the interplanar force constant between the surface and the next to the last plane.

We now turn to discuss the properties of the (111) surface of an FCC crystal. In this case for a general Q point inside the 2DBZ, depicted in Fig. 19, there are no particular symmetries to be invoked so that we cannot block-diagonalize Eq. (4.1). Only for Q along symmetry directions as the Σ (Q in the $[11\bar{2}]$ direction) and the T (Q in the $[1\bar{1}0]$ direction) line, it is possible to decouple the dynamical matrix. The details of this analysis which is very cumberstome are not presented here. We want only to mention that in the Σ direction one can divide the displacements in sagittal and shear horizontal modes. We remind that for the (111) surface are available very accurate measurements performed with the neutral atom scattering[5] for Q along the $[11\bar{2}]$ and $[1\bar{1}0]$ directions. As already mentioned, since the cross section data depend on the atom-surface potential, we pospone the presentation of our results to Section 7 after a detailed discussion of this problem.

5. ATOM SURFACE SCATTERING CROSS SECTION

Explicit formulae for the inelastic one-phonon scattering cross section in the DWBA have been derived some years ago by Cabrera et al[3]. In this section we present the essential points of the derivation. For a two body potential $V(\vec{r}-\vec{r}_1-\vec{u}_1)$ between the gas atom at $\vec{r}$ and the atoms in the solid at $\vec{r}_1+\vec{u}_1$ the interaction is:

$$\sum_1 \vec{u}_1 \cdot \nabla_{\vec{r}_1} V(\vec{r}-\vec{r}_1) \tag{5.1}$$

to first order in the phonon induced displacements $\vec{u}_1$. For the repulsive part of the interaction, only the atoms in the surface layer contribute appreciably to the sum in Eq. (5.1). The l-dependence of the $\vec{u}_1$, for a phonon of lateral momentum $\vec{Q}$, is then given by $\exp(i\vec{Q}\cdot\vec{R}_1)$ (capital letters denote vectors in the plane of the surface). The matrix element of the interaction in Eq. (5.1) must be taken between eigenstates of the static potential:

$$\sum_1 V(\vec{r}-\vec{r}_1) \tag{5.2}$$

Only the zero Fourier component of the potential $V_0(z)$ is appreciable for smooth metallic surfaces such as the (111) surface of the noble and transition metals. The eigenfunctions of $V_0(z)$ are then of the form $\chi_{k_z}(z)\exp(i\vec{K}\cdot\vec{R})$ where, for an atom of energy E and mass m:

$$k_z^2 + K^2 = 2mE \tag{5.3}$$

The matrix elements for transitions from $(\vec{K}_i, k_{iz})$ to $(\vec{K}_f, k_{fz})$ contain the integral:

$$\sum_1 e^{i\vec{Q}\cdot\vec{R}_1} \nabla_{\vec{r}_1} \int d^2R \; e^{i(\vec{K}_f-\vec{K}_i)\cdot\vec{R}} V(\vec{r}-\vec{r}_1) \tag{5.4}$$

which vanishes unless $\vec{Q}=\vec{K}_f-\vec{K}_i+\vec{G}$ where $\vec{G}$ is a reciprocal lattice vector. Ultimately then, we need to compute:

$$N \int dz \chi_{iz}^{*}(z)\chi_{fz}(z) \int d^2R \; e^{-i(\vec{Q}+\vec{G})\cdot\vec{R}} \nabla_{\vec{r}_1} V(\vec{r}-\vec{r}_1) \tag{5.5}$$

where N is the number of atoms for unit surface area and χ_{iz}, χ_{fz} are eigenfunctions of the laterally averaged potential:

$$V_0(z) = N \int d^2R \; V(\vec{r}) \tag{5.6}$$

The Fourier components of the potential then determine both the dynamic interaction and the static potential. $\nabla_{\vec{r}}$ can be replaced by $(i(\vec{Q}+\vec{G}), \partial/\partial z)$. By taking the square modulus of Eq. (5.5) and by summing all processes in the volume $d^2K_f dk_{fz}=(2mE_f)^{\frac{1}{2}} m \, dE_f$ and dividing by $dE_f d\Omega_f$ we obtain the differential reflection coefficient or cross section per unit surface area. In this manner we obtain:

$$\frac{d^2R}{dE_f d\Omega_f} = \frac{2}{MN} \sum_{\vec{Q},j} \frac{(2mE_f)^{\frac{1}{2}}k_{iz}}{\omega_j(\vec{Q})} n\{\omega_j(\vec{Q})\} \, e^{-2W} \; x$$

$$|\vec{e}_{(N-1)/2}(\vec{Q},j)\cdot\langle\chi_{iz}|(i\vec{Q},\partial/\partial z)V(\vec{Q},z)|\chi_{fz}\rangle|^2$$

$$\delta(\vec{K}_f-\vec{K}_i-\vec{Q})\delta\{(E_f-E_i-\omega_j(\vec{Q})\} \tag{5.7}$$

Here we have introduced the phonon frequencies $\omega_j(\vec{Q})$ and the polarization vectors $e_n(\vec{Q},j)$. $n(\omega)$ is the Bose factor, phonon creation (energy loss) corresponds to $\omega_j(\vec{Q}) < 0$; phonon annihilation (energy gain) corresponds to $\omega_j(\vec{Q}) > 0$. $V(\vec{Q},z)$ is the 2-dimensional Fourier transform of the potential and $\vec{Q}$ now is considered in the extended zone scheme. The Debye Waller factor 2W is given by:

$$2W = \sum_{\vec{Q},j} \{|k_{iz}|+k_{fz}\}^2 \lesssim |e_{(N-1)/2z}(\vec{Q},j)|^2 > \qquad (5.8)$$

6. ATOM-SURFACE POTENTIAL

The atom surface potential is generally divided as follows:

$$V(\vec{r}) = V_R(\vec{r}) + V_A(z) \qquad (6.1)$$

The short range repulsive part V_R arises from the overlap of the electronic cloud of the atom and the surface electrons while the long range attractive polarization potential V_A has the asymptotic form:

$$V_A(z) = -\frac{c}{z^3} \qquad (6.2)$$

To determine the repulsive part, which gives the major contribution to the matrix elements in Eq. (5.7) we use the Esbjerg-Norskov approximation[24]:

$$V_R(\vec{r}) = \alpha \, \overline{\rho(\vec{r})} \qquad (6.3)$$

where α can be determined from electron-He atom scattering data. $\rho(\vec{r})$ is the surface charge density averaged with the atom electrostatic potential. We have found that the averaging effects of the He atom on the surface charge density at the typical incident energies used in atom-scattering experiments (E_i=8-30 meV) are a minor correction so that in Eq.(6.3) we will take directly the charge density $\rho(\vec{r})$.

We than evaluate $\rho(\vec{r})$ with a superposition of atomic charges. The reason for using a simple superposition of atomic charges is that the impinging He atom experiences the charge density at large distances from the surface, where the surface states and the bonding effects are negligible. Eq. (6.3) is evaluated at the classical turning point for which:

$$V_R(\vec{R}, z_t) = E_i \qquad (6.4)$$

Around this region the potential can be written as:

$$V_R(\vec{R}, z) = E_i\, e^{-\beta\{z - z_t - D(\vec{R})\}} \qquad (6.5)$$

where β is the softness parameter and $D(\vec{R})$ is the corrugation function. The corrugation is measured with respect to z_t evaluated at $\vec{R}=0$ so that $D(0)=0$. To determine β and D we have evaluated $\rho(\vec{r})$ with a thirty atom cluster. Since in the evaluation of the cross section we need also the gradient of the potential we have

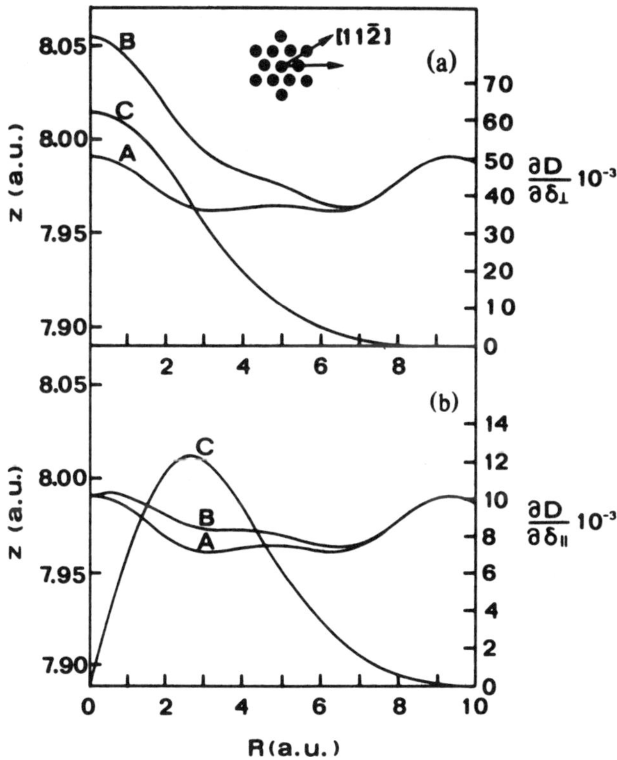

Fig. 20 Corrugation profiles and their derivatives $\partial D(\vec{R})/\partial\delta$. Curve A: $D(\vec{R})$. Curve B: $D(\vec{R})$ obtained by mouving an atom: in (a) along the z direction; in (b) along the $\vec{R}$ direction. Curve C represents the numerical derivative.

calculated numerically the normal and in-plane derivatives. In Fig. 20 are plotted the results for Ag(111) with $\vec{R}$ along the $[11\bar{2}]$ surface direction. As one can see the normal derivative of the potential has a Gaussian type behaviour while the lateral derivative is the derivative of the same Gaussian. We have found an analytic form of the potential which reproduces very accurately the results of the full calculation. This approximate expression can be cast in the form:

$$V_R(\vec{R},z) = V \sum_1 e^{-\beta|\vec{r}-\vec{r}_1|} = V \sum_1 e^{-\beta\{(\vec{R}-\vec{R}_1)^2+(z-z_1)^2\}^{\frac{1}{2}}} \tag{6.6}$$

In the last expression we have considered only surface atoms at $z_1=0$. From our calculations for noble metals it turns out that the classical turning point z_t is very large ($z_t > 4$ A). For this reason if the coordinate $\vec{R}$ happens to be in the 1-th surface cell $|\vec{R}-\vec{R}_1| \ll z_t$. In this cell we can expand the potential as:

$$e^{-\beta\{(\vec{R}-\vec{R}_1)^2+z^2\}^{\frac{1}{2}}} \simeq e^{-\beta z}\, e^{-\frac{\beta}{2z_t}(\vec{R}-\vec{R}_1)^2} \tag{6.7}$$

This expansion has to be applied to all the nearest neighbors. For all the other neighbors the contribution of the two-body potential to the total potential is negligible so that this approximation can be performed without altering the accuracy of the total potential. V_R can thus be written as:

$$V_R(\vec{R},z) = V\, e^{-\beta z} \sum_1 e^{-\frac{\beta}{2z_t}(\vec{R}-\vec{R}_1)^2} \tag{6.8}$$

We emphasize that also the gradient of the potential reproduce to within 1% the results of the full calculation. The 2-dimensional Fourier transform of this potential has therefore the form:

$$e^{-Q^2/Q_c^2} \tag{6.9}$$

This lateral cut-off gives a quantitative explanation of the Armand effect i.e. of the reduction in the cross section produced by the cooperative mouvement of neigboring surface atoms. The cut-off factor reduces the scattering intensities for large lateral momentum transfer. We want to point out that this cut-off is just associated with the deformation of charge produced by a single

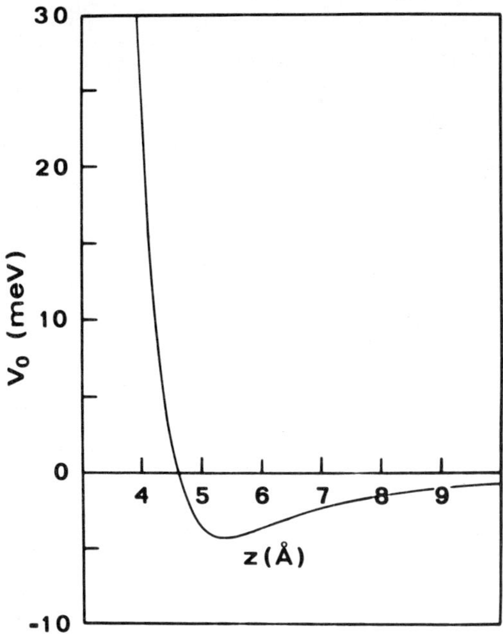

Fig. 21   The laterally averaged   potential   for   He:Ag(111).   The
softness   parameter $\beta=2.2$ A^{-1}, $k_c=1$ A^{-1} and V =296 eV.

mouving atom and exists because of the softness of the repulsive
potential.

We turn now to discuss the long range attractive dipolar
interaction. As shown by Zaremba and Kohn[22] an approximate form of
V_A is given by:

$$V_A(z) = - \frac{c_{VW}}{(z-z_{VW})^3} \qquad (6.10)$$

where for Ag c_{VW}=0.44 eV A and z_{VW}=1.38 A. The divergence of V_A
at $z=z_{VW}$ has been recently considered by Norlander and Harris[25].
These authors found that the inclusion of exchange and correlation
between the He atom and the surface regularize the divergence. In

fact according to these authors V_A has to be multiplied by the function:

$$f(x) = 1-\{2x(1+x)+1\}e^{-2x}$$

$$x = k_c(z-z_{VW}) \qquad (6.11)$$

that eliminates the divergence. k_c is a free parameter of the order of the inverse of the atomic radius. The laterally averaged total potential can then be written as:

$$V_0(z) = V\,e^{-\beta z} - \frac{c_{VW}}{(z-z_{VW})^3}\,f\{k_c(z-z_{VW})\} \qquad (6.12)$$

The value of k_c can be obtained by fitting for a given surface the physisorption well or the bound states of the He-metal potential. In Fig. 21 is drawn this potential for He/Ag(111).

7. CALCULATION FOR THE (111) SURFACE

The eigenvectors $\vec{e}\,(\vec{Q},j)$ and eigenvalues $\omega_j(\vec{Q})$ have been evaluated for a slab of 45 (111) ideal atomic planes, within a force constant parametrization of the bulk dynamics. Central and angular forces have been included up to second neighbors. In this model the surface interactions are determined by the nearest neighbor radial force

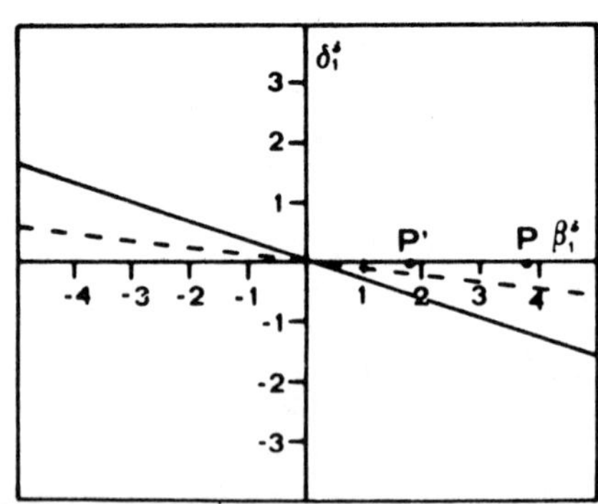

Fig. 22 Stability diagram of the Ag(111) surface as a function of the nearest neighbor surface force constants (THz²). β_1^s and δ_1^s at fixed $\alpha_1^s=\alpha_1^B$. P is for $\beta_1^s=\beta_1^B$ and P' is for $\beta_1^s=0.48\beta_1^B$.

constant β_1^S, the angular force constant δ_1^S and the tangential force constant α_1^S. By using surface force constants equal to those of the bulk, it is possible to explain the energy position and the fall off of the scattering intensity of the Rayleigh wave (RW) as a function of the momentum transfer $\vec{Q}$. In the bulk region, however, is absent any significant structure, in complete disagreement with the experiment [5] that shows the presence of a well defined peak in this region. By considering the effect due to the termination of the electron gas at the surface we can explain the origin of this new structure. The different screening produced by the electrons in the surface region can be introduced phenomenologically in our model by varying the surface force constants with respect to those of the bulk (hereafter denoted by the superscript B). By lowering the surface force constants we are able to reproduce the observed peak [26]. The variation of the surface force constants could cause the softening of some mode so that these variations cannot be performed arbitrarly. In Fig. 22 the surface stability diagram at fixed α_1^S is shown. The full line represents the values of the surface force constants for which the sagittal M mode becomes unstable. The dashed line corresponds to the instability of the M shear horizontal (SH) mode. As one can see a small lowering of δ_1^S makes the surface unstable against the M SH mode while the frequency of

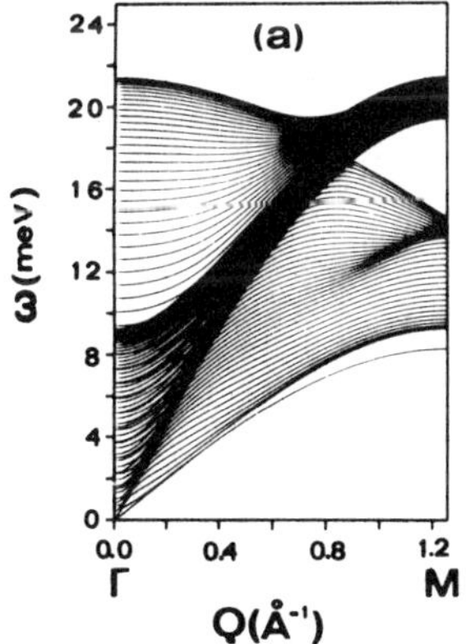
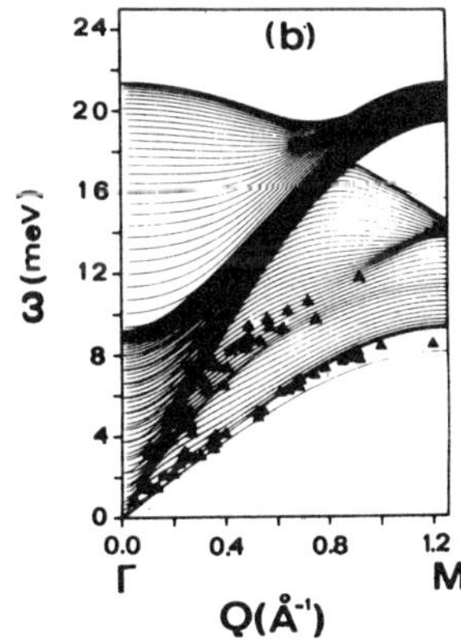

Fig. 23 a) calcualated phonon dispersion curves for Ag(111) with $\vec{Q}$ $[11\bar{2}]$. The surface force constants are taken to be equal to their bulk values. b) same calculations performed with $\beta_1^S=0.48\beta_1^B$. Experimental points (triangles) from ref. 5.

the M sagittal mode does not change appreciably. A significant lowering of the sagittal mode can be obtained by reducing the radial force constant β_1^S and keeping $\alpha_1^S = \alpha_1^B$ and $\delta_1^S = \delta_1^B$. For $\beta_1^S = 0.48\beta_1^B$ we found a well defined resonance of longitudinal character embedded in the transverse bulk continuum which explains the experimental TOF data. In Fig. 23 are drawn the sagittal phonons of Ag (111) with $\vec{Q}$ along $[11\bar{2}]$.

Fig 23a refers to the ideal calculation. As one can see there is no evidence of resonances which become evident by lowering β_1^S as is shown in Fig 23b. We want to stress that in both calculations the dispersion of the Rayleigh wave is practically the same, as is evident from Figs. 23a and 23b. The TOF spectra evaluated with Eq. (5.7) are presented in Fig. 24 for $\beta_1^S = 0.48\beta_1^B$. The calculated TOF spectra accurately reproduce the experimental line shape. Notice that also the behaviour of the scattering intensities versus $\vec{Q}$ is in agreement with the experimental data.

The calculated phonon spectrum for the Ag(111) surface with $\vec{Q}$ along $[1\bar{1}0]$ is shown in Fig. 25. The direction $[1\bar{1}0]$ is even more interesting than the $[11\bar{2}]$ because in the former case there is no decoupling of sagittal and shear horizontal (SH) modes, which instead are mixed together. This situation allows the appearance of a long wavelength pseudo Rayleigh wave (PRW) in this direction[27].

We recall that for $\beta_1^S = \beta_1^B$ this mode is present in the long wavelength limit and its velocity is very near to that of the upper transverse threshold [27]. Fig. 25b shows very clearly the existence of the PRW. The reduction of β_1^S lowers considerably this phonon branch for $Q > 0.2$ A^{-1}, so that it becomes very close to the true RW, which is only slightly affected by variations of β_1^S. Therefore the lowest set of experimental points is explained as due to the combined effect of the RW and of the PRW. Notice that although the PRW in the long wavelength limit is mainly polarized as SH (which is not detectable in atomic scattering) it becomes strongly polarized in the sagittal plane for large Q (> 0.4 A^{-1}). The upper set of experimental points coincides with the calculated dispersion curve of the longitudinal resonance (LR). Also this phonon branch is strongly affected by β_1^S and is lowered by reducing β_1^S. The LR gives the explanation for the upper set of experimental data, as in the $[112]$ direction.

An explanation of the softnening of the relevant force constant β_1^S can be qualitatively understood by considering the pseudopotential expression for the total energy in the case of a semi-infinite crystal[28]. In the Fermi Thomas limit, the total energy for bulk has the form $\Sigma \exp(-\lambda r_{ij})/r_{ij}$ where λ is the Fermi-Thomas screening length and r_{ij} is the distance between atom i and atom j. The inclusion of the surface produces new leading

436

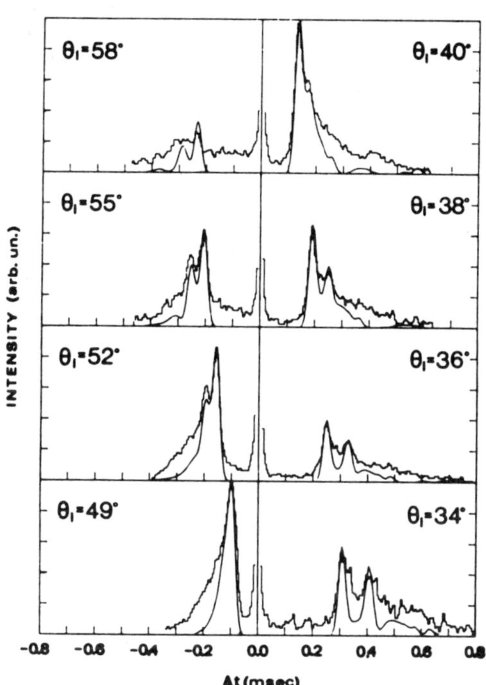

Fig. 24   The TOF spectra evaluated in the distorted Born approximation for Ag(111) and $\vec{Q}$ [11$\bar{2}$]. The light lines represent the experimental data. The incident atom energy E_i is equal to 17.5 meV and the angle between the incident and scattered beam is equal to 90^0.

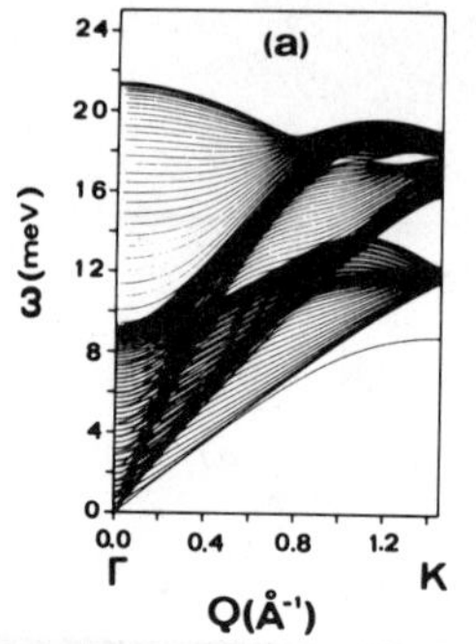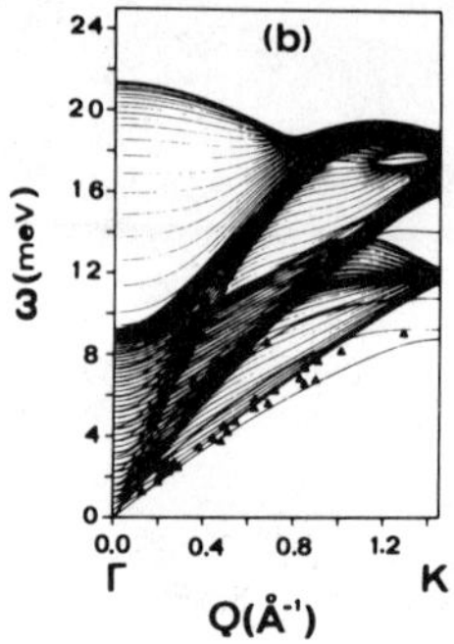

Fig. 25 same calculations as in Fig. 23 with $\vec{Q}$ along $[1\bar{1}0]$.

terms of the form:

$$- \frac{e^{-\lambda r_{ij}^+}}{r_{ij}^+} - \frac{|z_i + z_j|}{2\lambda} \frac{e^{-\lambda |z_i + z_j|}}{R_{ij}^3}$$

$$r_{ij}^+ = \{(R_i - R_j)^2 + (z_i + z_j)^2\}^{\frac{1}{2}} \tag{7.1}$$

where $\vec{R}_{ij}$ is the parallel component of $\vec{r}_{ij}$ and z_i and z_j are the normal components of the ionic position measured from the edge of the negative charge, located at $a_0/2$ in front of the outermost lattice plane. For the surface atoms $z_i = z_j = -a_0/2$ the bulk and surface terms in total energy strongly cancel, producing a reduction in the surface force constants. On the other hand for atoms in the bulk the surface term becomes negligible. Moreover, Moriarty[29] has shown that for noble metals the effect of the s-d interaction is very important in the evaluation of the band structure energy and its effect increases in going from Cu to Au. According to these arguments it is reasonable to expect that the lowering of surface force constants will be bigger in Au than in Ag, as we have found with our fitting procedure.

438

APPENDIX

In this appendix we report the analytical expressions of the forces relative to the (100) surface of an FCC crystal. In these expressions the force constants in the surface region are taken to be different from those of the bulk. The superscript "S" means that all the atoms involved in the interaction (two atoms for central and three for angular forces) are on the surface plane. The superscript "P" refers to interactions involving a surface atom and atoms that are on the first plane below the surface. The superscript "T" refers to interactions between surface atoms and atoms on the second internal plane. Finally the superscript "B" refers to bulk-like forces. u_{ni}, v_{ni} and w_{ni} are the cartesian components of the relative displacement u_{ni} between atom n and atom i.

The force F_n^S acting on the surface plane results to be:

$$
\begin{aligned}
F_{nx}^S = {}& -\{\tfrac{1}{2}(\beta_1^S+\alpha_1^S)+\tfrac{3}{2}\delta_1^P+2\delta_2^S\}\sum_{i=1,4} u_{ni} -\{\tfrac{1}{2}(\beta_1^P+\alpha_1^P)+3\delta_1^P+\tfrac{3}{2}\delta_2^P\}\sum_{i=5,6} u_{ni} \\[4pt]
& -(\alpha_1^P+3\delta_1^P)\sum_{i=9,10} u_{ni} -\beta_2^S\sum_{i=13,14} u_{ni} -\alpha_2^S\sum_{i=15,16} u_{ni} -\alpha_2^T u_{n18} \\[4pt]
& -\tfrac{1}{6}(\beta_3^T+5\alpha_3^T)\sum_{i=19,22} u_{ni} -\tfrac{1}{6}(\beta_3^P+5\alpha_3^P)\sum_{i=27,30} u_{ni} \\[4pt]
& -\tfrac{1}{3}(2\beta_3^P+\alpha_3^P)\sum_{i=35,38} u_{ni} -\tfrac{1}{2}(\beta_4^S+\alpha_4^S)\sum_{i=43,46} u_{ni} -\alpha_4^T\sum_{i=51,52} u_{ni} \\[4pt]
& -\tfrac{1}{2}(\beta_4^T+\alpha_4^T)\sum_{i=47,48} u_{ni} -\{\tfrac{1}{2}(\beta_1^S-\alpha_1^S)+\tfrac{3}{4}\delta_1^P\}\sum_{i=1,4}\frac{p_i}{q_i} v_{ni} \\[4pt]
& -\tfrac{1}{6}(\beta_3^T-\alpha_3^T)\sum_{i=19,22}\frac{p_i}{q_i} v_{ni} -\tfrac{1}{3}(\beta_3^P-\alpha_3^P)\sum_{\substack{i=27,30\\35,38}}\frac{p_i}{q_i} v_{ni} \\[4pt]
& -\tfrac{1}{2}(\beta_4^S-\alpha_4^S)\sum_{i=43,46}\frac{p_i}{q_i} v_{ni} -\{\tfrac{1}{2}(\beta_1^P-\alpha_1^P)+\tfrac{3}{2}\delta_1^P+\tfrac{1}{2}\delta_2^P\}\sum_{i=5,6}\frac{p_i}{r_i} w_{ni} \\[4pt]
& +\tfrac{9}{4}\delta_1^P\sum_{i=1,4} p_i w_{ni} +\tfrac{1}{2}\delta_2^P\sum_{i=13,14} p_i w_{ni} -\tfrac{1}{3}(\beta_3^T-\alpha_3^T)\sum_{i=19,22}\frac{p_i}{r_i} w_{ni} \\[4pt]
& -\tfrac{1}{3}(\beta_3^P-\alpha_3^P)\sum_{i=35,38}\frac{p_i}{r_i} w_{ni} -\tfrac{1}{6}(\beta_3^P-\alpha_3^P)\sum_{i=27,30}\frac{p_i}{r_i} w_{ni} \\[4pt]
& -\tfrac{1}{2}(\beta_4^T-\alpha_4^T)\sum_{i=47,48}\frac{p_i}{r_i} w_{ni}
\end{aligned}
$$

$$F_{ny}^S = -\{\tfrac{1}{2}(\beta_1^S-\alpha_1^S)+\tfrac{3}{4}\delta_1^P\}\sum_{i=1,4} u_{ni} - \tfrac{1}{6}(\beta_3^T-\alpha_3^T)\sum_{i=19,22}\frac{q_i}{p_i}u_{ni}$$

$$-\tfrac{1}{3}(\beta_3^P-\alpha_3^P)\sum_{\substack{i=27,30\\35,38}}\frac{q_i}{p_i}u_{ni}-\tfrac{1}{2}(\beta_4^S-\alpha_4^S)\sum_{i=43,46}\frac{q_i}{p_i}u_{ni}$$

$$-\{\tfrac{1}{2}(\beta_1^S+\alpha_1^S)+\tfrac{3}{2}\delta_1^P+2\delta_2^S\}\sum_{i=1,4}v_{ni}-(\alpha_1^P+3\delta_1^P)\sum_{i=5,6}v_{ni}$$

$$-\{\tfrac{1}{2}(\beta_1^P+\alpha_1^P)+3\delta_1^P+\tfrac{3}{2}\delta_2^P\}\sum_{i=9,10}v_{ni}-\alpha_2^S\sum_{i=13,14}v_{ni}-\beta_2^S\sum_{i=15,16}v_{ni}$$

$$-\alpha_2^T v_{n18}-\tfrac{1}{6}(\beta_3^T+5\alpha_3^T)\sum_{i=19,22}v_{ni}-\tfrac{1}{6}(\beta_3^P+5\alpha_3^P)\sum_{i=35,38}v_{ni}$$

$$-\tfrac{1}{3}(2\beta_3^P+\alpha_3^P)\sum_{i=27,30}v_{ni}-\tfrac{1}{2}(\beta_4^S+\alpha_4^S)\sum_{i=43,46}v_{ni}-\alpha_4^T\sum_{i=47,48}v_{ni}$$

$$-\tfrac{1}{2}(\beta_4^T+\alpha_4^T)\sum_{i=51,52}v_{ni}-\{\tfrac{1}{2}(\beta_1^P-\alpha_1^P)+\tfrac{3}{2}\delta_1^P+\tfrac{1}{2}\delta_2^P\}\sum_{i=9,10}\frac{q_i}{r_i}w_{ni}$$

$$+\tfrac{9}{4}\delta_1^P\sum_{i=1,4}q_i w_{ni}+\tfrac{1}{2}\delta_2^P\sum_{i=15,16}q_i w_{ni}-\tfrac{1}{3}(\beta_3^T-\alpha_3^T)\sum_{i=19,22}\frac{q_i}{r_i}w_{ni}$$

$$-\tfrac{1}{3}(\beta_3^P-\alpha_3^P)\sum_{i=27,30}\frac{q_i}{r_i}w_{ni}-\tfrac{1}{6}(\beta_3^P-\alpha_3^P)\sum_{i=35,38}\frac{q_i}{r_i}w_{ni}$$

$$-\tfrac{1}{2}(\beta_4^T-\alpha_4^T)\sum_{i=51,52}\frac{q_i}{r_i}w_{ni}$$

$$F_{nz}^S = -\tfrac{9}{4}\delta_1^P\sum_{i=1,4}p_i u_{ni}-\{\tfrac{1}{2}(\beta_1^P-\alpha_1^P)+\tfrac{3}{2}\delta_1^P-\tfrac{1}{2}\delta_2^P\}\sum_{i=5,6}\frac{r_i}{p_i}u_{ni}$$

$$-\tfrac{1}{2}\delta_2^P\sum_{i=13,14}p_i u_{ni}-\tfrac{1}{3}(\beta_3^T-\alpha_3^T)\sum_{i=19,22}\frac{r_i}{p_i}u_{ni}$$

$$-\tfrac{1}{3}(\beta_3^P-\alpha_3^P)\sum_{i=35,38}\frac{r_i}{p_i}u_{ni}-\tfrac{1}{6}(\beta_3^P-\alpha_3^P)\sum_{i=27,30}\frac{r_i}{p_i}u_{ni}$$

$$-\tfrac{1}{2}(\beta_4^T-\alpha_4^T)\sum_{i=47,48}\frac{r_i}{p_i}u_{ni}-\tfrac{9}{4}\delta_1^P\sum_{i=1,4}q_i v_{ni}-\tfrac{1}{2}\delta_2^P\sum_{i=15,16}q_i v_{ni}$$

$$-\{\tfrac{1}{2}(\beta_1^P-\alpha_1^P)+\tfrac{3}{2}\delta_1^P-\tfrac{1}{2}\delta_2^P\}\sum_{i=9,10}\frac{r_i}{q_i}v_{ni}-\tfrac{1}{3}(\beta_3^T-\alpha_3^T)\sum_{i=19,22}\frac{r_i}{q_i}v_{ni}$$

$$-\frac{1}{3}(\beta_3^P-\alpha_3^P)\sum_{i=27,30}\frac{r_i}{q_i}v_{ni}-\frac{1}{6}(\beta_3^P-\alpha_3^P)\sum_{i=35,38}\frac{r_i}{q_i}v_{ni}$$

$$-\frac{1}{2}(\beta_4^T-\alpha_4^T)\sum_{i=51,52}\frac{r_i}{q_i}v_{ni}-(\alpha_1^S+\frac{3}{2}\delta_1^P)\sum_{i=1,4}w_{ni}-\beta_2^P w_{n18}$$

$$-\left\{\frac{1}{2}(\beta_1^P+\alpha_1^P)+3\delta_1^P+\frac{3}{2}\delta_2^P\right\}\sum_{i=5,6\atop 9,10}w_{ni}-\alpha_2^S\sum_{i=13,16}w_{ni}$$

$$-\frac{1}{6}(\beta_3^P+5\alpha_3^P)\sum_{i=27,30\atop 35,38}w_{ni}-\frac{1}{3}(2\beta_3^T+\alpha_3^T)\sum_{i=19,22}w_{ni}$$

$$-\frac{1}{2}(\beta_4^T+\alpha_4^T)\sum_{i=47,48\atop 51,52}w_{ni}-\alpha_4^S\sum_{i=43,46}w_{ni}$$

The force F_n^P acting on the first internal plane results to be:

$$F_{nx}^P = -\left\{\frac{1}{2}(\beta_1^B+\alpha_1^B)+\frac{3}{2}(\delta_1^B+\delta_1^P)+2\delta_2^B\right\}\sum_{i=1,4}u_{ni}-(\alpha_1^P+3\delta_1^P)\sum_{i=11,12}u_{ni}$$

$$-\left\{\frac{1}{2}(\beta_1^P+\alpha_1^P)+3\delta_1^P+\frac{3}{2}\delta_2^P\right\}\sum_{i=7,8}u_{ni}-(\alpha_1^B+3\delta_1^B)\sum_{i=9,10}u_{ni}$$

$$-\left\{\frac{1}{2}(\beta_1^B+\alpha_1^B)+3\delta_1^B+\frac{3}{2}\delta_2^B+\frac{1}{2}\delta_2^P\right\}\sum_{i=5,6}u_{ni}-\beta_2^B\sum_{i=13,14}u_{ni}-\alpha_2^B u_{n18}$$

$$-\alpha_2^B\sum_{i=15,16}u_{ni}-\frac{1}{6}(\beta_3^B+5\alpha_3^B)\sum_{i=19,22\atop 27,30}u_{ni}-\frac{1}{3}(2\beta_3^B+\alpha_3^B)\sum_{i=35,38}u_{ni}$$

$$-\frac{1}{6}(\beta_3^P+5\alpha_3^P)\sum_{i=31,34}u_{ni}-\frac{1}{3}(2\beta_3^T+\alpha_3^P)\sum_{i=39,42}-\alpha_4^B\sum_{i=51,52}u_{ni}$$

$$-\frac{1}{2}(\beta_4^B+\alpha_4^B)\sum_{i=43,48}u_{ni}-\left\{\frac{1}{2}(\beta_1^B-\alpha_1^B)+\frac{3}{4}(\delta_1^B+\delta_1^P)\right\}\sum_{i=1,4}\frac{p_i}{q_i}v_{ni}$$

$$-\frac{1}{6}(\beta_3^B-\alpha_3^B)\sum_{i=19,22}\frac{p_i}{q_i}v_{ni}-\frac{1}{3}(\beta_3^B-\alpha_3^B)\sum_{i=27,30\atop 35,38}\frac{p_i}{q_i}v_{ni}$$

$$-\frac{1}{3}(\beta_3^P-\alpha_3^P)\sum_{i=31,34\atop 39,42}\frac{p_i}{q_i}v_{ni}-\frac{1}{2}(\beta_4^B-\alpha_4^B)\sum_{i=43,46}\frac{p_i}{q_i}v_{ni}$$

$$-\left\{\frac{1}{2}(\beta_1^P-\alpha_1^P)+\frac{3}{2}\delta_1^P-\frac{1}{2}\delta_2^P\right\}\sum_{i=7,8}\frac{p_i}{r_i}w_{ni}+\frac{1}{2}(\delta_2^B-\delta_2^P)\sum_{i=13,14}p_iw_{ni}$$

$$-\left\{\frac{1}{2}(\beta_1^B-\alpha_1^B)+\frac{3}{2}\delta_1^B+\frac{1}{2}(\delta_2^B-\delta_2^P)\right\}\sum_{i=5,6}\frac{p_i}{r_i}w_{ni}$$

$$+\frac{9}{4}(\delta_1^B-\delta_1^P)\sum_{i=1,4}p_iw_{ni}-\frac{1}{3}(\beta_3^B-\alpha_3^B)\sum_{\substack{i=19,22\\35,38}}\frac{p_i}{r_i}w_{ni}$$

$$-\frac{1}{6}(\beta_3^B-\alpha_3^B)\sum_{i=27,30}\frac{p_i}{r_i}w_{ni}-\frac{1}{3}(\beta_3^P-\alpha_3^P)\sum_{i=39,42}\frac{p_i}{r_i}w_{ni}$$

$$-\frac{1}{6}(\beta_3^P-\alpha_3^P)\sum_{i=31,34}\frac{p_i}{r_i}w_{ni}-\frac{1}{2}(\beta_4^B-\alpha_4^B)\sum_{i=47,48}\frac{p_i}{r_i}w_{ni}$$

$$F_{ny}^P=-\left\{\frac{1}{2}(\beta_1^B-\alpha_1^B)+\frac{3}{4}(\delta_1^B+\delta_1^P)\right\}\sum_{i=1,4}\frac{q_i}{p_i}u_{ni}-\frac{1}{6}(\beta_3^B-\alpha_3^B)\sum_{i=19,22}\frac{q_i}{p_i}u_{ni}$$

$$-\frac{1}{3}(\beta_3^B-\alpha_3^B)\sum_{\substack{i=27,30\\35,38}}\frac{q_i}{p_i}u_{ni}-\frac{1}{3}(\beta_3^P-\alpha_3^P)\sum_{\substack{i=31,34\\39,42}}\frac{q_i}{p_i}u_{ni}$$

$$-\frac{1}{2}(\beta_4^B-\alpha_4^B)\sum_{i=43,46}\frac{q_i}{p_i}u_{ni}-(\alpha_1^P+3\delta_1^P)\sum_{i=7,8}v_{ni}$$

$$-\left\{\frac{1}{2}(\beta_1^P+\alpha_1^P)+\frac{3}{2}\delta_2^P+3\delta_1^P\right\}\sum_{i=11,12}v_{ni}-(\alpha_1^B+3\delta_1^B)\sum_{i=5,6}v_{ni}$$

$$-\left\{\frac{1}{2}(\beta_1^B+\alpha_1^B)+\frac{3}{2}(\delta_1^B+\delta_1^P)+2\delta_2^B\right\}\sum_{i=1,4}v_{ni}-\alpha_2^B\sum_{i=13,14}v_{ni}$$

$$-\left\{\frac{1}{2}(\beta_1^B+\alpha_1^B)+3\delta_1^B+\frac{3}{2}\delta_2^B+\frac{1}{2}\delta_2^P\right\}\sum_{i=9,10}v_{ni}-\beta_2^B\sum_{i=15,16}v_{ni}$$

$$-\alpha_2^B v_{n18}-\frac{1}{6}(\beta_3^B+5\alpha_3^B)\sum_{\substack{i=19,22\\35,38}}v_{ni}-\frac{1}{3}(2\beta_3^B+\alpha_3^B)\sum_{i=27,30}v_{ni}$$

$$-\frac{1}{6}(\beta_3^P+5\alpha_3^P)\sum_{i=39,42}v_{ni}-\frac{1}{3}(2\beta_3^P+\alpha_3^P)\sum_{i=31,34}v_{ni}$$

$$-\frac{1}{2}(\beta_4^B+\alpha_4^B)\sum_{\substack{i=43,46\\51,52}}v_{ni}-\alpha_4^B\sum_{i=47,48}v_{ni}+\frac{9}{4}(\delta_1^B-\delta_1^P)\sum_{i=1,4}q_iw_{ni}$$

$$-\left\{\tfrac{1}{2}(\beta_1^P-\alpha_1^P)+\tfrac{3}{2}\delta_1^P-\tfrac{1}{2}\delta_2^P\right\}\sum_{i=11,12}\frac{q_i}{r_i}\,w_{ni}+\tfrac{1}{2}(\delta_2^B-\delta_2^P)\sum_{i=15,16}q_i w_{ni}$$

$$-\left\{\tfrac{1}{2}(\beta_1^B-\alpha_1^B)+\tfrac{3}{2}\delta_1^B+\tfrac{1}{2}(\delta_2^B-\delta_2^P)\right\}\sum_{i=9,10}\frac{q_i}{r_i}\,w_{ni}$$

$$-\tfrac{1}{3}(\beta_3^B-\alpha_3^B)\sum_{\substack{i=19,22\\27,30}}\frac{q_i}{r_i}\,w_{ni}-\tfrac{1}{6}(\beta_3^B-\alpha_3^B)\sum_{i=35,38}\frac{q_i}{r_i}\,w_{ni}$$

$$-\tfrac{1}{3}(\beta_3^P-\alpha_3^P)\sum_{i=31,34}\frac{q_i}{r_i}\,w_{ni}-\tfrac{1}{6}(\beta_3^P-\alpha_3^P)\sum_{i=39,42}\frac{q_i}{r_i}\,w_{ni}$$

$$-\tfrac{1}{2}(\beta_4^B-\alpha_4^B)\sum_{i=51,52}\frac{q_i}{r_i}\,w_{ni}$$

$$F_{nz}^P = -\tfrac{9}{4}(\delta_1^B-\delta_1^P)\sum_{i=1,4}p_i u_{ni}-\left\{\tfrac{1}{2}(\beta_1^P-\alpha_1^P)+\tfrac{3}{2}\delta_1^P+\tfrac{1}{2}\delta_2^P\right\}\sum_{i=7,8}\frac{r_i}{p_i}\,u_{ni}$$

$$-\left\{\tfrac{1}{2}(\beta_1^B-\alpha_1^B)+\tfrac{3}{2}\delta_1^B-\tfrac{1}{2}(\delta_2^B-\delta_2^P)\right\}\sum_{i=5,6}\frac{r_i}{p_i}\,u_{ni}$$

$$-\tfrac{1}{2}(\delta_2^B-\delta_2^P)\sum_{i=13,14}p_i u_{ni}-\tfrac{1}{3}(\beta_3^B-\alpha_3^B)\sum_{\substack{i=19,22\\35,38}}\frac{r_i}{p_i}\,u_{ni}$$

$$-\tfrac{1}{6}(\beta_3^B-\alpha_3^B)\sum_{i=27,30}\frac{r_i}{p_i}\,u_{ni}-\tfrac{1}{3}(\beta_3^P-\alpha_3^P)\sum_{i=39,42}\frac{r_i}{p_i}\,u_{ni}$$

$$-\tfrac{1}{6}(\beta_3^P-\alpha_3^P)\sum_{i=31,34}\frac{r_i}{p_i}\,u_{ni}-\tfrac{1}{2}(\beta_4^B-\alpha_4^B)\sum_{i=47,48}\frac{r_i}{p_i}\,u_{ni}$$

$$-\tfrac{9}{4}(\delta_1^B-\delta_1^P)\sum_{i=1,4}q_i v_{ni}-\left\{\tfrac{1}{2}(\beta_1^P-\alpha_1^P)+\tfrac{3}{2}\delta_1^P+\tfrac{1}{2}\delta_2^P\right\}\sum_{i=11,12}\frac{r_i}{q_i}\,v_{ni}$$

$$-\left\{\tfrac{1}{2}(\beta_1^B-\alpha_1^B)+\tfrac{3}{2}\delta_1^P-\tfrac{1}{2}(\delta_2^B-\delta_2^P)\right\}\sum_{i=9,10}\frac{r_i}{q_i}\,v_{ni}$$

$$-\tfrac{1}{2}(\delta_2^B-\delta_2^P)\sum_{i=15,16}q_i v_{ni}-\tfrac{1}{3}(\beta_3^B-\alpha_3^B)\sum_{\substack{i=19,22\\27,30}}\frac{r_i}{q_i}\,v_{ni}$$

$$-\tfrac{1}{6}(\beta_3^B-\alpha_3^B)\sum_{i=35,38}\frac{r_i}{q_i}\,v_{ni}-\tfrac{1}{3}(\beta_3^P-\alpha_3^P)\sum_{i=31,34}\frac{r_i}{q_i}\,v_{ni}$$

$$-\frac{1}{6}(\beta_3^P-\alpha_3^P)\sum_{i=39,42}\frac{r_i}{q_i}v_{ni}-\frac{1}{2}(\beta_4^B-\alpha_4^B)\sum_{i=51,52}\frac{r_i}{q_i}v_{ni}$$

$$-\{\alpha_1^B+\frac{3}{2}(\delta_1^B+\delta_1^P)\}\sum_{i=1,4}w_{ni}-\{\frac{1}{2}(\beta_1^P+\alpha_1^P)+3\delta_1^P+\frac{3}{2}\delta_2^P\}\sum_{\substack{i=7,8\\11,12}}w_{ni}$$

$$-\{\frac{1}{2}(\beta_1^B+\alpha_1^B)+3\delta_1^B+\frac{3}{2}\delta_2^B+\frac{1}{2}\delta_2^P\}\sum_{\substack{i=5,6\\9,10}}w_{ni}-\alpha_2^B\sum_{i=13,16}w_{ni}$$

$$-\beta_2^B w_{n18}-\frac{1}{6}(\beta_3^B+5\alpha_3^B)\sum_{\substack{i=29,30\\35,38}}w_{ni}-\frac{1}{3}(2\beta_3^B+\alpha_3^B)\sum_{i=19,22}w_{ni}$$

$$-\frac{1}{6}(\beta_3^P+5\alpha_3^P)\sum_{\substack{i=31,34\\39,42}}w_{ni}-\frac{1}{2}(\beta_4^B+\alpha_4^B)\sum_{\substack{i=47,48\\51,52}}w_{ni}-\alpha_4^B\sum_{i=43,46}w_{ni}$$

The force F_n^T acting on the second internal plane results to be:

$$F_{nx}^T=-\{\frac{1}{2}(\beta_1^B+\alpha_1^B)+3\delta_1^B+2\delta_2^B\}\sum_{i=1,6}u_{ni}-(\alpha_1^B+3\delta_1^B)\sum_{i=9,12}u_{ni}$$

$$-\{\frac{1}{2}(\beta_1^B+\alpha_1^B)+3\delta_1^B+\frac{3}{2}\delta_2^B+\frac{1}{2}\delta_2^P\}\sum_{i=7,8}u_{ni}-\beta_2^B\sum_{i=13,14}u_{ni}$$

$$-\alpha_2^B\sum_{i=15,16}u_{ni}-\alpha_2^T u_{n17}-\alpha_2^B u_{n18}-\frac{1}{6}(\beta_3^B+5\alpha_3^B)\sum_{\substack{i=19,22\\27,34}}u_{ni}$$

$$-\frac{1}{3}(2\beta_3^B+\alpha_3^B)\sum_{i=35,42}u_{ni}-\frac{1}{6}(\beta_3^T+5\alpha_3^T)\sum_{i=23,26}u_{ni}$$

$$-\frac{1}{2}(\beta_4^B+\alpha_4^B)\sum_{i=43,46}u_{ni}-\alpha_4^B\sum_{i=51,52}u_{ni}-\alpha_4^T\sum_{i=53,54}u_{ni}$$

$$-\frac{1}{2}(\beta_4^T+\alpha_4^T)\sum_{i=49,50}u_{ni}-\{\frac{1}{2}(\beta_1^B-\alpha_1^B)+\frac{3}{2}\delta_1^B\}\sum_{i=1,4}\frac{p_i}{q_i}v_{ni}$$

$$-\frac{1}{6}(\beta_3^B-\alpha_3^B)\sum_{i=19,22}\frac{p_i}{q_i}v_{ni}-\frac{1}{3}(\beta_3^B-\alpha_3^B)\sum_{i=27,42}\frac{p_i}{q_i}v_{ni}$$

$$-\frac{1}{6}(\beta_3^T-\alpha_3^T)\sum_{i=23,26}\frac{p_i}{q_i}v_{ni}-\frac{1}{2}(\beta_4^B-\alpha_4^B)\sum_{i=43,46}\frac{p_i}{q_i}v_{ni}$$

$$-\left\{\tfrac{1}{2}(\beta_1^B-\alpha_1^B)+\tfrac{3}{2}\delta_1^B-\tfrac{1}{2}(\delta_2^B-\delta_2^P)\right\}\sum_{i=7,8}\frac{p_i}{r_i}\,w_{ni}$$

$$-\left\{\tfrac{1}{2}(\beta_1^B-\alpha_1^B)+\tfrac{3}{2}\delta_1^B\right\}\sum_{i=5,6}\frac{p_i}{r_i}\,w_{ni}-\tfrac{1}{3}(\beta_3^B-\alpha_3^B)\sum_{\substack{i=19,22\\35,42}}\frac{p_i}{r_i}\,w_{ni}$$

$$-\tfrac{1}{6}(\beta_3^B-\alpha_3^B)\sum_{i=27,34}\frac{p_i}{r_i}\,w_{ni}-\tfrac{1}{3}(\beta_3^T-\alpha_3^T)\sum_{i=23,26}\frac{p_i}{r_i}\,w_{ni}$$

$$-\tfrac{1}{2}(\beta_4^B-\alpha_4^B)\sum_{i=47,48}\frac{p_i}{r_i}\,w_{ni}-\tfrac{1}{2}(\beta_4^T-\alpha_4^T)\sum_{i=49,50}\frac{p_i}{r_i}\,w_{ni}$$

$$F_{ny}^T = -\left\{\tfrac{1}{2}(\beta_1^B-\alpha_1^B)+\tfrac{3}{2}\delta_1^B\right\}\sum_{i=1,4}\frac{q_i}{p_i}\,u_{ni}-\tfrac{1}{6}(\beta_3^B-\alpha_3^B)\sum_{i=19,22}\frac{q_i}{p_i}\,u_{ni}$$

$$-\tfrac{1}{3}(\beta_3^B-\alpha_3^B)\sum_{i=27,42}\frac{q_i}{p_i}\,u_{ni}-\tfrac{1}{6}(\beta_3^T-\alpha_3^T)\sum_{i=23,26}\frac{q_i}{p_i}\,u_{ni}$$

$$-\tfrac{1}{2}(\beta_4^B-\alpha_4^B)\sum_{i=43,46}\frac{q_i}{p_i}\,u_{ni}-\left\{\tfrac{1}{2}(\beta_1^B+\alpha_1^B)+3\delta_1^B+2\delta_2^B\right\}\sum_{\substack{i=1,4\\9,10}}v_{ni}$$

$$-(\alpha_1^B+3\delta_1^B)\sum_{i=5,8}v_{ni}-\left\{\tfrac{1}{2}(\beta_1^B+\alpha_1^B)+3\delta_1^B+\tfrac{3}{2}\delta_2^B+\tfrac{1}{2}\delta_2^P\right\}\sum_{i=11,12}v_{ni}$$

$$-\alpha_2^B\sum_{\substack{i=13,14\\18}}v_{ni}-\beta_2^B\sum_{i=15,16}v_{ni}-\alpha_2^T v_{n17}-\tfrac{1}{6}(\beta_3^B+5\alpha_3^B)\sum_{\substack{i=19,22\\35,42}}v_{ni}$$

$$-\tfrac{1}{3}(2\beta_3^B+\alpha_3^B)\sum_{i=27,34}v_{ni}-\tfrac{1}{6}(\beta_3^T+5\alpha_3^T)\sum_{i=23,26}v_{ni}$$

$$-\tfrac{1}{2}(\beta_4^B+\alpha_4^B)\sum_{i=43,46}v_{ni}-\alpha_4^B\sum_{i=47,48}v_{ni}-\tfrac{1}{2}(\beta_4^T+\alpha_4^T)\sum_{i=53,54}v_{ni}$$

$$-\alpha_4^T\sum_{i=49,50}v_{ni}-\left\{\tfrac{1}{2}(\beta_1^B-\alpha_1^B)+\tfrac{3}{2}\delta_1^B-\tfrac{1}{2}(\delta_2^B-\delta_2^P)\right\}\sum_{i=11,12}\frac{q_i}{r_i}\,w_{ni}$$

$$-\left\{\tfrac{1}{2}(\beta_1^B-\alpha_1^B)+\tfrac{3}{2}\delta_1^B\right\}\sum_{i=9,10}\frac{q_i}{r_i}\,w_{ni}-\tfrac{1}{3}(\beta_3^B-\alpha_3^B)\sum_{\substack{i=19,22\\27,34}}\frac{q_i}{r_i}\,w_{ni}$$

$$-\tfrac{1}{6}(\beta_3^B-\alpha_3^B)\sum_{i=35,42}\frac{q_i}{r_i}\,w_{ni}-\tfrac{1}{3}(\beta_3^T-\alpha_3^T)\sum_{i=23,26}\frac{q_i}{r_i}\,w_{ni}$$

$$- \frac{1}{2}(\beta_4^B - \alpha_4^B) \sum_{i=51,52} \frac{q_i}{r_i} w_{ni} - \frac{1}{2}(\beta_4^T - \alpha_4^T) \sum_{i=53,54} \frac{q_i}{r_i} w_{ni}$$

$$
\begin{aligned}
F_{nz}^T = & -\left\{ \frac{1}{2}(\beta_1^B - \alpha_1^B) + \frac{3}{2}\delta_1^B + \frac{1}{2}(\delta_2^B - \delta_2^P) \right\} \sum_{i=7,8} \frac{r_i}{p_i} u_{ni} \\
& -\left\{ \frac{1}{2}(\beta_1^B - \alpha_1^B) + \frac{3}{2}\delta_1^B \right\} \sum_{i=5,6} \frac{r_i}{p_i} u_{ni} - \frac{1}{3}(\beta_3^B - \alpha_3^B) \sum_{\substack{i=19,22 \\ 35,42}} \frac{r_i}{p_i} u_{ni} \\
& - \frac{1}{6}(\beta_3^B - \alpha_3^B) \sum_{i=27,34} \frac{r_i}{p_i} u_{ni} - \frac{1}{3}(\beta_3^T - \alpha_3^T) \sum_{i=23,26} \frac{r_i}{p_i} u_{ni} \\
& - \frac{1}{2}(\beta_4^B - \alpha_4^B) \sum_{i=47,48} \frac{r_i}{p_i} u_{ni} - \frac{1}{2}(\beta_4^T - \alpha_4^T) \sum_{i=49,50} \frac{r_i}{p_i} u_{ni} \\
& -\left\{ \frac{1}{2}(\beta_1^B - \alpha_1^B) + \frac{3}{2}\delta_1^B + \frac{1}{2}(\delta_2^B - \delta_2^P) \right\} \sum_{i=11,12} \frac{r_i}{q_i} v_{ni} \\
& -\left\{ \frac{1}{2}(\beta_1^B - \alpha_1^B) + \frac{3}{2}\delta_1^B \right\} \sum_{i=9,10} \frac{r_i}{q_i} v_{ni} - \frac{1}{3}(\beta_3^B - \alpha_3^B) \sum_{\substack{i=19,22 \\ 27,34}} \frac{r_i}{q_i} v_{ni} \\
& - \frac{1}{6}(\beta_3^B - \alpha_3^B) \sum_{i=35,42} \frac{r_i}{q_i} v_{ni} - \frac{1}{3}(\beta_3^T - \alpha_3^T) \sum_{i=23,26} \frac{r_i}{q_i} v_{ni} \\
& - \frac{1}{2}(\beta_4^B - \alpha_4^B) \sum_{i=51,52} \frac{r_i}{q_i} v_{ni} - \frac{1}{2}(\beta_4^T - \alpha_4^T) \sum_{i=53,54} \frac{r_i}{q_i} v_{ni} \\
& -(\alpha_1^B + 3\delta_1^B) \sum_{i=1,4} w_{ni} - \left\{ \frac{1}{2}(\beta_1^B + \alpha_1^B) + 3\delta_1^B + \frac{3}{2}\delta_2^B + \frac{1}{2}\delta_2^P \right\} \sum_{\substack{i=7,8 \\ 11,12}} w_{ni} \\
& -\left\{ \frac{1}{2}(\beta_1^B + \alpha_1^B) + 3\delta_1^B + 2\delta_2^B \right\} \sum_{\substack{i=5,6 \\ 9,10}} w_{ni} - \alpha_2^B \sum_{i=13,16} w_{ni} - \beta_2^B w_{n18} \\
& -\beta_2^T w_{n17} - \frac{1}{6}(\beta_3^B + 5\alpha_3^B) \sum_{i=27,42} w_{ni} - \frac{1}{3}(2\beta_3^B + \alpha_3^B) \sum_{i=19,22} w_{ni} \\
& - \frac{1}{3}(2\beta_3^T + \alpha_3^T) \sum_{i=23,26} w_{ni} - \frac{1}{2}(\beta_4^B + \alpha_4^B) \sum_{\substack{i=47,48 \\ 51,52}} w_{ni} - \alpha_4^B \sum_{i=43,46} w_{ni} \\
& - \frac{1}{2}(\beta_4^T + \alpha_4^T) \sum_{\substack{i=49,50 \\ 53,54}} w_{ni}
\end{aligned}
$$

446

The force F_n^B acting on a bulk-like plane results to be:

$$
\begin{aligned}
F_{nx}^B = &-\{\tfrac{1}{2}(\beta_1^B+\alpha_1^B)+3\delta_1^B+2\delta_2^B\}\sum_{i=1,8} u_{ni}-(\alpha_1^B+3\delta_1^B)\sum_{i=9,12} u_{ni}\\
&-\beta_2^B\sum_{i=13,14} u_{ni}-\alpha_2^B\sum_{i=15,18} u_{ni}-\tfrac{1}{6}(\beta_3^B+5\alpha_3^B)\sum_{i=19,34} u_{ni}\\
&-\tfrac{1}{3}(2\beta_3^B+\alpha_3^B)\sum_{i=35,42} u_{ni}-\tfrac{1}{2}(\beta_4^B+\alpha_4^B)\sum_{i=47,50} u_{ni}\\
&-\alpha_4^B\sum_{i=51,54} u_{ni}-\{\tfrac{1}{2}(\beta_1^B-\alpha_1^B)+\tfrac{3}{2}\delta_1^B\}\sum_{i=i,4}\frac{p_i}{q_i} v_{ni}\\
&-\tfrac{1}{6}(\beta_3^B-\alpha_3^B)\sum_{i=19,26}\frac{p_i}{q_i} v_{ni}-\tfrac{1}{3}(\beta_3^B-\alpha_3^B)\sum_{i=51,54}\frac{p_i}{q_i} v_{ni}\\
&-\tfrac{1}{2}(\beta_4^B-\alpha_4^B)\sum_{i=43,46}\frac{p_i}{q_i} v_{ni}-\{\tfrac{1}{2}(\beta_1^B-\alpha_1^B)+\tfrac{3}{2}\delta_1^B\}\sum_{i=5,8}\frac{p_i}{r_i} w_{ni}\\
&-\tfrac{1}{3}(\beta_3^B-\alpha_3^B)\sum_{i=19,42}\frac{p_i}{r_i} w_{ni}+\tfrac{1}{6}(\beta_3^B-\alpha_3^B)\sum_{i=27,34}\frac{p_i}{r_i} w_{ni}\\
&-\tfrac{1}{2}(\beta_4^B-\alpha_4^B)\sum_{i=47,50}\frac{p_i}{r_i} w_{ni}
\end{aligned}
$$

$$
\begin{aligned}
F_{ny}^B = &-\{\tfrac{1}{2}(\beta_1^B-\alpha_1^B)+\tfrac{3}{2}\delta_1^B\}\sum_{i=1,4}\frac{q_i}{p_i} u_{ni}-\tfrac{1}{6}(\beta_3^B-\alpha_3^B)\sum_{i=19,26}\frac{q_i}{p_i} u_{ni}\\
&-\tfrac{1}{3}(\beta_3^B-\alpha_3^B)\sum_{i=27,42}\frac{q_i}{p_i} u_{ni}-\tfrac{1}{2}(\beta_4^B-\alpha_4^B)\sum_{i=43,46}\frac{q_i}{p_i} u_{ni}\\
&-\{\tfrac{1}{2}(\beta_1^B+\alpha_1^B)+3\delta_1^B+2\delta_2^B\}\sum_{\substack{i=1,4\\9,12}} v_{ni}-(\alpha_1^B+3\delta_1^B)\sum_{i=5,8} v_{ni}\\
&-\alpha_2^B\sum_{\substack{i=13,14\\17,18}} v_{ni}-\beta_2^B\sum_{i=15,16} v_{ni}-\tfrac{1}{6}(\beta_3^B+5\alpha_3^B)\sum_{\substack{i=19,26\\35,42}} v_{ni}\\
&-\tfrac{1}{3}(2\beta_3^B+\alpha_3^B)\sum_{i=27,34} v_{ni}-\tfrac{1}{2}(\beta_4^B+\alpha_4^B)\sum_{\substack{i=43,46\\51,54}} v_{ni}-\alpha_4^B\sum_{i=47,50} v_{ni}\\
&-\{\tfrac{1}{2}(\beta_1^B-\alpha_1^B)+\tfrac{3}{2}\delta_1^B\}\sum_{i=9,12}\frac{q_i}{r_i} w_{ni}-\tfrac{1}{3}(\beta_3^B-\alpha_3^B)\sum_{i=19,34}\frac{q_i}{r_i} w_{ni}
\end{aligned}
$$

$$-\frac{1}{6}(\beta_3^B-\alpha_3^B)\sum_{i=35,42}\frac{q_i}{r_i}w_{ni}-\frac{1}{2}(\beta_4^B-\alpha_4^B)\sum_{i=51,54}\frac{q_i}{r_i}w_{ni}$$

$$\begin{aligned}
F_{nz}^B = &-\{\frac{1}{2}(\beta_1^B-\alpha_1^B)+\frac{3}{2}\delta_1^B\}\sum_{i=5,8}\frac{r_i}{p_i}u_{ni}-\frac{1}{3}(\beta_3^B-\alpha_3^B)\sum_{\substack{i=19,26\\35,42}}\frac{r_i}{p_i}u_{ni}\\
&-\frac{1}{6}(\beta_3^B-\alpha_3^B)\sum_{i=27,34}\frac{r_i}{p_i}u_{ni}-\frac{1}{2}(\beta_4^B-\alpha_4^B)\sum_{i=47,50}\frac{r_i}{p_i}u_{ni}\\
&-\{\frac{1}{2}(\beta_1^B-\alpha_1^B)+\frac{3}{2}\delta_1^B\}\sum_{i=9,12}\frac{r_i}{q_i}v_{ni}-\frac{1}{3}(\beta_3^B-\alpha_3^B)\sum_{i=19,34}\frac{r_i}{q_i}v_{ni}\\
&-\frac{1}{6}(\beta_3^B-\alpha_3^B)\sum_{i=35,42}\frac{r_i}{q_i}v_{ni}-\frac{1}{2}(\beta_4^B-\alpha_4^B)\sum_{i=51,54}\frac{r_i}{q_i}v_{ni}\\
&-(\alpha_1^B+3\delta_1^B)\sum_{i=1,4}w_{ni}-\{\frac{1}{2}(\beta_1^B+\alpha_1^B)+3\delta_1^B+2\delta_2^B\}\sum_{i=5,12}w_{ni}\\
&-\alpha_2^B\sum_{i=13,16}w_{ni}-\beta_2^B\sum_{i=17,18}w_{ni}-\frac{1}{6}(\beta_3^B+5\alpha_3^B)\sum_{i=27,42}w_{ni}\\
&-\frac{1}{3}(2\beta_3^B+\alpha_3^B)\sum_{i=19,26}w_{ni}-\frac{1}{2}(\beta_4^B+\alpha_4^B)\sum_{i=47,54}w_{ni}-\alpha_4^B\sum_{i=43,46}w_{ni}
\end{aligned}$$

In these expressions we have used the notation $p_i=R_{ix}/|R_{ix}|$, $q_i=R_{iy}/|R_{iy}|$ and $r_i=R_{iz}/|R_{iz}|$.

Acknowledgements

We like to thank V. Celli for usefull discussions. The calculations where performed with the support of Centro di Calcolo, Universita' di Modena.

References

1. V. Bortolani, F. Nizzoli and G. Santoro, Phys.Rev.Lett. __41__, 39 (1978).
2. V. Bortolani, F. Nizzoli, G. Santoro, A. Marvin and J.R.. Sandercock, Phys.Rev.Lett. __43__, 224 (1979).
3. N. Cabrera, V. Celli and R. Manson, Phys.Rev.Lett __22__, 346 (1969);
4. M. G. Brusdeylins, R.B. Doak and J.P. Toennies, Phys.Rev.Lett. __16__, 937 (1981); and to be published

5. R.B. Doak, U. Harten and J.P. Toennies, Phys. Rev. Lett.
    $\underline{51}$, 578 (1983).
6. S. Lehwald, J.M. Szeftel, H. Ibach, T.S. Rahman and D.L.
    Mills, Phys.Rev.Lett. $\underline{5}$, 518 (1983).
7. B.N. Brockhouse, "Phonons and neutron scattering" in Phonons
    and Phonon Interactions, Edited by T.A. Bak, W.A.Benjamin,
    Inc. New York, pag. 221 (1964);
8. L. Van Hove, Phys. Rev. $\underline{95}$, 249 (1954).
9. R.J. Glauber, Phys. Rev. $\underline{98}$, 1092 (1955).
10. M. Born and K. Huang, Dynamical Theory of Crystal Lattices,
    Oxford Univ. Press (1954).
11. D. Castiel, L. Dobrzynski and D. Spanjaard, Surf. Sci. $\underline{59}$,
    252 (1976).
12. D.H. Dutton, B.N. Brockhouse and A.P. Miller, Canad.J.Phys.,
    $\underline{50}$, 2915 (1972).
13. S.K. Sinha, Phys. Rev. $\underline{143}$, 422 (1966).
14. W.A. Kamitakahra and B.N. Brockhouse, Phys.Lett., $\underline{29A}$, 639
    (1969).
15. J.W. Lynn, H.G. Smith and R.N.Nicklow, Phys. Rev. $\underline{B8}$, 3493
    (1973).
16. R.J. Birgenau, J. Cordes, G. Dolling and A.D.B. Woods,
    Phys. Rev. $\underline{136}$, A1359 (1964).
17. J.F. Cornwell, Group Theory and Electronic Energy Bands in
    Solids, North-Holland Publ.Comp. Amsterdam (1969).
18. W.R. Lambert, P.L. Trevol, R.B. Doak and M.J. Cardillo,
    J.Vac.Sci.Technol. $\underline{A2}$, 1066 (1984).
19. R. Manson and V. Celli, Surf. Sci.$\underline{26}$, 695 (1971)
20. V. Bortolani, A. Franchini, N. Garcia, F. Nizzoli and G.
    Santoro, Phys. Rev. $\underline{B28}$, 7358 (1983).
21. V. Celli, in Dynamics of Gas-Surface Interaction, ed.
    G.Benedek and U.Valbusa (Springer, Berlin, 1982) p. 1
22. E. Zaremba and W. Kohn, Phys. Rev.$\underline{B15}$, 1769 (1977)
23. V. Bortolani, A. Franchini, F. Nizzoli and G. Santoro , in
    Dynamics of Gas-Surface Interaction, ed. G.Benedek and
    U.Valbusa (Springer, Berlin, 1982) p. 196
24. N. Esbjerg and J.K. Norskov, Phys. Rev. Lett. $\underline{45}$, 807
    (1980)
25. P.Norlander and J.Harris, J. Phys. $\underline{C}$: Solid St. Phys. $\underline{17}$,
    1141 (1984)
26. V. Bortolani, A. Franchini, F. Nizzoli, G. Santoro, Phys.
    Rev. Lett. $\underline{52}$, 429 (1984).
27. G.W. Farnell, in "Physical Acoustics", vol. VI, ed.
    W.P.Mason and R.N. Thurston (Academic Press, New York,
    1970) p. 109.
28. D.E. Beck, V. Celli,G. Lo Vecchio and A. Magnaterra, Nuovo
    Cim. $\underline{B17}$, 230 (1970)
29. A. Moriarty, Phys. Rev. $\underline{B6}$, 1239 (1972).
30. J.P.Toennies and coworkers, data presented at the Modena
    Meeting of the Surface Group of GNSM, December 1983,
    unpublished and to be published

INTERVALLEY ELECTRON-PHONON AND HOLE-PHONON INTERACTIONS

IN SEMICONDUCTORS: EXPERIMENT AND THEORY

Fred H. Pollak[*] and Orest J. Glembocki[**]

Department of Physics
Brooklyn College of The City University of New York
Brooklyn, N.Y. 11210

INTRODUCTION

Intervalley electron-phonon (EP) and hole-phonon(HP) interactions play an important role in many optical[1-5], time-dependent
optical[6], and transport[7-10] properties of semiconductors. However,
relatively little is known about them either experimentally or
theoretically. This deficiency can be traced to the difficulty in
obtaining reliable experimental values for the matrix elements of
these interactions. In multivalley indirect semiconductors the
fundamental absorption process is phonon-assisted. This process
can proceed by two mechanisms involving EP as well as HP scattering matrix elements and in such a way that they can interfere
either constructively or destructively. However, the nature of
this interference phenomena inhibits the evaluate of these matrix
elements directly by measuring only one parameter, e.g. the
absorption coefficient or luminescence. This difficulty can be
overcome by the application of a uniaxial stress along appropriate
crystallographic axes which reduces the symmetry of the valence and/
or conduction bands. We will discuss various piezospectroscopic
experiments that have been performed on Si $(\Gamma-\Delta)$,[3] GaP$(\Gamma-X)$[1,11] and
Ge $(\Gamma-L)$[4]. These studies, combined with previously measured
values of the absorption coefficient, can be used to evaluate the
EP and HP scattering matrix elements for the TO phonon in Si[2] as
well as the LA and TA phonons in GaP[2,11]. For Ge the situation is

____
[*]Also at Physics Dept., Graduate School and University Center of
  CUNY, New York, N.Y. 10036
[**]Present address: Naval Research Laboratory,
  Washington D.C. 20375

somewhat more complex but valuable information about the LA phonon assisted process can be obtained.[4] Information about the EP interactions in GaAs have been obtained from transport[9,10] and time-dependent Raman experiments.[6] Theoretical calculations for the Γ-Δ and Γ-L EP and HP matrix elements of Si[12,13] and Ge[14,15] using the "rigid-pseudoion" model will be discussed. This model represents the lattice displacements through the rigid-ion model and the potential and electronic states by local pseudopotential theory. The comparison of the theoretical values with experimental results shows, in general, good agreement.

EXPERIMENTAL RESULTS

In a multivalley indirect semiconductor the fundamental absorption process is phonon-assisted and proceeds by different scattering mechanisms involving EP as well as HP interactions. The "oscillator strength" of the absorption coefficient is proportional to the square of the sum of terms involving EP and HP matrix elements and thus constructive or destructive interferences can occur between the two scattering mechanisms.[1,3,4,16] This is a result of the delicate balance between both the magnitude and relative phase of the EP and HP matrix elements of a given process. We denote these matrix elements as S_{e-ph} and S_{h-ph}, respectively.

The sensitivity of the indirect transitions to the EP and HP matrix elements makes this area ideal for the evaluation of S_{e-ph} and S_{h-ph}. However, the nature of the interference phenomena does not allow us to determine the matrix elements directly by measuring only one parameter, e.g., the absorption coefficient or luminescence. This difficulty has been overcome by various workers, who realized that the application of uniaxial stress along appropriate crystallographic axes can be used to reduce the symmetry of the valence and/or conduction bands in such a way that would produce additional transitions for each phonon-assisted process that takes place.[1,3,4,16,17] It was recognized that this extra information can be utilized in evaluating the ratio of the EP and HP scattering matrix elements. This concept has been applied successfully to evaluate the values of S_{e-ph} and S_{h-ph} for the TO phonon-assisted transition in Si (Γ-Δ)[3,16] and the LA and TA phonon-assisted transition in GaP (Γ-X).[1,11,17] Information has also been obtained about the LA phonon-assisted transition in Ge (Γ-L) although the interpretation of this experiment is somewhat more complex.[4,18] Shown in Figs. 1 - 3 are the various phonon-assisted indirect transitions for Si, GaP and Ge, respectively.

We shall describe in some detail the ideas behind the piezo-spectroscopy of the absorption coefficient of the indirect gap.

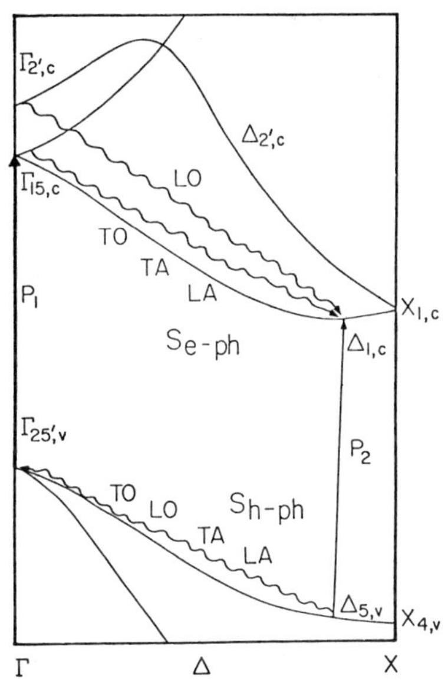

Fig. 1. Schematic representation of the band structure of Si showing the allowed Γ-Δ transitions.

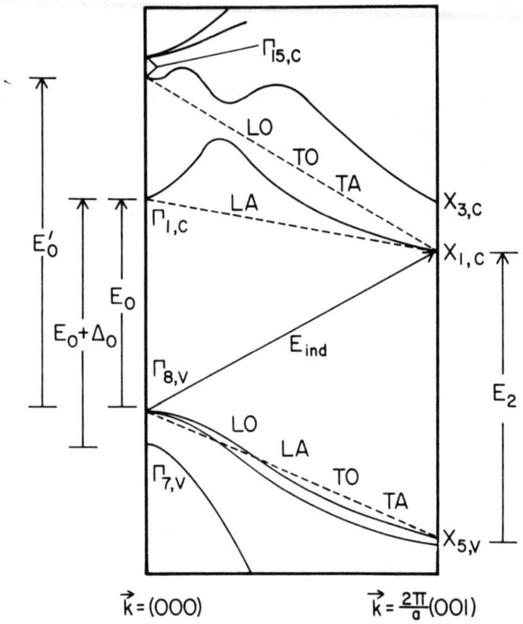

Fig. 2. Schematic representation of the band structure of GaP showing the allowed Γ-X transitions.

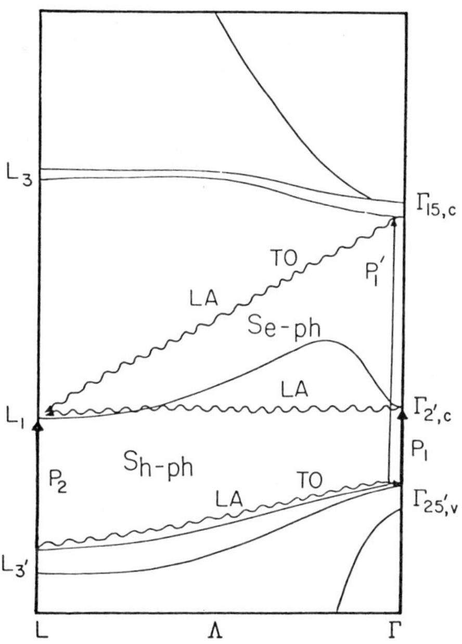

Fig. 3. Schematic representation of the band structure of Ge showing the allowed Γ-L transitions.

<u>Absorption Coefficient</u>

The absorption coefficient α_ℓ of an indirect-gap semi-conductor, resulting from an electronic transition between states differing in energy by E_g and accompanied by the creation or annihilation of a phonon of the ℓth branch is given by [1,5]

$$\alpha_\ell = Af_\ell \left\{ L \left[\hbar\omega - E_g \mp \hbar\omega_\ell(Q) \right] / \hbar\omega \right\} (n_Q + \tfrac{1}{2} \pm \tfrac{1}{2}) \tag{1}$$

where, $\hbar\omega$ is the photon energy, $\vec{Q}$ the wave vector of a phonon of frequency $\omega_\ell(Q)$, where the upper and lower signs of $\mp$ and $\pm$ refer to phonon emission and absorption, respectively, and n_Q is the phonon occupation number. The frequency independent (only over small energy ranges) term, Af_ℓ of Eq. (1) is related to the "strength" of the transition. The parameter A is a density-of-states constant involving certain materials parameters such as the index of refraction, electron and hole effective masses, etc. In Eq. (1) f_ℓ is the "oscillator strength" of the indirect transition between the valence state $\Psi_{v,k}$ and a conduction state $\Psi_{c,k'}$ and proceeding via an intermediate state $\Psi_{i,k}$ (or j,k) and is given by[1,5]

$$f_\ell = \sum_{c,v\,t} \left| \sum_i \frac{\left\langle \Psi_{v,\vec{k}} | \hat{e}\cdot\vec{p} | \Psi_{i,\vec{k}} \right\rangle \left\langle \Psi_{i,\vec{k}} | H_\ell^t(Q) | \Psi_{c,\vec{k}'} \right\rangle}{E_c(k') - E_i(k) \mp \hbar\omega_\ell(Q)} \right.$$

$$\left. + \sum_j \frac{\left\langle \Psi_{v,\vec{k}} | H_\ell^t(Q) | \Psi_{j,\vec{k}'} \right\rangle \left\langle \Psi_{j,\vec{k}'} | \hat{e}\cdot\vec{p} | \Psi_{c,\vec{k}'} \right\rangle}{E_v(k) - E_j(k') \pm \hbar\omega_\ell(Q)} \right|^2 \tag{2}$$

where $H_\ell^t(Q)$ is the effective electronic perturbation due to the creation or annihilation of a phonon of polarization t and wave vector Q, $\hat{e}$ is the unit polarization vector of the incident electronic field and $\vec{p}$ is the linear momentum of the electron.

Equation (2) contains the matrix elements of the EP and HP interactions. These are given by:

$$S_{e\text{-ph}}^\ell = \left\langle \Psi_{i,\vec{k}} | H_\ell^t(Q) | \Psi_{c,\vec{k}'} \right\rangle \tag{3a}$$

$$S_{h\text{-ph}}^\ell = \left\langle \Psi_{v,\vec{k}} | H_\ell^t(Q) | \Psi_{j,\vec{k}'} \right\rangle \tag{3b}$$

where for $S_{e\text{-ph}}$ of DZB materials $\Psi_{i,\vec{k}}$ is $\Gamma_{1,c}$ ($\Gamma_{2',c}$) or $\Gamma_{15,c}$, depending on which phonon is involved (see Fig. 1 - 3).

Excluding Ge, the valence-band intermediate state $\Psi_{j,k'}$ is $X_{5,v}$ $(\Delta_{5,v})$ for all phonon modes of these semiconductors (see Figs. 1 and 2). For Ge, the LA phonon-assisted transition proceeds via both $\Gamma_{2',c}$ and $\Gamma_{15,c}$ (see Fig. 3). This fact complicates the interpretation of the piezospectroscopy experiments on this material in relation to Si or GaP. For TO phonon scattering to L the conduction band intermediate state is $\Gamma_{15,c}$. The valence band intermediate state for both LA and TO phonon scattering is $L_{3,v}$ for Ge.

Theoretical expressions for f_ℓ have been derived for the Γ-L (Ge) and Γ-X(Δ) phonon assisted transitions in DZB indirect gap semiconductors.[1,5] Equation (2) is evaluated by using the symmetry of the phonon and the various states which are involved in the process. The selection rules for the optical transitions and intervalley phonon scattering along with the details of the calculation are presented in Refs. 1 and 5. In Tables I and II we display the equations for the Γ-X(Δ) and Γ-L processes.[16,17] Also listed in these tables are the expressions for f_ℓ for transitions from the spin-orbit split bands, $\Gamma_{7,v}^+$ $(\Gamma_{7,v})$.

The form of the line-shape factor in Eq. (1) depends upon whether excitons or free electron-hole pairs are formed. In the excitonic regions, it is given by

$$L_{ex}\left[\hbar\omega - E_g \mp \hbar\omega_\ell(Q)\right] = \left[\hbar\omega - E_g \mp \hbar\omega_\ell(Q) + \frac{R}{n^2}\right]^{\frac{1}{2}} \tag{4}$$

where R is the exciton Rydberg, and n is the principal quantum number. The coefficient A_{ex} which involves materials parameters such as effective masses, index of refraction, etc. is defined in Eq. (5) of Ref. 1.

When the photon energy is such that $\left[\hbar\omega - E_g \mp \hbar\omega_\ell(Q)\right] \gg R$, where R is the exciton Rydberg, the absorption is due to the creation of free electron-hole pairs. In this case the line-shape factor is given by

$$L\left[\hbar\omega - E_g \mp \hbar\omega_\ell(Q)\right] = \left[\hbar\omega - E_g \mp \hbar\omega_\ell(Q)\right]^2 \tag{5}$$

and the coefficient A is given by Eq. (12) of Ref. 1.

The relative intensities of the phonon assisted transitions in a given material are determined by the product f_ℓ $(n_Q + \frac{1}{2} \mp \frac{1}{2})$. The effects of the oscillator strength are clearly manifested in phonon emission processes, for which $n_Q \ll 1$. For temperatures up to room temperature this conditions is satisfied and f_ℓ becomes the principal factor governing the relative strengths. For phonon absorption processes, the phonon occupation number has a large effect on the relative intensities. In this case the role of

458

Table I. Theoretical expressions for the "oscillator strengths" of the LA, LO, TA, and TO phonon-assisted transitions in $\langle 001 \rangle$ indirect materials. The subscript ℓ refers to the phonon branch being considered. The spin orbit split band, $\Gamma_{7,v}^{+}$ is considered.

Phonon Branch (ℓ)	Oscillator Strength (f_ℓ)	
	$\Gamma_{8,v}^{+}$	$\Gamma_{7,v}^{+}$
TA, TO (Si, GaP, AlSb, AlAs)	$\frac{8}{3}\left[U_\ell^2 + (U_\ell + V_\ell)\right]^2$	$\frac{2}{3}\left[U_\ell^2 + (U_\ell + V_\ell)\right]^2$
LA (GaP, AlSb) LO (Si, AlAs)	$\frac{4}{3}\left[W_\ell^2 + 2(W_\ell + V_\ell)\right]^2$	$\frac{1}{3}\left[W_\ell^2 + 2(W_\ell + V_\ell)\right]^2$
LO (GaP, AlSb) LA (Si, AlAs)	$\frac{8}{3}\left[(U_\ell + V_\ell)\right]^2$	$\frac{2}{3}\left[(U_\ell + V_\ell)\right]^2$

$$U_\ell = \frac{\left\langle \Gamma_{8,v}^{x} \mid P_x \mid \Gamma_{15,c}^{y} \right\rangle \left\langle \Gamma_{15,c}^{y} \mid H_\ell^{y} \mid X_{1,c} \right\rangle^{a,b}}{E(X_{1,c}) - E(\Gamma_{15,c}) - (\hbar\omega)_\ell} \qquad V_\ell = \frac{\left\langle \Gamma_{8,v}^{z} \mid H_\ell^{y} \mid X_{5,v}^{x} \right\rangle \left\langle X_{5,v}^{x} \mid P_x \mid X_{1,c} \right\rangle^{a,b}}{E(\Gamma_{8,v}) - E(X_{5,v}) + (\hbar\omega)_\ell}$$

$$W_\ell = \frac{\left\langle \Gamma_{8,v}^{x} \mid P_x \mid \Gamma_{1,c} \right\rangle \left\langle \Gamma_{1,c} \mid H_\ell \mid X_{1,c} \right\rangle^{a,b}}{E(X_{1,c}) - E(\Gamma_{1,c}) - (\hbar\omega)_\ell}$$

a. These expressions apply to the zincblende materials. For Si, $\Gamma_{8,v}^{z}$ becomes $\Gamma_{25',v}^{z}$, $\Gamma_{1,c}$ is replaced by $\Gamma_{2',c}$, $X_{5,v}^{x}$ becomes $\Delta_{5,v}^{x}$ and $X_{1,c}$ is replaced by $\Delta_{1,c}$.

b. For the spin orbit split band, $\Gamma_{7,v}$ replaces $\Gamma_{8,v}$.

Table II. Theoretical expressions for the oscillator strength of the Γ–L LA and TO phonons. For the LA phonon, scattering from $\Gamma_{15,c}$ is included.

Phonon	Initial state $\Gamma_{8,v}^{+}$	$\Gamma_{7,v}^{+}$ [a]
LA	$\frac{4}{3}\left[2\left(W_{LA} + \frac{5}{3} V_{LA} - U_{LA}\right)^2 + \left(U_{LA} - \frac{2}{3} V_{LA}\right)^2\right]$	$\frac{1}{3}\left(W_{LA} + U_{LA}\right)^2 + \left(W_{LA} + \frac{8}{9} V_{LA} - \frac{1}{3} U_{LA}\right)^2 + \frac{2}{9}\left(\frac{2}{3} V_{LA} - U_{LA}\right)^2$
TO	$\frac{1}{3}\left[U_{TO}^2 + \frac{16}{3}\left(U_{TO} - \frac{5}{3} X_{TO}\right)^2 + \frac{5}{3}\left(U_{TO} + \frac{4}{3}\sqrt{2}\, V_{TO}\right)^2\right]$	$\frac{1}{6}\left[U_{TO}^2 + \frac{16}{3}\left(U_{TO} - \frac{5}{3} X_{TO}\right)^2 + \frac{5}{3}\left(U_{TO} + \frac{4}{3}\sqrt{2}\, V_{TO}\right)^2\right]$

$$W_{LA} = \frac{\left\langle \Gamma_{8,v}^{\bar{x}} \mid p_{\bar{x}} \mid \Gamma_{2',c}\right\rangle\left\langle \Gamma_{2',c} \mid H_{LA} \mid L_{1,c}\right\rangle}{E(L_{1,c}) - E(\Gamma_{2',c}) - \hbar\omega_{LA}}$$

$$U_{TO} = \frac{\left\langle \Gamma_{8,v}^{\bar{y}} \mid p_{\bar{y}} \mid \Gamma_{15,c}^{\bar{y}}\right\rangle\left\langle \Gamma_{15,c}^{\bar{y}} \mid H_{TO}^{\bar{y}} \mid L_{1,c}\right\rangle}{E(L_{1,c}) - E(\Gamma_{15,c}) - \hbar\omega_{TO}}$$

$$U_{LA} = \frac{\left\langle \Gamma_{8,v}^{\bar{y}} \mid p_{\bar{x}} \mid \Gamma_{15,c}^{\bar{z}}\right\rangle\left\langle \Gamma_{15,c}^{\bar{z}} \mid H_{LA} \mid L_{1,c}\right\rangle}{E(L_{1,c}) - E(\Gamma_{15,c}) - \hbar\omega_{LA}}$$

$$V_{TO} = \frac{\left\langle \Gamma_{8,v}^{\bar{x}} \mid H_{TO}^{\bar{y}} \mid L_{3',v}^{\bar{x}}\right\rangle\left\langle L_{3',v}^{\bar{x}} \mid p_{\bar{x}} \mid L_{1,c}\right\rangle}{E(\Gamma_{8,v}^{+}) - E(L_{3',v}) + \hbar\omega_{TO}}$$

$$V_{LA} = \frac{\left\langle \Gamma_{8,v}^{\bar{x}} \mid H_{LA} \mid L_{3',v}^{\bar{x}}\right\rangle\left\langle L_{3',v} \mid p_{\bar{x}} \mid L_{1,c}\right\rangle}{E(\Gamma_{8,v}^{+}) - E(L_{3',v}) + \hbar\omega_{LA}}$$

$$X_{TO} = \frac{\left\langle \Gamma_{8,v}^{\bar{z}} \mid H_{TO}^{\bar{x}} \mid L_{3',v}^{\bar{x}}\right\rangle\left\langle L_{3',v}^{\bar{x}} \mid p_{\bar{x}} \mid L_{1,c}\right\rangle}{E(\Gamma_{8,v}^{+}) - E(L_{3',v}) + \hbar\omega_{TO}}$$

$\bar{x} = 1/\sqrt{2}(x-y)$, $\vec{y} = 1/\sqrt{6}(x+y-2z)$, $\vec{z} = 1/\sqrt{3}(x+y+z)$

a. For the spin orbit split band, $\Gamma_{7,v}^{+}$ replaces $\Gamma_{8,v}^{+}$

the oscillator strength is not apparent. In all of our work with
relative intensities, we will include the phonon occupation number.
However, only phonon creation processes will be considered when we
study the role of f_ℓ in indirect absorption.

Stress Effects

In the previous section, expressions were presented for the
oscillator strengths of the various phonon-assisted indirect
transitions. From these equations (see Tables I and II) it is
evident that a measurement of only the absorption coefficient is
inadequate for obtaining values of the EP and HP matrix elements.
However, it is well known that uniaxial stress, applied along
appropriate axes can be used to reduce the symmetry of the
crystal and in such a way as to provide enough additional information so that an evaluation of the ratio S_{e-ph}/S_{h-ph} can be
performed.[1-5]

In this section we will consider the effect of stress along
[001] and [111] on the indirect absorption edge of DZB type
materials. From the theory, we will be able to obtain expressions
for the "oscillator strength", f_ℓ of the strained crystal for the
Γ-L transitions in Ge and the Γ-Δ(X) processes in the other DZB
materials. For the Γ-Δ(X) transitions we will also include the
stress-induced coupling between the top of the valence-band and
its spin-orbit split component. In Ge, the spin-orbit splitting
is large (300 meV) in comparison to the stress induced splittings
generally attainable. Therefore, in Ge this effect will be
neglected. The expressions for the oscillator strength in the
strained crystal will make it possible to obtain the ratio of the
EP and HP matrix elements.

The total Hamiltonian, H, for the $\Gamma_{25'}(\Gamma_{15})$ valence band can
be written as[19]

$$H_1 = H_{so} + H_1 + H_2 \tag{6}$$

where H_{so} is the spin-orbit Hamiltonian without stress, H_1
and H_2 are the orbital-strain and stress-dependent spin-orbit
components of the Hamiltonian respectively.

It has been shown that H_1 can be written as[19]

$$H_1 = -a_1(\varepsilon_{xx} + \varepsilon_{yy} + \varepsilon_{zz}) - 3b_1\left[(L_x^2 - 1/3\, L^2)\,\varepsilon_{xx} + cp\right]$$
$$- \sqrt{3}d_1\left[(L_xL_y + L_xL_y)\,\varepsilon_{xy} + cp\right] \tag{7}$$

where ε_{ij} is the ijth component of the strain tensor, $\vec{L}$ is the
angular momentum operator, a_1 is the deformation potential for

460

the hydrostatic pressure shift, while b_1 and d_1 are the uniaxial stress deformation potentials. In Eq. (7) cp denotes cyclic permutations with respect to the indices x, y, z.

The stress-dependent spin-orbit Hamiltonian, H_2 is given by[19]

$$H_2 = -a_2(\varepsilon_{xx} + \varepsilon_{yy} + \varepsilon_{zz}) (\vec{L} \cdot \vec{\sigma}) - 3b_2 [(L_x \sigma_x -$$

$$1/3 \; \vec{L} \cdot \vec{\sigma})\varepsilon_{xx} + cp] - \sqrt{3} d_2 [(L_x \sigma_y + L_y \sigma_x)\varepsilon_{xy} + cp] \qquad (8)$$

where a_2, b_2 and d_2 are deformation potentials describing the effects of a strain on the spin-orbit interaction and σ is the Pauli spin matrix.

In addition, uniaxial stress may cuase a shift and splitting of any equivalent conduction band minima. This effect can be expressed as[19]

$$\delta E = \hat{n} \cdot \{ \mathcal{E}_1 (\varepsilon_{xx} + \varepsilon_{yy} + \varepsilon_{zz})\overleftrightarrow{1}$$

$$+ \mathcal{E}_2 [\underset{\approx}{\varepsilon} - 1/3 \; (\varepsilon_{xx} + \varepsilon_{yy} + \varepsilon_{zz})\overleftrightarrow{1}] \} \cdot \hat{n} \qquad (9)$$

where $\hat{n}$ is a unit vector in the direction of the critical point in k space, 1 is the unit diadic and $\mathcal{E}_1$ and $\mathcal{E}_2$ are the hydrostatic and shear deformation potentials.

A uniaxial stress may also result in a coupling between the conduction band minima and the next highest (in energy) state at (or near) the zone edge, i.e., $\Delta_{1,c} - \Delta_{2',c}$ in Si[7] and $X_{1,c} - X_{3,c}$ in GaP.[1,11] However, this effect is expected to be small.

Stress Along [111]

⟨001⟩ Conduction Bands

For stress along [111], no intervalley splitting of the ⟨001⟩ conduction bands takes place. However, they experience a hydrostatic pressure shift which can be obtained from Eq. (9) and is given by[19]

$$\delta E_H^c = \mathcal{E}_1 (S_{11} + 2S_{12})X \qquad (10)$$

where S_{11} and S_{12} are elastic compliance constants and X is the applied stress. Figure 4 illustrates this effect schematically.

⟨111⟩ Conduction Bands

For the case of the ⟨111⟩ conduction bands, there is both a hydrostatic and shear effect. The (111) valley splits away

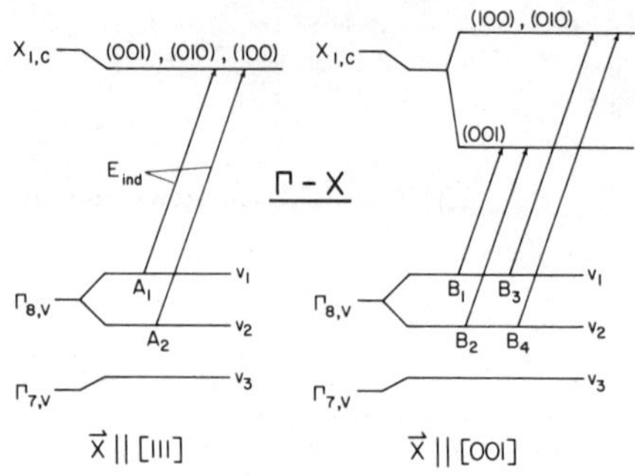

Fig. 4. Schematic representation of the effects of stress X
along [111] and [001] in the Γ-X(Δ) transitions.

from the $(\bar{1}\bar{1}1)$, $(1\bar{1}\bar{1})$ and $(\bar{1}1\bar{1})$ equivalent valleys. The energy
shift of the (111) valley is given by

$$\delta E_{1,c}^{(111)} = 1/3 \ (S_{44} \mathcal{E}_2 X) + \delta E_H^c \tag{11a}$$

and for the $(\bar{1}1\bar{1})$, $(1\bar{1}\bar{1})$, $(\bar{1}1\bar{1})$ valleys

$$\delta E_{1,c}^{(\overline{\overline{111}})} = - 1/9 \ (S_{44} \mathcal{E}_2 X) + \delta E_H^c \tag{11b}$$

where X is the applies stress, S_{44} is an elastic compliance
constant and $\mathcal{E}_2$ has been previously defined. Fig. 5 illustrates
this splitting schematically.

The Valence Band

The valence band, $\Gamma_{25',v}$ ($\Gamma_{15,v}$) is p-like (J=3/2) and six-
fold degenerate. The spin-orbit interaction splits the $|1/2,1/2\rangle$
state $\Gamma_{7,v}^+$ ($\Gamma_{7,v}$) from the $|3/2,3/2\rangle$ and $|3/2,1/2\rangle$ states,
$\Gamma_{8,v}^+$ ($\Gamma_{8,v}$). A uniaxial stress along [111] removes the
degeneracy of the $\Gamma_{8,v}^+$ ($\Gamma_{8,v}$) states and produces v_2 which is
purely $|3/2,3/2\rangle$ and v_1 which is mainly $|3/2,1/2\rangle$, with a
stress-induced admixture of $|1/2,1/2\rangle$.   However, the Kramers

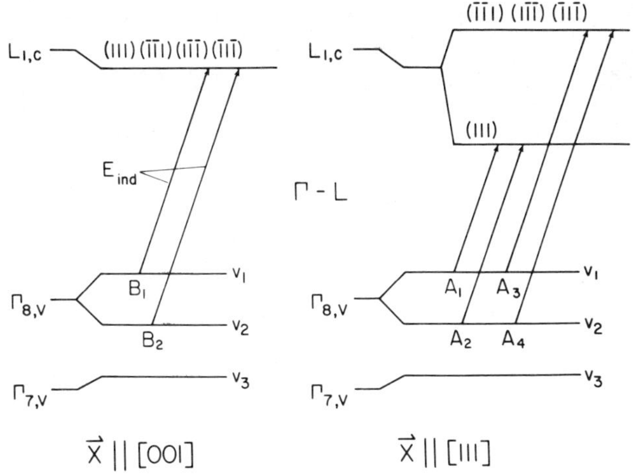

Fig. 5. Schematic representation of the effects of stress X along [001] and [111] on the Γ-L transitions.

degeneracy still remains. Figures 4 and 5 depict this splitting schematically.

With the valence band wave functions in the (J, M_J) representation, the Hamiltonian matrix of Eqs. (6) - (8) becomes:[19]

$$
\begin{array}{ccc}
\left|\frac{3}{2}, \frac{3}{2}\right\rangle_\beta & \left|\frac{3}{2}, \frac{1}{2}\right\rangle_\beta & \left|\frac{1}{2}, \frac{1}{2}\right\rangle_\beta
\end{array}
$$

$$
\begin{bmatrix}
-\delta E_H - \frac{1}{2}\delta E_\beta & 0 & 0 \\[2ex]
0 & -\delta E_H + \frac{1}{2}\delta E_\beta & \delta E'_\beta/\sqrt{2} \\[2ex]
0 & \delta E'_\beta/\sqrt{2} & -\Delta_0 - \delta E'_H
\end{bmatrix} \tag{12}
$$

where $\beta = [111]$, refers to the stress direction and

$$\delta E_H = (a_1 - 2a_2)(S_{11} = 2S_{12})X = a(S_{11} + 2S_{12})X \qquad (13a)$$

$$\delta E_H' = (a_1 + a_2)(S_{11} + 2s_{12})X = a'(S_{11} + 2S_{12})X \qquad (13b)$$

$$\delta E_{[111]} = (1/\sqrt{3})(d_1 + 2d_2)S_{44}X = (d/\sqrt{3})S_{44}X \qquad (13c)$$

$$\delta E_{[111]} = (1/\sqrt{3})(d_1 - d_2)S_{44}X = (d'/\sqrt{3})S_{44}X \qquad (13d)$$

where S_{11}, S_{12} and S_{44} are elastic compliance constants, X is the applied stress and Δ_o is the spin-orbit splitting in the absence of stress.

Diagonalizing the matrix of Eq. (12) yields the following eigenvalues for the v_1 band

$$\delta E_1 = -\ 1/2\ (\bar{\Delta}_o - 1/2\ \delta E_\beta) - \delta E_H$$
$$+\ 1/2\left[(\bar{\Delta}_o + 1/2\ \delta E_\beta)^2 + 2(\delta E_\beta')^2\right]^{\frac{1}{2}} \qquad (14)$$

while, the v_2 band energy eigenvalue is

$$\delta E_2 = 1/2\ \delta E_\beta - \delta E_H \qquad (15)$$

and for the spin-orbit split band, v_3, we find

$$\delta E_3 = -\ 1/2\ (\bar{\Delta}_o - 1/2\ \delta E_\beta) - \delta E_H$$
$$-\ 1/2\left[(\bar{\Delta}_o + 1/2\ \delta E_\beta)^2 + 2(\delta E_\beta')^2\right]^{\frac{1}{2}} \qquad (16)$$

where, $\bar{\Delta}_o = \Delta_o - 3a_2(S_{11} + S_{12})X$.

It has been previously demonstrated that the stress dependence of Δ_o is very small and thus we can take $\bar{\Delta}_o = \Delta_o$ in our calculations.

The stress dependent eigenfunctions are given by:

$$v_2 = |3/2,\ 3/2\rangle_\beta \qquad (17a)$$

$$v_1 = \frac{1}{q_1}\left[(\sqrt{2}\delta E_\beta')\ |3/2,\ 1/2\rangle_\beta + (n_i - m_i)\ |1/2,\ 1/2\rangle_\beta\right] \qquad (17b)$$

$$v_3 = \frac{1}{q_i}\left[(\sqrt{2}\delta E_\beta')\ |1/2,\ 1/2\rangle_\beta - (n_i - m_i)\ |3/2,\ 1/2\rangle_\beta\right] \qquad (17c)$$

where $i=1$ and $\beta = [111]$ for stress along $[111]$. These equations have been given in general because, for stress along $[001]$, they will have a similar form. In Eqs. (17) m_i, n_i and q_i are given by:

$$m_i = \Delta_o + 1/2 \, \delta E_\beta \tag{18a}$$

$$n_i = \left[m_i^2 + 2(\delta E_\beta')^2 \right]^{\frac{1}{2}} \tag{18b}$$

$$q_i = \left[2n_i(n_i - m_i) \right]^{\frac{1}{2}} \tag{18c}$$

where we have used the approximation $\overline{\Delta}_o = \Delta_o$.

Stress Along [001]

$\langle 001 \rangle$ Conduction Bands

For this stress direction, the $\langle 001 \rangle$ valleys experience a splitting, i.e., the (001) valley splits away from the (100) and (010) valleys as shown schematically in Fig. 4. From Eq. (9), the energy shift of the singlet, (001) valley is given by:

$$\delta E_{1,c}^{(001)} = -2/3 \, \mathcal{E}_2 (S_{11} - S_{12})X + \delta E_H^{\,c} \tag{19}$$

and for the (010) and (100) valleys (doublet)

$$\delta E_{2,c}^{(010)} = 1/3 \, \mathcal{E}_2 (S_{11} - S_{12})X + \delta E_H^{\,c} \tag{20}$$

This splitting is shown schematically in Fig. 4.

$\langle 111 \rangle$ Conduction Bands

For this case, there is no splitting of the $\langle 111 \rangle$ conduction bands and only a hydrostatic pressure shift occurs. The shift is obtained from Eq. (9) and is given by Eq. (10). Figure 5 illustrates this effect.

The Valence Band

The degeneracy of the valence band, $\Gamma_{8,v}^{+}$ ($\Gamma_{8,v}$) is also lifted producing two states v_1 and v_2. The Hamiltonian matrix is just Eq. (12) with $\beta = [001]$ and

$$\delta E_{[001]} = 2(b_1 + 2b_2)(S_{11} - S_{12})X = 2b(S_{11} - S_{12})X \tag{21a}$$

$$\delta E'_{[001]} = 2(b_1 - b_2)(S_{11} - S_{12})X = 2b'(S_{11} - S_{12})X \tag{21b}$$

Equation (12) with $\beta = [001]$ and Eqs. (21) can be diagonalized to obtain the eigenvalues and eigenvectors of the Hamiltonian. These turn out to be just Eqs. (14) - (18) with $\beta = [001]$ and $i = 0$.

Thus the application of a uniaxial stress along [001] and [111] reduced the degeneracy of the valence and (or) conduction

bands. Table III summarizes the splittings which are found for
both stresses for the various polarizations.

The polarization dependent relative as well as absolute
intensities of the above transitions can be calculated by using
Eq. (2) for the oscillator strength, Eq. (17) for the stress
dependent valence band wave functions and the optical and phonon
scattering selection rules.[1,2,5,11,17] The results of this calculation for the transverse modes in Si and GaP are presented in
Table IV for the electric field vector of the incident light, $\vec{E}$,
polarized parallel ($\parallel$) perpendicular ($\perp$) to $\vec{X}$.

The terms η_i^j represent the contribution of the stress-induced
mixing of the $|3/2,1/2\rangle$ and $|1/2,1/2\rangle$ states. If this effect is
neglected, $\eta_i^j = 1$. The fact that the ratio U_ℓ/V_ℓ is real allows
us to write these expressions in the simplified form presented in
Table IV. Results for the longitudinal modes in Si and GaP are
presented in Table I and II of Ref. 1. For Ge the correct
results for the LA phonon assisted Γ-L transitions including the
$\Gamma_{15,c}$ intermediate state is given in Table V.

The experimental procedure involves measuring the wavelength
modulated absorption (WMA) spectrum[12] $d\alpha/dE$, of the strained
crystal for stresses along [001] and [111] with the electric
field vector of the incident light polarized parallel and perpendicular to the stress axis. For a given phonon, ratios of the
relative intensities of the extra transitions resulting from the
stress will be compared to the theoretical expressions derived
in the previous section. By performing this analysis for both
stress directions and light polarizations, it is possible to
obtain the ratio $K = S_{e-ph}/S_{h-ph}$ for the phonon assisted transition
under consideration for Si and GaP.

From these ratios the experimental values of the absorption
coefficient and the theoretical expressions for Af_ℓ from the
previous section it is possible to evaluate individually S_{e-ph}
and S_{h-ph} in these materials. Germanium will be left out of
this analysis because the LA transition proceeds via two intermediate conduction band states and therefore, the ratio, K_{LA}
cannot be determined uniquely.

The experimental details will not be discussed here but can
be found in Refs. 1 and 3.

Experimental Results

Shown in Fig. 6 is the normalized WMA spectrum of the TO
phonon-assisted indirect exciton in silicon[3] at 77 K for stress
X = 0. Also plotted in Fig. 6 is the spectrum for

Table III. Summary of the various allowed Γ-L and Γ-Δ(X) transitions for stresses along [111] and [001]. The label inside the parenthesis refers to the Γ-L transitions, while the one outside is used for Γ-Δ(X).

Transition	Γ-Δ(X)		Γ-L	
Γ-Δ (Γ-L)	X ∥ [001]	X ∥ [111]	X ∥ [001]	X ∥ [111]
$A_1(B_1)$		$v_1 \rightarrow (1,0,0),\ (010),\ (001)$	$v_1 \rightarrow (111),\ (\bar{1}\bar{1}1),\ (1\bar{1}\bar{1}),\ (\bar{1}1\bar{1})$	
$A_2(B_1)$		$v_2 \rightarrow (100),\ (010),\ (001)$	$v_2 \rightarrow (111),\ (\bar{1}\bar{1}1),\ (1\bar{1}\bar{1}),\ (\bar{1}1\bar{1})$	
$B_1(A_1)$	$v_1 \rightarrow (001)$			$v_1 \rightarrow (111)$
$B_2(A_2)$	$v_2 \rightarrow (001)$			$v_2 \rightarrow (111)$
$B_3(A_3)$	$v_1 \rightarrow (010),\ (001)$			$v_1 \rightarrow (\bar{1}\bar{1}1),\ (1\bar{1}\bar{1}),\ (\bar{1}1\bar{1})$
$B_4(A_4)$	$v_2 \rightarrow (010),\ (001)$			$v_2 \rightarrow (\bar{1}\bar{1}1),\ (1\bar{1}\bar{1}),\ (\bar{1}1\bar{1})$

Table IV. Theoretical expressions for the relative intensities of the TO- and TA-phonon-assisted indirect transition in GaP and Si as a function of stress for $X \parallel [001]$ and $X \parallel [111]$ and light polarized parallel and perpendicular to the stress axis. In the expressions below, the subscript T refers to either TO or TA phonon. The matrix elements of H_T are referred to the (001) valley. Note at the η_3^1 and η_4^1 of this table correspond, respectively, to η_1^1 and η_2^1 in Ref. 3. For the TO phonon H_T^y becomes H_T^x.

	$\vec{E} \parallel \vec{X}$	$\vec{E} \perp \vec{X}$
$X \parallel [001]$		
$B_1 [v_1 \rightarrow (001)]$	$1/3\eta_1^o U_T^2$	$2/3\eta_2^o (U_T+V_T)^2$
$B_2 [v_2 \rightarrow (001)]$	U_T^2	0
$B_3 [v_1 \rightarrow (001), (010)]$	$1/3\eta_1^o (U_T+V_T)^2$	$(1/6\eta_1^o+2/3\eta_2^o)U_T^2+1/6\eta_1^o(U_T+V_T)^2$
$B_4 [v_2 \rightarrow (100), (010)]$	$(U_T+V_T)^2$	$1/2 \ U_T^2+(U_T+V_T)^2$
$X \parallel [111]$		
$A_1 [v_1 \rightarrow (001), (010), (100)]$	$2\eta_3^1(U_T^2+U_TV_T)+2/3V_T^2$	$\eta_4^1(U_T^2+U_TV_T)+2/3V_T^2$
$A_2 [v_2 \rightarrow (001), (010), (100)]$	$2/3(U_T^2+V_T^2+U_TV_T)$	$1/3(5U_T^2+2V_T^2+U_TV_T)$

$$U_T = \langle \Gamma_{8,v}^z \mid P_x \mid \Gamma_{15,c}^y \rangle \langle 15,c \mid H_T^y \mid 1,c \rangle / E(X_{1,c})-E(\Gamma_{15,c})-(\hbar\omega)_T$$

$$V_T = \langle \Gamma_{8,v} \mid H_T^y \mid X_{5,v}^x \rangle \langle X_{5,v}^x \mid P_x \mid X_{1,c} \rangle / E(\Gamma_{8,v})-E(X_{5,v})+(\hbar\omega)_T$$

The terms η_i^j are defined in Refs. 1 and 3.

Table V. Polarization dependent relative intensities of the LA phonon assisted Γ-L transition for $X \parallel [111]$ and $X \parallel [001]$ and $\vec{E} \parallel \vec{X}$ and $\vec{E} \perp \vec{X}$. Scattering from $\Gamma_{15,c}$ is included. The quantities W_{LA}, U_{LA} and V_{LA} are defined in Table II.

$X \parallel [111]$	$\vec{E} \parallel \vec{X}$	$\vec{E} \perp \vec{X}$
A_1	$\dfrac{2}{3}\left(W_{LA} + U_{LA}\right)^2$	$\dfrac{1}{6}\left(V_{LA} - \dfrac{1}{2}U_{LA}\right)^2$
A_2	0	$\dfrac{1}{2}\left(V_{LA} - \dfrac{1}{2}U_{LA}\right)^2$
A_3	$\dfrac{2}{9}\left(U_{LA} - \dfrac{17}{3}V_{LA}\right)^2 + \dfrac{1}{9}\left(U_{LA} - \dfrac{2}{3}V_{LA}\right)^2$	$\dfrac{1}{12}\left(U_{LA} + \dfrac{8}{3}V_{LA}\right)^2 + \dfrac{1}{6}\left(V_{LA} - \dfrac{1}{2}U_{LA}\right)^2 + \dfrac{1}{3}\left(U_{LA} - \dfrac{2}{3}V_{LA}\right)^2$
A_4	$\dfrac{1}{3}\left(U_{LA} - \dfrac{2}{3}V_{LA}\right)^2$	$\dfrac{1}{4}\left(U_{LA} + \dfrac{8}{3}V_{LA}\right)^2 + \dfrac{1}{2}\left(U_{LA} + \dfrac{8}{3}V_{LA}\right)^2 + \dfrac{1}{3}\left(U_{LA} - \dfrac{2}{3}V_{LA}\right)^2$
$X \parallel [001]$	$\vec{E} \parallel \vec{X}$	$\vec{E} \perp \vec{X}$
B_1	$\dfrac{8}{3}\left(W_{LA} + \dfrac{2}{3}V_{LA}\right)^2 + \dfrac{4}{27}\left(V_{LA} - \dfrac{3}{2}U_{LA}\right)^2$	$\dfrac{2}{3}\left(W_{LA} + \dfrac{2}{3}V_{LA}\right)^2 + \dfrac{10}{27}\left(V_{LA} - \dfrac{3}{2}U_{LA}\right)^2$
B_2	$\dfrac{4}{9}\left(V_{LA} - \dfrac{3}{2}U_{LA}\right)^2$	$2\left(W_{LA} + \dfrac{2}{3}V_{LA}\right)^2 + \dfrac{2}{9}\left(V_{LA} - \dfrac{3}{2}U_{LA}\right)^2$

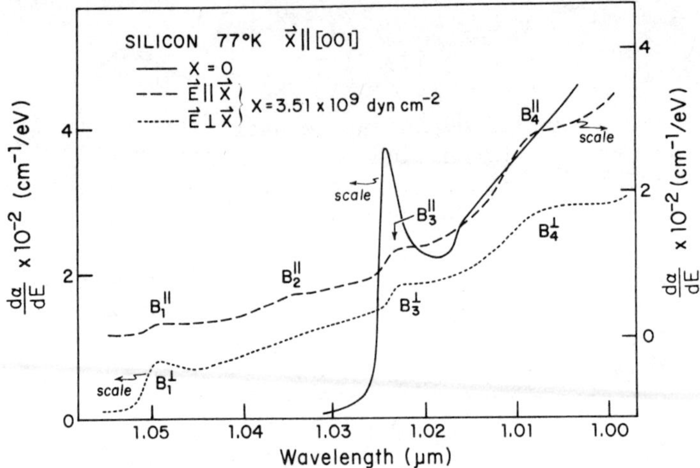

Fig. 6. Wavelength-modulated absorption spectra of the TO phonon assisted indirect exciton of silicon at 77K for X = 0 and X = 3.51 x 10⁹ dyn cm⁻² along [001] for the electric field vector of the light $\vec{E}$ polarized parallel and perpendicular to the stress axis.

for X = 3.51 x 10⁹ dyn -cm⁻² along [001] for the $\vec{E} \parallel \vec{X}$ and $\vec{E} \perp \vec{X}$. Note that four peaks, B_1 – B_4 are observed that correspond to the transitions denoted in Fig. 4. Similar results have been observed for X $\parallel$ [111] where only two peaks, A_1 and A_2, are observed as indicated schematically in Fig. 4.

It has been shown that the lineshape of the derivative absorption spectrum of the indirect exciton in Si (and Ge) can be fitted by an expression of the form:

$$\frac{d\alpha}{dE} \sim F(W) \tag{22a}$$

where α is the absorption coefficient. The function F(W) is given by:

$$F(W) = \left[(W^2 + 1)^{\frac{1}{2}} + W \right]^{\frac{1}{2}} / \left[W^2 + 1 \right]^{\frac{1}{2}} \tag{22b}$$

where $W = (E - E_{exciton})/\Gamma$ and Γ is the broadening parameter. By appropriate subtraction of the background it is possible to fit

470

the lineshapes of the various peaks to this form and hence obtain
a quantitative determination of the amplitude of the modulated
exciton spectra. In Fig. 7 we have plotted the experimental
values (solid line) of the $B_1^{||}$ peak of Fig. 6 and the theoretical
fit (dashed line) from Eqs. (22). There is good agreement in the
lineshapes. Similar results have been obtained for the other B
peaks as well as the A structures. In order to obtain a measure
of integrated intensity, $d\alpha$, we have multiplied the value of
$[d\alpha/dE]$ by Γ for the various B and A peaks at the indicated
stresses for $\vec{E} \parallel \vec{X}$ and $\vec{E} \perp \vec{X}$. Similar results have been obtained
at several other applied stresses for both stress directions.[3]

The general theoretical expression for the intensities of the
excitonic lines in the optical spectra, corresponding to TO phonon
assisted indirect transitions between the stress-split $\Gamma_{8,v}$
valence and $\Delta_{1,c}$ conduction bands via $\Gamma_{15,c}$ conduction and $\Delta_{5,v}$
valence band intermediate states are listed in Table IV.

Comparison of the theoretical expressions of Table IV and the
experimental values listed in Table VI, enables the ratio of

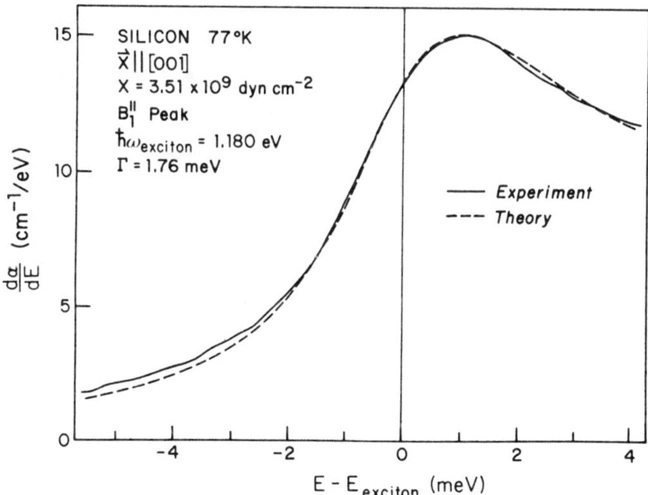

Fig. 7. Experimental values (solid lines) of the wavelength
modulated absorption spectra of the $B_1^{||}$ peak of Fig. 6 after
appropriate background subtraction and the theoretical fit
(dashed line) obtained from Eqs. (22).

Table VI. Experimental and theoretical values of the intensities for the TO-phonon assisted indirect transitions in silicon. The relative and actual (in parentheses in units of cm^{-1}) experimental values were obtained by multiplying $d\alpha/dE$ by the broadening parameter Γ. The theoretical values were calculated using $U_{TO}/V_{TO} = 1$.

$x \parallel [001]$ ($X = 3.51 \times 10^9$ dyn-cm^{-2})	$\vec{E} \parallel \vec{x}$		$\vec{E} \perp \vec{x}$	
	Exp.	Theory	Exp.	Theory
B_1	3% (0.026)	3%	25% (0.139)	46%
B_2	16% (0.135)	17%	0 (0.00)	0
B_3	18% (0.156)	11%	12% (0.068)	18%
B_4	63% (0.551)	69%	63% (0.360)	36%

$x \parallel [111]$ ($X = 7.59 \times 10^9$ dyn-cm^{-2})	$\vec{E} \parallel \vec{x}$		$\vec{E} \perp \vec{x}$	
	Exp.	Theory	Exp.	Theory
A_1	74% (0.244)	65%	44% (0.087)	45%
A_2	26% (0.086)	35%	56% (0.112)	55%

U_{TO}/V_{TO} to be determined. The value of $B_4^{\parallel}/B_2^{\parallel} = 4$ yields that $U_{TO}/V_{TO} = 1$ or $-1/3$. The above ambiguity is resolved by an examination of the $A_2^{\parallel}/A_1^{\parallel}$ ratio, for which the agreement between experiment and theory is good only for $U_{TO}/V_{TO} = 1$. Using this value of the ratio we have calculated the relative theoretical values listed in Table VI. There is in general good agreement between experiment and theory. Not only can comparisons be made between peak intensities of a given polarization but between the same peak for the two observed polarizations, thus eliminating any effects due to different line broadenings. For example, the theoretical ratio of $A_2^{\perp}/A_2^{\parallel} = 1.34$ is in good agreement with the experimental ratio of $(0.112/0.086) = 1.3$. Similar correspondences are found for $A_1^{\perp}/A_1^{\parallel}$ and $B_4^{\perp}/B_4^{\parallel}$. We find, however, that any ratio of intensities that involves η_1^0 or η_2^0 does not yield as good an agreement with theory.

In a similar manner it has been possible to obtain the ratio EP and HP scattering matrix elements for the LA and TA phonons in GaP.[1,11] It is found in Ref. 1 that for the former phonon $W_{LA}/V_{LA} = -4.5$ while for the latter phonon $U_{TA}/V_{TA} = -1.6$.

Evaluation of the Electron–Phonon and Hole–Phonon Scattering Matrix Elements

Having obtained values for the ratio of the EP and HP scattering matrix elements for the LA and TA phonons of GaP and for the TO phonon of Si, we now turn our attention to evaluating the matrix elements individually. This can be accomplished by comparing the ratio K together with experimental values of the absorption coefficient to the theoretical expressions for the absorption coefficient.

Silicon

The case of Si will be considered first. The piezospectroscopic investigation discussed above found that for (001) scattering, $K_{TO} = U_{TO}/V_{TO} = 1.0$, where

$$U_{TO} = \frac{\langle \Gamma_{25',v}^{z} \mid P_x \mid \Gamma_{15,c}^{y} \rangle \langle \Gamma_{15,c}^{y} \mid H_{TO}^{x} \mid \Delta_{1,c} \rangle}{E(\Delta_{1,c}) - E(\Gamma_{15,c}) - (\hbar\omega)_{TO}} \tag{23a}$$

and

$$V_{TO} = \frac{\langle \Gamma_{25',v}^{z} \mid H_{TO}^{x} \mid \Delta_{5,v}^{x} \rangle \langle \Delta_{5,v}^{x} \mid P_x \mid \Delta_{1,c} \rangle}{E(\Gamma_{25',v}) - E(\Delta_{5,v}) + (\hbar\omega)_{TO}} \tag{23b}$$

The study of Macfarlane, McLean, Quarrington and Roberts
(MMQR) provides absorption data in both the excitonic and free-
pair energy regions.[20] MMQR were able to decompose the absorption
coefficient into a sum of components, each having the same line
shape and corresponding to the creation or annihilation of TA and
TO phonons. In Fig. 8, where $(\alpha \hbar \omega)^{1/2}$ (this function is convenient
for the analysis of the Si data) is plotted against $\hbar \omega$, we exhibit
the decomposition which MMQR obtained for the absorption coefficient at 4.2K. Since at this temperature $n_Q \approx 0$, we do not observe
the phonon annihiliation processes and thus the data of Fig. 8
corresponds to transitions accompanied by phonon creation. The
low energy component is a result of the emission of TA phonons,
while the higher energy one is a consequence of TO phonon
creation. A further examination of the figure reveals that each
component begins with a well defined knee and ultimately rises
linearly with energy. In the "knee" region excitons are created,
while in the linear region free electron-hole pairs are formed.

In the knee (or excitonic) region of Fig. 8, MMQR were able
to obtain empirical expressions for the components of the
absorption coefficient.[20] For the TO phonon, they found the
following relationship:

$$\alpha_{TO} = \frac{B_{TO}}{\hbar \omega} \left[\hbar \omega - Eg - (\hbar \omega)_{TO} + R \right]^{1/2} \tag{24}$$

where $B_{TO} = 26.9 \times 10^{-9}$ $Ry^{1/2}/a_0$ and a_0 is the Bohr radius
($= 5.29 \times 10^{-9}$ cm).

The value of B_{TO} enables us to evaluate S^{TO}_{e-ph} and S^{TO}_{h-ph}.
The theoretical expression for the absorption coefficient in
this energy region is given by Eqs. (1), (2) and (4), and hence
$B_{TO} = A_{ex}f_{TO}$. The constant A_{ex} can be evaluated from known
materials parameters. Thus we can obtain a value for
f_{TO} ($=B_{TO}/A_{ex}$). Using this value, the expression for the
oscillator strength listed in Table I and the experimentally
determined ratio $K_{TO} = 1.0$ we compute U_{TO} and V_{TO} individually.
From these values, the expressions for U_{TO} and V_{TO} from Eqs. (23),
optical matrix elements (obtained from $k \cdot p$ calculations), energy
denominators obtained from the optical transitions and known
phonon frequencies we find:[2,21]

$$S^{TO}_{e-ph} = 17.4 \pm 2.6 \times 10^{-3} \text{ Ry}$$

$$S^{TO}_{h-ph} = -22.9 \pm 3.4 \times 10^{-3} \text{ Ry}$$

Since the quantity Af_ℓ is an absolute magnitude of the sum
of electron and hole scattering terms, it is sensitive to the
phase of the ratio S_{e-ph}/S_{h-ph}. Therefore, the phases of the
individual matrix elements are arbitrary to within a common

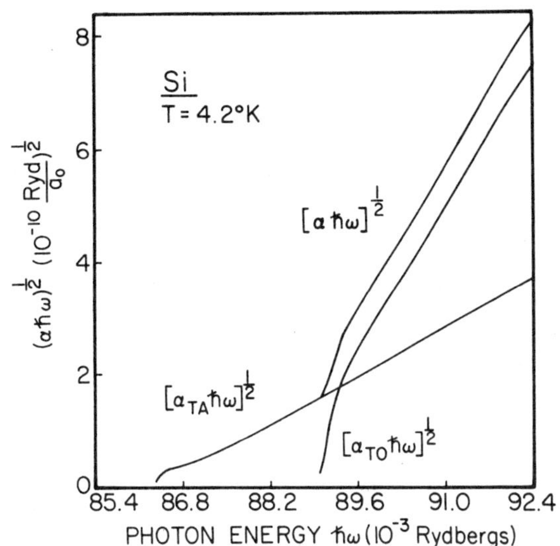

Fig. 8.   Decomposition of the absorption coefficient of Si at 4.2°K.

factor. We have chosen this factor such that S^{TO}_{e-ph} is real and positive, making S^{TO}_{h-ph} real and negative. This sign convention will be followed throughout this article.

In the linear region of Fig. 8 the theoretical expression for the absorption coefficient is given by Eqs. (1), (2) and (5). From here, proceeding as we did in the excitonic case we find:[2,21]

$$S^{TO}_{e-ph} = 17.9 \pm 2.7 \times 10^{-3} \text{ Ry}$$

$$S^{TO}_{h-ph} = -23.6 \pm 3.5 \times 10^{-3} \text{ Ry}$$

This illustrates the consistency of the parameters in the different spectral regions.

Averaging over a number of different spectral regions (excitonic, free electron-hole), different temperature ranges etc. and estimating the error at 15% we obtain for silicon:[2,21]

$$S^{TO}_{e-ph} = 17.3 \pm 2.6 \times 10^{-3} \text{ Ry}$$

$$S^{TO}_{h-ph} = -22.9 \pm 3.4 \times 10^{-3} \text{ Ry}$$

Details of the various regions are given in Table III of Ref. 2.

GaP

A similar type of analysis has been performed for GaP. For this material we find for the average value of the LA and TA phonon matrix elements:[2]

LA Phonon (GaP)

$$S^{LA}_{e-ph} = 7.6 \pm 1.1 \times 10^{-3} \text{ Ry}$$

$$S^{LA}_{h-ph} = 13.0 \pm 2.0 \times 10^{-3} \text{ Ry}$$

TA Phonon (GaP)

$$S^{TA}_{e-ph} = 19.2 \pm 2.8 \times 10^{-3} \text{ Ry}$$

$$S^{TA}_{h-ph} = 17.4 \pm 2.5 \times 10^{-3} \text{ Ry}$$

We shall return to the case of Ge later.

GaAs

For GaAs values for the EP scattering matrix element from Γ-X and Γ-L have been deduced by various experiments such as transferred electron effects[9,10] and time-dependent Raman scattering.[6] In general, these workers find that:[22]

$$S_{e-ph} \ (\Gamma\text{-X}) \approx 1 \times 10^9 \text{ eV/cm}$$

$$S_{e-ph} \ (\Gamma\text{-L}) \approx 1.5 \times 10^8 \text{ eV/cm}$$

THEORETICAL CALCULATIONS

The calculation of the EP and HP scattering matrix elements is, in general, a difficult problem. Microscopic theories have been discussed by a number of authors including Zimam,[23] Sham and Zimam[24] and Vogl.[25,26] These authors have pointed out the fact

that many degrees of freedom are involved, i.e., a quasi-infinite
lattic of ions is embedded in a sea of electrons. The ions and
electrons interact with each other through Coulomb forces. This
is a many body problem, in which the actions of a given electron
or ion will affect the whole ensemble, thereby introducing a
complicated screening of the bare particle potential. For example,
the displacement of an ion in such a system is expected to cause
not only a shift of the core potential, but also a redistribution
of the outer-electrons which are not tightly bound to the ion.
Thus, the calculation of the EP and HP matrix elements should be
done self consistently.

Various approximations, such as the "rigid-ion" and
"deformable-ion" models and long wavelength approaches have been
developed.[23-26] Of particular interest to us is an application
of the "rigid-ion" model, the "rigid-pseudoion" model.[27,28] This
approach combines the rigid-ion description of the lattice
displacements and local pseudopotential theory to provide the first
step in calculating EP and HP interactions. In short, it neglects
such effects as charge redistribution and the nonlocal nature of
the pseudopotential. These effects should ultimately be included.
However, their neglect seems to provide a reasonable starting
point for calculating these interactions.

The Rigid-Pseudoion Model

The foundation of the "rigid-pseudoion" model is the "rigid-
ion" description of the lattice dynamics, which was introduced by
Nordheim.[29] In this approximation, the ion-core potential of an
atom moves rigidly when the atom is displaced from equilibrium.
Thus, the potential seen by an electron due to a lattic of such
atoms is just:[12,24,27,28]

$$
V(\vec{r}) = \sum_{j,\beta} V_{j,\beta}(\vec{r} - \vec{R}_{j,\beta}{}^{o}) \tag{26}
$$

where $V_{j,\beta}(\vec{r} - \vec{R}_{j,\beta}{}^{o})$ is the atomic potential of an atom in its
equilibrium position, $\vec{R}_{j,\beta}{}^{o}$, having a lattice index, j, and basis
index β. Within this approximation, the EP (HP) interaction
Hamiltonian due to small displacements from equilibrium resulting
from a phonon of mode ℓ and polarization, t, is given by:[12,28]

$$
H_{\ell}^{t} = \sum_{j,\beta} \delta\vec{R}_{j,\beta}^{\ell,t} \cdot \vec{\nabla} V_{j,\beta}(\vec{r} - \vec{R}_{j,\beta}{}^{o}) \tag{27}
$$

where $\delta\vec{R}_{j,\beta}^{\ell,t}$ is the displacement of the atom from equilibrium and
can be written in terms of phonon coordinates as:[12,28]

$$
\delta\vec{R}_{j,\beta}^{\ell,t} = \left[\frac{\hbar}{2M_{\beta}N\omega_{\ell}(Q)} \right]^{\frac{1}{2}} \hat{e}_{\ell,\beta}^{t}\, e^{i\vec{Q}\cdot\vec{R}_{j,\beta}{}^{o}} \tag{28}
$$

where the phonon creation and annihilation operators are implied.
In Eq. (28), $\vec{Q}$ is the phonon wave vector, $\hat{e}^t_{\ell,\beta}$ is the polarization
vector of the βth atom of mass M_β and $\omega_\ell(Q)$ is the phonon
frequency. The index t is used to distinguish between any
degenerate phonons of a given mode, ℓ. In the above equation, N is
the number of primitive cells in the crystal. Since this quantity
is ultimately cancelled out in calculating measureable quantities
such as absorption coefficients, it will be left out of the square
root in any subsequent discussions.

The potential can be Fourier analyzed to obtain

$$V_{j,\beta}(\vec{r}-\vec{R}^o_{j,\beta}) = \frac{\Omega}{(2\pi)^3}\int d\vec{q}\ V_{\vec{q},\beta}\ e^{i\vec{q}\cdot(\vec{r} - \vec{R}^o_{j,\beta})} \tag{29}$$

where Ω is the volume of a primitive cell. Writing $\vec{R}^o_{j,\beta} =$
$\vec{R}^o_j + \vec{\tau}_\beta$ where R^o_j is a lattice vector and $\vec{\tau}_\beta$ is a basis vector of
the βth atom. By putting Eqs. (28) and (29) into (27) we obtain
for the EP (HP) interaction:

$$H^t_\ell = \sum_{\beta,\vec{G}} i\left[\frac{\hbar}{2M_\beta\omega_\ell(Q)}\right]^{\frac{1}{2}} \hat{e}^t_{\ell,\beta} \cdot (\vec{Q} - \vec{G})x$$

$$V_{\vec{Q} - \vec{G},\beta}\ e^{i\vec{G}\cdot\vec{\tau}_\beta}\ e^{i(\vec{Q} - \vec{G})\cdot\vec{r}} \tag{30}$$

where $\vec{G}$ is a reciprocal lattice vector. We have used Bloch's
theorem in the form:

$$\sum_j e^{i\vec{R}^o_j \cdot (\vec{Q} - \vec{q})} = N\delta_{\vec{G},\vec{Q} - \vec{q}} \tag{31}$$

and changed the integral over q into a sum by the following
relationship:

$$\frac{\Omega N}{(2\pi)^3}\int d\vec{q} \longrightarrow \sum_q \tag{32}$$

In order to calculate the matrix elements of Eq. (30), which
is a general result within the context of the "rigid-ion" model,
the atomic potential and electronic states must be known. In
principle, this can become a very difficult program. However,
pseudopotential theory provides an elegant method of obtaining the
electronic band structure of a solid.[28] It takes advantage of the
fact that between lattice sites, the potential is weak and thus
the electron can be considered as being nearly-free. Under these
conditions, the electronic wave functions can be chosen to be plane
waves, whose overlap with the core states is removed through the
use of an effective potential, the pseudopotential. If we assume

478

that the pseudopotential is just a function of $|\vec{r} - \vec{R}^0_j|$ then we are within the regime of local pseudopotential theory.[30] In the following discussion, we will use this approximation. With this choice, the wave functions are pseudo-plane-waves and can be written as:[30]

$$\Psi_{\vec{k},s}(\vec{r}) = \sum_{\vec{G}} c^{\vec{k},s}_{\vec{G}} \, e^{i(\vec{k} + \vec{G}) \cdot \vec{r}} \tag{33}$$

By using Eq. (33) in Eqs. (3), (30) and (31) we find the matrix element of the electron (hole)-phonon Hamiltonian to be:

$$S^{\ell}_{e\text{-ph}} = \left\langle \Psi_{\vec{k}',s'} \,\Big|\, H^t_{\ell} \,\Big|\, \Psi_{\vec{k},s} \right\rangle$$

$$S^{\ell}_{e\text{-ph}} = i \sum_{\beta,\vec{G}} \left[\frac{\hbar}{2M_{\beta}\omega_{\ell}(Q)} \right]^{\frac{1}{2}} V_{\vec{Q}-\vec{G},\beta}\, e^{i\vec{G}\cdot\vec{\tau}_{\beta}} \; x$$

$$\hat{e}^t_{\ell,\beta} \cdot (\vec{Q} - \vec{G}) \sum_{\vec{G}'} (c^{\vec{k}',s'}_{\vec{G}'+\vec{G}})^* \, c^{\vec{k},s}_{\vec{G}'} \tag{34}$$

where $\vec{k}$ and $\vec{k}'$ are related through $\vec{Q} = \vec{k}' - \vec{k}$. For hole scattering, the subscript h-ph replaces e-ph. Equation (34) is the general form of the EP(HP) matrix elements within the "rigid-pseudoion" model. We will now apply it to the calculation of the EP and HP matrix elements in the diamond-type semiconductors.

<u>Application to Diamond-Type Materials</u>

In Eq. (34), the phonon polarization vectors are complex numbers. However, in diamond-type materials, such as Si and Ge, with two similar atoms within the basis ($\beta = 1,2$) a simplification can be made. If we choose the origin to lie midway between the two atoms, we find that $\vec{\tau}_2 = -\vec{\tau}_1 = \vec{\tau} = (a/8)(111)$. By employing time reversal invariance and inversion symmetry, we find that the polarization vectors $\hat{e}^t_{\ell,1}$ and $\hat{e}^t_{\ell,2}$ are related in the following manner:[5,12,31]

$$\hat{e}^t_{\ell,1} = (\hat{e}^t_{\ell,2})^* \tag{35}$$

This relationship between the polarization vectors allows us to make a transformation to a real set of vectors, $\hat{e}^t_{\ell,+}$ and $\hat{e}^t_{\ell,-}$, which are given by:[5,31,32]

$$\hat{e}^t_{\ell,+} = \frac{1}{\sqrt{2}} (\hat{e}^t_{\ell,1} + \hat{e}^t_{\ell,2}) \tag{36a}$$

$$\hat{e}^t_{\ell,-} = \frac{1}{\sqrt{2}} (\hat{e}^t_{\ell,1} - \hat{e}^t_{\ell,2}) \tag{36b}$$

where $(\hat{e}^t_{\ell,+})^2$ and $(\hat{e}^t_{\ell,-})^2$ correspond to the fraction of the mode which is "acoustic" and "optic", respectively. Using Eq. (36) in Eq. (34) and expanding $e^{iG\cdot\tau_1}$ and $e^{iG\cdot\tau_2}$ we obtain:

$$S^\ell_{e-ph} = i \left[\frac{\hbar}{2M\omega_\ell(Q)} \right]^{\frac{1}{2}} \sum_{\vec{G}} V_{\vec{Q}-\vec{G}} \; x$$

$$(\vec{Q} - \vec{G}) \cdot (\hat{e}^t_{\ell,+} \cos \vec{G}\cdot\vec{\tau} + \hat{e}^t_{\ell,-} \sin \vec{G}\cdot\vec{\tau}) \; x$$

$$\sum_{\vec{G}'} (C^{\vec{k}',s'}_{\vec{G}'+\vec{G}})^* \; C^{\vec{k},s}_{\vec{G}'} \tag{37}$$

where we used the fact that for a diamond-type material $M_1 = M_2 = M$ and $V_{Q-G,1} = V_{Q-G,2} = V_{0-G}$. The computation is further simplified because the plane wave expansion coefficients can be chosen to be real numbers for these types of semiconductors, i.e., the Hamiltonian matrix is real. Thus, Eq. (37) is just two sums of real numbers.

In the following section, we will apply Eq. (37) to Si and Ge. For both materials, we will consider the fundamental indirect gap.

In Si, the conduction band minimum is at $\Delta \left[\vec{k} = (2\pi/a)(0,0,0.85) \right]$ and $\Gamma-\Delta$ indirect transitions, assisted by TO, LO, TA and LA phonons are allowed (see Fig. 1). A subsidiary conduction band minimum is found at L and thus at a photon energy of 1.65 eV at 1.6K, indirect processes ($\Gamma-L$) can occur with the aid of LA and TO phonons. In this section, we will concentrate on the EP and HP interactions for the $\Gamma-\Delta$ scattering, which are associated with the primary gap.

The evaluation of Eq. (37) was performed by using a 67 plane wave expansion for the wave functions and a local pseudopotential form factor V_q, in the range $V_{0.85} < V_q < V_\infty$[11]. V_q was interpolated through the values of Pantelides,[33] which were used to calculate the band structure. The form of Pantelides reproduces the Cohen-Bergstresser[34] form factors V_3, V_8 and V_{11}. Convergence was checked by comparing the wave functions from 59 and 74 plan wave expansions and by examining the value of a given matrix element as the calculation was performed, i.e., the program would print out the value of the matrix element for each plane wave component which was used. Computational difficulties with the subroutine, which diagonalized the band structure Hamiltonian limited the number of plane waves to 89.

In evaluating Eq. (37) we used experimental values of the
phonon frequencies.[5] The phonon polarization vectors were
obtained from a second nearest neighbor "rigid-ion" model
calculation, which utilized experimentally determined phonon
frequencies at $\vec{k} = 0$ and at the Brillouin zone edges at X and L.[30]
Such an approach to the lattice dynamics is inadequate for
calculating dispersion curves and elastic constants. However, it
was found to yield phonon polarization vectors which were within
2% of those derived from the more exact Weber "bond-charge"
model.[35] This surprising result was obtained for both Si and Ge
and seems to indicate that for diamond-type material, the phonon
polarization vectors may be model independent and not a major
source of error in our calculation. Thus within the context of
the model, the most uncertain quantities are the interpolation
of the form factor and the pseudowave functions.

Now we will consider the wave function and form factors.
These two parameters are intimately related, since the form
factors are the potential in the Fourier transformed Schroedinger
equation, whose solution ultimately determines the wave functions.
However, because of the structure factor, only a few select values
of V_q are needed to determine the band structure.[30] The values of
q^2 which are important for diamond type materials are 3, 8, and
11 $\left[\text{in units of } (2\pi/a)^2\right]$. Larger values of q^2 do not enter,
because the form factor approaches zero at large values of the
wave vector. It is not uncommon to take $V_q = 0$ beyond $q = 4$.[30]
On the other hand as $q \to 0$, it has been deduced that the form
factor should approach a value of $(-2/3)E_f$, where E_f is the Fermi
energy.[30,36,37] In the calculation of the EP and HP matrix
elements, the relevant values of q are reciprocal lattice vectors
shifted by the wave vector of the phonon. In addition to this,
the structure factor is different $\left[\text{see Eq. (37)}\right]$. Consequently,
the form factors which we use are not those of the band structure
calculations. In particular, the values of $q^2 < 3$ and $q^2 > 11$ are
important.[38] Since the form factor is not known in these regions,
there is a certain amount of freedom in interpolating V_q. This
$q = 0$ limit is exact in the case of metals, but can be relaxed for
semiconductors. It has recently been proposed[13,32,39] that
$V_q \to 0$ as $q \to 0$. We shall return to this point later.

From the above discussion, we see that there are a number of
form factors which can be used and that we have no way of telling
of which one is right for our purposes. However, the resulting
S_{e-ph} and S_{h-ph} matrix elements must be in a delicate balance in
order to accound for the relative as well as absolute intensities
of the indirect transitions in Si. In addition, the values of the
matrix elements for the TO phonon should be in accord with the
experimental ones. A further test of the calculation comes from
the quantity of Af_ℓ. The large and small q limits and the above

requirements were used as guidelines in choosing the interpolation of the form factor.

At this point, it must be noted that the phase of the ratio S_{e-ph}/S_{h-ph} is fixed. However, the phases of the individual matrix elements are arbitrary to within a common factor. This factor has been chosen so that the S_{e-ph}^{TO} is real and positive, thereby making S_{h-ph}^{TO} real and negative.

The best overall fit to S_{e-ph}^{TO}, S_{h-ph}^{TO} and Af_ℓ was obtained by fitting the form factor of Fig. 9 to a cubic equation of the form:

$$V_q = Aq^3 + Bq^2 + Cq + D \tag{38}$$

where A, B, C and D are adjustable parameters. This form is shown by curve a in Fig. 9. However, for $q^2 > 11$ we find that the best fit is obtained if V_q is given by:[5]

$$V_q = \frac{V_o}{q^2} \cos(Rq - \theta) \tag{39}$$

where V_o, R and θ are also adjustable parameters. This oscillatory tail is shown by curve b in Fig. 9. Note that for $q^2 < 11$ curves a and b coincide.

The results for S_{e-ph} and S_{h-ph} using the form factors of Eqs. (38) and (39) are listed in Table VII together with the experimental values of S_{e-ph}^{TO} and S_{h-ph}^{TO}. Listed in Table VIII are the results for $A_{ex}f_\ell$. As can be seen there is in general quite good agreement.

The reason why a form factor with an oscillating tail works better than the usual one, which is cut off near q = 4, is not known. This may be an indication that nonlocality is important in Si, i.e., the core potential is not properly cancelled out. It may also simulate effects due to charge redistribution.

In a similar calculation, Bednarek and Rossler[13] also had to use an unusual potential in order to get good agreement between experiment and theory for the LA and TA phonon assisted absorption. In their case, the form factor was cut off at q = 5 and approached zero for small values of q. They felt that their potential simulated nonlocal and charge redistribution effects. The results of their calculations are also listed in Tables VII and VIII. Except for the LA phonon there is little difference in the two works.

A more detailed analysis is required to identify the source of the difficulty in obtaining good agreement between theory and

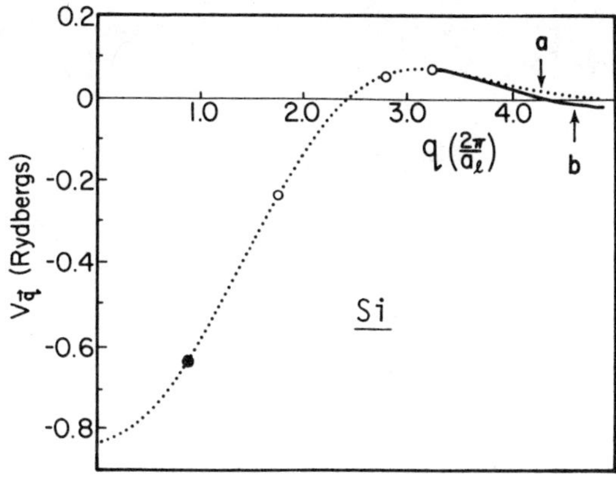

Fig. 9.   Form factor interpolation for Si.

experiment for the LA and TA phonons. In particular, a nonlocal
calculation would be a good starting point for such an analysis.

However, the agreement which we have obtained through our
potential is good and can be taken as a demonstration of the
validity of the "rigid-pseudoion" model for calculating EP and HP
matrix elements.

Bassani and Guidotti[15] have deduced a value of $S^{TA}_{e-ph}/S^{TA}_{h-ph} =$
0.45 for Si from polarization ratio of luminescence measurements
for X ‖ [001]. These results are in sharp disagreement with the
theoretical results of Table VII, which shows a ratio of −12.
Examination of Table IV shows that either ratio would account for
the results of Ref. 16. However, as pointed out in Ref. 3 measurements for X along only [001] are not unique. Thus a study for
X ‖ [111] should be undertaken to resolve this ambiguity.

Germanium, just like Si, is a diamond type material and thus
Eq. (37) can be used to calculate its EP and HP matrix elements.
However, in Ge, the fundamental absorption results from Γ-L
indirect processes assisted by LA and TO phonons. The LA
transition has two intermediate conduction band states, $\Gamma_{2',c}$ and

Table VII. Theoretical and experimental values of S_{e-ph} and S_{h-ph} (units of 10^{-3}Ry) for the $\Gamma-\Delta$ transition (TO, LO, TA, LA) of Si.

| | TO | | LO | | TA | | LA | |
	S_{e-ph}	S_{h-ph}	S_{e-ph}	S_{h-ph}	S_{e-ph}	S_{h-ph}	S_{e-ph}	S_{h-ph}
THEORY	19.0[a] 19.0[b] 13.0[c]	−23.0[a] −23.0[b] −18.0[c]	−12.0[a] −13.0[b] −	−36.0[a] −36.0[b] −	−3.3[a] −4.4[b] −	0.3[a] 0.3[b] −	25.0[a] 25.0[b] −	28.0[a] 28.0[b] −
EXPERIMENT	17.3±2.6	−22±3.4						

[a]From curve a of Fig. 9
[b]From curve b of Fig. 9
[c]From Ref. 13

Table VIII. Theoretical and Experimental values of the quantity Af_ℓ (units of 10^{-9} Ry$^{\frac{1}{2}}$ a_o^{-1}) for the indirect exciton ($\Gamma-\Delta$) of Si. This quantity is presented for the TO, LO, TA and LA phonons. The LA transition is so weak that it is not observed experimentally.

	TO	LO	TA	LA
THEORY	30.0[a] 30.0[b] 29.4[c]	5.1[a] 4.8[b] 4.4[c]	0.43[a] 0.76[b] 0.89[c]	0.27[a] 0.27[b] 0.03[c]
EXPERIMENT	26.9[d] 21.8[e] 30.9[f]	− 3.23[e] 3.40[f]	0.72[d] 0.64[e] 0.80[f]	− −

[a]From curve a of Fig. 9
[b]From curve b of Fig. 9
[c]Ref. 13
[d]Ref. 20
[e]Ref. 40
[f]Ref. 41

$\Gamma_{15,c}$ while the TO phonon proceeds via $\Gamma_{15,c}$. Both phonons have $L_{3',v}$ as their valence band intermediate state (see Fig. 3).

We have evaluated Eq. (37) by using a 70 plane wave expansion for the wave function and a form factor given by:

$$V_q = \frac{u_o \cos (r_1 q)}{q^2 + k_o^2} \; \exp \left[k_1^2 (3-q^2) + k_2^2 (q-3.34)^2 \right] \qquad (40)$$

where the parameters u_o, r_1 and k_o^2 are determined by the known form factors V_3, V_8 and V_{11}. The other quantities are used to control the form factor in the regions $q^2 > 11.2$ and $q^2 < 3$. The phonon polarization vectors for Ge were obtained in the same manner as they were for Si.

In Table IX, we list the calculated values of S_{e-ph} and S_{h-ph} for the LA and TO phonons (Γ-L). For the LA phonon, EP matrix elements are presented for the two allowed conduction band intermediate states, $\Gamma_{15,c}$ and $\Gamma_{2',c}$.

The TO phonon hole scattering requires two matrix elements (see Table II). This is a consequence of the fact that $\Gamma_{25',v}$ is compatible with $\Lambda_{3,v} + \Lambda_{1,v}$ and each component couples to $L_{3',v}$ through the TO phonon. Our results for the two components are consistent with the luminescence study of Smith and McGill who found that $S_{h-ph}^{z,TO}/S_{h-ph}^{x,TO} < 1.0$ in order to explain their experimental data[42]. From Table IX we see that this ratio is 1/2.6.

For Ge reliable experimental data exists for Af_{LA} and the polarization dependent relative intensities of the LA phonon in the strained crystal, as shown in Fig. 10. Reference 41 could not resolve the TO transition and thus there is no known value for the quantity Af_{TO}. The lack of data for the TO phonon can be remedied by making use of the WMA spectra of Fig. 10 at zero stress. A line is found approximately 9 meV higher in energy than the strong LA peak and is a result of TO phonon assisted indirect transitions. This conclusion is based on the fact that at L the LA and TO phonons differ in energy by 9 meV.

By using the procedures described in the previous section it was possible to fit the TO peak to the function F(W) of Eqs. (22). From the data and Eqs. (1) and (4) and by accounting for broadening, a value for the quantity Af_{TO} was obtained. As a check of the procedure, we also evaluated Af_{LA} from the data of Fig. 10 and found it to be in good agreement with the value of Ref. 42. This result gave us confidence in this method, which had previously been sucessful in Si as discussed above. Consequently, we now have a value of Af_{TO} with which we can compare the theoretical number. The results of this analysis are presented along with

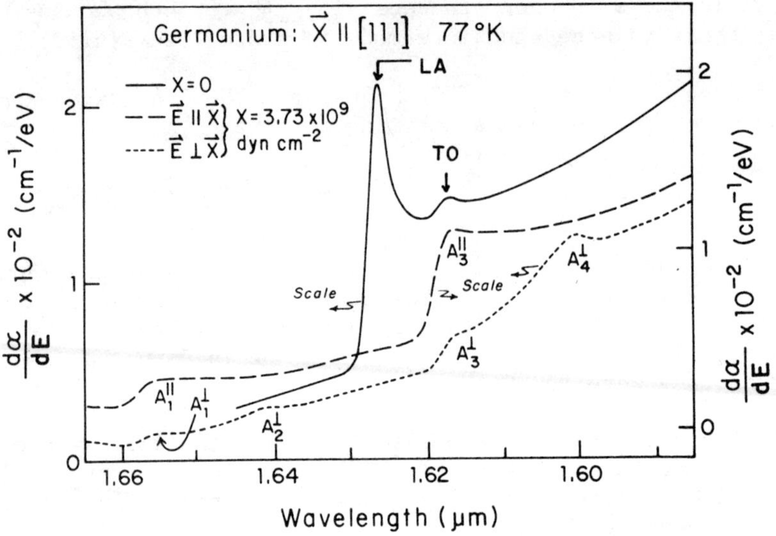

Fig. 10. Wavelength-modulated absorption spectra of the LA-phonon assisted indirect transition in Ge at 77K for X = 0 and 3.73 x 10^9 dyn-cm^{-2} along [111] for $\vec{E} \parallel \vec{X}$ and $\vec{E} \perp \vec{X}$. For the X = 0 spectra the TO-phonon assisted peak is also observed as marked.

those of Ref. 42 and the theory in Table X.

We have made a theoretical evaluation of the quantity Af_ℓ for the LA and TO phonon assisted transitions of Ge. In the case of the LA phonon, the conduction band energy difference, $E(L_{1,c})-E(\Gamma_{2',c})+\hbar\omega_{LA}$, is on the order of magnitude of S_{e-ph}^{LA} for the $\Gamma_{2',c} \rightarrow L_{1,c}$ scattering and thus the usual first-order perturbation term of Eq. (2) (S_{e-ph} divided by an energy denominator) is not adequate in calculating Af_{LA}. In our work, we used an exact expression for the $\Gamma_{2',c} \rightarrow L_{1,c}$ coupling coefficient, which is based on a two band model and given by:[5]

$$d(y_\ell) = \frac{\left[1 - (1 + 4y_\ell^2)^{\frac{1}{2}}\right]}{\left\{4y_\ell^2 + \left[1 - (1 + 4y_\ell^2)\right]^2\right\}^{\frac{1}{2}}} \tag{41}$$

where

486

Table IX. Theoretical values of S_{e-ph} and S_{h-ph} (in units of 10^{-3} Ry) for the Γ-L transitions in Ge.

$$S_{e-ph}$$

Intermediate State \ Phonon	LA	TO	LO	LA
$\Gamma_{2',c}$	-9.0	0.0	0	0
$\Gamma_{15,c}$	21.0	-4.0	0	0

$$S_{h-ph}$$

Final State \ Phonon	LA	TO	LO	LA
$\Gamma_{25'}^{\bar{x}},_{v}$ or $\Gamma_{25'}^{\bar{y}},_{v}$	21.0	-26.0	0	0
$\Gamma_{25'}^{\bar{z}},_{v}$	0.0	10.0	0	0

$\bar{x} = 1/\sqrt{2}\ (x-y)$; $\quad \bar{y} = 1/\sqrt{6}\ (x+y-2z)$; $\quad \bar{z} = 1/\sqrt{3}\ (x+y+z)$

$$y_\ell = \frac{S^\ell_{e-ph}}{E(L_{1,c}) - E(\Gamma_{2',c}) - \hbar\omega_\ell}$$

$$\simeq S^\ell_{e-ph} / \Delta E_c \tag{42}$$

with this expression, W_ℓ is replaced by

$$W_\ell = d(y_\ell) \langle \Gamma_{25,v}^{\bar{z}} \mid p_{\bar{z}} \mid \Gamma_{2',c} \rangle \tag{43}$$

It is easy to see that as $\Delta E_c \to 0$, $d(\infty) \to -1/\sqrt{2}$ and that as ΔE_c becomes much larger (in magnitude) than S_{e-ph}, $d(y_\ell) \to y_\ell$ and the result of Eq. (2) is recovered. Thus the transition intensity increased asymptotically as $\Delta E_c \to 0$, instead of becoming infinity large. This phenomena is also important in other materials such as ternary compounds, whose fundamental gap can be changed from direct to indirect by alloying or pressure.

Listed in Table X are the experimental and theoretical values
of Af_ℓ for the LA and TO phons of Ge. The agreement between
experiment and theory is quite good even though the theory overestimates Af_{TO} somewhat.

As a further test of the calculation, we have evaluated the
polarization dependent intensities of the LA phonon assisted
transitions in the strained crystal for stresses along [001] and
[111].[4] This was accomplished through the use of the equations
listed in Table V along with Eqs. (41) – (43). The results of
this calculation along with the experimental values of Ref. 4 are
presented in Table XI. The agreement between experiment and
theory is quite good for both of the stresses for $\vec{E} \parallel \vec{X}$ and
$\vec{E} \perp \vec{X}$.

Based on the above considerations a number of other EP and HP
matrix elements have been calculated for Ge and Si. For example
Ref. 14 compares theoretical calculations of S_{e-ph} and S_{h-ph} for
the Γ-Δ phonons (TO, LO, LA and TA) in Si and Ge. Reference 5
lists S_{e-ph} and S_{h-ph} for the "second" indirect transition in
Si ($\Gamma_{15,c} - L_{1,c}$) and Ge ($\Gamma_{2',c} - \Delta_{1,c}$) as well as electron (hole)
scattering matrix elements halfway across the zone for a variety
of transitions.

Herbert et al.[27] have used a similar model to calculate
electron-phonon matrix elements (but not hole-phonon) for a number
of transitions in Ge, Si, GaAs and InP. These values have been
compared to experiment, particularly transport measurements. In
general there is good agreement. However, such comparisons are
not as sensitive as those made for the indirect gap measurements,
both stressed and unstressed, since in this case there is a
delicate balance in the interference between EP and HP terms.

Although Herbert et al.[27] have calculated a number of intervalley electron-phonon matrix elements there is overlap with our
theoretical determinations in the case of Γ-L of Ge (LA-phonon).
However, from Herbert's work it is not clear whether he has
considered both intermediate states (i.e., $\Gamma_{2',c}$ and $\Gamma_{15,c}$). If
we assume that he has taken into account only $\Gamma_{2',c}$ then there is
fairly good agreement between his values and ours (see Table IX).

SUMMARY

Piezospectroscopy experiments in Si and GaP have made it
possible to determine the ratio of S_{e-ph}/S_{h-ph} for the Γ-Δ (X)
transitions for several intervalley phonons, i.e. TO in Si and
LA and TA in GaP. Comparing these results with experimental
values and theoretical expressions for the absorption coefficient
related to these phonon-assisted processes we have been able to

Table X. Theoretical and experimental values of the coefficient Af_ℓ for the LA and TO (Γ-L) phonons of Ge. The numbers are listed in atomic units and a commonly used mixed unit.

Af_ℓ \ Phonon	LA		TO	
	Theory	Expt.	Theory	Expt.
$(10^{-9}Ry^{\frac{1}{2}}a_o^{-1})$	11.2	11.6^a 12.2^b	1.3	$-a$ 1.8^b
$(eV^{\frac{1}{2}}cm^{-1})$	7.8	8.1^a 8.5^b	0.9	$-a$ 0.7^b

[a]Ref. 42
[b]From the analysis of the TO peak in the WMA spectrum of Fig. 10.

Table XI. Theoretical and experimental values for the intensities of the LA-phonon assisted transition in Ge for stress, X, along $[111]$ and $[001]$ for the electric field vector, $\vec{E}$, of the incident light polarized parallel and perpendicular to $\vec{X}$.

X $\parallel$ $[111]$	$\vec{E} \parallel \vec{X}$		$\vec{E} \perp \vec{X}$	
	Expt.	Theory	Expt.	Theory
A_1	16%	9%	6%	8%
A_2	0	0	24%	24%
A_3	84%	89%	17%	18%
A_4	0	2%	54%	50%

X $\parallel$ $[001]$	$\vec{E} \parallel \vec{X}$		$\vec{E} \perp \vec{X}$	
	Expt.	Theory	Expt.	Theory
B_1	100%	95%	25%	27%
B_2	0	5%	75%	73%

evaluate individually S_{e-ph} and S_{h-ph} for these transitions.
Although in Ge a piezospectroscopy experiment has been performed
on the LA phonon-assisted transition its interpretation is more
complex due to the fact that there are two intermediate electron
states. However, from this work as well as the WMA spectrum of
the TO-phonon assisted transition at zero stress valuable
information about S_{e-ph} and S_{h-ph} for both these transitions was
obtained.

The success of piezospectroscopy experiments in the above
materials strongly suggests the application to other indirect gap
binary semiconductors such as AlSb,[44] AlAs, etc. as well as
indirect alloy materials such as GaAlAs, GaAsP, etc. In addition,
information about intervalley EP and HP scattering matrix elements
could also be obtained from ultra high pressure experiments (such
as in a diamond anvil cell) where a direct gap material, such as
GaAs, can be made indirect.[45]

Motivated by the experimental results we then performed
theoretical calculations of S_{e-ph} and S_{h-ph} in Si and Ge, using
the "rigid-pseudoion" model. The Γ-Δ and Γ-L processes were
calculated in both materials. From the results of this work we
were able, for the first time, to account for the relative as well
as absolute intensities of the indirect transition occurring at
the fundamental gaps of both Si (Γ-Δ) and Ge (Γ-L). The results
of Ref. 13 for Si, using the "rigid-pseudoion" model but a
different extrapolation of the form factor as $q \to 0$, are, also,
in good agreement with experiment. In Ge, we were able to
explain the polarization dependent relative intensities of the LA
transition in the strained crystal by considering both intermediate electron states.

Based on the success of both our calculations and those of
Ref. 12 we conclude that the "rigid-pseudoion" model provides a
good description of the EP and HP intervalley scattering matrix
elements in Ge and Si. One of the questions to be resolved is
the difference in the extrapolation of the form factor between our
work and that of Ref. 13. This may be due to the empirical nature
of the pseudopotential that has been used and suggests that a more
self-consistent approach should be tried. Also the extension of
the present work to zincblend-type materials is an area of
considerable interest.

ACKNOWLEDGEMENTS

We acknowledge the support of the Office of Naval Research
through Contract No. N00014-78-C0890 for portions of this work.
Also we wish to express our thanks to Dr. George Wright for his

help and encouragement during some difficult phases of this
investigation.

REFERENCES

1. See, for example, O.J. Glembocki and F.H. Pollak, Phys. Rev.
    B25, 1179 (1982) and references therein.
2. See, for example, O.J. Glembocki and F.H. Pollak, Phys. Rev.
    B25, 1193 (1982) and references therein.
3. F.H. Pollak, A. Feldblum, H.D. Park and P.E. Vanier, Solid
    State Comm. 28, 161 (1978).
4. F.H. Pollak, A. Feldblum, H.D. Park and P.E. Vanier, Proceed-
    ings of the Fourteenth International Conference on the
    Physics of Semiconductors, ed. by B.L. H. Wilson (Institute
    of Physics, Bristol and London, 1979), p. 867.
5. O.J. Glembocki, Ph.D. Thesis, City University of New York,
    1982 (unpublished).
6. C.L. Collins and P.Y. Yu, Phys. Rev. B27, 2602 (1983).
7. See, for example, B.R. Nag, Electron Transport in Compound
    Semiconductors (Springer-Verlag, New York, 1980).
8. See, for example, J.D. Wiley, in Semiconductors and Semi-
    metals, ed. by R.K. Willardson and A.C. Beer (Academic
    Press, New York, 1975), Vol. 10, p. 91.
9. S. Kratzer and J. Frey, J. Appl. Phys. 49, 4064 (1978).
10. P.J. Vinson, C. Pickering, A.R. Adams, W. Fawcett and G.D.
    Pitt, in Proceedings of the Thirteenth International
    Conference on the Physics of Semiconductors, edited by
    F.G. Fumi (Tipogravia, Rome, 1976), p. 1243.
11. U. Rossler, E. Fornoff and R.G. Humphreys, J. Phys. C: Solid
    State Phys. 16, 259 (1983).
12. O.J. Glembocki and F.H. Pollak, Phys. Rev. Lett. 48, 415 (1982).
13. S. Bednarek and U. Rossler, Phys. Rev. Lett. 48, 1296 (1982).
14. O.J. Glembocki and F.H. Pollak, Phys. Rev. B25, 7863 (1982).
15. O.J. Glembocki and F.H. Pollak, Physica 117B and 118B, 546
    (1983).
16. F. Bassani and D. Guidotti, Physica 117B and 118B, 549 (1983).
17. S. Bednarek and U. Rossler, Phys. Stats. Sol. (b) 110, 565
    (1982).
18. In Ref. 4 LA phonon scattering via the $\Gamma_{15,c}$ intermediate
    state was omitted.
19. See, for example, F.H. Pollak, Surface Science 37, 863 (1973)
    and references therein; L.D. Laude, F.H. Pollak and M.
    Cardona, Phys. Rev. B3, 2623 (1971); G.L. Bir and G.E.
    Pikus, Symmetry and Strain-Induced Effects in Semiconduc-
    tors (John Wiley and Sons, New York, 1974).
20. G.G. MacFarlane, T.P. McLean, J.E. Quarrington and V. Roberts,
    Phys. Rev. 111, 1245 (1957).

21. The values of S_{e-ph} and S_{h-ph} quoted here are slightly different from those of Ref. 2 due to the use of somewhat more accurate values of some of the energy gaps and masses.

22. In many references, including Refs. 6-10 the electron (hole)-phonon coupling constant D is given in units of eV/cm. For a given phonon mode ℓ the parameters D and S are connected by the relationship $S^\ell = (\hbar/2M\omega_\ell)^{\frac{1}{2}}D\ell$.

23. J.M. Ziman, _Electrons and Phonons_ (Oxford University Press, London and New York, 1960).

24. L.J. Sham and J.M. Zimam in _Solid State Physics_, ed. by F. Seitz and D. Turnbull (Academic Press, New York, 1963), Vol. 15, p. 223.

25. P. Vogl, Phys. Rev. B$\underline{13}$, 694 (1976).

26. P. Vogl in _Physics of Nonlinear Transport in Semiconductors_, ed. by D.K. Ferry, J.R. Barker and C. Jacoboni (Plenum Press, New York, 1980) p. 75.

27. D.C. Herbert, J. Phys. C. $\underline{6}$, 2788 (1973); D.C. Herbert, W. Fawcett, A.H. Lettington and D. Jones, _Proceedings of the Eleventh International Conference on the Physics of Semiconductors_ (Polish Scientific Publishers, Warsaw, 1972), p. 1221.

28. M.L. Cohen and Y.W. Tang, _Proceedings of the Conference on the Physics of Semi-Metals and Narrow-Gap Semiconductors_, ed. by D.L. Carter and R.T. Bate (Pergamon Press, New York, 1971), p. 303.

29. L. Nordheim, Ann. Physik $\underline{9}$, 607 (1931).

30 See, for example, M.L. Cohen and V. Heine, _Solid State Physics_, ed. by H. Ehrenreich, F. Seitz and D. Turnbull (Academic Press, New York, 1970), Vol. 24.

31. M. Lax, _Symmetry Principles in Solid State and Molecular Physics_ (Wiley, New York, 1974).

32. P.B. Allen and M. Cardona, Phys. Rev. $\underline{27}$B, 4760 (1983).

33. S. Pantelides, private communication; J. Bernholc, N.O. Lipari and S.T. Pantelides, Phys. Rev. B$\underline{21}$, 3545 (1980).

34. M.L. Cohen and T.K. Bergstresser, Phys. Rev. $\underline{141}$, 789 (1966).

35. P.B. Allen, private communication; W. Weber, Phys. Rev. B$\underline{15}$, 4789 (1977).

36. W.A. Harrison, _Solid State Physics_, ed. by H. Ehrenreich, F. Seitz and D. Turnbull (Academic, New York, 1970), Vol. 24.

37. V. Heine, _Solid State Physics_, ed. by H. Ehrenreich, F. Seitz and D. Turnbull (Academic, New York, 1970), Vol. 24.

38. In Ref. 30 the importance of the large q region is noted for phonon calculations.

39. J.A. Verges, D. Glatzel, M. Cardona and O.K. Anderson, Phys. Stat. Sol. (b) $\underline{113}$, 519 (1982)

40. P.J. Dean, Y. Yafet and J.R. Haynes, Phys. Rev. $\underline{184}$, 837 (1969).

41. T. Nishino, M. Takeda and Y. Hamakawa, Solid State Comm. $\underline{12}$, 1137 (1973).

42. G.G. MacFarlane, T.P. McLean, J.E. Quarrington and V. Roberts, Phys. Rev. 111, 1245 (1958).
43. D.L. Smith and T.C. McGill, Phys. Rev. B14, 2448 (1976).
44. L. Laude, M. Cardona and F.H. Pollak, Phys. Rev. 1B, 1436 (1970).
45. D. Wolford, private communication.

NONLINEAR ELECTRONIC AND DYNAMICAL RESPONSE OF

SOLIDS IN THE ULTRASHORT TIME DOMAIN

W.E. Bron

Department of Physics
Indiana University
Bloomington, IN 47405

GENERAL INTRODUCTION

This lecture departs from many of the others of this volume
mostly because it concentrates on a comparison of experimental
results with theory and because it is being presented by an experimentalist rather than by a theorist. Actually, the material which
I present involves experimental results which have outpaced their
theoretical basis, not in the sense that a theoretical basis does
not exist, but rather that its application to the experimental
situation is either too cumbersome to carry out or some components
of the analysis, such as crystal potential or electronic wavefunctions, are not sufficiently known. Hence, a secondary task of mine
is to indicate areas in the electronic and dynamical response of
solids in which renewed efforts toward detailed, quantitative,
theoretical approaches appear to be appropriate.

Over the past decade it has become apparent that the nonlinear
response of solids offers a rich field of inquiry which often yields
more detailed information on the basic properties of solids than
does the linear response. In this lecture the nonlinear response is
to be presented in terms of measurements of two fundamental properties of the solid, namely, the lifetime of vibrational excitations
(phonons) and the third-order nonlinear (electronic) susceptibility,
$\chi_E^{(3)}$. Traditionally, these quantities have been determined in the
frequency domain through, respectively, measurements of the half-
width of the Raman scattering intensity, and the frequency dependence
of the three-wave mixing intensity. These same two quantities can
be determined by a single measurement in the time domain, namely,
the temporal response of the three-wave-mixing signal when the
frequency difference of two of the excitation fields is set to a

Raman resonance frequency. The two measuring methods complement
each other, in that the measurements in the frequency domain work
best in the range of short phonon lifetimes ($\lesssim$ 5 ps) and vica versa.

In any event, experimental results for these quantities are now
available for a series of solids. Although, qualitative agreement
is found between experiment and existing (for the most part ad hoc)
theoretical approaches, quantitative comparisons are poor.

OPTICAL PHONON LIFETIMES

Theoretical Background

The main focus of this section of the lecture is on the lifetimes of optical phonons. It is for these phonons that a new
experimental effort has produced new definitive and interesting
results. I will approach the theoretical basis of phonon lifetimes
in terms of very simple introductory ideas of phonons in general,
plus a short review of what is known about their lifetimes.

It is assumed that the reader has sufficient knowledge of the
general theoretical description of phonons so that only that part
of the description, which is pertinent to the discussion of later
sections, needs to be briefly reviewed here. (For a general
introductory treatment of the subject see Ashcroft and Mermin (1976).
For a more detailed discussion see, for example, Born and Huang
(1956), Ziman (1960), or Tucker and Rampton (1972).)

If the ions at one end of a crystal lattice are somehow suddenly
uniformly displaced from their equilibrium positions, as a consequence of interionic forces, information on the sudden displacement
propagates through the lattice in the form of a displacement wave.
Such a displacement wave can always be described in terms of linear
combinations of plane wave Fourier components of the form

$$\vec{\eta}_\alpha(\vec{k}) = \vec{A}_\alpha(\vec{k})\exp[i(\vec{k}\cdot\vec{r}_\alpha - \omega t)] \tag{1}$$

in which $\vec{\eta}_\alpha$ is the displacement of the αth ion (located at $\vec{r}_\alpha$ in the
lattice) due to the Fourier component with frequency ω and wave-
vector $\vec{k}$ after the elapsed time t. Since the lattice masses are not
continuous but discrete, the vector $\vec{k}$ must be one of the discrete
set of allowed vectors of the reciprocal lattice. The amplitudes
$|\vec{A}_\alpha|$ are at this point arbitrary except that the sum over all
Fourier components must yield the initial displacement wave.

It is often more convenient to describe the displacement wave
instead in terms of an orthonormal set of displacement vectors of
the form

$$\eta_\alpha^i(\vec{k},\lambda) = N^{-3/2} \varepsilon_\mu^i(\vec{k},\lambda) \exp[i(\vec{k}\cdot\vec{r}_\alpha - \omega t)] \tag{2}$$

in which i refers to the ith Cartesian component, ε is called the
polarization vector, it is assumed that there are N^3 unit cells in
the lattice, and we have introduced the branch index λ. (The total
number of branches equals three times s, the number of ions in the
unit cell, and μ runs over 1, 2,..., s.) For crystals with inversion
symmetry at all sites the polarization vectors ε may be chosen to be
real and to obey the following orthogonality and closure conditions:

$$\sum_i \varepsilon_\mu^i(\vec{k}\lambda) \varepsilon_\mu^i(\vec{k}\lambda') = \delta_{\lambda\lambda'} \tag{3}$$

and

$$\sum_\lambda \varepsilon_\mu^i(\vec{k}\lambda) \varepsilon_{\mu'}^{i'}(\vec{k}\lambda) = \delta_{\mu\mu'}\delta_{ii'} \quad . \tag{4}$$

The relation (2) can also be obtained by solving the 3s-dimensional
dynamical equations of motion of a harmonic lattice and are thereby
identified as the normal modes of vibrations of the lattice. As is
the case in electromagnetic displacements, it is often convenient to
replace the normal modes by an equivalent corpuscular description
of the lattice displacements. As a result it is usual to associate
a phonon with each of the normal modes.

The solutions of the dynamical equation also yield the relationship between the phonon frequency ω and $\vec{k}$ along the 3s branches, the
so-called dispersion relations. The density of allowed phonon
states, $g(\omega)$, with the same energy $\hbar\omega$ associated with any one branch
s turns out to be

$$g^s(\omega) = \frac{1}{2(\pi)^3} \int \frac{dS}{\vec{\nabla}_k \omega^s} \tag{5}$$

in which $\vec{\nabla}_k \omega^s$ is the gradient of ω with respect to $\vec{k}$ on the branch
s, and the integration is over all points in the dispersion relation
for which the phonon frequency equals ω. For an idealized isotropic
solid with linear dependence of ω on k (a dispersionless solid) and
with a Debye cut-off frequency ω_D, the total density of phonon
states are:

$$g(\omega) = \frac{3}{2\pi^2} \frac{\omega^2}{(\vec{\nabla}_k \omega)^3} \qquad \text{for } \omega < \omega_D$$

and $$\tag{6}$$

$$g(\omega) = 0 \qquad \text{for } \omega > \omega_D .$$

The starting point for most theoretical approaches to calculations of phonon lifetimes is the Hamiltonian of the harmonic

lattice in which freely propagating, but none interacting, phonons exist. To this are added perturbative terms describing the anharmonic components of the dynamical potential which lead to interactions among the phonons. In the standard Landau-Rumer (1937) approach the interactions are assumed to be weak and are taken into account perturbatively only to first or second lowest order.

Following the notation of Tucker and Rampton (1972) the elastic potential energy density in the crystal is then

$$W = \frac{1}{2} C_{ijkl}\, \eta_{ij}\eta_{kl} + C_{ijklmn}\eta_{ij}\eta_{kl}\eta_{mn} + \text{h.o. terms} \tag{7}$$

in which C_{ijkl} and C_{ijklmn} are, respectively, the second and third order elastic constants and the η_{ij} are the components of the deformation tensor. The elastic constants, of course, reflect the crystal in which the elastic wave described by the η_{ij} is propagating.

It is generally more convenient to transform to normal mode coordinates and into second quantized form. Then the third order term in the elastic potential energy expansion becomes

$$\Phi_3 = -1/6(\hbar^{(3/2}/2^{3/2}N^{1/2}) \sum_{\substack{q_1 q_2 q_3 \\ j_1 j_2 j_3}} M^{q_1 q_2 q_3}_{j_1 j_2 j_3} A^{q_1 q_2 q_3}_{j_1 j_2 j_3} \tag{8}$$

where

$$M^{q_1 q_2 q_2}_{j_1 j_2 j_3} = \phi^{q_1 q_2 q_3}_{j_1 j_2 j_3} [\omega_{j_1}(\vec{q}_1)\omega_{j_2}(\vec{q}_2)\omega_{j_3}(\vec{q}_3)]^{-1/2} , \tag{9}$$

$$A^{q_1 q_2 q_3}_{j_1 j_2 j_3} = [a_{j_1}(-\vec{q}_1)-a^*_{j_1}(\vec{q}_1)][a_{j_2}(-\vec{q}_2)-a^*_{j_2}(\vec{q}_2)]$$
$$[a_{j_3}(-\vec{q}_2)-a^*_{j_3}(\vec{q}_3)] \tag{10}$$

and $\phi^{q_1 q_2 q_3}_{j_1 j_2 j_3}$ is related to the third order derivative of the potential energy with respect to the lattice displacements as indicated in Eq. 4.9 of Tucker and Rampton (1972). In Eqs. (8) through (10) above, N is the number of normal modes, $\vec{q}_i$ the wavevector associated with the ith mode, j_i the mode branch index, and the a_i, a^*_i are creation and annihilation operators.

It is now possible to determine, through standard lowest-order time dependent perturbation theory, the rate at which Φ_3 may create or destroy a phonon in the mode $(j_1 q_1)$ with the proviso that energy and crystal momentum are conserved; the latter of which atleast to within a reciprocal lattice vector.

At this point it becomes necessary to choose a model for the
crystal so that the derivatives, ϕ, of the potential energy can be
calculated and so that the density of states available to the
phonons produced by the decay can be determined. For this purpose
one needs either the lattice dynamics or the dispersion relations
for the crystal, plus either the anharmonic components of the
vibrational potential or certain of its derivatives.

It is at this point that most of the treatments diverge. By
far the simplest, though least physical, assumption is to choose an
elastically isotropic, dispersionless Debye solid without a cut-off
frequency. In this case, it is possible to write down analytical
expressions for the decay of a longitudinal acoustic (LA) phonon
into another LA and a transverse acoustic (TA) phonon conserving
energy and wavevector along the way. Similar expressions for higher
order phonon-phonon interactions can also be obtained (Orbach and
Vredevoe (1964), Tucker and Rampton (1972), and Kwok and Miller
(1966)). For example, one finds for an ambient temperature $T_A \sim 0K$

$$\tau^{-1}_{\ell \to \ell+t} = \frac{\hbar[(C_1+C_4) + \frac{3}{2}(C_2+C_5)]^2}{16\pi\rho^3 v_\ell^6 v_t^3}(1-\beta^2)(\beta^2-\beta+\frac{3}{8})\omega^5 \qquad (11)$$

in which ℓ=LA, t=TA, ρ is the mass density, v_ℓ and v_t the sound
velocities, $\beta = v_t/v_\ell$, and the C_i's are linear combinations of
second and third order elastic constants.

A related expression has been formulated by Klemens (1967) for
the spontaneous decay of a LA phonon into two TA phonons. In this
case it is found, again for $T_A \sim 0K$, that

$$\tau^{-1}_{\ell \to t+t} = \frac{3\pi}{4\sqrt{2}} \gamma^2 \omega \left(\frac{\hbar\omega}{Mv_\ell^2}\right) \left(\frac{\omega}{\omega_o}\right)^3 \beta \qquad (12)$$

in which M is the atomic mass, and γ is the Grueneisen constant.
It is important to note at this point that the LA phonon lifetime
against spontaneous decay varies as ω^{-5}, i.e., high frequency LA
phonons are predicted to decay much more rapidly than low frequency
phonons. In simple systems near $T_A \sim 0$, the spontaneous three-phonon
decay of LA phonons, described above, should dominate the observed
temporal evolution of nonequilibrium phonon distributions (Orbach
and Vredevoe, 1964).

The above expressions describe the decay of nonequilibrium LA
phonons in an ambient environment in which the temperature $T_A \sim 0K$.
In a technique, to be discussed later, it is possible to lift this
restraint on T_A which can then take on any value. Moreover, in
this case it is the spontaneous decay of optical phonons into pairs
of lower energy acoustic phonons which is probed. An approximate
expression for the decay time for this case has been obtained by

Klemens (1966), and is

$$\tau^{-1} \sim \omega_o \frac{J}{24\pi} \gamma^2 \frac{\hbar\omega_o}{Mv_o^2} \frac{a^3\omega_o^3}{v_g^3} [(N_o+1)N'N''-N_o(N'+1)(N''+1)] \tag{13}$$

in which the subscript o stands for the optical phonon which decays into two acoustic phonons denoted by superscript prime and double prime, v_g is the group velocity of the acoustic phonons, a^3 is the volume per atom, the Ns are phonon occupation numbers, and J is a number between 1 and 6. Expression (13) was derived by Klemens for the simple case of the Si lattice. The approximations used to account for the strain gradients associated with optical phonons limit the application of Klemen's result to monoatomic lattices.

The approaches discussed so far are not self-consistent in a number of ways. Perhaps the most important of these is the failure to correct the vibrational eigenstates of the lattice for the anharmonic component of the potential. This point has been addressed by a number of authors; we follow here the treatment by Maradudin and Fein (1962) who derive the scattering cross section for coherent inelastic scattering of thermal neutrons by an anharmonic Bravais lattice. The problem is solved again perturbatively considering only one-phonon interactions with the incident neutron. These phonons, however, interact with other phonons through third and fourth order anharmonic terms in the vibrational potential. The interactions lead to a complex shift of the neutron scattering resonances, which is calculated here in a high temperature (near room temperature) and a low temperature ($\sim 0°K$) regime. Since the low temperature regime favors three phonon processes over other phonon interactions, I will limit further comments to this case. The real part of the complex shift, of course, describes the renormalization of the vibrational states of the lattice, whereas the imaginary part of the shift is related to the reciprocal of the lifetime of the phonon which interacts with the neutron.

The problem is solved using an elegant phonon propagator and diagrammatic technique. The general solution for the reciprocal of the phonon lifetime at the absolute zero of temperature is

$$\tau^{-1}(\omega,\vec{q}j) = \frac{\pi\hbar}{16N\omega_3(\vec{q}j)} \sum_{\substack{q_1q_2q_3 \\ j_1j_2j_3}} \Delta(-\vec{q}_3+\vec{q}_1+\vec{q}_3)$$

$$\times \left| \phi^{q_1q_2q_3}_{j_1j_2j_3} \right|^2 [\omega(\vec{q}_1j_1)\omega(\vec{q}_2j_2)]^{-1} \tag{14}$$

$$\times \{\delta(\omega_3-\omega_1-\omega_2) - \delta(\omega_3+\omega_1\omega+_2)\} \, .$$

In. Eq. (14) the Δ and δ functions guarantee conservation of crystal momentum and energy, respectively.

At this point it is necessary, as in the previous discussion, to introduce a model for the anharmonic crystal. The authors choose a face centered cubic lattice with central forces between nearest neighbor ions only. The necessary derivatives of the anharmonic potential are then calculated for a lead crystal and fit to experimental values of the thermal expansion, compressibility, and the lattice parameter.

The numerical evaluation of the frequency shifts and the (reciprocal) lifetimes is carried out over a representative set of q_3-vectors and includes both "normal" and "umklapp" processes. Numerical results on lifetimes are obtained only for the high temperatures (425°K) for which fairly good agreement with experimental neutron widths is obtained. No comparison with low temperature results are reported.

Phonons once propagating in a crystal system undergo various other scattering interactions. Such scattering events, which cause a change in the phonon wavevector or phase, occur at crystal boundaries and as a result of interactions with lattice imperfections or with conduction electrons. It is possible experimentally to limit interactions with surfaces and electrons; the latter by concentrating on insulators and semiconductors with low carrier concentrations. That leaves imperfection scattering to consider.

Phonons may scatter inelastically from dislocations and scatter elastically from crystal imperfections such as vacancies, and isotopic and foreign-ion impurities. Theoretical expressions for the scattering rates at imperfections are in general quite complicated (see, for example, Klein (1968)). A qualitative insight into the scattering probability can, however, be gained through the assumption that the scattering is of the Rayleigh type (Klemens 1958), that the crystal is an isotropic dispersionless Debye solid, and that the disturbance is limited to the local change in mass, ΔM. Under these assumptions the elastic scattering rate, τ^{*-1}, becomes

$$\tau^{*-1} = (n/\eta)(\Delta M/M)^2 \omega^4 (4\pi v^3)^{-1} \tag{15}$$

in which ω is the phonon frequency, n is the defect concentration, η is the number of atoms per unit volume, and v is the average group velocity in the acoustic branches. Note should be taken that high-frequency phonons are predicted to be scattered much more strongly than low-frequency phonons.

It is worth noting that, according to Eq. (15), elastic impurity scattering of acoustic phonons can be approximated by

$\tau^* \sim A^*\omega^{-4}$, whereas Eqs. (7) and (8) predict that spontaneous anharmonic scattering of LA phonons can be approximated by $\tau \sim A\omega^{-5}$. Average values of A^*, obtainable from the literature for crystals containing ~100 ppm imperfections, are of the order of 10^{39} s^{-3}. Similarly, A is found in insulators to be typically of the order of 10^{54} s^{-4}. On the assumption that the phonon propagation velocity is approximately constant and equal to 5×10^3 ms^{-1}, one concludes that the mean free path against impurity and anharmonic scattering is much less than the dimensions of typically sized samples only if the phonon frequency $\nu \gtrsim 1$ THz.

Optical phonons, in contrast, normally possess low group velocities, so that impurity dominated elastic scattering times are long compared to those of most acoustic phonons. On the other hand, lifetimes against anharmonic decay typically lie in the picosecond regime, and thereby dominate the observable transport properties of these phonons.

<u>Nonlinear Response of Solids</u>

One of the most exciting prospects in the study of phonon dynamics is the use of nonlinear optical phenomena to generate monochromatic, coherent phonon distributions and to detect their subsequent evolution.

Of the various methods which fall under the category of non-linear optical techniques (Demtroder 1981, Levenson and Song 1980, Laubereau and Kaiser 1978, von der Linde 1981), the one which so far appears to have the most potential is the generation of short duration (subnanosecond) coherent phonon packets through coherent excitation, and the detection of the packet by coherent antiStokes Raman scattering (CARS).

According to Placzek (1934), in a Raman active medium the important terms in the optical polarizability, α, can be written as

$$\alpha = \alpha_o + (\partial\alpha/\partial Q_v)Q_v \qquad\qquad (16)$$

in which Q_v is the coordinate corresponding to some normal mode of vibration (phonon) of the solid, and α_o refers to a static polarizability if one is present.

An electromagnetic field, E, can interact with this polarization. The pertinent term in the interaction Hamiltonian, H_I, corresponding to the lowest order Raman process, is

$$H_I = -\frac{1}{2} (\partial\alpha/\partial Q_v)Q_v E^2 \; . \qquad\qquad (17)$$

It follows that as a result a force

$$F = - \frac{\partial H_I}{\partial Q_v} = \frac{1}{2} (\partial \alpha / \partial Q_v) E^2 \tag{18}$$

acts on the vibrational system. Moreover, under the action of both the vibrational and the em field, a polarization, P, is induced with

$$P = - \frac{\partial H_I}{\partial E} = N(\partial \alpha / \partial Q_v) Q_v E \tag{19}$$

in which N is the number density of the vibrational modes. For simplicity we associate with each vibrational mode a damped oscillator of mass m, and write according to Newton's second law an equation of motion,

$$\ddot{Q}_v + 2\Gamma \dot{Q}_v + \omega_v^2 Q_v = \frac{1}{2m} (\partial \alpha / \partial Q_v) E^2 \tag{20}$$

in which Γ describes the damping of Q. Under coherent excitation Q becomes the coherent amplitude and Γ^{-1} is related to the dephasing time T_2. As we shall see, it is actually T_2 which is measured.

In the actual experiment we chose an electromagnetic field which contains two frequency components, ω_ℓ and ω_s, produced by two well defined coherent laser beams propagating with wavevector $\vec{k}_\ell$ and $\vec{k}_s$. The em field amplitude of these beams is

$$E(\vec{x},t) = \frac{1}{2} E_\ell \exp[i(\omega_\ell t - \vec{k}_\ell \cdot \vec{x})] \; +$$

$$\frac{1}{2} E_s \exp[i(\omega_s t - \vec{k}_s \cdot \vec{x})] + cc \; , \tag{21}$$

in which $\vec{x}$ is the position vector and t is the temporal coordinate, respectively. It is to be understood that the laser beam frequencies ω_ℓ and ω_s will be in the visible region, whereas ω_v is in the far infrared. Hence $\omega_\ell, \; \omega_s \gg \omega_v$.

In order to solve (20) for Q_v, we use the trial function

$$Q_v = \frac{1}{2} q_v \exp(i\omega t) + cc \tag{22}$$

and pick the terms which are consistent with the inequalities in ω_ℓ, ω_s and ω_v, i.e.,

$$(\omega_v^2 - \omega^2 - i2\Gamma\omega) q_v \exp(i\omega t) = \frac{1}{2m} (\partial \alpha / \partial Q_v) E_\ell E_s \exp[i(\omega_\ell - \omega_s)$$

$$t - (\vec{k}_\ell - \vec{k}_s) \cdot x] \tag{23}$$

Note should be taken that the temporal terms in Eq. (23) require that $\omega = \omega_\ell - \omega_s$, i.e., that the vibrational system is being driven at the difference frequency of the two lasers.

Solving Eq. (23) for Q_v yields

$$Q_v = \frac{1}{4m} \ \frac{\partial\alpha/\partial Q_v}{[\omega_v^2-(\omega_\ell-\omega_s)^2+i2\Gamma(\omega_\ell-\omega_s)]} \tag{24}$$

$$\exp[-i(\omega_\ell-\omega_s)t-(\vec{k}_\ell-\vec{k}_s)\cdot x]$$

It is clear from (24) that a strong resonant excitation can occur if $\omega_\ell-\omega_s$ equals the frequency of a Raman mode, ω_v, providing that the phase matching condition $\vec{k}_v=\vec{k}_\ell-\vec{k}_s$ can be met.

If the durations of the exciting laser pulses Δt_ℓ and Δt_s are short compared to the relaxation time of the phonon modes, then the excitation undergoes the equivalence of "free fall decay." This decay process can be monitored by a Raman interaction between the vibrational mode and a time delayed third pulsed laser beam. This interaction can also be driven coherently as described below.

The presence of the coherent phonon packet causes an oscillatory polarizability at frequency ω_v, as indicated in Eq. (16), which can in turn interact with the third laser at frequency ω_3 to yield a nonlinear polarization, P^{NL}, (see Eq. (19))

$$P^{NL} = \chi^{(3)}E_\ell E_s E_3 \tag{25}$$

where $\chi^{(3)}$ is the so called total third-order nonlinear susceptibility which contains (as we shall see) contributions from the response of the electronic and the lattice dynamical system. The polarization acts as a source term in Maxwell's equation (see e.g., Armstrong et al. 1962 and Maker and Terhune 1965) to produce a strong coherent output beam with $\omega_4 = \omega_3+(\omega_\ell-\omega_s)$ provided that $\vec{k}_4 = \vec{k}_\ell-\vec{k}_s+\vec{k}_3$ can be satisfied. It is usual to set $\omega_3=\omega_\ell$ by taking a small amplitude component of the incident laser beam. Therefore, in practice, $\omega_4 = 2\omega_\ell-\omega_s$. The corresponding energy and wavevector diagrams are shown in figure 1. It is clear from the figure that the interaction leads to coherent antiStokes Raman scattering, CARS. Accordingly, we identify ω_{AS} with ω_4 and $\vec{k}_{AS}$ with $\vec{k}_4$.

It follows from the above discussion that the intensity of the CARS signal as a function of the delay time between the pump and the probe laser can be written as

$$I(\Delta t) = AS(\Delta k)\int_{-\infty}^{+\infty}dt\left|\vec{E}_p(t-\Delta t)[NR_aQ(t)+3\chi_E^{(3)}\vec{E}_\ell(t)\vec{E}_s^*(t)]\right|^2 \tag{26}$$

where

$$S(\Delta k) = \sin^2(\Delta kL/2)/(\Delta kL/2)^2 \tag{27}$$

and

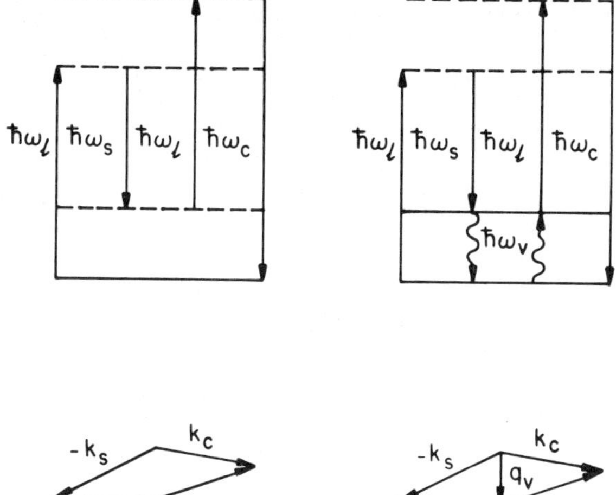

Fig. 1. Energy level system and electromagnetic and lattice transitions for the nonresonant (1a) and resonant (1b) CARS technique.

$$\Delta\vec{k} = \vec{k}_c - [\vec{k}_\ell + \vec{k}_p - \vec{k}_s] \tag{28}$$

and

$$\ddot{Q} + \Gamma\dot{Q} + \omega_v^2 Q = R_a m^{-1} \vec{E}_\ell(t)\vec{E}_s^*(t) \tag{29}$$

In these expressions R_a is the Raman tensor, and the coherent amplitude, Q, of the excited phonon packet is given by Eq. (29). The wavevector mismatch is given by Δk; it is set experimentally to approximately zero (hence, $S(\Delta k)\sim 1$). Finally, Δt is the temporal delay between the pulses of the probe laser, p, and those of the lasers ℓ and s; m is the reduced lattice mass, N is the number of primitive cells per cm^3, and $A = 2\pi\omega_c^2 L^2/c\varepsilon_c$. In the expression for A, c is the speed of light, L is the effective length within the medium over which spatial overlap of the three-wave mixing exists, and ε_c is the dielectric constant of the medium at ω_c.

According to Eq. (26), $I(\Delta t)$ has two components. If $\omega_\ell - \omega_s$ is detuned from the LO phonon frequency, ω_v, Q becomes very small and only that part of Eq. (26) controlled by $\chi_E^{(3)}$ remains. Thus a purely "electronic" part of $I(\Delta t)$ (i.e., independent of any Raman excitation) appears under these circumstances. If, on the other hand, $\omega_v = \omega_\ell - \omega_s$, the curve with crosses appears. The exponentially decreasing part of the signal is controlled by Γ and corresponds to the dephasing of the coherent LO phonon packet referred to earlier.

Experimental Results

Application of the nonlinear response of solids to the determination of the lifetimes of optical phonons from time domain studies has received recent attention (Gale and Laubereau (1983), Kuhl and Bron (1984), Schosser and Dlott (1984)). I present here as model cases results for high purity GaP and ZnSe. These measurements are facilitated by recent rapid advances in commercially available mode-locked lasers and synchronously driven dye lasers. This system generates picosecond duration pulses at high repetition rates, thus it becomes possible to observe decay processes which are typically in the 1 to 10 psec range.

The experimental apparatus is depicted in figure 2. A mode-locked, argon-ion laser is used to pump synchronously two matched dye lasers. The dye laser autocorrelation pulse typically has durations of 2 to 3 ps FWHM, a repetition rate of 80 MHz and an average power in each output beam of 30-40 mW. The dye laser outputs are tuned to the frequencies ω_ℓ and ω_s such that $\omega_\ell - \omega_s$ equals the frequency of the LO phonons, $\omega_v = \omega_{LO}$. When these two beams are spatially and temporally focused in the GaP crystals, such that the wavevector conservation is obeyed, a coherent packet of LO phonons propagating along $\vec{k}_v$ is generated.

The subsequent temporal variation of the coherent LO phonon packet is measured through the CARS technique. For this purpose a small part of the ω_ℓ beam is extracted and delayed by an optical delay line, so that pulses from it appear at the focal point at any delayed time, Δt, of between 0 and 0.3 ns, with a resolution of fractions of picoseconds. As described above, the delayed beam scatters from the component of the packet which remains coherent after Δt and produces a coherent beam propagating along $\vec{k}_{AS}$.

Figure 3 is an illustration of the typical temporal variation of the CARS signal intensity, $I(\Delta t)$, from LO phonons with $\omega_v = 403$ cm^{-1} and an ambient temperature of 5 K for a GaP crystal with a carrier concentration of less than 10^{16} cm^{-3}. The component of the signal from ~ -40 to ~ $+20$ ps is primarily the result of nonlinear excitation of the electronic system. The full shape of this signal (dashed curve) can be obtained by slightly detuning $\omega_\ell - \omega_s$ away from ω. This part of the signal is controlled by $\chi_E^{(3)}$ and will be discussed in the next section.

The signal beyond 20 ps is a measure of the temporal evolution of the coherent intensity $<Q>^2 \alpha [-2t/T_2]$, in which $<Q>$ is the coherent amplitude of the excited Raman mode (Laubereau and Kaiser,

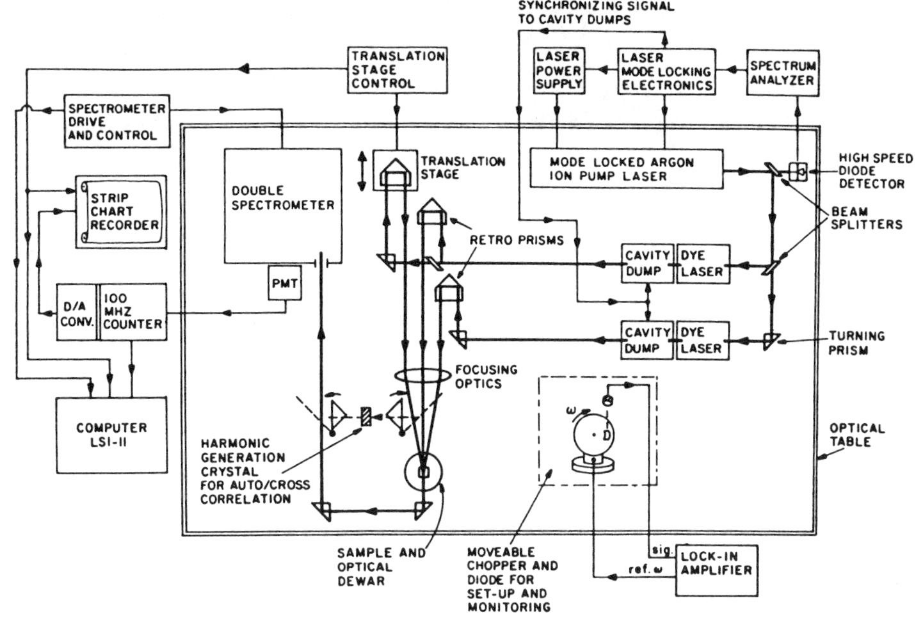

Fig. 2. CARS apparatus.

1978). The intensity is seen from figure 2 to decay exponentially over several orders of magnitude with a time constant $\tau = T_2/2$. Here T_2 is the traditional <u>dephasing time</u>. The temperature dependence of τ (and the uncertainty in its measurement), as observed over a range from 5 to 300 K, is shown in figure 4 (full circles).

It is generally accepted that the linewidth $\delta\nu$ (in cm^{-1}) of the same Raman mode, ω_v, as observed from spontaneous Raman scattering is related to T_2 by $\delta\nu = (\pi c T_2)^{-1}$ provided that the Raman line is homogeneously broadened (see, e.g., Laubereau and Kaiser, 1978).

Linewidth measurements of the LO phonon (at $\vec{k}\sim0$) have previously been reported by Bairamov, et al. (1974, 1979) who used a high resolution Fabry-Perot interferometer. The corresponding dephasing times are designated as BPTU in figure 4 (open circles). We have repeated these measurements using our sample and a standard Raman double monochromator providing 0.15 cm^{-1} resolution. Phonon dephasing times are obtained by deconvoluting the instrument profile from the measured Raman spectrum assuming a Gaussian profile for the spectrometer function and a Lorentzian for the Raman line. The results are displayed as crosses in figure 4.

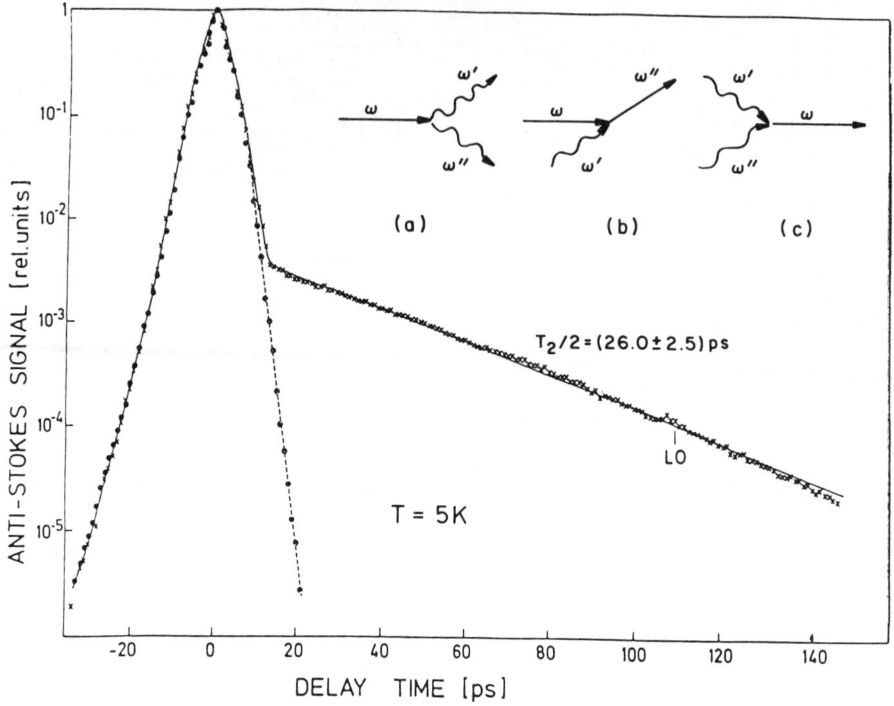

Fig. 3.   Time dependence of the CARS signal for GaP at 5K.

The time, T_2, includes contributions from all dephasing phenomena such as elastic scattering from surfaces, from impurities, from imperfections, and from electronic carriers, etc., and also includes phonon–phonon scattering which leads to phonon population decay. The population decay time is conventionally referred to as T_1. Since T_2 contains T_1 it follows that $T_2 \leq T_1$.

In the absence of any detailed knowledge of the strength of the various scattering processes it is not possible to calculate the magnitude of T_2, although it is possible to predict theoretically its temperature dependence. Contributions to the temperature dependence of T_2 can arise from scattering from thermally activated carriers and from the temperature dependence of the various depopulation processes. We concentrate here on the latter processes. At low ambient temperatures, such that $\hbar\omega_V \gg k_B T$, the spontaneous three-phonon decay of $\vec{q} \approx 0$, optical phonons into two acoustic phonons (plus the corresponding recombination process) should dominate all other phonon–phonon interactions (Orbach and Vredevoe 1964).

The possible three-phonon decay channels are illustrated in the inset to figure 3. Channel (b) is not active since it would produce a phonon of higher energy than the LO phonon which is not

508

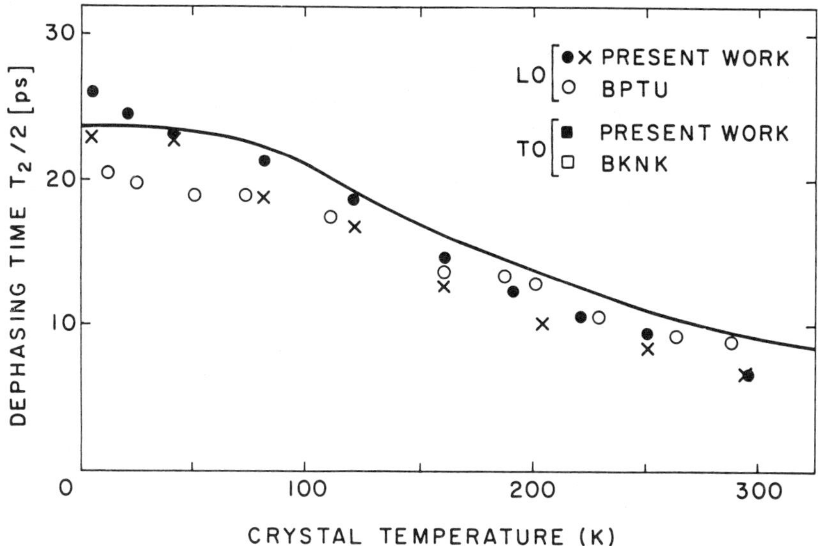

Fig. 4.   Temperature dependence of $T_2/2$ for GaP.   Symbols—data; solid line—theory.

possible in GaP. Accordingly, the temperature dependent part of the expression for the inverse lifetime of the LO phonons is (Klemens 1966)

$$\Gamma^{-1} = \Gamma_0^{-1}[1+N'+N'']$$

(30)

in which Γ_0 is the lifetime at $T=0°K$ and the N's are the phonon occupation numbers of the acoustic phonons. Eq. (30) also reflects the fact that the topography of the dispersion relations of GaP (see for e.g., Bilz and Kress, 1979) suggest that the decay proceeds primarily as $LO(\omega,o)\rightarrow LA(\omega/2,k)+LA(\omega/2,-k)$. Expression 30, with its magnitude fitted to the low temperature data, appears as the solid line in figure 4. Comparison with the data shows indeed that the low temperature dependence of T_2 is dominated by three-phonon inter- actions, i.e., by T_1. Beyond this temperature additional processes, including higher order phonon-phonon interactions and scattering from thermally activated carriers, must also be considered.

Quite similar results and agreement have been obtained for the temperature dependence of the LO phonon in ZnSe eventhough the topography of the dispersion relations for this crystal allow decay to various sets of acoustic phonons. The fit of Eq. (30) for this

509

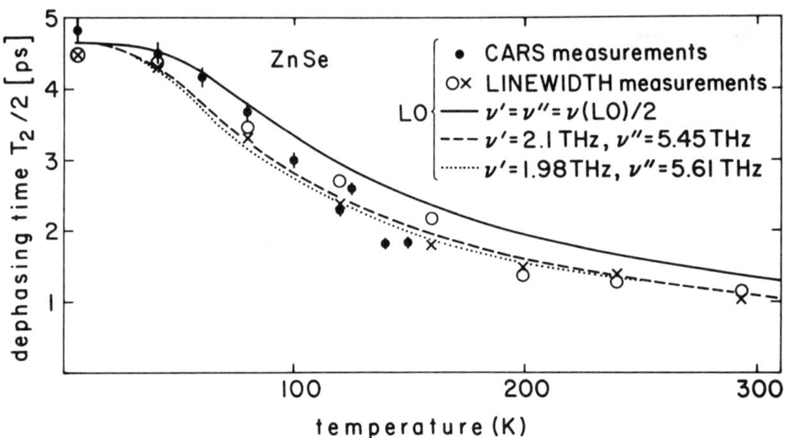

Fig. 5. Temperature dependence of $T_2/2$ for ZnSe.
Symbols--data; lines--theory.

510

case is given in figure 5 for each of the possible decay routes
(see insert to figure 5).

We conclude, therefore, that it is possible to measure the
temperature dependence of the dephasing time of LO phonons in GaP
and ZnSe and to account for the observation in terms of the simple
Bose-Einstein dependence on phonon occupation numbers. Although
this procedure does indeed yield quite good agreement between experiment and theory, no theoretical apparatus exists for the prediction
of the magnitude of T_2 at T=0K.

The only readily usable formulation for T_1 is that by Klemens
(1966) which does not directly apply to diatomic lattices, and if
forced to apply to such lattices yields values of T_1 one to two
orders of magnitude smaller than those observed.

THIRD ORDER NONLINEAR ELECTRONIC SUSCEPTIBILITY

Theoretical Background

The third order nonlinear susceptibility, $\chi^{(3)}$, has already
been defined in terms of Eq. (25). I will not attempt to arrive at
a theoretical description of $\chi^{(3)}$ to the same level as was done
above for T_2. This is not due to any special difficulty in the
derivation but rather due to the fact that the nomenclature becomes
rather cumbersome and, moreover, a number of excellent reviews of
the theory appear elsewhere (see, for example, Flytzanis, 1978, or
Flytzanis and Bloembergen, 1976).

Accordingly, I will present only the result which is applicable
to the experimental conditions to be presented below. The pertinent
expression for the case of three interacting electromagnetic fields
($\vec{E}_\ell(\omega_\ell)$, $\vec{E}_s(\omega_s)$, $\vec{E}_p(\omega_p=\omega_\ell)$) which produce a CARS signal at
$\omega_{AS}=2\omega_\ell-\omega_s$ is

$$\chi^{(3)}_{xyxy}(-\omega_c,\omega_\ell,\omega_\ell,-\omega_s) = \chi^{(3)}_{xyxyE}$$

$$-\frac{4\pi(\Delta\hat{k}_z)^2}{3\varepsilon(\Delta\omega)}\,[\chi^{(2)}_{xyzE}]^2 + \frac{4\pi(1-(k_z/k'))^2}{3[(ck/\omega)^2-\varepsilon(2\omega_\ell)]}\,[\chi^{(2)}_{xyzE}]^2$$

$$+ (\Delta\hat{k}_z)^2\,\frac{N\alpha_{LO}^2}{3M(\omega_{LO}^2-\Delta\omega^2+i\Delta\omega\Gamma)}$$

$$(31)$$

$$+ \frac{4\pi(1-\hat{\Delta k}_z^2)}{3[(c\Delta k/\Delta\omega)^2-\varepsilon(\Delta\omega)]} \{[\chi_{xyzE}^{(2)}]^2$$

$$+ \frac{N\chi_{xyzE}^{(2)}\ \ell_{TO}^*\alpha_{TO}^{(1)}}{M(\omega_{TO}^2-\Delta\omega^2+i\Delta\omega\Gamma)}$$

$$+ \frac{N\alpha_{TO}^2}{4M(\omega_{TO}^2-\Delta\omega^2+i\Delta\omega\Gamma)}\ [(c\Delta k/\Delta\omega)^2-\varepsilon(\Delta\omega)]\} \ .$$

In this expression $\chi_{ijk\ell E}^{(i)}$ refers to the purely electronic contribution, $\vec{\Delta k} = \vec{k}_\ell - \vec{k}_s$, $\hat{\Delta k}_z$ is a unit vector in the direction of the phonon wavevector, $\varepsilon(\nu)$ is the dielectric constant at the frequency ν, $[\chi_{xyzE}^{(2)}]^2$ are contributions to the total $\chi_E^{(3)}$ from so-called two-step order nonlinear processes, k' is the wavevector for the frequency $2\omega_\ell$, $\Delta\omega=\omega_\ell-\omega_s$, α_{TO}, α_{LO} are, respectively, the Raman tensor at the TO and LO frequencies, M is the reduced lattice mass, $\vec{k}'$ is the wavevector at $2\omega_\ell$.

When $\omega_\ell-\omega_s=\omega_{LO}$, and in the absence of any contributions from two-step second order nonlinear processes, Eq. (31) reduces to

$$\chi_{xyxy}^{(3)}(-\omega_c,\omega_\ell,\omega_\ell,-\omega_s) = \chi_{xyxyE}^{(3)}$$

$$- (\hat{\Delta k}_z)^2 \frac{iN\alpha_{LO}^2}{3M\Delta\omega\Gamma} \tag{32}$$

$\chi_{xyxyE}^{(3)}$ is the purely electronic contribution to the third-order nonlinear response, whereas the second term describes the damped response of the LO phonons. The relation of these terms to the two components of $I(\Delta t)$ as given in Eq. (26) is straightforward. Consequently, if by experimental means, the two contributions to $I(\Delta t)$ can be separated from each other, the contributions from the lattice and that of the electronic response can be separately determined. This is the procedure we have carried out for the example case of GaP.

Experimental Results

It follows from Eqs. (26) and (29) that, once Γ is determined, by the method described in the earlier sections of this paper, a further analysis of the full $I(\Delta t)$ becomes possible in terms of the integration over the temporal profiles of the laser fields $E_i(t)$, plus an independent evaluation of the Raman tensor elements $\bar{R}_a$. The result of this procedure leads to a determination of the components of $\chi^{(3)}_{ijk\ell E}$. In the analysis presented here (Rhee, Bron, and Kuhl, 1984), the actual form of the temporal profiles of the laser pulses is not directly obtained from $I(\Delta t)$ but is extracted from the auto- and cross-correlation signals which in the experiment are obtained by mixing the ℓ and s lasers in a KDP crystal. The observed autocorrelation signal can be fitted well by the slightly asymmetric exponential function $f(t)=\exp(\delta\gamma t)$ for $t<0$ and $f(t)=\exp(-\gamma t)$ for $t\geq 0$. We further assume that the observed broadening of the cross-, compared to the auto-correlation signal can be accounted for by a Gaussian temporal fluctuation (jitter) between the ℓ and s lasers. Since the final $I(\Delta t)$ involves the product of the three fields E_ℓ, E_s, E_p, we shift for convenience the Gaussian broadening totally on to E_s and write its temporal profile as $f'(t)=[\exp(\delta\gamma t)-(1/z)\exp(z\delta\gamma t)]/(1-1/z)$ for $t<0$ and $f'(t)=[\exp(-\gamma t)-(1/z)\exp(-z\gamma t)]/(1-1/z)$ for $t\geq 0$. Here we have taken advantage of the expansion for small values of α of the Gaussian profile $\exp(-\alpha t^2)$.

Every effort is made before each experimental run to maximize the temporal overlap (at $\Delta t=0$) between the ℓ and s laser. All the same, a temporal mismatch, $\Delta\tau$, of the order of $\lesssim 0.5$ ps is often found between the two lasers. (The mismatch can be extracted from the slight asymmetry which it causes in the CARS signal in the vicinity of the peak at $\Delta t=0$). Finally, the relative contributions of the two components of $I(\Delta t)$ can be experimentally specified by a quantity, Y, defined as the difference between the peak of the normalized CARS intensity at zero time delay, $I(\Delta t=0)$ and the intensity of the phonon dephasing component extrapolated back to $\Delta t=0$. This procedure is illustrated in figure 6 for a subset of the laser field polarizations indicated in Table I. (In Table I the polarization direction is given in terms of an angle measured relative to a direction in the plane of the optical table and which is coplanar with the (010) and (100) crystal directions which lie in a plane perpendicular to the optical table.)

At this point the entire function, $I(\Delta t)$, can be specified (for each polarization condition) in terms of Y, Γ, γ, δ, z, and $\Delta\tau$. The first four quantities are obtained directly from $I(\Delta t)$. For the asymmetry factor, δ, we choose 1.3, which is the average value obtained from a set of $I(\Delta t)$ measured with the $\omega_\ell-\omega_s$ detuned from ω_v. The remaining quantities are determined from a best fit to each $I(\Delta t)$ as shown by the solid lines in figure 6.

It is important to recognize that R_a and $\chi_E^{(3)}$ as they appear in Eqs. (26) and (29) are each effective quantities whose magnitudes depend on the actual polarization (relative to the crystal axes) of the fields $\vec{E}_\ell$, $\vec{E}_s$, and $\vec{E}_p$. At resonance the phonon part of $I(\Delta t)$ is purely imaginary (see Eq. (32)). Moreover, $\chi_E^{(3)}$ is in general complex; particularly, in the present case (Levenson and Bloembergen, 1974) for which $2\hbar\omega_\ell$ (or $2\hbar\omega_s$) exceeds the fundamental

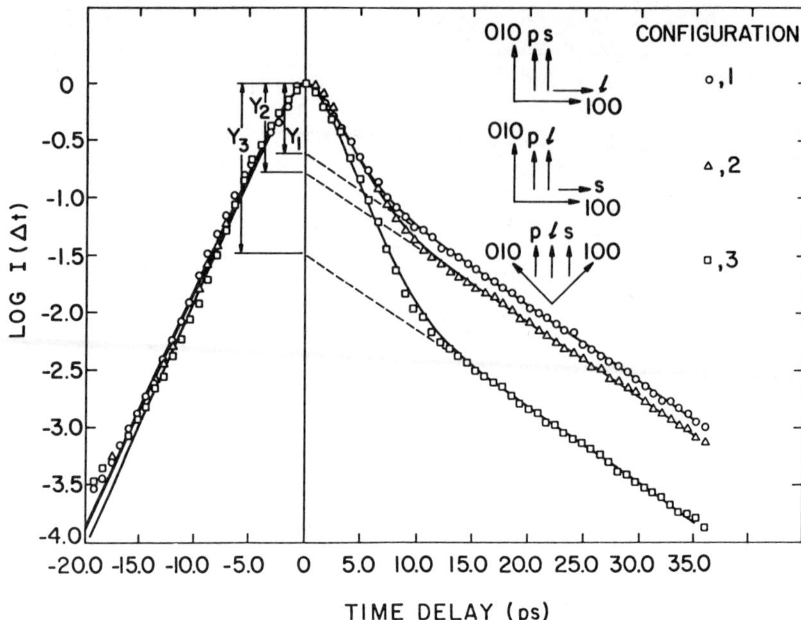

Fig. 6. Experimental data (symbols) and fitted I(Δt) (solid lines)
for each of the polarization configurations indicated in
the inset to the figure and defined in Table I.

gap energy, i.e., in the case that two photon absorptive processes
may occur.

The complex effective $\chi_E^{(3)}$ for the various polarization conditions are listed in Table I. Note that four out of the seven
configurations lead to redundant results. A grand average over 32
runs, distributed over all seven configurations, together with the
published value of R_a for GaP (Calleja, et al., 1982), leads to the
following values of the magnitude of the real, χ', and of the
imaginary, χ'', parts of the active components of $\chi_E^{(3)}$ (in units of
10^{-10} e.s.u.):

$$\chi'_{1221}=2.1\pm0.15, \quad \chi''_{1221}= -0.48\pm0.24; \quad \chi'_{1122}=1.8\pm0.44,$$

$$\chi''_{1122}= -0.7\pm0.48; \quad \chi'_{1111}=2.1\pm0.7, \quad \chi''_{1111}= -0.73\pm0.81.$$

The imaginary part of $\chi_E^{(3)}$ has its main effect on I(Δt) through
crossterms with the phonon part. For values of $\chi'' < 10^{-10}$ e.s.u.
these crossterms effect only slightly the curvature of I(Δt) in the
region (Zinth, et al., 1978) $0 < \Delta t < 10$ ps; hence, the relatively
large uncertainty in χ''. It is observed that the sign of χ'' is
opposite to that of χ' and to that of the (positive definite)
phonon part.

TABLE I

Laser Polarization, Crystal Axis Orientation and Effective $\chi^{(3)}$

Configuration	Polarization Condition (in degrees)					Effective $\chi^{(3)(b)}$
	Lasers			Analyser[a]	Crystal Axis	
	p	ℓ	s		010	100	
1	90	90	180	180	90	180	χ_{1221}
2	90	180	90	180	90	180	χ_{1122}
3	90	90	45	135	45	135	$2\chi_{1122}$
4	90	90	90	90	45	135	$\frac{1}{2}(\chi_{1111} + 2\chi_{1122} + \chi_{1221})$
5	90	90	45	45	45	135	$\chi_{1111} + \chi_{1221}$
6	90	45	90	45	45	135	$\chi_{1111} + \chi_{1122}$
7	90	45	90	135	45	135	$\chi_{1122} \quad \chi_{1221}$

(a) The analyser is placed into the CARS beam.
(b) See e.g. reference 1 for an explanation of the subscript convention.
     The effective $\chi^{(3)}$ are derived here such that the phonon part of $I(\Delta t)$
     is normalized to unity.

In order to avoid complications arising from high lying conduction bands, most previous experimental determinations of $\chi_E^{(3)}$, and most theoretical evaluations (see e.g., Flytzanis, 1978) have been limited to the "zero frequency limit," i.e., to the case that $2\hbar\omega_\ell$ (and $2\hbar\omega_s$) is much less than the gap energy, E_g, of the medium so that two-photon absorptive processes can be neglected. This constraint means that $\chi_E^{(3)}$ is real, and also leads to the so-called Kleinman symmtery (Kleinman, 1962) for which $\chi'_{1122}=\chi'_{1221}$. This condition is not met in the present experiment on GaP for which $E_g=2.26$ eV whereas $2\hbar\omega_\ell=4.30$ eV ($2\hbar\omega_s=4.2$ eV). It is, accordingly, not unexpected that we find $\chi_E^{(3)}$ to be complex and that $\chi'_{1122}\neq\chi'_{1221}$; the observed ratio $\chi'_{1221}/\chi'_{1122}=1.2$. Moreover, for an isotropic medium we expect the anisotropy factor $\sigma = [2\chi'_{1122}+\chi'_{1221}-\chi'_{1111}]/\chi'_{1111}$ to equal zero whereas 1.7 is observed here. This implies coupling to considerably more anisotropic electronic states than has previously been reported for other tetrahedrally bonded crystals such as diamond ($\sigma=0.17$), Ge ($\sigma=0.83$), Si ($\sigma=0.44$), and GaAs ($\sigma=0.56$) (see Levenson and Bloembergen, 1974,

515

Wynne, 1969, and Yablonovitch, et al., 1972).

Unfortunately, the need to take higher excited electronic
bands into account makes it unlikely that the mostly phenomenologically theoretical approaches used so far are applicable to the
results reported here (Levenson and Bloembergen, 1974). It is
particularly unlikely that these models would be able to account
for the observed anisotropy in $\chi_E^{(3)}$. On the other hand, it is
possible that the more microscopic, ab initio, calculations of
semiconductor systems currently being developed (Martin, 1981) will
in time become theoretical bases for a comparison to the experimentally observed components of $\chi_E^{(3)}$ reported here.

CONCLUSION

I have demonstrated that it is possible to determine experimentally the dephasing times of optical phonons and their temperature
dependence, and that it is equally possible to determine experimentally all components of the third order nonlinear electronic
susceptibility. The measurements reported here were taken in the
temporal domain. Similar measurements of T_2 and $\chi_E^{(3)}$ taken in the
spectral domain had previously been reported for other crystal
systems.

Although it has been possible to account for the temperature
dependence of T_2 for LO phonons in GaP and ZnSe, full quantitative
comparisons between experiment and theory of the magnitude of T_2
and $\chi_E^{(3)}$ has not been possible. In the case of T_2, the effects of
elastic scattering events can be limited by suitable experimental
conditions, such that the measured T_2 reflects mainly the effects
of anharmonic phonon-phonon scattering time T_1. At very low temperatures three-phonon interactions should dominate anharmonic
processes. Indeed, experimental and theoretical evidence on the
temperature dependence of T_1 substantiate this claim. However, the
only theoretical formulations for T_1 are based on crude simplifications of the anharmonic crystal; namely, a crystal for which the
anharmonic component of the vibrational potential is given by the
Grueneisen parameter or by higher order elastic constants, and for
which the density of phonon states stems from an isotropic, dispersionless Debye solid. Clearly, such formulations can be relied
upon to yield only qualitative comparisons to experimental results.
More detailed, and more self-consistent, formulations exist but
their application to real crystal systems require detailed knowledge
of the anharmonic potential and a self-consistent lattice density
of states. Again these requirements often lead to simplifying
approximations. In any event, attempts along these lines have been
limited to only a few crystal systems. Clearly, as the wealth and
detail of the experimental results accumulate, an effort toward a
fully self-consistent, ab initio, treatment will become necessary.

516

A quite similar case can be made for the comparison of the observed components of $\chi_E^{(3)}$ and theory. The theoretical basis for $\chi_E^{(3)}$ (see e.g., Flytzanis and Bloembergen, 1976) is straightforward though rather cumbersome, and requires for its application to real cases knowledge of various orders of the crystal polarizabilities. Traditionally, these parameters have been obtained in terms of the response of simplifying models of the interionic bonds. This approach has been qualitatively successful for simple cases providing the nonlinear excitation energy $2\hbar\omega_\ell$ is much less than the energy of the fundamental gap, E_g, of the crystal (Levenson and Bloembergen, 1974). In the experimental case presented here, $2\hbar\omega_\ell > E_g$, so that the response of electronic states both at and above E_g need to be taken into account. Moreover, our observation of the considerable anisotropy in the observed $\chi_E^{(3)}$, suggests that the theoretical treatment will need to take the microscopic details of the participating wavefunctions into account. Again, it is hoped that the ab initio, large scale, calculations currently being attempted will in time yield a sufficiently detailed theoretical basis with which a comparison to the experimental results can be made.

ACKNOWLEDGMENT

The experimental work reported here was supported in part through ARO DAAG29-83-K-91 and was carried out through the collaboration with J. Kuhl and B.K. Rhee.

REFERENCES

Armstrong, J.A., Bloembergen, N., Ducuing, J., and Pershan, P.S., 1962, Phys. Rev. 127:1918.
Ashcroft, N.W. and Mermin, N.D., 1976, "Solid State Physics," Holt, Rinehardt, and Winston, New York, Chap. 25.
Bairamov, B.Kh., Kitaev, Yu.E., Negoduiko, V.K., and Kashkotzev, Z.M., 1974, Fiz. Tverd. Tela. 16:2036.
Bairamov, B.Kh., Parshin, V.V. Toporov, and Ubaidullav, Sh.B., 1979, Pis'ma Zh. Tekh. Fiz. 5:1116, Engl. translation: Sov. Tech. Phys. Lett. 5:466.
Born, M. and Huang, H., 1956, "Dynamical Theory of Crystal Lattices," Clarendon, Oxford.
Calleja, J.M., Vogt, H., and Cardona, M., 1983, Phil. Mag. A45:239.
Demtroder, W., 1981, "Laser Spectroscopy," Springer, Berlin.
Flytzanis, C., 1978, "Quantum Electronics: A. Treatise," vol. 1, part A, ed. Rabin, H. and Tang, C.L., Academic, New York, pg. 9.
Flytzanis, C. and Bloembergen, N., 1976, Proc. Quant. Elect. 4:271.
Gale, G.M. and Laubereau, A., 1983, Opt. Comm. 44:273.

Klein, M.V., 1968, "Physics of Color Centers," ed. Fowler, W.B.,
    Academic, New York, pg. 429.
Kleinman, D.A., 1969, Phys. Rev. 126:1977.
Klemens, P.G., 1958, "Solid State Physics," vol. 1, ed. Seitz, F.
    and Turnbull, D., Academic, New York.
Klemens, P.G., 1966, Phys. Rev. 148:845.
Klemens, P.G., 1967, J. Appl. Phys. 38:4573.
Kuhl, J. and Bron, W.E., 1984, Sol. State Comm. 48:935.
Kwok, P.C. and Miller, P.B., 1966, Phys. Rev. 146:592.
Landau, L. and Rumer, G., 1937, Phys. Zeit. Sovietunion 11:18.
Laubereau, A. and Kaiser, W., 1978, Rev. Mod. Phys. 50:607.
Levenson, M.D. and Bloembergen, N., 1974, Phys. Rev. B10:4447.
Levenson, M.D. and Song, J.J., 1980, "Coherent Nonlinear Optics,"
    ed. Feld, M.S. and Letokhov, V.S., Springer, Berlin.
von der Linde, D., 1981, "Defects in Insulating Crystals," ed.
    Tuckevich, V.M. and Shvartz, Springer, Berlin, pg. 542.
Maker, P.D. and Terhune, R.W., 1965, Phys. Rev. 137:A801.
Maradudin, A.A. and Fein, A.E., 1962, Phys. Rev. 168:2589.
Martin, R.M., 1981, J. Physique 42:C6-617.
Orbach, R. and Vredevoe, R., 1964, Physics 1:91.
Placzek, G., 1934, "Handbuch der Radiologie," ed. Marx, E.,
    Akademische Verlagsgesellschaft, Leipzig.
Rhee, B.K., Bron, W.E., and Kuhl, J., 1984, to be published.
Schosser, C.L. and Dlott, D.D., 1984, submitted to Physical Review.
Tucker, J.W. and Rampton, V.W., 1972, "Microwave Ultrasonics in
    Solid State Physics," North Holland, Amsterdam.
Wynn, J.J., 1969, Phys. Rev. 178:1295.
Yablonovitch, E., Flytzanis, C., and Bloembergen, N., 1972, Phys.
    Rev. Lett. 29:865.
Ziman, J., 1960, "Electrons and Phonons," Oxford Univ. Press,
    Oxford.
Zinth, W., Laubereau, A., and Kaiser, W., 1978, Opt. Comm. 26:457.

III. MOLECULAR DYNAMICS

MOLECULAR-DYNAMIC SIMULATIONS OF MANY-PARTICLE SYSTEMS:

NEW FACES ON OLD PROBLEMS

Duane C. Wallace

Los Alamos National Laboratory

Los Alamos, New Mexico 87545

I. INTRODUCTION

There used to be two realities in the world of physics: Experiment and Theory. Now there are three, and the third one is
The Computer. In the community of physicists outside of the
computer-simulation enclave, there is a good deal of skepticism
about computer simulations of many-particle systems. This skepticism is certainly justified at the present time. Nevertheless,
in my opinion, computer simulation will affect the progress of
physics in a profound way. The many-body problems that we worked
on for decades will finally yield to computer simulation. This
does not mean that many-body problems will suddenly become
simple; the complications of these problems will appear in a new
form. The question will be, how are computer simulations to be
interpreted in terms of the mathematically posed problem, or in
terms of physical reality. A computer-generated many-particle
process contains an enormous amount of information, and the challenge will be to extract the information we want, leaving computer artifacts behind. In other words, the computer may have
the answer, but we will have to figure out the question.

As an introductory example of this general situation, I will
discuss the computer simulation technique of molecular dynamics
(MD), with application to real metals in solid and fluid phases.

II. DESCRIPTION OF EQUILIBRIUM MOLECULAR DYNAMICS

The Ordinary MD System

Imagine a 3-dimensional space filled with identical boxes, the box walls being merely mathematical surfaces. In each box is a set of particles, whose positions and velocities are functions of time. With respect to the set of particles, the boxes are all identical for all time. We consider a specific box, and call it the computational cell; the remaining boxes are the image cells. Let us say that the particles interact through a central pairwise potential. What we do in MD is to integrate the classical equations of motion for the particles in the computational cell, and simultaneously move the particles in the image cells along the same trajectory, to keep the infinite array perfectly periodic. In present-day calculations, the number of particles in the computational cell is usually in the range 10^2 to 10^3. An extensive discussion of finite-size effects is presented in Section III below.

When we speak of an MD system, we mean the set of particles in the computational cell, and the dynamical variables which belong to that set. For example, the total energy is a dynamical variable, and one should note that a portion of the energy of the particles in the computational cell is due to interactions with particles in neighboring image cells. This is why a system is different from a cell. The computational cell has volume V and contains N particles. For the present work we take the particles to be identical, label them with the index $K = 1,2,\cdots,N$, and introduce the following notations.

$$\vec{x}_K = \text{particle position}\quad,$$

$$\vec{p}_K = \text{particle momentum}\quad,$$

$$m = \text{particle mass}\quad,$$

$$r_{KL} = \left| \vec{x}_K - \vec{x}_L \right| \quad . \tag{1}$$

The system Hamiltonian is

$$H = KE + \Phi \quad , \tag{2}$$

where KE is the total kinetic energy,

$$KE = \Sigma_K \frac{\vec{p}_K^2}{2m} \quad , \tag{3}$$

and Φ is the total potential energy. For Φ, we take the general form

$$\Phi = \Omega(V) + \frac{1}{2} \Sigma'_{KL} \; \phi(r_{KL};V) \quad , \tag{4}$$

where $\Omega(V)$ is the volume dependent (or density dependent) "one-body" potential of a metal, and $\phi(r;V)$ is the volume-dependent effective potential between two ions separated by a distance r. In (4), the Σ_K is over all ions in the system, while the Σ'_L is over all neighbors of K, within the range of the potential ϕ. Evaluation of these potentials for pseudopotential metals is discussed in Section IV. The notable properties of $\Omega(V)$ and ϕ are that $\Omega(V)$ is negative and much larger in magnitude than the $\Sigma\phi$, hence $\Omega(V)$ accounts for most of the binding energy, and that the effective ion-ion interaction ϕ is volume dependent. These properties characterize the physics of metals in general, and they have to be accounted for in any calculation which is intended to represent a real metal. Finally, the total linear momentum of the MD system will be denoted by $\vec{J}$:

$$\vec{J} = \Sigma_K \vec{p}_K \quad . \tag{5}$$

523

Regarding the techniques for numerically integrating the classical equations of motion, and for implementing periodic boundary conditions, the student is referred to the earliest MD papers by Alder and Wainwright,[1] Erginsoy et. al.,[2] Rahman,[3] and Hoover and Alder,[4] and also to the recent reviews by Wood,[5] Erpenbeck and Wood,[6] Kushick and Berne,[7] and by Hockney and Eastwood.[8]

Exercise. Draw a picture of a two-dimensional array, showing the computational cell and its neighboring image cells. Confirm that the potential energy (4) belongs to the computational cell, even though some of the particles involved are in image cells. If the cell width is w, and if the range of ϕ is $\geq \frac{1}{2}$ w, show that a particle and its image can simultaneously interact with the same third particle.

Classical Canonical Ensemble

Let us switch our thinking to the area of formal statistical mechanics. A system consists of N particles in a volume V, but now the system is so large that boundary conditions and/or surface effects can be neglected. An ensemble is a very large number of systems in contact with the appropriate reservoir, and with the appropriate constraints applied to each system. The ensemble weight function $W(\{\vec{x}_K,\vec{p}_K\})$ is the distribution over the systems at a given instant of time of the phase $\{\vec{x}_K,\vec{p}_K\}$ of a system. The ensemble average of a system variable is the average over all systems at a given instant of time.

For the canonical ensemble, the systems are constrained to constant N and constant V, and are in contact with a reservoir at constant temperature T. Hence it is an N,V,T ensemble. Since all the ensembles we will consider are at constant N and V, the notation of constant N and V will be suppressed. The canonical weight function is

$$W(\beta) \propto e^{-\beta H} \quad , \tag{6}$$

where $\beta = 1/kT$, with k = Boltzmann's constant. The classical canonical partition function is

$$Z(\beta) = \frac{1}{N!h^{3N}} \int \cdots \int e^{-\beta H} \Pi_K d\vec{x}_K d\vec{p}_K \quad . \tag{7}$$

The normalization constant in (7) requires careful attention. For a fluid, the integration domain of each $\vec{x}_K$ is the entire volume V, so all permutations of the particles are contained in the integrals. For a crystal, if one restricts each $\vec{x}_K$ to a region near the lattice site K, then Z has to be multiplied by N! to count the particle permutations. The momentum integrations in (7) can be carried out, transforming Z to the form

$$Z(\beta) = \frac{V^N}{N!\Lambda^{3N}}\, Q(\beta) \quad , \tag{8}$$

where Λ is the thermal de Broglie wavelength of the particles,

$$\Lambda = \sqrt{\frac{2\pi\hbar^2}{mkT}} \quad , \tag{9}$$

and $Q(\beta)$ is the configuration integral,

$$Q(\beta) = \frac{1}{V^N} \int \cdots \int e^{-\beta\Phi}\Pi_K d\vec{x}_K \quad . \tag{10}$$

The connection of statistical mechanics to thermodynamics is most commonly made through the canonical ensemble. This is the procedure we will adopt. Further, the connection must be made through one and only one equation, which we will take to be the Helmholtz free energy F:

$$F = - kT \ln Z(\beta) \quad . \tag{11}$$

All remaining thermodynamic functions are then obtained from F, according to the customary relations. The entropy S is

$$S = - \left.\frac{\partial F}{\partial T}\right)_V \quad , \tag{12}$$

the thermodynamic internal energy U is

$$U = F - TS \quad , \tag{13}$$

and the pressure P is

$$P = - \left.\frac{\partial F}{\partial V}\right)_T \quad . \tag{14}$$

The thermodynamic functions, as defined above, can be shown to be given by ensemble averages. For a dynamical variable $A = A(\{\vec{x}_K,\vec{p}_K\})$, the canonical-ensemble average $\langle A|\beta\rangle$ is given by

$$\langle A|\beta\rangle = \frac{\int \cdots \int A e^{-\beta H}\, \Pi_K d\vec{x}_K d\vec{p}_K}{\int \cdots \int e^{-\beta H}\, \Pi_K d\vec{x}_K d\vec{p}_K} \quad . \tag{15}$$

It follows that the ensemble average of H is equal to U:

$$\langle H \,|\, \beta \rangle = - \frac{\partial \ln Z(\beta)}{\partial \beta} = U \quad , \tag{16}$$

where the first equality follows from (15), and the second follows from (11)-(13). It is important to note that we cannot merely <u>assume</u> the relation $\langle H \,|\, \beta \rangle = U$; we have to show that it follows from the connection (11). Components of the particle momenta satisfy

$$\langle p_{Ki} \,|\, \beta \rangle = 0 \quad ;$$

$$\langle p_{Ki} p_{Lj} \,|\, \beta \rangle = \langle p_{Ki}^2 \,|\, \beta \rangle \delta_{KL} \delta_{ij} = mkT \delta_{KL} \delta_{ij} \quad , \tag{17}$$

where the subscripts i,j indicate Cartesian components. Hence the canonical average of the total momentum vanishes,

$$\langle J_i \,|\, \beta \rangle = 0 \quad , \tag{18}$$

and the canonical average of the total kinetic energy exhibits the well-known relation to the temperature,

$$\langle KE \,|\, \beta \rangle = \frac{3}{2} NkT \quad . \tag{19}$$

Finally, the pressure is expressed as an ensemble average by means of

$$PV = - V\frac{d\Omega}{dV} + NkT - \langle Y \,|\, \beta \rangle \quad , \tag{20}$$

where the generalized virial function Y is

$$Y = \frac{1}{2} \Sigma'_{KL} \left(V\frac{\partial \phi_{KL}}{\partial V} + \frac{1}{3} r_{KL} \frac{\partial \phi_{KL}}{\partial r_{KL}} \right) \quad . \tag{21}$$

<u>Exercise.</u> Prove equations (16) and (17). Derive (20) for a metal whose Hamiltonian is given by (2)-(4).

The dynamical variable representing a fluctuation is $\delta A = A - \langle A \rangle$, where $\langle A \rangle$ is the average in whatever ensemble we are concerned with. Correspondingly, $\langle \delta A \rangle$ is always zero, and the average fluctuation squared is $\langle (\delta A)^2 \rangle$. The word fluctuation is used rather carelessly in the literature, to mean for example δA, or $(\delta A)^2$, or $\langle (\delta A)^2 \rangle$. Generally, except for critical phenomena, one says that fluctuations are of relative order N^{-1}, which means specifically $\langle \delta A \delta B \rangle \sim N^{-1} \langle AB \rangle$.

<u>Exercise.</u> For a classical canonical ensemble, derive the fluctuation formula $\partial \langle A \,|\, \beta \rangle / \partial \beta = - \langle \delta A \delta H \,|\, \beta \rangle$.

526

The Logic of Calculating Thermodynamics from MD

The equations of motion cannot be integrated exactly; in
practice they are evaluated as finite-difference equations, with
a finite timestep. This means the total energy H, and the total
linear momentum $\vec{J}$, will not be exact constants of the motion, but
will have small variations in the course of time, due to the
finite differencing. However, with any good integration algorithm, these variations are extremely small, and have no statistical significance in terms of phase space sampling, so that H
and $\vec{J}$ can be thought of as exact constants of the motion. On the
other hand, the angular momentum of an MD system is not conserved, because occasionally a particle will move out through one
wall of the computational cell, while simultaneously one of its
image particles moves in through the opposite wall. Hence to
construct an MD ensemble, we imagine a large number of systems in
contact with an angular momentum reservoir, with each system
having constant N and V, with H held constant at the value E, and
with $\vec{J}$ held constant at the value $\vec{M}$. As usual, the parameters N
and V are suppressed, so that the MD ensemble is specified by the
parameters E and $\vec{M}$. The MD ensemble weight function is

$$W(E,\vec{M}) \propto \delta(H-E)\delta(\vec{J}-\vec{M}) \quad . \tag{22}$$

The classical partition function is

$$Z(E,\vec{M}) \propto \int \cdots \int \delta(H-E)\delta(\vec{J}-\vec{M})\Pi_K d\vec{x}_K d\vec{p}_K \quad , \tag{23}$$

and the ensemble average of the dynamical variable A is

$$\langle A \,|\, E,\vec{M}\rangle = \frac{\int \cdots \int A\delta(H-E)\delta(\vec{J}-\vec{M})\Pi_K d\vec{x}_K d\vec{p}_K}{\int \cdots \int \delta(H-E)\delta(\vec{J}-\vec{M})\Pi_K d\vec{x}_K d\vec{p}_K} \quad . \tag{24}$$

Historically, the MD ensemble was defined by Lebowitz, Percus,
and Verlet,[9] and by Wood.[5] In keeping with the customary procedure of statistical mechanics, the ensemble is defined, by the
above equations, without any reference to boundary conditions
and/or surface effects. It should be noted that the MD ensemble
is not a microcanonical ensemble, since the microcanonical ensemble has no restriction to a constant total linear momentum.

When we want to calculate thermodynamics from MD, the first
question which arises is, how are thermodynamic functions to be
defined in the MD ensemble? It is natural to define thermodynamic functions in the MD ensemble in terms of averages of dynamic variables, exactly as in equations (16)-(21) for the canonical
ensemble. However, in different ensembles, the averages of the
same dynamical variable are different, the difference being of

relative order N^{-1}. Hence if we simply applied the canonical
equations to the MD ensemble, the interrelations among our thermodynamic variables would have errors of relative order N^{-1}. To
eliminate such errors, we have to find the relations between averages in the MD ensemble, and averages in the canonical ensemble. The situation is much more striking in the case of fluctuations, because in different ensembles, the same dynamical fluctuations are different in relative order 1. For example, the average of $(\delta H)^2$ is zero in the MD ensemble, but it is nonzero in the
canonical ensemble. Since it is often convenient to work in
terms of fluctuations, we also need to find the relations between
fluctuation averages in the MD and canonical ensembles. All of
these needed ensemble relations will be derived in Section III.

The next step in preparing to calculate thermodynamic functions from MD is to invoke the quasiergodic hypothesis. This is
well described by Reif,[10] and for a general ensemble composed of
systems in contact with a reservoir, it may be stated as follows.

<u>Quasiergodic hypothesis.</u> The time average of the dynamical
variable A for a single system, when averaged for a time sufficiently long that the average becomes independent of the averaging
time, is the same as the ensemble average <A>.

Speaking pictorially, the quasiergodic hypothesis implies
that the trajectory of a single system, while it does not cover
phase space, uniformly samples phase space. Questions regarding
ergodic properties of Hamiltonian systems are extremely complicated, and the subject is under active study today. But in practice an MD system is not quite Hamiltonian, because solving the
equations of motion as finite difference equations introduces a
certain randomness into the phase space trajectory. This property of the trajectory, which will be discussed in detail in Section III, presumably justifies the quasiergodic hypothesis for an
MD system.

<u>Summary.</u> The logical foundation for using MD to calculate
equilibrium thermodynamic properties is based on the following
three steps.
  (1) Express thermodynamic functions as canonical ensemble
       averages.
  (2) Relate the canonical ensemble averages to MD ensemble
       averages.
  (3) Replace the MD ensemble average by the time average for
       a single MD system.

III. INTERPRETATION OF EQUILIBRIUM MOLECULAR DYNAMICS

The Appearance of MD Data

At Los Alamos, we have carried out extensive MD calculations
for metallic sodium in solid and fluid phases.[11-14] A calculation is usually started with the sodium ions located at the
lattice sites of a bcc crystal, and with a random Gaussian distribution of velocities. Intensive variables of the MD system,
such as the kinetic energy per particle KE/N, or the generalized
virial per particle Y/N, are graphed as functions of time. These
are fluctuating signals which evolve to a steady state, the
steady state being characterized by the appearance of a constant
mean and a constant bandwidth of the signal. We will refer to
the steady state as "equilibrium," and will make this definition
more precise later on. In approaching equilibrium, the mean and/
or the variance of the signal may change, as shown in Figures 1
and 2. We have found that when the system remains in the crystalline phase, the kinetic energy increases to its equilibrium
level, but when the system becomes a fluid, the kinetic energy
decreases to its equilibrium level.

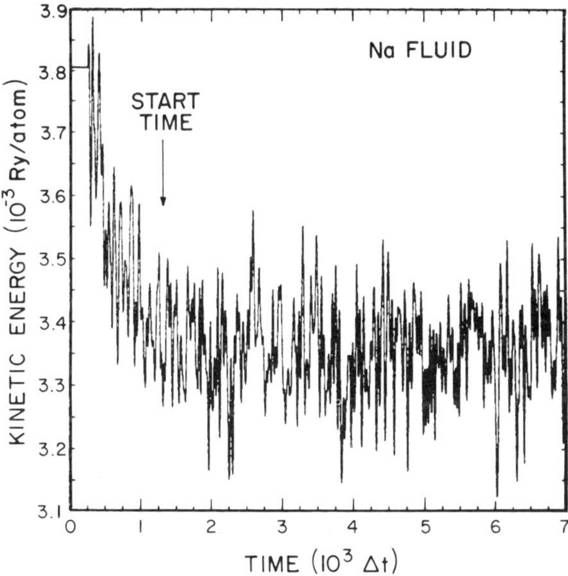

Fig. 1. Equilibration of MD data, showing variation in the mean
of the signal as the system approaches equilibrium.

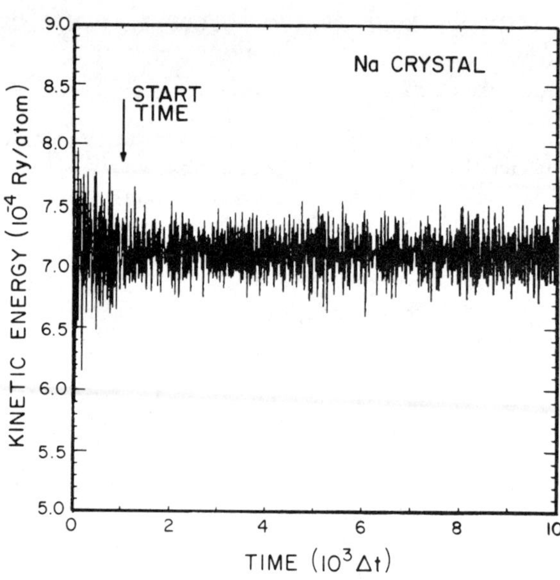

Fig. 2.   Equilibration of MD data, showing variation in the band-
width of the signal as the system approaches equilibrium.

To discuss the MD data in detail, we need to make some
definitions.

$$t = \text{time},$$
$$X(t) = \text{system intensive variable (e.g. KE/N, etc.)},$$
$$\Delta t = \text{timestep for the numerical integration},$$
$$\Delta t_f = \text{mean fluctuation time.} \tag{25}$$

The mean fluctuation time is determined by counting the local
maxima in a section of the equilibrium graph; a section containing 20 or so local maxima produces a reliable evaluation of Δt_f.
For our calculations in solid and fluid sodium, fluctuation times
are in the range $30\Delta t$ to $60\Delta t$. Important properties of the signals $\dot{X}(t)$, for different dynamical variables in the same MD calculation, are as follows.
   (1)   Different $X(t)$ approach equilibrium on different time
        scales.
   (2)   The Δt_f are different for different $X(t)$.

These observations, which seem surprising at first, tell us that
different dynamical variables of an MD system are "statistically

independent" variables. Another property we have observed is that
the equilibrium can be metastable. For example, on cooling the
fluid phase, the system usually passes into an amorphous or glassy
phase, having internal energy and entropy well above the crystalline phase. In the present work, we will always use the term
equilibrium in the most general sense, including metastable
states.

We can discuss MD data on a more sophisticated level by
introducing the probability distribution $P(X)$, for each variable
X. For the energy E, for example, and for a general ensemble
whose statistical weight function is W,

$$P(E) = \frac{Tr\ W\delta(H-E)}{Tr\ W} \quad , \tag{26}$$

where Tr stands for the phase-space integral. $P(X)$ is in fact a
probability, since it is normalized:

$$\int P(X)dX = 1 \quad . \tag{27}$$

The ensemble average of X, or of any function $f(X)$, is given by

$$\langle f(X) \rangle = \int f(X)P(X)dX \quad . \tag{28}$$

A helpful discussion of joint probabilities of several variables
is given by Callen.[15] For the present purposes, it is convenient
to use a separate probability $P(X)$ for each variable X.

In any practical MD calculation, we are interested in only a
small number of system variables X, say three or four. A formal
definition of equilibrium may be stated as follows.

<u>Definition.</u> Equilibrium for an MD system is the state in
which $\overline{P(X)}$ is independent of time, for all variables X of
interest.

Then how is $P(X)$ related to the actual MD data, as represented by a graph of the function $X(t)$? The answer to this is
found in the quasiergodic hypothesis, which may be restated in the
following form.

<u>Quasiergodic hypothesis restated.</u> In the course of a suffi-
ciently long time, the variable X for a single equilibrium MD
system is distributed according the the probability distribution
$P(X)$.

In other words, when the system is in equilibrium, the graph
of $X(t)$ actually traces out the probability distribution $P(X)$. It
should be remembered that $P(X)$ depends on the ensemble parameters

E and $\vec{M}$, or in other words, P(X) depends parametrically on the
thermodynamic state of the system.

In practice it is convenient to work with the moments of the
distribution P(X), and in fact we need only the low-order
moments. A thermodynamic function can be expressed as a first
moment (a mean), and a mean fluctuation squared is a second
moment (a variance). The mean of an intensive quantity should be
independent of N, while the variance should go as N^{-1}. In fact,
it is _extremely difficult_ to actually compute moments higher than
second order. From our calculations on metallic sodium, we found
that the length of data required to calculate a standard deviation to a given accuracy, say to 1%, is 100 times greater than
that required to calculate the mean to the same accuracy. A further complication lies in the fact that the computer generated
distribution P(X) depends on the timestep Δt, and the dependence
will become stronger as one moves to regions where P(X) is smaller. This leads us to record the following observation.

Notice. To calculate an accurate result for a _higher_ moment
will require using a _smaller_ Δt, compared to the same accuracy
for a lower moment.

Statistical Analysis of MD Data

We will consider only the most basic questions which arise
in the analysis of MD data; namely, given the fluctuating signal
X(t) for some variable X, when can one start the averaging, and
what are the confidence limits on the mean? The discussion is
based on the work of Schiferl and Wallace.[16] Regarding the first
question, we have already noted that different variables approach
equilibrium on different time scales. Since the calculation of
the mean of a given variable is independent of the behavior of
the other variables, we will further specialize our definition of
equilibrium, as follows.

Definition. An MD system is in equilibrium with respect to
the variable X when the mean and variance of X(t) are independent
of time.

The condition of equilibrium can be estimated by looking at
a graph of X(t), as shown for example in Figures 1 and 2. But
the visual determination of equilibrium is highly subjective, so
we have developed a statistical procedure for this determination. The "start times" marked in Figures 1 and 2 were determined from our statistical procedure. Regarding the question of
confidence limits, we note that the statistical uncertainty in
the mean of X(t) is due to the fluctuations. While this uncer-

tainty decreases as the averaging time increases, the averaging
time will always be finite. Hence, in order to compare MD calculations with experiment, or with other theoretical results, it is
important to estimate confidence limits for the MD mean values.

For the function $X(t)$, the sequence of MD data points at
successive timesteps is called the "raw data," and is denoted by

$$X_r = X(r\Delta t), \quad r = 1,2,\cdots \quad . \tag{29}$$

The essential complication in trying to estimate confidence
limits for the mean is the presence of strong correlation in the
raw data. An expression for the variance of means for a stationary time series, which includes the effects of correlation, has
been given by Davenport.[17] However, to evaluate this expression
requires knowledge of the autocorrelation function of the data.
This function is not known _a priori_, but has to be computed from
the data that we want to analyze. From an examination of a large
amount of MD data on metallic sodium in solid and liquid phases,
we find that calculation of the autocorrelation function requires
an enormous amount of data, and hence we reach the following conclusion.

Observation. For a given amount of data $X(t)$, from a particular MD calculation, trying to calculate the autocorrelation
function from this data, and then calculating the variance of
means from the autocorrelation function, is an inefficient method
for estimating confidence limits for the mean.

We therefore take an alternate approach, and try to construct "coarse-grained data," which will have negligible correlation. To construct coarse-grained data, we divide the series of
points X_r into sequential nonoverlapping segments k, with k =
$1,2,\cdots,n$. Each segment has m points, and the mean and variance
of the raw data for segment k are denoted by x_k and S_k^2, respectively. The length of each segment is $\tau = m\Delta t$. The n values of
x_k constitute a sample, drawn from a population whose mean is the
mean of the MD data. The sample mean and variance are $\bar{x}$ and s^2,
respectively:

$$\bar{x} = \frac{1}{n} \sum_{k=1}^{n} x_k \quad ; \tag{30}$$

$$s^2 = \frac{1}{n-1} \sum_{k=1}^{n} (x_k - \bar{x})^2 \quad . \tag{31}$$

To analyze MD data, one should first construct a graph of
$X(t)$ for the entire MD calculation. A "start time" t_s is then

chosen, after which the mean and the bandwidth of the graph
appear nominally constant. The mean fluctuation time is calculated from a portion of the data after t_s, and a coarse-grained
sample is constructed, starting at t_s, and with $\tau \approx \Delta t_f$ and
with $n \geq 24$. Four statistical tests are then performed on the
sample.

    (a)   Test for lack of trend in the x_k (Mann-Kendall test).
    (b)   Test for lack of trend in the S_k (Mann-Kendall test).
    (c)   Test for normality of the x_k (W test, or shape test).
    (d)   Test for lack of positive correlation in the x_k (one-
            tailed von Neumann test).

If test (a) and/or (b) fails, the indication is to start
later (increase t_s), because $X(t)$ has not reached equilibrium.
If both (a) and (b) pass, and if (c) and/or (d) fails, the indication is to increase τ. When all four tests pass, the mean of
the MD data, which is the same as the ensemble average $\langle X \rangle$, can
be assigned the following confidence interval:

$$\langle X \rangle = \bar{x} \pm \alpha_n \frac{s}{\sqrt{n}} \quad , \tag{32}$$

where α_n is the 0.975-fractile of "Student's" t-distribution
with n-1 degrees of freedom. For $n \geq 24$, $\alpha_n \approx 2$.

The Mann-Kendall test for trend, or Kendall test for correlation, is described by Bradley,[18] and by Conover.[19] In testing
for normality, the W test is recommended for sample sizes $n \leq 50$;
the test is described by Shapiro and coworkers.[20,21] For
larger samples, the shape test can be used, as described by
Snedecor and Cochran,[22] and by Pearson and Hartley.[23] The von
Neumann test for serial correlation is described by Hald.[24] Further details on the analysis of MD data can be found in the paper
of Schiferl and Wallace.[16]

We have constructed a rule-of-thumb estimate for confidence
limits, which is consistent with the more accurate analysis of
our sodium data. With the notation

    S = standard deviation of the raw data,
    N_f = number of fluctuations in the data averaged,

then our rule-of-thumb estimate is

$$\langle X \rangle \approx \bar{x} \pm \frac{2S}{\sqrt{N_f}} \quad . \tag{33}$$

This result should be used only for a quick estimate, and is in
no way intended to replace the more reliable procedure based on
the tests (a)-(d), as outlined above.

$\underline{\text{Exercise.}}$ Use the known N-dependence of the fluctuation $\langle(\delta X)^2\rangle$, and reasonable hypotheses about the dependences of s^2, to transform (32) to (33).

Generalized Canonical Ensemble

We now turn to the problem of how to relate MD-ensemble averages to canonical-ensemble averages. The MD ensemble has two constants of the motion (besides N and V), namely H and $\vec{J}$. The ordinary canonical ensemble has only one parameter, namely β which is conjugate to H, and it has zero for the average of $\vec{J}$. To construct an ensemble conjugate to the MD ensemble, we have to introduce a second parameter $\vec{b}$, which gives $\vec{J}$ a nonzero average value. This is the generalized canonical ensemble, whose statistical weight function is

$$W(\beta,\vec{b}) \propto e^{-\beta H} e^{-\vec{b}\cdot\vec{J}} \quad . \tag{34}$$

The generalized canonical partition function is

$$Z(\beta,\vec{b}) = \frac{1}{N!h^{3N}} \int \cdots \int e^{-\beta H} e^{-\vec{b}\cdot\vec{J}} \Pi_K d\vec{x}_K d\vec{p}_K \quad . \tag{35}$$

The generalized canonical ensemble was previously studied by Grad,[25] and by Lado.[26] The discussion here is based on the paper of Wallace and Straub.[27]

To relate generalized- and ordinary-canonical ensembles, it is useful to define the velocity $\vec{u}$, which turns out to be the mean center-of-mass velocity:

$$\vec{b} = -\beta\vec{u} \quad . \tag{36}$$

The center-of-mass momentum of particle K is $\vec{p}_K - m\vec{u}$. $Z(\beta,\vec{b})$ can be evaluated in terms of $Z(\beta)$, by completing squares of the center-of-mass momenta in equation (35), to give

$$Z(\beta,\vec{b}) = e^{Nm\vec{b}^2/2\beta} Z(\beta) \quad . \tag{37}$$

Thus when $\vec{b} = 0$, the generalized-canonical ensemble reduces to the ordinary-canonical ensemble. The generalized-canonical average of A is $\langle A|\beta,\vec{b}\rangle$, which we will abbreviate by $\bar{A}$:

$$\bar{A} = \langle A|\beta,\vec{b}\rangle = \frac{\int \cdots \int A e^{-\beta H} e^{-\vec{b}\cdot\vec{J}} \Pi_K d\vec{x}_K d\vec{p}_K}{\int \cdots \int e^{-\beta H} e^{-\vec{b}\cdot\vec{J}} \Pi_K d\vec{x}_K d\vec{p}_K} \quad . \tag{38}$$

By using (35) and (37), $\bar{H}$ and $\bar{J}_i$ can be related to the ordinary-canonical averages of H and J_i, equations (16) and (18), as follows:

$$\bar{H} = \frac{-\partial \ln Z(\beta,\vec{b})}{\partial\beta} = U + \frac{Nm\vec{b}^2}{2\beta^2} \quad , \tag{39}$$

$$\bar{J}_i = \frac{-\partial \ln Z(\beta,\vec{b})}{\partial b_i} = \frac{-Nmb_i}{\beta} \quad , \tag{40}$$

where the subscript i indicates Cartesian components. From the last equation, the mean momentum per particle is seen to be $m\vec{u}$.

Differentiating (38) with respect to the ensemble parameters gives the fluctuation formulas,

$$\partial\bar{A}/\partial\beta = - \langle \delta A \delta H | \beta,\vec{b} \rangle \quad , \tag{41}$$

$$\partial\bar{A}/\partial b_i = - \langle \delta A \delta J_i | \beta,\vec{b} \rangle \quad . \tag{42}$$

With these results, and the equations for $\bar{H}$ and $\bar{J}_i$, the following important relations can be established.

$$\frac{\partial\bar{H}}{\partial\beta} = - \langle (\delta H)^2 | \beta,\vec{b} \rangle = \frac{\partial U}{\partial\beta} - \frac{Nm\vec{b}^2}{\beta^3} \quad , \tag{43}$$

$$\frac{\partial\bar{H}}{\partial b_i} = \frac{\partial\bar{J}_i}{\partial\beta} = - \langle \delta H \delta J_i | \beta,\vec{b} \rangle = \frac{Nmb_i}{\beta^2} \quad , \tag{44}$$

$$\frac{\partial\bar{J}_i}{\partial b_j} = - \langle \delta J_i \delta J_j | \beta,\vec{b} \rangle = - \frac{Nm\delta_{ij}}{\beta} \quad . \tag{45}$$

The average kinetic energy now contains, in addition to the temperature term, the mean translational energy,

$$\overline{KE} = \frac{3}{2} NkT + \frac{1}{2} Nm\vec{u}^2 \quad . \tag{46}$$

Because Φ does not depend on the particle momenta, we can write

$$\langle \delta\Phi\delta KE | \beta,\vec{b} \rangle = 0 \quad . \tag{47}$$

Then the mean squared fluctuations of KE and Φ can be evaluated with the aid of (41), to find

$$\langle (\delta KE)^2 | \beta,\vec{b} \rangle = \frac{3}{2} N(kT)^2 + Nm\vec{b}^2/\beta^3 \quad , \tag{48}$$

$$\langle (\delta\Phi)^2 | \beta, \vec{b} \rangle = N(kT)^2(c - \tfrac{3}{2}) \quad , \tag{49}$$

where c is the constant-volume heat capacity in units of k per particle,

$$c = \frac{1}{Nk} \left. \frac{\partial U}{\partial T} \right)_V \quad . \tag{50}$$

<u>Exercise.</u> By straightforward evaluation of momentum integrals in equations of the form (38), verify the results (40), (44), (45), (47), and (48). From the definition of H, and the equation (39) for $\bar{H}$, verify (49).

For any dynamical variable that depends only on the volume and the relative positions of the particles, as for example Φ and Y, the generalized canonical average is the same as the canonical average:

$$\langle Y | \beta, \vec{b} \rangle = \langle Y | \beta \rangle \quad , \tag{51}$$

$$\langle \Phi | \beta, \vec{b} \rangle = \langle \Phi | \beta \rangle \quad . \tag{52}$$

This follows from (35) and (38), and is equivalent to the equation $\partial \bar{Y}/\partial b_i = 0$, and the same for $\bar{\Phi}$. Let us define KE_{CM} as the kinetic energy in the center-of-mass frame,

$$KE_{CM} = \frac{1}{2m} \Sigma_K (\vec{p}_K - m\vec{u})^2 \quad . \tag{53}$$

The generalized-canonical average of KE_{CM} is

$$\overline{KE}_{CM} = \frac{3}{2} NkT \quad . \tag{54}$$

With these results, the ordinary-canonical expressions (16) and (20) for U and PV can be written in terms of generalized-canonical averages:

$$U = \overline{KE}_{CM} + \bar{\Phi} \quad , \tag{55}$$

$$PV = - V \frac{d\Omega}{dV} + \frac{2}{3} \overline{KE}_{CM} - \bar{Y} \quad . \tag{56}$$

It is now possible to construct a generalized canonical ensemble from a collection of MD ensembles. The MD statistical weight function is $W(E, \vec{M})$, the partition function is $Z(E, \vec{M})$, and the average of A is $\langle A | E, \vec{M} \rangle$. These quantities are expressed in equations (22)-(24). In terms of these, the corresponding generalized-canonical quantities are as follows.

$$W(\beta,\vec{b}) \propto \iint W(E,\vec{M}) \exp(-\beta E - \vec{b}\cdot\vec{M}) \, dE\,d\vec{M} \quad , \tag{57}$$

$$Z(\beta,\vec{b}) \propto \iint Z(E,\vec{M}) \exp(-\beta E - \vec{b}\cdot\vec{M}) \, dE\,d\vec{M} \quad , \tag{58}$$

$$\langle A|\beta,\vec{b}\rangle = \frac{\iint \langle A|E,\vec{M}\rangle Z(E,\vec{M}) \exp(-\beta E - \vec{b}\cdot\vec{M}) \, dE\,d\vec{M}}{\iint Z(E,\vec{M}) \exp(-\beta E - \vec{b}\cdot\vec{M}) \, dE\,d\vec{M}} \quad . \tag{59}$$

The integrations on E and $\vec{M}$ are formally over all space for these variables, but the integrations are effectively restricted, because the MD functions $W(E,\vec{M})$ and $Z(E,\vec{M})$ vanish for certain regions of E and $\vec{M}$.

<u>Exercise.</u> Do the integrals on E and $\vec{M}$, and show that (58) and (59) are identical to (35) and (38), respectively.

Ensemble Corrections for the MD Ensemble

We can use fluctuation theory to relate different ensemble averages of the same dynamical variable. In the MD ensemble, with H and J_i held constant at the values E and M_i, the average of A is $\langle A|E,M_i\rangle$. To relate this to the generalized-canonical average $\bar{A}$, let us fix the values of H and J_i at $\bar{H}$ and $\bar{J}_i$, respectively. The corresponding MD average is $\langle A|\bar{H},\bar{J}_i\rangle$, and is related to $\bar{A}$ according to

$$\langle A|\bar{H},\bar{J}_i\rangle = \bar{A} + \frac{1}{2}\frac{\partial}{\partial\beta}\frac{\partial\bar{A}}{\partial\bar{H}} + \frac{1}{2}\Sigma_i\frac{\partial}{\partial b_i}\frac{\partial\bar{A}}{\partial\bar{J}_i} + \cdots \quad , \tag{60}$$

where the correction terms, which are those in addition to $\bar{A}$, are of relative order N^{-1}, and the $+\cdots$ represents terms of lowest relative order $N^{-3/2}$. Similarly, for the ensemble averages of fluctuations, we have the relation

$$\langle \delta A\delta B|\bar{H},\bar{J}_i\rangle = \overline{\delta A\delta B} + \frac{\partial\bar{A}}{\partial\beta}\frac{\partial\bar{B}}{\partial\bar{H}} + \Sigma_i\frac{\partial\bar{A}}{\partial b_i}\frac{\partial\bar{B}}{\partial\bar{J}_i} + \cdots \quad , \tag{61}$$

where the correction terms, which are those in addition to $\overline{\delta A\delta B}$, are now of relative order 1, and the $+\cdots$ represents terms of relative order $N^{-1/2}$ and higher.

<u>Exercise.</u> In the paper of Lebowitz, Percus, and Verlet,[9] check the algebra in equations (2.1)-(2.12), and verify the equations (60) and (61) above. Or you can do the derivation directly from (57)-(59) above.

It is useful to eliminate the extensive quantities $\bar{H}$ and

$\bar{J}_i$ from the right sides of (60) and (61). In this way, the
ensemble corrections are expressed directly in terms of the
ensemble parameters β and b_i. To carry out this change of
variables, it is necessary to invert the equations (43)-(45).
The results are conveniently expressed in terms of the
differential operator D,

$$D = \frac{\partial}{\partial\beta} + \Sigma_i \frac{b_i}{\beta} \frac{\partial}{\partial b_i} \quad . \tag{62}$$

Equation (60) becomes

$$<A|\bar{H},\bar{J}_i> = \bar{A} - \frac{1}{2N} \left[D \frac{\beta^2}{c} D\bar{A} + \frac{3\beta}{c} D\bar{A} + \frac{\beta}{m} \Sigma_i \frac{\partial^2\bar{A}}{\partial b_i^2} \right] \quad , \tag{63}$$

and equation (61) becomes

$$<\delta A\delta B|\bar{H},\bar{J}_i> = \overline{\delta A\delta B} - \frac{1}{N} \left[\frac{\beta^2}{c} (D\bar{A})(D\bar{B}) + \frac{\beta}{m} \Sigma_i \frac{\partial\bar{A}}{\partial b_i} \frac{\partial\bar{B}}{\partial b_i} \right] \quad , \tag{64}$$

where we have specifically neglected the higher-order terms. In
the last equation, D does not operate through the round
brackets. These results hold for all $\beta > 0$ and $\vec{b}^2 \geq 0$.

<u>Exercise.</u> Do the algebra to verify the last two equations.

It is instructive to work out a few examples of the MD
ensemble corrections. For the kinetic energy per particle, KE/N,
equation (63) gives

$$\frac{<KE|\bar{H},\bar{J}_i>}{N} = \frac{3}{2} kT \left[1 - \frac{1}{N} \left(1 - \frac{3}{2c} - \frac{T}{2c^2} \frac{\partial c}{\partial T} \right) \right] + \frac{m\vec{u}^2}{2} \quad . \tag{65}$$

The last term is the kinetic energy due to the mean translational
motion. The term in N^{-1} on the right side expresses the fact
that the MD average of the center-of-mass kinetic energy is not
quite $\frac{3}{2}$ kT. In fact, the ensemble correction for the center-of-
mass kinetic energy is just this term:

$$\frac{<KE_{CM}|\bar{H},\bar{J}_i>}{N} = \frac{3}{2} kT \left[1 - \frac{1}{N} \left(1 - \frac{3}{2c} - \frac{T}{2c^2} \frac{\partial c}{\partial T} \right) \right] \quad . \tag{66}$$

<u>Notice.</u> Extreme caution must be observed when calculating
the ensemble correction for a dynamical variable, such as KE_{CM},
which depends on the ensemble parameters β and/or $\vec{b}$.

The quantity PV is given by (56) in terms of generalized-canonical averages. Let us compute the same averages in the MD ensemble, and call it PV_{MD}:

$$PV_{MD} = -V\frac{d\Omega}{dV} + \langle \frac{2}{3} KE_{CM} - Y | \bar{H}, \bar{J}_i \rangle \quad . \tag{67}$$

A straightforward calculation gives the relation

$$PV_{MD} = PV - kT\left(1 - \frac{3}{2}\gamma + \frac{1}{2}T\frac{\partial\gamma}{\partial T}\right) \quad , \tag{68}$$

where γ is the Grüneisen parameter,

$$\gamma = V\frac{\partial P}{\partial U}\bigg)_V \quad . \tag{69}$$

Since PV is of order N, the correction term in (68) is of relative order N^{-1}. In the generalized canonical ensemble, fluctuation averages are related to thermodynamic functions, as in (48) and (49). Such results, together with the fluctuation correction equation (64), can be used to evaluate fluctuation averages in the MD ensemble. Two results of interest are,

$$\frac{\langle(\delta KE)^2 | \bar{H}, \bar{J}_i \rangle}{N} = \frac{3}{2}(kT)^2\left(1 - \frac{3}{2c}\right) \quad , \tag{70}$$

$$\frac{\langle \delta Y \delta \Phi | \bar{H}, \bar{J}_i \rangle}{N} = \frac{3}{2}(kT)^2\left(\frac{1}{c} - \gamma\right) \quad . \tag{71}$$

For a further discussion of ensemble corrections, the student is referred to the paper of Wallace and Straub.[27] We have also derived the MD ensemble corrections for isothermal elastic constants computed from fluctuations.[28] The important point about ensemble corrections which should always be kept in mind is the following.

Important point. The ensemble correction for a fluctuation average is of the same order as the fluctuation average itself.

Exercise. Derive equations (65), (68), (70), and (71).

Exercise. Show that the MD ensemble corrections vanish for H and for J_i.

Exercise. Use equation (64) to show that the MD averages of $(\delta H)^2$ and of $(\delta J_i \delta J_j)$ are zero.

Exercise. For an ideal gas, where $U = \frac{3}{2}NkT$, prove:

$$\langle KE_{CM} | \bar{H}, \bar{J}_i \rangle = \frac{3}{2} NkT \quad , $$

$$PV_{MD} = PV = NkT \quad , $$

$$\langle (\delta KE)^2 | \bar{H}, \bar{J}_i \rangle = 0 \quad . $$

Remaining Finite-Size Effects

To help us in understanding finite-size effects in molecular dynamics, it is useful to separate them into two contributions, defined as follows.

1. <u>Ensemble corrections.</u> These result from the fact that different ensembles have different statistical weight functions, as discussed in the preceding pages. Since ensembles are defined without reference to boundary conditions, the same is true of ensemble corrections.

2. <u>Boundary condition effects.</u> This category is meant to include all finite-size effects not contained in the ensemble corrections. This is abbreviated BC effects, and PBC effects in the case of periodic boundary conditions. In the remainder of this subsection, we will discuss what is currently known (to the author) about BC effects.

First, Fisher and Lebowitz[29] have proved that the thermodynamic limit exists for periodic boundary conditions, and is identical with that for systems contained in normal domains with walls or boundaries. A normal domain is a regularly-shaped volume with surface-to-volume ratio going to zero in the thermodynamic limit. Next, Lebowitz and Percus[30] identified two types of N-dependence in the pressure and other intensive properties of small systems. They did this specifically in terms of cluster integrals and virial coefficients, so their analysis applies only when the cluster expansion converges. Their "simple dependence," which goes as a power series in N^{-1}, is apparently the ensemble correction for the canonical ensemble relative to the grand ensemble. The other dependence is a PBC effect; Lebowitz and Percus showed that this does not go as N^{-1}, and in fact is not even a regular function of N as $N \to \infty$. The constant pressure ensemble (N,P,T ensemble) was studied by Salsburg.[31] His calculations show how the virial coefficients approach their limits as $N \to \infty$, but there is no information here about BC effects. In Salsburg's paper, boundary condition effects are eliminated when the $\tilde{A}_k$ are set zero in equation (2.8). Finally, N-dependence of computer simulations has been studied numerically. In the latest such effort, and the most accurate to date, Slattery, Doolen, and DeWitt[32] found that their Monte Carlo results for energies and pressures of the one-component plasma are consistent with N-dependence of the form N^{-1} as N increases.

Let us pause to consider more deeply the nature of periodic boundary conditions. As we will see, periodic boundary conditions have exceptional properties. Once again, imagine the infinite periodic array of identical boxes, each box containing an identical set of particles, and with the box walls being merely mathematical surfaces.

First observation. The network of surfaces can be translated by an arbitrary amount, and the infinite system is still periodic. In other words, the periodicity of the infinite system is invariant under continuous translation of the walls.

Second observation. Consider the infinite periodic array at a given time, say t = 0. Remove the walls. Now allow all the particles to move under the equations of motion, and the infinite system will remain periodic forever. In other words, periodic boundary conditions are equivalent to an initial condition for the infinite system.

A conclusion can be drawn from these observations. First, imagine a real N-atom system with a surface, as for example a cubic centimeter of gold resting on the bench top, or some water in a beaker. Material in the surface region is different from the bulk material. Since the number of atoms in the surface region relative to the total is of order $N^{-1/3}$, surface effects are of relative order $N^{-1/3}$. In contrast, in the infinite periodic array there is no surface. More precisely, no surface is associated with the box walls; a "real" physical surface can still be constructed within the computational cell. Hence, there are no effects of order $N^{-1/3}$ associated with periodic boundary conditions, and PBC contributions to dynamical variables are presumably of lowest order N^{-1} in general.

Beyond this N-dependence, periodic boundary conditions can introduce other nonphysical effects into an MD or Monte Carlo calculation. It is generally believed that periodic boundary conditions will affect the relative stability of different structures, as for example amorphous and crystalline structures, and will inhibit phase separation in the evolution of an MD system. The effect of periodic boundary conditions on the calculation of equilibrium fluctuations, and of time-dependent correlation functions, is a problem that needs to be studied, especially in systems of particles which interact through continuous potentials.

Summary. In the calculation of thermodynamic functions for an equilibrium MD system, for gas, fluid, or solid phases, the present knowledge of PBC effects can be summarized as follows.

(a) The N-dependence as $N \to \infty$ is not known from theory.
(b) A reasonable assumption is the N-dependence going as
     N^{-1} as $N \to \infty$.
(c) Numerical results are consistent with the assumption
     (b).

Phase-Space Trajectory of an MD System

Suppose the evolution of an MD system is calculated, under
any specified boundary conditions. We inquire about the motion
of the phase-space point of the system, starting at any time, say
t = 0. The "true trajectory" is the phase-space trajectory the
system would follow if the equations of motion could be integrated exactly. Note in making this definition, we ignore the
presence of instabilities, and possible bifurcations of the
system motion. The "actual trajectory" is the trajectory calculated on the computer. For particles interacting through a continuous potential, as e.g. the central potential $\phi(r)$, the actual
trajectory departs from the true one, because the equations of
motion are replaced by finite-difference equations. Note that
machine round-off error is not important here; it is finite differencing that causes divergence of the actual trajectory from
the true one. On the other hand, for hard-core particles, as
e.g. hard spheres, there are no equations of motion to integrate,
there are only collisions. However, since the scattering angle
for hard spheres is extremely sensitive to the impact parameters,
machine round-off error is now important. Hence for both real
particles, with continuous potentials, and for hard-core particles, the actual MD trajectory diverges from the true trajectory,
but the source of this divergence is different in the two cases.
This situation has been noted by Wood.[5] The question arises, how
does this "divergence of trajectories" affect the way in which MD
calculations are to be interpreted?

First, suppose we want to calculate thermodynamic functions
for an equilibrium MD system. It is always possible that we
could average over a period of time long compared to the trajectory divergence time. In this case, the actual trajectory is unrelated to the true trajectory. The actual long-time trajectory
is essentially a statistical sampling of phase space, with the
equations of motion operating merely to constrain the actual trajectory to satisfy constants of the motion. As a matter of fact,
this is not a problem. The first thing we learn in statistical
mechanics is that the actual phase-space trajectory of a many-
particle system is not important, because it consists of precisely the microscopic information that we don't want to worry
about. Almost any trajectory consistent with the ensemble weight
function will do. Since the actual MD trajectory is consistent
with $W(E,\vec{M}) \propto \delta(H-E)\delta(\vec{J}-\vec{M})$, then our calculation of equilibrium

thermodynamic properties is correct, as far as the trajectory is
concerned.

Exercise. There exists a theorem and proof consisting of
the following three statements; which of these statements are
incorrect?
(1) Linear response theory cannot possible be correct,
     because:
(2) Linear response assumes that, when a perturbation is
     applied to a dynamical system, the phase-space trajec-
     tory changes only by a perturbation;
(3) Whereas in fact when a perturbation is applied to a dy-
     namical system, the phase-space trajectory diverges ex-
     ponentially.

An extremely valuable property of molecular dynamics is that,
in principle, this technique can describe a real physical evolutionary process. This cannot be done by statistical sampling techniques, as e.g. Monte Carlo, because they do not calculate physical
evolution. However, as we have seen from the preceding discussion
of the MD trajectory, on a sufficiently long time scale, molecular
dynamics is in fact a statistical sampling technique. Does this
mean that molecular dynamics has to give up its claim of being able
to calculate real evolutionary processes? To address this question, we need to recognize that a phase-space trajectory carries
two classes of information, namely microscopic and macroscopic.
Suppose the microscopic state of the system, i.e. the precise
phase-space point, is known at t = 0. This information is lost by
an actual MD calculation, in a time order of τ_μ, which is the
time required for the actual trajectory to diverge significantly
from the true one. The relaxation time τ_μ is expected to be
quite short, perhaps of the order of $10^3 \Delta t$, for particles interacting through continuous potentials. Now consider macroscopic
information. This is usually in the form of field variables, which
give local average values of density, fluid velocity, energy density, and so on. If the macroscopic state of the system is known
at t = 0, the information is lost in an actual MD calculation in a
time of order τ_M, which is the time required for the macroscopic
state on the actual trajectory to diverge significantly from the
macroscopic state on the true trajectory. In other words, because
the equations of motion are solved only approximately, macroscopic
causal relations are washed out in a time of order τ_M. But for
times less than τ_M, molecular dynamics is capable of following a
physical evolutionary process, such as the decay of a local density
fluctuation. In fact, a more general statement can be made, namely
that molecular dynamics can properly describe processes which are
governed by rates down to the order of τ_M^{-1}. We expect $\tau_M \gg \tau_\mu$
to hold in general. Beyond this, τ_M must have a complicated
dependence on the details of an MD calculation.

IV. MOLECULAR-DYNAMIC CALCULATIONS FOR METALLIC SODIUM

Pseudopotential Theory

The pseudopotential is an approximate potential for the interaction between a single electron and a single ion. It is designed to give a reasonably good value for the electron energy, and for the electron wave function in the region outside the ion core. To construct a metallic system, we assume each ion carries its pseudopotential rigidly. The conduction electrons screen the pseudopotential, and the screened-pseudopotential is treated as a perturbation. The groundstate energy of the conduction electrons, E_G, then contributes to the effective potential among the ions, according to the Born-Oppenheimer approximation. For a system of N identical ions, each with a net charge z, together with zN conduction electrons, the total effective potential Φ is

$$\Phi = NI_z + E_G + \Omega_I \quad , \tag{72}$$

where I_z is the ionization energy for one ion, and Ω_I is the direct potential among the ions.

In the old days, the energies E_G and Ω_I were each split into real-space and Fourier contributions, to obtain better convergence in the numerical calculations.[33] However, to do molecular-dynamic calculations, we want to write Φ completely in terms of real-space potentials, specifically in the form of equation (4). In our pseudopotential formulation, Φ contains I_z explicitly, and the zero of Φ corresponds to N free atoms:

$$\Phi = NI_z + \Omega_z(V) + \frac{1}{2} \Sigma'_{KL} \phi(r_{KL};V) \quad . \tag{73}$$

Comparison with (4) shows that $\Omega(V) = NI_z + \Omega_z(V)$.

We will outline the transformation of (72) to (73), relying heavily on Reference 33, pages 305-325. The potential Ω_I is already a sum of direct ion-ion interactions:

$$\Omega_I = \frac{1}{2} \Sigma'_{KL} \frac{z^2 e^2}{r_{KL}} + \frac{1}{2} \Sigma'_{KL} \alpha e^{-\gamma r_{KL}} \quad , \tag{74}$$

where the first term is the Coulomb interaction among point ions, and the second term is the short-range Born-Mayer repulsion between ion cores, with α and γ being parameters. The groundstate energy can be written in the form

$$E_G = zN[\frac{3}{5} \varepsilon_F + \varepsilon_X + \varepsilon_C + \frac{1}{2} W(q=0) - \frac{1}{2} W_X(q=0)]$$

$$+ N\Sigma_{\vec{q}} S(\vec{q})S(-\vec{q})F(q) \quad . \tag{75}$$

Here $\frac{3}{5} \varepsilon_F + \varepsilon_X + \varepsilon_C$ is the kinetic plus exchange plus correlation energy per electron of a uniform electron gas. $W(q=0)$ and $W_X(q=0)$ are $q=0$ Fourier components of the total pseudopotential, and of the exchange contribution to the total pseudopotential, respectively, and $F(q)$ is the energy-wavenumber characteristic, which is quadratic in the pseudopotential. All the dependence of E_G on the ion positions $\vec{x}_K$ is contained in the structure factor $S(\vec{q})$:

$$S(\vec{q}) = N^{-1}\Sigma_K e^{i\vec{q}\cdot\vec{x}_K} \quad . \tag{76}$$

We evaluate $W(q=0)$ and $W_X(q=0)$ as limits for $q\rightarrow 0$, and find

$$W(q=0) = -\frac{2}{3}\varepsilon_F \quad , \tag{77}$$

$$W_X(q=0) = -\frac{2e^2 f}{3\pi\xi} \quad , \tag{78}$$

where f is the free-electron Fermi wavevector, and ξ is the parameter in the Hubbard form of the screening correction. Both f and ξ depend on the electronic density, or on V. We transform the Fourier sum in (75) to an integral, and express Φ in the form of (73), where

$$\Omega_z(V) = zN[\frac{4}{15}\varepsilon_F + \varepsilon_X + \varepsilon_C + \frac{e^2 f}{3\pi\xi}]$$

$$+ N\frac{V_A}{2\pi^2}\int_0^\infty F(q)q^2 dq \quad ; \tag{79}$$

$$\phi(r;V) = \frac{z^2 e^2}{r} + \alpha e^{-\gamma r} + \frac{V_A}{\pi^2}\int_0^\infty F(q)\frac{\sin qr}{qr}q^2 dq \quad ; \tag{80}$$

where $V_A = V/N$ is the volume per atom.

Exercise. Derive equation (75). This is the lengthy part of the above derivation. It can be done by starting with equation (27.1) of Reference 33, and making multiple-counting corrections and keeping all $q\rightarrow 0$ terms. Or it can be done by starting with the results of Reference 33, equations (27.14) and (27.17), and by restoring the $q\rightarrow 0$ terms (whose sum is identically zero), to obtain the sum of (74) and (75) above.

For our pseudopotential calculations, we took the pseudopotential model for sodium that was determined many years ago in some work on lattice dynamics.[34] The model has four parameters, two in the Harrison-type pseudopotential, and α and γ in the

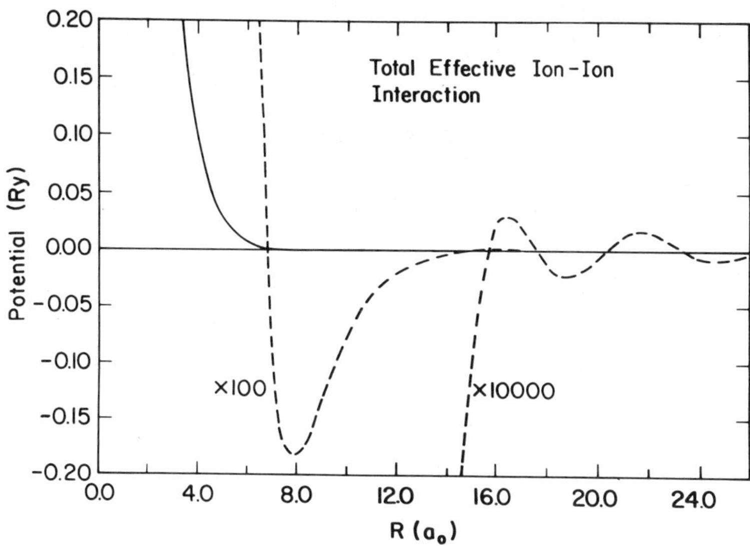

Fig. 3. The effective ion-ion potential $\phi(R;V_A)$ vs R for our
pseudopotential model of sodium, at $V_A = 256\ a_o^3$.

Born-Mayer repulsion. The parameter γ was fixed to agree with
the findings of Tosi on the alkali halides.[35] The three remaining parameters were adjusted so that theory and experiment would
agree for the crystal energy and its first two volume derivatives, at zero temperature and pressure. This model gives a good
account of the phonon dispersion curves of bcc sodium, and of the
thermodynamic properties of sodium up to room temperature.[34]
Hence we can use the model to see how well molecular dynamics
does in calculating thermodynamic properties of solid and fluid
metallic sodium. The effective ion-ion potential $\phi(r;V)$ at a
volume of 256 a_o^3/atom, which is close to the experimental volume
at T=0 and P=0, is shown in Figure 3. As in all metals, $\phi(r;V)$
is not responsible for the binding, but instead the large negative term $NI_z + \Omega_z(V)$ dominates the energy under normal
conditions.

Crystalline Sodium and Anharmonicity
<u>   </u>

Our computational cell is a rectangular parallelopiped, or
sometimes a cube, with N in the range 672-2048. Our MD run is
normally started by putting the ions at the lattice sites of a
perfect bcc crystal, with a random Gaussian distribution of velocities. If the total kinetic energy of the ions is below a cer-

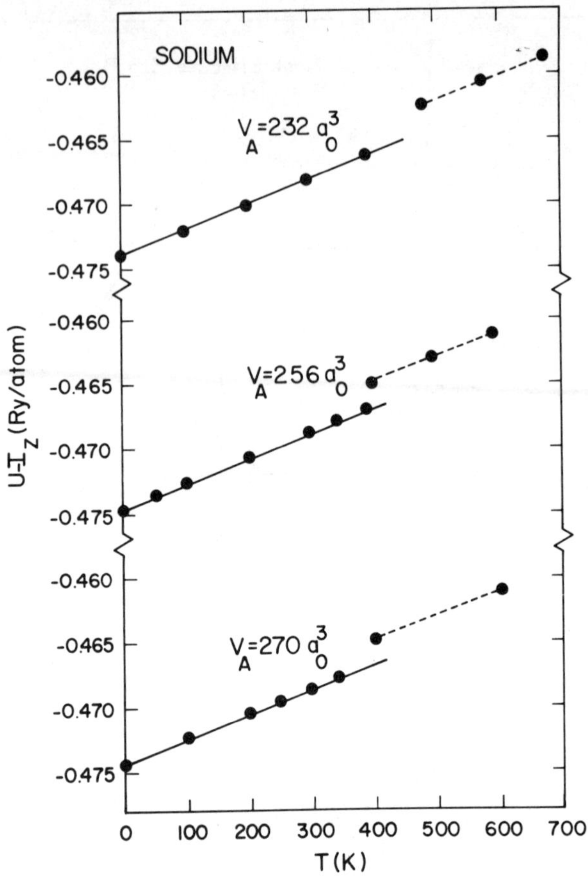

Fig. 4. Solid points are the MD results for the total energy
per atom minus the ionization energy. Solid lines are
the quasiharmonic contributions, and the dashed lines
are merely fits to the data for the fluid phase.

tain level, the system comes to equilibrium as a bcc crystal at
some temperature. If the total kinetic energy is above a certain
level, the crystal melts and the system comes to equilibrium as a
fluid, at a temperature above the "dynamic melting temperature."
This behavior is illutrated in Fig. 4, which shows the internal
energy as a function of temperature, at equilibrium, for three
different volumes. The solid points are the MD results, and they
fall on two separate curves for each volume, corresponding to
solid and fluid phases. Dynamic melting at constant volume
occurs between the highest point on the solid curve and the
lowest point on the fluid curve. The dynamic melting temperature
is a function of volume. At the middle volume of $V_A = 256\ a_0^3$, we

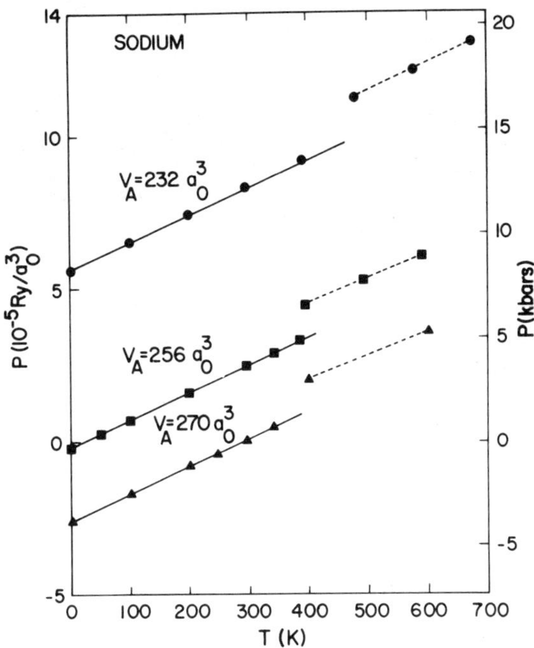

Fig. 5. Solid points are the MD results for the pressure, solid
lines are the quasiharmonic contributions, and dashed
lines are merely fits to the fluid-phase data.

made runs close to the dynamic melting temperature. The highest
stable crystal we observed is at 385 K, and the lowest stable
fluid we observed is at 396 K. An intermediate run proved to be
unstable, in that the mean of the kinetic energy continued to
drift for as long as the calculation continued. The pressure for
the same set of runs is shown in Figure 5. Again for the
pressure, at each volume, the data points fall on two separate
curves, according to whether the system has come to equilibrium
as a bcc crystal, or as a fluid.

The MD calculation of thermodynamic functions is valid at
temperatures for which the ion motion is classical. This is in
the lattice-dynamics high-temperature region $T \gtrsim \theta_{H\infty}$, where
$\theta_{H\infty}$ is the harmonic high-temperature Debye temperature, given
by equation (19.32) of Reference 33:

$$k\theta_{H\infty} = [\tfrac{5}{3}\langle \hbar^2 \omega_\kappa^2 \rangle_{BZ}]^{1/2} \quad . \tag{81}$$

Here ω_κ is a phonon frequency, and the brackets $\langle \ \rangle_{BZ}$ represent a Brillouin-zone average. $\theta_{H\infty}$ is volume dependent, and is 168K for pseudopotential sodium at $V_A = 256\ a_o^3$. The lattice-dynamics high-temperature expansion of the Helmholtz free energy F is given by equations (16.7) and (19.26) of Reference 33:

$$F = \Phi_0 - 3NkT\ \ln\ (T/\theta) + F_A \quad . \tag{82}$$

Here Φ_0 is the total potential Φ evaluated at the perfect crystal configuration, and θ is the classical characteristic temperature, given by

$$\ln\ k\theta = \langle \ln\ \hbar\omega_\kappa \rangle_{BZ} \quad . \tag{83}$$

An important point is that F_A, the anharmonic free energy, is arbitrary here, not necessarily a perturbation. From the above expression for F, the internal energy is

$$U = \Phi_0 + 3NkT + U_A \quad , \tag{84}$$

and the pressure is

$$P = -\frac{d\Phi_0}{dV} - 3NkT\ \frac{d\ \ln\ \theta}{dV} + P_A \quad , \tag{85}$$

where the anharmonic contributions are given by

$$U_A = F_A - T\ \frac{\partial F_A}{\partial T}\bigg|_V \quad , \tag{86}$$

$$P_A = -\ \frac{\partial F_A}{\partial V}\bigg|_T \quad . \tag{87}$$

Again the anharmonic contributions U_A and P_A are not necessarily small, and are not necessarily expressed by perturbation theory.

We now want to use our MD results to determine the anharmonic thermodynamic functions for bcc sodium. The first step is to calculate Φ_0 and θ, and their volume derivatives, for pseudo-potential sodium at the volumes of interest. This is done by straight-forward lattice-dynamic techniques, as described in Reference 33. The quasiharmonic contributions to U and P are then subtracted from the MD results for U and P, to find the anharmonic parts U_A and P_A, respectively. The anharmonic energy is shown in Figure 6, and the anharmonic pressure is shown in

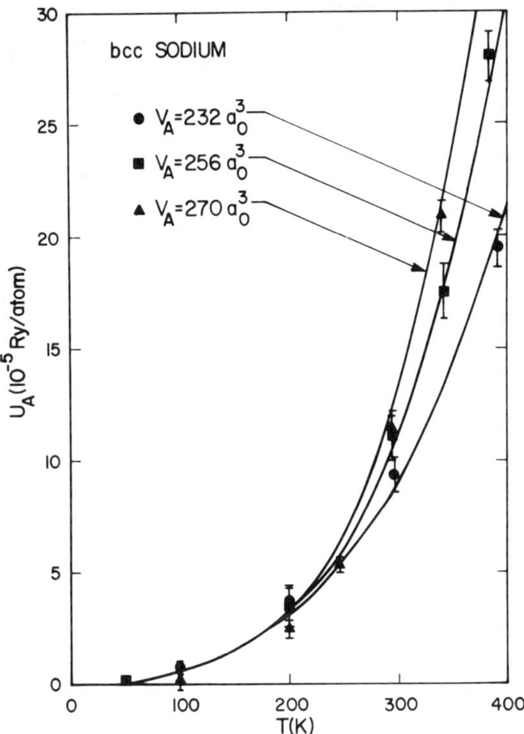

Fig. 6.   Anharmonic internal energy vs temperature. Solid lines
represent the fit to all the anharmonic data, through
equation (90) for F_A.

Figure 7. The striking property of these curves is that they are
not of the form prescribed by leading-order anharmonic perturbation theory. From leading-order anharmonic perturbation theory,
in the high temperature region, both U_A and P_A should go as
T^2, with volume-dependent coefficients. Our results for U_A and
P_A cannot be fitted to T^2 curves, within the estimated statistical errors of our MD averages.

From an examination of numerical results, Hoover[36] suggested
that the function $\ln k\theta$ may converge as $\Gamma^{-1}\ln\Gamma$, as $\Gamma\to\infty$, where Γ
is the number of points used in calculating the Brillouin-zone
average. This result has sometimes been misinterpreted as a

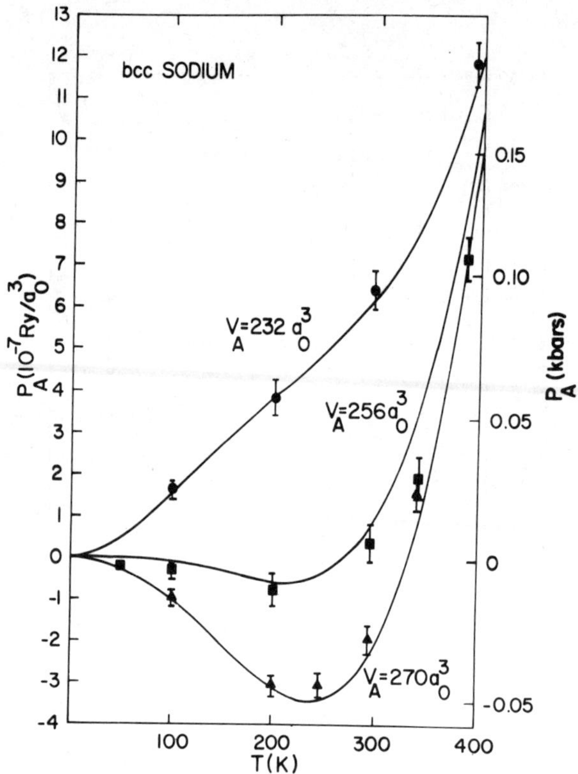

Fig. 7.   Anharmonic pressure vs temperature. Solid lines
represent the fit to all the anharmonic data, through
equation (90) for F_A.

finite-N effect in molecular dynamics. It is important to realize
that the lattice dynamics calculations are entirely separate from
the MD calculations, and that Brillouin-zone averages can be made
as accurate as desired, independently of the MD calculations.

The quantity contained in F, which is not directly accesible
to MD calculation, is the entropy S. From the lattice-dynamics
equation (82) for F, it follows

$$S = 3Nk[1 + \ln (T/\theta)] + S_A \tag{88}$$

where

$$
S_A = - \left. \frac{\partial F_A}{\partial T} \right)_V \quad . \tag{89}
$$

The quasiharmonic part of S contains the temperature-independent
term $3Nk(1 - \ln\theta)$. To determine S_A, we <u>assume</u> that S_A contains no temperature-independent term. Such a term would follow
from a contribution to F_A of the form $A_1 T$, where A_1 may be
volume-dependent. From our MD calculations of U_A and P_A, we
would not be able to determine A_1. However, anharmonic perturbation theory suggests that F_A in the high-temperature region
should be a power series in T, of lowest order T^2. Therefore, to
analyze the sodium calculations, we write

$$
F_A = A_2 T^2 + A_3 T^3 + A_4 T^4 \quad , \tag{90}
$$

where each of the coefficients A_2, A_3, A_4 is taken as a quadratic
function of the volume. This form gives a good fit to all the anharmonic data, as shown by the solid lines in Figures 6 and 7. In
this way, S_A and F_A are completely determined in the high temperature region.

Our MD results are compared with experimental data for sodium
up to the melting temperature in Reference 12. What is compared is
the volume at zero pressure, the heat capacity, the bulk modulus,
and the Grüneisen parameter. The comparison with experiment is
good for all the data. From these comparisons, together with the
lattice-dynamics studies of many years ago, we can draw two general
conclusions.

<u>First conclusion.</u> The combination of quasiharmonic lattice
dynamics in the quantum regime, for $T < \theta_{H\infty}$, together with molecular dynamics in the classical regime, $T \gtrsim \theta_{H\infty}$, provides a
simple and reasonably accurate representation of the vibrational
thermodynamics of a nonquantum solid.

<u>Second conclusion.</u> At last, the problem of calculating anhar-
monic thermodynamic functions at classical temperatures is solved,
by a combination of molecular dynamics and quasiharmonic lattice
dynamics.

Fluid Sodium and Melting

We carried out MD calculations for fluid sodium at several
volumes and at temperatures up to 2200K.[14] The ion-ion potential
ϕ becomes deeper as the volume increases, as shown in Figure 8,
while at the same time the negative term $\Omega_z(V)$ increases alge-
braically. We also included contributions to the energy and the

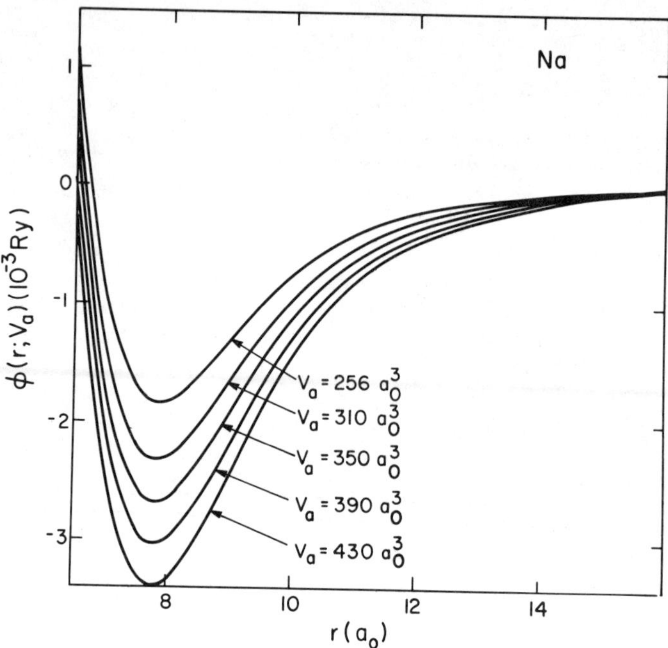

Fig. 8.   The effective ion-ion potential $\phi(r;V_a)$ for our pseudopotential model of sodium, for the range of volumes
discussed in the present work.

pressure which arise from thermal excitation of conduction electrons. For temperatures up to 1200 K, where experimental data
exist for the heat capacity, the bulk modulus, and the Grüneisen
parameter, our calculations are in good agreement with experiment.[14] However, for the volume at zero pressure, our calculations
diverge from experiment for temperatures above 1200 K. As shown in
Figure 9, our pseudopotential sodium does not expand as rapidly as
does real sodium in the high-temperature region. This discrepancy
is not as serious as it may appear from Figure 9, because at 2200 K
it would take only 2 kbar of pressure to compress real sodium to
our theoretical zero-pressure volume. In terms of the best theoretical equation-of-state calculations of the present time, 2 kbar
is essentially zero. But the discrepancy in Figure 9 is still
meaningful. We do not believe it represents a failure of molecular
dynamics, nor does it represent a failure of pseudopotential theory
in general. The discrepancy in Figure 9 is probably due to extrapolation of our pseudopotential model, with its fixed parameters,
over too large a range of conduction-electron density.

Now that we have a good theory for crystalline sodium, and for
fluid sodium, we should be able to calculate the melting temperature. Of course, dynamic melting occurs spontaneously on the com-

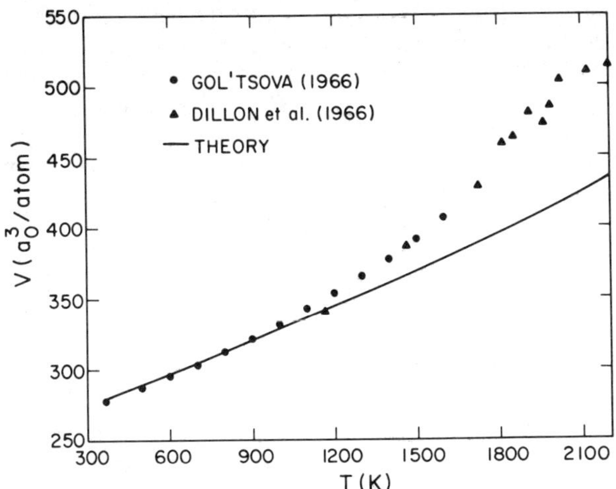

Fig. 9.   Calculated and measured values of the zero-pressure
volume vs temperature, for fluid sodium.

puter, and in fact the dynamic melting temperature is close to the
experimental melting temperature, but we don't know to what extent
dynamic melting might be a computer artifact. We therefore have to
calculate the free energies of the solid and fluid phases, and determine the "equilibrium melting temperature." Specifically, the
Gibbs free energy G is

$$G = F + PV \quad , \tag{91}$$

and this quantity is the same for solid and fluid phases at the
equilibrium melting temperature and pressure, T_m and P_m respectively:

$$G_f(P_m,T_m) = G_s(P_m,T_m) \quad . \tag{92}$$

We have already determined the free energy of bcc sodium in the
high temperature region. We now have to devise a procedure for
calculating the free energy of the fluid phase.

Let us fix the volume, hence fix the potentials $\Omega_z(V)$ and
$\phi(r;V)$, and go to the high-temperature limit. Of course, in a

555

real metal, thermal ionization of core electrons will take place at sufficiently high temperatures, and this will change the potentials $\Omega_z(V)$ and $\phi(r;V)$. But here we can ignore this effect, since our object is merely to determine the entropy in the fluid phase at moderate temperatures, where the operating potentials are $\Omega_z(V)$ and $\phi(r;V)$. From the classical canonical partition function (7), with the Hamiltonian given by equations (2)-(4), the cluster expansion of the Helmholtz free energy is[37]

$$F = \Omega(V) + NkT \left[\ln(\rho\Lambda^3) - 1 + \sum_{n=2}^{\infty} \frac{B_n \rho^n}{n-1} \right] \quad , \tag{93}$$

where $\rho = V_A^{-1}$ is the atomic density, and Λ is the deBroglie wavelength, given by equation (9). The B_n are virial coefficients, and usually $B_n = B_n(T)$, but for metals we have $B_n = B_n(V,T)$. The first coefficient is

$$B_2 = 2\pi \int_0^{\infty} (1 - e^{-\beta\phi(r;V)}) \, r^2 dr \quad . \tag{94}$$

From equation (93), the following expansions are found for the internal energy and the pressure:

$$U = \Omega + NkT \left[\frac{3}{2} - \sum_{n=2}^{\infty} \left(\frac{T\partial B_n}{\partial T} \right)_V \frac{\rho^{n-1}}{n-1} \right] \quad , \tag{95}$$

$$P = -\frac{d\Omega}{dV} + kT\rho + kT \sum_{n=2}^{\infty} \left[B_n - \frac{V_A}{n-1} \left(\frac{\partial B_n}{\partial V_A} \right)_T \right] \rho^n \quad . \tag{96}$$

The important point of the cluster expansion is that it converges at sufficiently high temperature.[38] This gives us a way to determine the free energy of the fluid phase, as follows. At a fixed volume, for a sequence of increasing temperatures, we calculate B_2 and its volume- and temperature-derivatives, and we also carry out MD calculations of U and P. At a sufficiently high temperature, the MD results for U and P will agree with the expansions (95) and (96), keeping only terms to order n=2. At this same temperature, the free energy is then assumed to be given by (93) to order n=2. Now from the MD data for U(T) at the fixed volume, the free energy can be calculated down to low temperatures, by integrating the relation

$$d\left(\frac{F}{T} \right) = -\frac{U}{T^2} \, dT, \text{ at constant V.} \tag{97}$$

From the MD data for the fluid at moderate temperatures, the free energy can also be determined as a function of volume, by integrating the relation

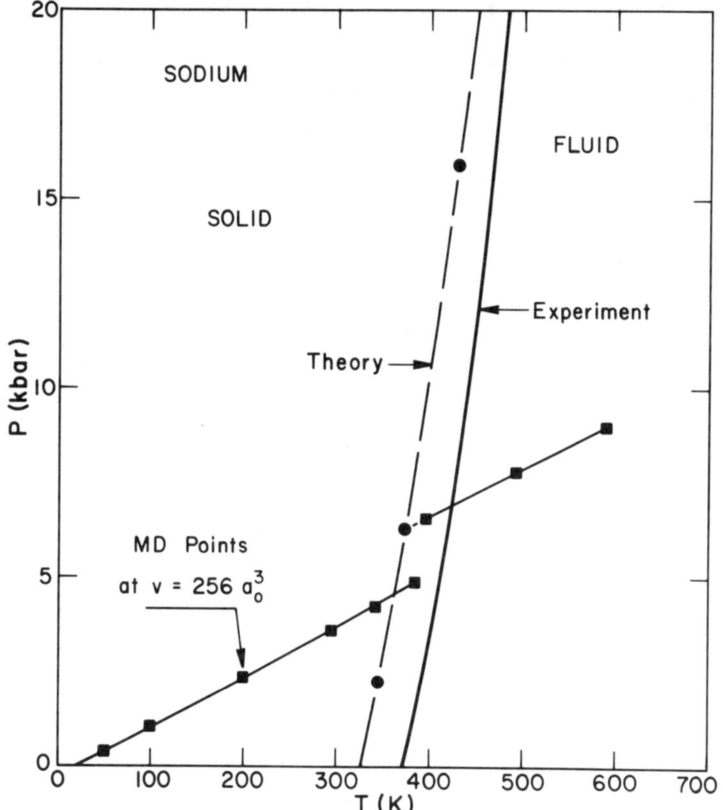

Fig. 10.   P-T phase diagram for sodium. Solid line is the experimental melting curve, and solid circles are the calculated equilibrium melting points. Solid squares are the
MD results at the fixed volume of 256 a_o^3/atom; dynamic
melting takes place between the highest point on the
solid branch (385 K) and the lowest point on the fluid
branch (396 K).

$$dF = - PdV, \text{ at constant T.} \tag{98}$$

This procedure was first described in Reference 13. In this way
we calculated G for fluid pseudopotential sodium, and hence determined the equilibrium melting temperature from the condition
(92).

Our calculated melting temperature as a function of pressure
is compared with experiment in Figure 10. The theoretical temperatures are 8-12% lower than experiment, which is excellent agreement
for a melting calculation. The discrepancy is not caused by a

failure of our entropy calculation for either the solid or fluid
phase, since these calculations are both in good agreement with experiment. As can be seen from Figure 10, the dynamic melting temperature is a little higher than the equilibrium melting temperature. We believe the dynamic melting process is sensitive to the
MD cell size, more so than equilibrium melting, and for large N we
expect the dynamic melting temperature to approach the equilibrium
value. It has been stated that, in melting of the alkali metals,
the potential $\Omega(V)$ can be ignored, since melting is merely a rearrangement of the atoms.[39] However, when sodium melts at zero pressure, the volume increases by 2.4%, and because $\Omega(V)$ is the dominant part of the total potential, it actually contributes more than
half of our calculated increase ΔU of the internal energy. This is
shown in the following list for melting of sodium at zero pressure.

$$
\begin{aligned}
\Delta\Omega \ (\text{calculated}) &= 296 \ \text{cal/mole} \\
\Delta\Sigma\phi \ (\text{calculated} &= 252 \ \text{cal/mole} \\
\Delta U \ (\text{total theory}) &= 548 \ \text{cal/mole} \\
\Delta U \ (\text{experiment}) &= 621 \ \text{cal/mole}
\end{aligned}
$$

Further details of our melting calculations can be found in
Reference 13.

V. MORE EXOTIC APPLICATIONS OF MOLECULAR DYNAMICS

We will attempt only a brief listing of MD applications beyond
ordinary equilibrium thermodynamics, merely to direct the student's
attention to interesting topics in this rapidly expanding field.

Let us first consider nonequilibrium properties of dense
fluids. Linear response theory relates transport coefficients to
the decay of position and velocity correlations among the particles
in an equilibrium fluid. For example, the shear viscosity η can be
expressed in Green-Kubo formalism as a time integral of a
particular correlation function:

$$
\eta = \frac{1}{kTV} \int_0^\infty \langle \dot{A}(t)\dot{A}(0)\rangle \, dt \quad , \tag{99}
$$

where the bracket $\langle \ \rangle$ denotes a canonical ensemble average, and
for N particles in a volume V,

$$
A = \Sigma_K x_{Ki} P_{Kj} \quad , \text{ for } i \neq j. \tag{100}
$$

The equivalent expression in terms of an Einstein relation is

$$
\eta = \frac{1}{kTV} \lim_{t\to\infty} \frac{1}{2t} \langle [A(t) - A(0)]^2\rangle \quad . \tag{101}
$$

The basic ideas of linear response theory are summarized by Zwanzig,[40] and detailed derivations are presented by Helfand,[41] and by
Hansen and McDonald.[42] One should keep in mind that these equations are derived for correlations which are found locally in a
fluid of infinite extent. In equilibrium MD, the motion of particles surrounding the computational cell is correlated <u>in a special
way</u> with the motion of particles within the computational cell, and
the presence of this <u>periodically correlated motion</u> will affect the
time development of correlation functions.

<u>Exercise.</u> How does the use of periodic boundary conditions
affect the evaluation of the Green-Kubo formula (99)? The Einstein
formula (101)?

<u>Exercise.</u> Linear response theory is derived in the canoni-
cal ensemble. If equation (99) or (101) is evaluated in the MD
ensemble, is the relative error of order N^{-1}, or of order 1?

Beyond linear response theory, molecular dynamics has the
capability in principle of simulating processes which are well
away from equilibrium. This capability has been exploited in the
development of nonequilibrium molecular dynamics, as described by
Hoover and Ashurst,[43] and recently reviewed by Hoover.[44] The
technique is to modify the equations of motion, which in effect
couples the system to momentum and energy reservoirs, so that the
computer can simulate a nonequilibrium steady state. Applications include viscous flows, heat flows, and chemical reactions.
We note that Hoover defines $\frac{3}{2}$ kT as the instantaneous kinetic
energy per particle, KE_{CM}/N, and for a "constant temperature
simulation," Hoover constrains the kinetic energy to be strictly
constant. This produces an ensemble whose weight function contains
$\delta(KE_{CM}$ - constant), and hence it is different from the MD
ensemble. Several articles on nonequilibrium fluids appear in the
January, 1984, issue of <u>Physics Today</u>, and of particular interest
for nonequilibrium molecular dynamics are the articles by Evans,
Hanley, and Hess,[45] by Hoover,[46] and by Alder and Alley.[47]

Several variations of ordinary MD have been developed.
Andersen[48] has proposed "molecular dynamics at constant temperature," in which an MD system is made to represent a canonical
system, by altering the momentum of random particles at sequential
random instants of time. The new momentum is picked from a Boltzmann distribution, with a given parameter β. Since the motion of
the system is no longer Hamiltonian, this procedure is a statistical sampling method. A combination technique was used by Wood and
Erpenbeck,[5,6] who ran a set of independent MD calculations,
with the initial phase of each calculation picked from a canonical,
or microcanonical, distribution. Andersen[48] also described "molecular dynamics at constant pressure," in which the pressure is a
parameter of the Lagrangian, and the system volume fluctuates.

Like the ordinary MD, this is an evolutionary method. However, the physical meaning of the Lagrangian is not clear, hence the time-scale of the motion is also of uncertain meaning. The technique was extended to the case of anisotropic stress, with the MD cell having both shape and volume fluctuations, by Parrinello and Rahman.[49] They used the procedure to induce phase changes from one crystal structure to another.[50] Since this theory is so far developed for volume-independent potentials, it is not in principle applicable to metals, because of the strong volume dependence of the potentials in metals.

Abraham has carried out an extensive study of the phases of two-dimensional systems of particles interacting through Lennard-Jones potentials.[51,52] His simulations include solid-fluid interfaces, fluid-vapor interfaces, and processes of melting and vaporization. We note that Abraham's "molecular dynamics at constant temperature and/or constant pressure" is different from that of Andersen. Abraham's[51,52] method is to rescale all the particle velocities, or all the particle positions, at certain instants during the course of the MD calculation. The rescaling transforms the calculations to a sampling technique. An important point to keep in mind is that each different MD procedure produces its own ensemble weight function, and that fluctuation phenomena are essentially different in different ensembles.

Density fluctuations in liquid rubidium were studied by Rahman and coworkers,[53,54] and stability of supercooled liquid rubidium was studied by Mountain and coworkers.[55,56] Collective modes in crystalline and fluid metals were studied by Jacucci and Klein,[57] and by Jacucci and McDonald.[58] Because of the volume dependence of the potentials, the problem of density fluctuations contains subtle and complicated difficulties. Indeed, one of the most intriguing unsolved problems in metals physics today is how to construct effective ion-ion potentials when a local density gradient exists, since these potentials operate nonlocally through the conduction electrons.

In an interesting series of papers, Stillinger and Weber[59-61] have used molecular dynamics to find the "inherent structure" in an atomic liquid, and to study the potential energy hypersurface underlying the motion of atoms in the liquid state. Abraham has carried out MD calculations of the adsorption of rare gas atoms onto graphite, and has simulated the nucleation and growth of incommensurate monolayer regions. These calculations are surveyed, and an extensive bibliography is also given, in a forthcoming publication of Abraham.[62]

ACKNOWLEDGMENTS

This work was carried out in collaboration with Galen
Straub, Sheila Schiferl, Richard Swanson, Brad Holian, and James
D. Johnson. The typescript was prepared by Barbara Forrest.

REFERENCES

1. B. J. Alder and T. E. Wainwright, Phys. Rev. 127, 359
    (1962).
2. C. Erginsoy, G. H. Vineyard, and A. Englert, Phys. Rev. 133,
    A595 (1964).
3. A. Rahman, Phys. Rev. 136, A405 (1964).
4. W. G. Hoover and B. J. Alder, J. Chem. Phys. 46, 686 (1967).
5. W. W. Wood, in Fundamental Problems in Statistical
    Mechanics, ed. by E. D. G. Cohen (North-Holland,
    Amsterdam, 1975), Vol. 3, p. 331.
6. J. J. Erpenbeck and W. W. Wood, in Statistical Mechanics,
    ed. by B. J. Berne (Plenum, New York, 1977), Pt. B, p. 1.
7. J. Kushick and B. J. Berne, in Statistical Mechanics, ed. by
    B. J. Berne (Plenum, New York, 1977), Pt. B, p. 41.
8. R. W. Hockney and J. W. Eastwood, Computer Simulation Using
    Particles (McGraw-Hill, New York, 1981).
9. J. L. Lebowitz, J. K. Percus, and L. Verlet, Phys. Rev. 153,
    250 (1967).
10. F. Reif, Fundamentals of Statistical and Thermal Physics
    (McGraw-Hill, New York, 1965).
11. G. K. Straub, R. E. Swanson, B. L. Holian, and D. C.
    Wallace, in Ab Initio Calculation of Phonon Spectra, ed.
    by J.T. Devreese, V.E. Van Doren and P.E. Van Camp (Plenum,
    New York, 1983), p. 137.
12. R. E. Swanson, G. K. Straub, B. L. Holian, and D. C.
    Wallace, Phys. Rev. B25, 7807 (1982).
13. B. L. Holian, G. K. Straub, R. E. Swanson, and D. C.
    Wallace, Phys. Rev. B27, 2873 (1983).
14. G. K. Straub, S. K. Schiferl, and D. C. Wallace, Phys. Rev.
    B28, 312 (1983).
15. H. B. Callen, Thermodynamics (Wiley, New York, 1960).
16. S. K. Schiferl and D. C. Wallace, J. Chem. Phys. (to be pub-
    lished).
17. W. B. Davenport, Jr., Probability and Random Processes
    (McGraw-Hill, New York, 1970); see Ch. 9, §8 and 9.
18. J. V. Bradley, Distribution-Free Statistical Tests
    (Prentice-Hall, Englewood Cliffs, New Jersey, 1968).
19. W. J. Conover, Practical Nonparametric Statistics (Wiley,
    New York, 1971).
20. S. S. Shapiro and M. B. Wilk, Biometrika 52, 591 (1965).
21. S. S. Shapiro, M. B. Wilk, and H. J. Chen, J. Am. Stat.
    Assoc. 63, 1343 (1968).

22. G. W. Snedecor and W. G. Cochran, _Statistical Methods_ (Iowa University Press, Ames, Iowa, 1967).

23. E. S. Pearson and H. O. Hartley, _Biometrika Tables for Statisticians_ (Cambridge University Press, Cambridge, 1954), Vol. I.

24. A. Hald, _Statistical Theory with Engineering Applications_ (Wiley, New York, 1952).

25. H. Grad, Comm. Pure and Appl. Math. $\underline{5}$, 455 (1952).

26. F. Lado, J. Chem. Phys. $\underline{75}$, 5461 (1981).

27. D. C. Wallace and G. K. Straub, Phys. Rev. A27, 2201 (1983).

28. D. C. Wallace, S. K. Schiferl, and G. K. Straub, Phys. Rev. A (to be published).

29. M. E. Fisher and J. L. Lebowitz, Comm. Math. Phys. $\underline{19}$, 251 (1970).

30. J. L. Lebowitz and J. K. Percus, Phys. Rev. $\underline{124}$, 1673 (1961).

31. Z. W. Salsburg, J. Chem. Phys. $\underline{44}$, 3090 (1966).

32. W. L. Slattery, G. D. Doolen, and H. E. DeWitt, Phys. Rev. A26, 2255 (1982).

33. D. C. Wallace, _Thermodynamics of Crystals_ (Wiley, New York, 1972).

34. D. C. Wallace, Phys. Rev. $\underline{176}$, 832 (1968).

35. M. P. Tosi, in _Solid State Physics_, edited by F. Seitz and D. Turnbull (Academic, New York, 1964), Vol. 16, p. 1.

36. W. G. Hoover, J. Chem. Phys. $\underline{49}$, 1981 (1968).

37. J. O. Hirschfelder, C. F. Curtiss, and R. B. Bird, _Molecular Theory of Gases and Liquids_ (Wiley, New York, 1954).

38. D. C. Wallace, B. L. Holian, J. D. Johnson, and G. K. Straub, Phys. Rev. A26, 2882 (1982).

39. I. N. Makarenko, A. M. Nikolaenko, and S. M. Stishov, in _High-Pressure Science and Technology_, edited by K. D. Timmerhaus and M. S. Barber (Plenum, New York, 1979), Vol. I, p. 347.

40. R. Zwanzig, Ann. Rev. Phys. Chem. $\underline{16}$, 67 (1965).

41. E. Helfand, Phys. Rev. $\underline{119}$, 1 (1960).

42. J. P. Hansen and I. R. McDonald, _Theory of Simple Liquids_ (Academic, New York, 1976).

43. W. G. Hoover and W. T. Ashurst, in _Theoretical Chemistry_, edited by H. Eyring and D. Henderson (Academic, New York, 1975), Vol. 1, p. 1.

44. W. G. Hoover, Ann. Rev. Phys. Chem. $\underline{34}$, 103 (1983).

45. D. Evans, H. J. M. Hanley, and S. Hess, Physics Today $\underline{37}$, 26 (1984).

46. W. G. Hoover, Physics Today $\underline{37}$, 44 (1984).

47. B. J. Alder and W. E. Alley, Physics Today $\underline{37}$, 56 (1984).

48. H. C. Andersen, J. Chem. Phys. $\underline{72}$, 2384 (1980).

49. M. Parrinello and A. Rahman, J. Appl. Phys. $\underline{52}$, 7182 (1981).

50. M. Parinello, A. Rahman, and P. Vashishta, Phys. Rev. Lett. $\underline{50}$, 1073 (1983).

51. F. F. Abraham, in Proc. Intern. Conf. on Ordering in Two
    Dimensions, edited by S. K. Sinha (North-Holland, New
    York, 1980), p. 155.
52. F. F. Abraham, Physics Reports 80, 339 (1981).
53. A. Rahman, Phys. Rev. A9, 1667 (1974).
54. S. W. Haan, R. D. Mountain, C. S. Hsu, and A. Rahman, Phys.
    Rev. A22, 767 (1980).
55. R. D. Mountain and P. K. Basu, Phys. Rev. A28, 370 (1983).
56. A. C. Brown and R. D. Mountain, J. Chem. Phys. 80, 1263
    (1984).
57. G. Jacucci and M. L. Klein, in Liquid and Amorphous Metals,
    edited by E. Lüscher and H. Coufal (Sigthoff and
    Nordhoff, Holland, 1980), p. 131.
58. G. Jacucci and I. R. McDonald, in Liquid and Amorphous
    Metals, edited by E. Lüscher and H. Coufal (Sigthoff and
    Nordhoff, Holland, 1980), p. 143.
59. F. H. Stillinger and T. A. Weber, Phys. Rev. A25, 978
    (1982).
60. F. H. Stillinger and T. A. Weber, Phys. Rev. A28, 2408
    (1983).
61. T. A. Weber and F. H. Stillinger, J. Chem. Phys. 80, 2742
    (1984).
62. F. F. Abraham, "Computer Simulations of Surfaces,
    Interfaces, and Physisorbed Films," to appear in J. Vac.
    Sci. Techn.

AUTHOR INDEX

Abarenkov, I.V., 11, 53,
    59-61, 65, 67
Abraham, F.F., 560, 562,
    563
Adams, A.R., 451, 452,
    476, 491, 492
Adler, S.L., 160, 172,
    302, 303, 312
Alber, R., 215, 225
Alder, B.J., 45, 110,
    185, 186. 189,
    197-199, 219, 220,
    337, 338, 390,
    395, 524, 559,
    561, 562
Aldinger, F., 364, 396
Allan, D.C., 183, 184,
    222
Allen, P.B., 46, 111,
    479, 481, 492
Allen, J.W., 185, 222
Alley, W.E., 559, 562
Alonso, J.A., 388, 397
Andersen, H.C., 559, 562
Andersen, O.K., 114, 120,
    153, 155, 181,
    190-192, 194, 205-
    207, 222-225, 324,
    332, 378, 397,
    481, 492
Anderson, P.W., 184, 222
Andrei, N., 185, 222
Animalu, A.O.E., 46, 53-
    55, 59, 60, 85,
    110, 111

Appelbaum, J.A., 46, 111,
    177, 193, 195, 205,
    221, 224, 368,
    378, 396, 397
Aravind, P.K., 44, 110
Arlinghaus, F.J., 378, 397
Armstrong, J.A., 504, 517
Ashcroft, N.W., 46, 51, 52,
    111, 190, 191,
    193, 197, 223,
    224, 496, 517
Ashurst, W.T., 559, 562
Austin, B.J., 49, 53, 54,
    56, 59, 67, 111
Averbach, 367
Averill, F.W., 190, 223,
    388, 397

Bachelet, G.B., 53-55, 61,
    112, 122, 153, 156,
    190, 192, 205, 207,
    210, 232, 280, 310,
    316, 320, 323, 332,
    388, 398
Bairamov, B.K., 507, 517
Bak, T.A., 402, 449
Baldereschi, A., 158, 183,
    184, 194, 222, 224,
    227, 230, 299, 303,
    304, 312
Baraff, G.A., 190, 192,
    205, 207, 210,
    218, 223, 225,
    323, 330, 332
Barber, M.S., 558, 562

Barsky, A., 368, 370, 376
Barthelet, J.C., 46, 110
Bar-Yam, Y., 205, 207,
        218, 225
Bassani, F., 452, 457,
        483, 491
Basu, P.K., 560, 563
Bate, R.T., 477, 478, 492
Batson, P.E., 36, 43, 108
Baym, G., 114, 115, 123-
        126, 128, 137, 141,
        149, 154, 156
Beck, D.E., 436, 449
Bednarek, S., 452, 457,
        466, 481, 482,
        484, 490, 491
Beer, A.C., 451, 491, 492
Bendt, P., 322, 332
Benedek, G., 449
Bennett, J., 137, 156
Bergstresser, T.K., 164,
        172, 480, 491
Berne, B.J., 524, 559, 561
Bernholc, J., 205, 207,
        218, 225, 480, 491
Bertoni, C.M., 96, 112,
        122, 153, 156,
        388, 398
Bethe, H., 185, 222
Bilz, H., 241, 305, 311,
        509
Bir, G.L., 460, 461, 463,
        491
Bird, R.B., 556, 562
Birgenau, R.J., 417, 449
Birman, J.L., 353, 395
Biswas, R., 177, 205, 214-
        216, 219-221, 360,
        395
Bloembergen, N., 504, 511,
        513-515, 518
Boettger, J.C., 193, 200,
        205, 224
Bohm, D., 36, 108
Bohnen, K.P., 150, 152,
        156
Born, M., 176, 179, 182,
        220, 222, 271, 312,
        314, 333, 350, 395,
        403, 449, 496, 517

Bortolani, V., 96, 112,
        401, 402, 435, 448,
        449
Boyle, 321, 333
Boyter, J.K., 259, 267,
        311
Bradley, J.V., 534, 561
Brener, N.E., 44, 110
Brinkman, W., 137, 156
Brittin, W.E., 38, 109
Brockhouse, B.N., 402, 413,
        416, 449
Bron, W.E., 495, 506, 513,
        518
Brosens, F., 9, 37-39, 41-
        44, 109, 110, 115,
        130, 135, 155, 157,
        167
Brown, A.C., 560, 563
Brueckner, K., 16, 108
Brusdeylins, M.G., 402,
        448
Brust, D., 232, 310, 323,
        333
Burstein, E., 279, 285, 312
Bylander, D.M., 320, 333

Cabrera, N., 402, 428, 448
Calandra, G., 96, 112, 122,
        153, 156, 388, 398
Calleja, J.M., 514, 517
Callen, H.B., 531, 561
Car, R., 205, 207, 218,
        225, 299, 312
Cardillo, M.J., 426, 449
Cardona, M., 117, 130, 150,
        155, 156, 239, 258,
        311, 328, 331, 460,
        461, 463, 479, 481,
        490-493, 514, 517
Carlsson, A.G., 197, 224
Carter, D.L., 477, 478, 492
Castiel, D., 403, 449
Celli, V., 402, 428, 436,
        448, 449
Ceperley, D.M., 45, 110,
        185, 186, 189, 197-
        199, 219, 223, 224,
        337, 338, 390, 395

Chadi, D.J., 157, 175, 194,
220, 224, 229, 310,
320, 333, 371, 374,
376, 396
Chakraverty, S., 197, 224
Chan, C.T., 342, 347, 349,
395
Chandesris, D., 380, 397
Chang, K.J., 236, 310
Chang, R.K., 215, 225
Chappell, W.E., 38, 109
Chelikowsky, J.R., 46, 111,
139, 156, 163, 164,
172, 181, 205-207,
217, 222, 231, 232,
310, 337, 342-347,
349-351, 368-370,
377-380, 395-397
Chen, C.H., 36, 43, 108
Chen, H.J., 534, 561
Chen, R., 348, 395
Chiang, C., 53, 112, 190,
192, 205, 206, 223,
232, 280, 310, 316,
320, 332, 337, 339,
340, 394
Chou, M.Y., 205, 206, 225,
347, 348, 364,365,
367, 395, 396
Christensen, N.E., 260,
311, 328, 333
Churcher, N., 305
Cochran, W.G., 534, 561
Cohen, E.D.G., 524, 527,
559, 561
Cohen, M.H., 19, 108, 157,
160, 172, 176, 179,
183, 220, 222, 284,
312, 353, 395
Cohen, M.L., 46, 97, 110-
112, 114, 120, 139,
155, 156, 163, 164,
172, 177, 180, 187,
191-195, 200, 202,
205-207, 209, 210,
213, 215-218, 221-
225, 229-232, 234,
236, 241-243, 246,
310, 311, 316, 318-
320, 324, 326, 328,

Cohen, M.L., (continued)
331-333, 337, 341-
348, 350-354, 359,
361-365, 367-373,
376-380, 388, 394-
397, 477-481, 492
Collins, C.L., 451, 452,
476, 491, 492
Connoly, J.W., 136, 156
Conover, W.J., 534, 561
Conrad, H., 382, 383, 397
Cordes, J., 417, 449
Cornwell, J.F., 420, 449
Coufal, H., 560, 563
Cousins, C.S.G., 260, 311,
328, 333
Curtiss, C.F., 556, 562

d'Amour, H., 328, 331
Davenport, W.B., Jr., 533,
561
Dayak, B., 259, 267, 311
Dean, P.J., 484, 492
Deb, B.M., 180, 181, 193,
222
Dekeyser, R., 37, 109
Delley, B., 178, 221
Del Sole, R., 371, 396
Demuth, J.E., 382-384, 386,
397
Demtroder, W., 502, 517
Denner, W., 328, 331
Dennis, J.E., Jr., 321,
322, 332
Dc Raedl, B., 38, 109
De Raedt, H., 38, 109
Devreese, J.T., 9, 35, 37-
39, 41-44, 98, 108-
110, 112, 114, 130,
132, 135, 139, 143,
144, 146, 155, 157-
159, 163, 168, 172,
175, 177, 180, 181,
183, 184, 194, 205,
220-222, 227, 230,
238, 239, 250, 252,
303, 310, 312, 321,
333, 529, 561
DeWitt, H.E., 541, 562
Blott, D.D., 506, 518

Doak, R.E., 402, 403, 426-428, 435, 448, 449
Dobrzynski, L., 403, 449
Dolling, G., 239, 241, 310, 417, 449
Donahue, J., 167, 170, 172, 209, 214, 215, 225
Doolen, G.D., 541, 562
Ducuing, J., 504, 517
Duke, C.B., 371, 396
Durand, P., 46, 110
Dutton, D.H., 413, 449

Eastman, D.E., 368, 370, 372, 374, 396
Eastwood, J.W., 524, 561
Eberhardt, W., 380, 382, 387, 397
Eden, R.C., 390, 391, 398
Ehrenfest, P., 179, 222, 314, 332
Ehrenreich, H., 19, 36, 44, 46, 56, 108, 110-112, 160, 172, 229, 243, 246, 310, 311, 479, 481, 492
Eisenberger, P., 36, 108
Ellis, D.E., 178, 221
Englert, A., 524, 561
Era, K., 12, 107
Erginsoy, C., 524, 561
Erpenbeck, J.J., 524, 559, 561
Ertl, G., 382, 383, 397
Esbjerg, N., 430, 449
Euwema, R.N., 12, 107, 151, 156
Evans, D., 559, 562
Eyring, H., 559, 562

Fahy, S., 363, 395
Falter, L., 183, 184, 222
Farnell, G.W., 436, 449
Fawcett, W., 451, 452, 476, 477, 488, 491, 492
Feibelman, P., 193, 205, 224
Fein, A.E., 500, 518

Feld, M.S., 502, 518
Feldblum, A., 451, 452, 460, 466, 468, 471, 483, 488, 491
Felton, R.H., 193, 223
Ferrell, R.A., 36, 108
Ferry, D.K., 476, 477, 492
Feynman, R.P., 180, 221, 243, 311, 314, 332, 366, 396
Fields, J.R., 36, 108
Fischer, C.F., 12, 107
Fisher, M.E., 541, 562
Fleszar, A., 304, 312
Flodstrom, S.A., 368, 370, 396
Flytzanis, C., 511, 515-518
Fock, V., 182, 222, 314, 332
Ford, W.K., 371, 396
Fornoff, E., 451, 461, 466, 491
Fowler, W.B., 137, 156, 501, 518
Franchini, A., 401, 435, 449
Franci, M.M., 178, 185, 193, 221
Freeman, A.J., 121, 153-156, 178, 193, 194, 197, 205, 207, 221, 223, 224, 378, 396
Frey, J., 451, 452, 477, 492
Froyen, S., 177, 205, 207, 221, 236, 242, 310, 316, 324, 328, 332, 337, 341, 344, 347, 348, 364, 388, 393, 394, 397
Fry, J.L., 44, 110
Fu, C.L., 180, 193, 194, 202-205, 222, 224, 242, 260, 311, 320, 332, 348, 350, 351, 395
Fuchs, K., 192, 223
Fulde, P., 150, 155
Fumi, F.G., 451, 452, 476, 491, 492
Furuya, K., 185, 222

Gale, G.M., 505, 517
Garcia, N., 449

Gay, J.G., 378, 397
Gelatt, C.D., Jr., 190-192, 200, 223
Geldart, D.J.W., 37, 44, 108, 110
Gell'Mann, M., 16, 108
Gerward, L., 328, 332
Gibbons, P.C., 36, 108
Giessen, B.C., 209, 212, 225
Girifalco, L.A., 388, 397
Glatzel, D., 481, 492
Glauber, R.J., 403, 449
Glembocki, O.J., 451, 452, 456, 457, 460, 461, 466, 468, 473, 474, 476, 479, 481-483, 488, 490-492
Glick, A.J., 36, 108
Glötzel, D., 114, 119, 120, 153, 155, 205, 207, 225, 324, 332
Glotzer, T., 115, 116, 130, 155
Goddard, W.A., III, 46, 53, 110, 111, 178, 221
Gomes Dacosta, P., 241, 246, 248, 250-255, 257-259, 263, 267, 269, 270, 311
Goldstein, H., 243, 311
Goodgame, M.M., 178, 221
Goodman, B., 137, 156
Gordon, R.G., 181, 222
Grad, H., 535, 562
Green, F.R., Jr., 388, 397
Greenside, H.S., 190, 192, 205, 207, 210, 223, 316, 323, 330-332
Guichar, G., 368, 370, 396
Guidotti, D., 452, 457, 483, 491
Gunnarsson, O., 114, 120-122, 155, 178, 189, 221, 223, 337, 338, 388-390, 395, 398
Gustafsson, T., 397

Haan, S.W., 560, 563
Hald, A., 534, 562
Hamakawa, Y., 484, 485, 492
Hamann, D.R., 46, 53, 54, 55, 61, 111, 112, 120, 121, 153, 155, 177, 190, 192, 193, 195, 205, 206, 208, 217, 224, 225, 229, 232, 242, 260, 280, 310, 311, 316, 320, 323, 331, 332, 337, 339, 340, 348, 350, 351, 368, 378, 388, 394-397
Haneman, D., 368, 370, 374, 396
Hanke, W., 113-116, 118, 130-137, 139-146, 150, 153, 155, 156, 388, 397
Hanley, H.J.M., 559, 562
Hansen, J.P., 559, 562
Hansson, G.V., 368, 370, 396
Harmon, B.N., 180, 193, 194, 202-205, 208, 217, 222, 224, 229, 242, 260, 310, 311, 348, 350, 351, 395
Harris, J., 178, 221, 433, 449
Harrison, J.G., 156
Harrison, W.A., 46, 51, 53, 54, 56, 58, 75-77, 93, 97, 101, 110-112, 481, 492
Harten, U., 402, 403, 427, 428, 435, 449
Hartley, H.O., 534, 562
Haynes, J.R., 484, 492
Hazi, A.U., 46, 110
Heaton, R.A., 156, 388, 397
Hedin, L., 37, 44, 108, 110, 114-117, 122, 137, 141, 149, 154, 155, 189, 223, 337, 338, 394

Heimann, P., 368, 370,
        374, 376, 396
Heine, V.L., 46, 49, 50,
        52-56, 58-61, 65,
        67, 85, 110-112,
        177, 181, 187, 221,
        229, 305, 310, 316,
        320, 332, 479, 481,
        492
Heisenberg, W., 182, 222,
        314, 332
Helfand, E., 559, 562
Hellman, H., 184, 221,
        314, 332
Henderson, D., 559, 562
Herbert, D.C., 477, 488,
        492
Herman, F., 259, 306,
        311,312
Herring, W.C., 48, 111,
        191, 220, 223,
        316, 332
Hess, S., 559, 562
Himpsel, F.J., 368, 370,
        374, 376, 396
Hirsch, J.E., 185, 223
Hirschfelder, J.O., 556,
        562
Ho, K.M., 46, 110, 111,
        184, 193-195, 202-
        205, 222, 224, 242,
        260, 311, 320, 321,
        332, 337, 342, 348,
        350, 351, 377-380,
        395, 397
Hockney, R.W., 520, 561
Höhberger, H.J., 36, 108
Hohenberg, P., 12, 79,
        107, 114, 155, 176,
        186, 221, 229, 310,
        313, 337, 388, 394
Holas, A., 43, 44, 110
Holian, B.L., 529, 553,
        556-558, 561, 562
Holzapfel, W.B., 209-211,
        213, 225
Holzschuh, E., 205, 206,
        225,236, 311
Holzwarth, N.A.W., 348,
        349, 380, 395, 397

Hoover, W.G., 524, 551, 559,
        561, 562
Hopfield, J., 163, 164, 169,
        172
Horsch, P., 150-152, 155,
        156
Houzay, F., 368, 370, 396
Hsu, C.S., 560, 563
Hu, J.Z., 209-211,225
Huang, K., 176, 179, 220,
        271, 312, 403, 449,
        496, 517
Hubbard, J., 37, 108
Hulthen, R., 390, 391, 398
Humphreys, R.G., 451, 461,
        466, 491
Huntingdon, H.B., 197, 224
Hybertsen, M.S., 388-393,
        397

Ibach, H., 402, 426, 428,
        449
Ichimaru, S., 45, 110
Ihm, J., 180, 187, 191,
        192, 195, 205, 206,
        217, 222, 224, 242,
        243, 311, 318, 319,
        321, 324, 332, 337,
        343, 370, 371, 376,
        377, 388, 395-397
Inkson, J.C., 137, 149,
        156
Ipatova, I.P., 244, 246,
        260, 261, 311

Jacoboni, C., 476, 477, 492
Jacucci, G., 560, 563
Janak, J.F., 114, 120, 155,
        177, 187, 190-193,
        200-202, 219, 221,
        223, 224, 229, 310
Jayaranan, A., 177, 221
Jepsen, O., 197, 224, 378,
        397
Jezequel, G., 368, 370, 396
Joannopoulos, J.D., 195,
        205, 207, 215, 217,
        218, 224, 225, 321,
        324, 332, 337, 376,
        377, 388, 394, 396,
        397

Johansson, B., 181, 222
Johnson, F.A., 157, 172
Johnson, J.D., 556, 562
Johnson, K.W., 190, 200,
     223
Johnson, P.D., 380, 382,
     397
Jona, F., 374, 396
Jones, D., 477, 488, 492
Jones, R.O., 114, 120,
     155, 178, 221,
     390, 398
Jonson, M., 114, 121, 122,
     155, 388, 389, 398
Jordan, P., 182, 222, 314,
     332
Jullian, R., 185, 222

Kadanoff, L.P., 123, 124,
     156
Kaiser, W., 502, 506,
     507, 514, 518
Kahn, R., 53, 111
Kallio, A., 45, 110
Kamitakahra, W.A., 416,
     449
Kane, E.O., 138, 139, 154,
     156
Kaschotzev, Z.M., 507,
     517
Kaspar, J.S., 214, 215,
     225
Kelly, P.S., 205, 207,
     218, 225
Kerker, G.P., 46, 111,
     190, 192, 205, 223,
     232, 260, 310, 316,
     320, 332, 337, 339,
     341, 380, 394, 397
Kim, Y.S., 181, 222
Kimball, J.C., 37, 109
Kitaev, Y.E., 507, 517
Kittel, C., 58
Klein, B.M., 153, 156
Klein, M.L., 560, 563
Klein, M.V., 501, 518
Kleinman, D.A., 515, 518
Kleinman, L., 37, 47, 54,
     108, 111, 320,
     326, 331, 332

Klemens, P.G., 499-501, 509,
     511, 518
Kobliska, R.J., 215, 225
Koch, E.E., 156
Koda, T., 12, 107
Kohn, W., 12-16, 79, 107,
     114-116, 120-122,
     125, 126, 154, 156,
     160, 161, 164, 172,
     176, 186, 187, 190,
     221, 223, 229, 310,
     313, 337, 338, 388,
     394, 433, 449
Korringa, J., 190, 223
Krakauer, H., 378, 397
Kratzer, S., 451, 452, 476,
     491, 492
Kress, W., 509
Kübler, J., 190, 191, 192,
     200, 223
Kugler, A.A., 37, 109
Kuhl, J., 506, 513, 518
Kunc, K., 119, 155, 158,
     161, 163, 175, 177,
     180, 181, 184, 187,
     191, 194, 195, 205,
     217-220, 225, 227,
     229-231, 238, 239,
     241, 246, 248, 250-
     255, 257-260, 263,
     265, 267, 269, 270,
     272, 276, 280, 285,
     289, 297, 299, 300,
     302, 304, 310-312,
     315, 316, 321, 331,
     333, 336, 350, 351,
     394, 395
Kunz, A.B., 119, 155
Kuppers, J., 382, 383, 397
Kushick, J., 524, 561
Kwok, P.C., 499, 518

Lado, F., 535, 562
Lam, P.K., 97, 112, 194,
     200, 205, 206, 224,
     225, 242, 311, 314,
     332, 347, 348, 350,
     351, 364, 365, 367,
     395, 396
Lambert, W.R., 419, 426

Land, R.H., 37, 108
Landau, L., 242, 312,
        498, 518
Landolt-Bornstein, 235,
        310, 390, 391, 398
Langer, D.W., 12, 107
Langreth, D.C., 37, 108,
        115, 123, 127, 155,
        388, 397
Lannoo, M., 119, 155, 371,
        396
Lantto, L., 45, 110
Latta, E.E., 382, 383, 397
Laubereau, A., 502, 506,
        507, 514, 517, 518
Laude, L.D., 460, 461, 463,
        490, 491, 493
Lax, M., 479, 492
Lebowitz, J.L., 527, 538,
        541, 561, 562
Lecante, J., 380, 397
Lee, D.H., 376, 377, 396
Lehwald, S., 402, 426,
        428, 449
Lemmens, L.F., 37-39, 41,
        43, 109, 110
Letokhov, V.S., 502, 518
Lettington, A.H., 477,
        488, 492
Levenson, M.D., 502, 513,
        515-518
Levine, Z., 338, 389,
        390, 395
Levy, M., 120, 156, 176,
        221
Ley, L., 117, 130, 150,
        155
Liberman, D.A., 190-192,
        196, 200, 223
Lieb, E.H., 185, 222
Lifshitz, E.M., 243, 311
Lin, C.C., 156, 388, 397
Lindhard, J., 36, 108
Lipari, N.O., 137, 156,
        205, 207, 218, 225,
        480, 492
London, G.J., 364, 396
Loucks, T.L., 190, 223

Louie, S.G., 46, 111, 158,
        163, 164, 172, 181,
        193, 202, 205-207,
        217, 222, 224, 225,
        231, 232, 310, 316,
        332, 334, 337, 338,
        341-354, 359, 363,
        364, 368-370, 373-
        397
Lo Vecchio, G., 436, 449
Lowenstein, H., 185, 222
Löwdin, P.O., 232, 310,
        323, 332
Lucovsky, G., 279, 285, 312
Ludwig, W., 183, 184, 222
Lundqvist, B.I., 114, 121,
        122, 149, 155, 156,
        337, 338, 388, 389,
        394, 395, 398
Lundqvist, S., 37, 44, 108,
        110, 114-117, 121,
        122, 125, 126, 149,
        154, 155, 156, 176,
        186, 187, 189, 221,
        223, 313, 332
Lüscher, E., 560, 563
Lynn, J.W., 417, 449

MacDonald, I.R., 559, 560,
        562, 563
MacFarlane, G.G., 474, 484-
        486, 489, 491-493
MacGill, T.C., 46, 110,
        493
MacGinn, G., 46, 110
Mackintosh, A.R., 181, 222
MacKitterick, J.G., 273,
        279, 303, 304, 312
MacLean, T.P., 474, 484-
        486, 489, 491-493
MacLellan, A.G., 182, 222,
        315, 333
MacMahan, A.K., 200, 202,
        207, 223, 224, 337,
        396
Madelung, O., 235, 311
Madsen, J., 378, 397
Magnaterra, A., 436, 449

Mahan, G.D., 16, 108
Makarenko, I.N., 558, 562
Maker, P.D., 504, 518
Mandal, S.S., 42, 43, 109
Manghi, F., 122, 153, 156,
        388, 398
Mann, J.B., 12, 107
Manson, R., 421,428,448,449
Maradudin, A.A., 244, 246,
        260, 261, 311,
        500, 518
March, N.H., 114, 121,
        122, 125, 126, 155,
        156, 176, 186, 187,
        189, 221, 313, 332
Marcus, A.M., 190, 191,
        193, 223
Martin, R.M., 157, 158,
        163, 172, 174-176,
        178-185, 187, 191,
        194-196, 205-224,
        229-231, 234, 238,
        239, 242, 243, 246,
        250, 252, 254, 258,
        260, 267, 272, 276,
        279, 280, 284, 285,
        289, 310-312, 313-
        315, 318, 321, 323,
        326, 328, 331, 333,
        336, 350, 351, 363,
        394, 395, 516, 518
Marvin, A., 402, 448
Marx, E., 502, 518
Mason, W.P., 436, 449
Mattausch, H.J., 114-116,
        130, 131, 134-137,
        139-144, 146, 150,
        153, 155, 156,
        388, 397
Mauger, A., 119, 155
Maysenholder, W., 374, 396
Mehl, M.J., 388, 397
Mele, E., 183, 184, 222
Melius, C.F., 53, 111
Mermin, N.D., 125, 156,
        190, 191, 193,
        223, 496, 517
Meskini, N., 113, 134, 156
Mickish, D.J., 119, 155

Miliotis, D.M., 36, 108
Miller, A.P., 413, 449
Miller, P.D., 499, 518
Mills, D.L., 402, 426, 428,
        449
Min, B.I., 197, 224
Monkhorst, H.J., 194, 224,
        232,310,320,323,333
Montroll, E.W., 244, 246,
        260, 311
Mooradian, A., 239, 310
Moré, J.J., 322, 332
Mori, H., 38, 109
Moriarty, J.A., 200, 224,
        364, 396, 438, 449
Moruzzi, V.L., 114, 120,
        155, 177, 187, 190,
        200-202, 219, 221,
        224, 229, 310
Moskowitz, J.W., 46, 110
Mountain, R.D., 560, 563
Mukhopadhyaya, G., 38, 109
Muller, H., 239, 311
Murnaghan, F.D., 206, 210
        215, 225, 234, 310,
        344, 345, 395
Musgrave, M.J.P., 259, 267,
        311

Nachtegaele, H., 44, 110
Nag, B.R., 451, 461, 491,
        492
Needs, R.J., 177, 194, 205,
        209-216, 219-221,
        224
Negoduiko, V.K., 507, 517
Nelin, G., 167-171, 172,
        241, 311
Newns, D.M., 385, 397
Neyer, H.R., 260, 311
Nicholls, J.M., 368, 370,
        396
Nicklow, R.N., 417, 449
Nielsen, O.H., 175, 177,
        182, 183, 187, 191,
        194-196, 205, 207-
        221, 234, 260, 310,
        313, 314, 318, 328,
        331, 333, 350, 395

Niklasson, G., 25, 37, 39,
    44, 108, 109
Nikolaenko, A.M., 558, 562
Nilsson, G., 79, 112, 167-
    172, 241, 311, 390,
    391, 398
Nishino, T., 484, 485, 492
Nizzoli, F., 96, 112, 402,
    435, 448, 449
Nordheim, L., 477, 492
Norlander, P., 433, 449
Norshov, J.V., 430, 449
Northrup, J.E., 177, 195,
    205, 221, 316, 333,
    370-373, 376, 396
Nozières, P., 36, 108
Nye, J.F., 325, 333

Olijnyk, H., 209-211, 213,
    225
Oppenheimer, R., 350, 395
Orbach, R., 499, 508, 518
Oshiyama, A., 205, 207,
    218, 225
Ostgaard, E., 197, 199,
    224
Otto, A., 36, 108
Overhauser, A.W., 37, 108

Pack, J.D. 194, 224, 320,
    323, 333
Pandey, K.C., 177, 195,
    205, 221, 371, 372,
    374-376, 396
Pantelides, S.T., 119, 155,
    205, 207, 218, 225,
    480, 492
Park, H.D., 451, 452, 460,
    466, 468, 471, 483,
    488, 491
Parrinello, M., 560, 562
Parshin, V.V., 507, 517
Pathak, K.N., 37, 109
Pauli, W., 182, 222
Pearson, E.S., 534, 562
Pepper, S.V., 374, 396
Percus, J.K., 527, 538,
    541, 561, 562

Perdew, J.P., 120, 156,
    176, 189, 221, 223,
    337, 338, 388, 390,
    395, 397
Pershan, P.S., 504, 517
Petri, E., 36, 108
Petroff, Y., 368, 370, 380,
    396, 397
Pettifor, D.G., 181, 196,
    222
Petzow, G., 364, 396
Phariseau, P., 242, 311
Phillips, J.C., 47, 54,
    111, 316, 332
Pick, R.M., 157, 172, 176,
    179, 183, 220, 222,
    284, 312
Pickering, C., 451, 452,
    476, 491
Pickett, W.E., 115, 122,
    155, 176, 221, 231,
    232, 310, 388, 397
Pietiläinen, P., 45, 110
Pikus, G.E., 460, 461, 463,
    491
Pinchaux, R., 368, 370,
    380, 396, 397
Pines, D., 36, 108
Piseri, L., 259, 267, 311
Pitt, G.D., 451,452,476,491,49
Placzek, G., 502, 518
Platzman, P., 36, 108
Plummer, E.W., 380, 382,
    387, 397
Pollak, F.H., 205, 218,
    225, 451, 452, 456,
    457, 460, 461, 463,
    466, 468, 471, 473-
    477, 483, 488, 491-
    493
Pollak, R.A., 117, 130,
    150, 155
Pople, J.A., 259, 267, 311
Posternak, M., 378, 397
Pulay, P., 180, 193, 222

Quarrington, J.E., 474, 484-
    486, 489, 491-493

Rabii, S., 348, 349, 395
Rabin, H., 511, 515, 517
Rahman, A., 524, 560-563
Rahman, T.S., 402, 426,
       428, 449
Rajan, V.T., 197, 224
Ramdas, A.K., 353, 395
Rampton, V.W., 496, 498,
       499, 518
Rao, B.K., 43, 109
Rath, J., 194, 224
Ratner, M.A., 46, 53,
       110, 112
Redondo, A., 46, 110
Reif, F., 365, 396, 528,
       561
Reissland, J.A., 243, 311
Resta, R., 158, 183, 184,
       205, 218, 222, 227,
       230, 273, 297, 299,
       300, 304, 312
Rhee, B.K., 513, 518
Rice, S.A., 46, 110
Rice, T.M., 131, 156
Richards, S.M., 214, 215,
       225
Riegler, G., 122, 153,
       156, 388, 398
Ritsko, J.J., 36, 108
Roberts, V., 474, 484-486
       489, 491-493
Rose, J.H., 197, 224
Ross, M., 190, 200, 202,
       223
Rossler, U., 451, 452,
       457, 461, 466, 481,
       482, 484, 490, 491
Rostocker, N., 190, 223
Rumer, G., 498, 518
Ruvalds, J., 353, 395

Salsburg, Z.W., 541, 562
Sambe, H., 193, 223
Sandercock, J.R., 402, 448
Santoro, G., 401, 402, 435,
       448, 449
Scalapino, D.J., 185, 223
Schiferl, S.K., 529, 532,
       534, 540, 553, 554,
       561, 562

Schlüter, M., 46, 53-55, 61,
       110-112, 115, 120,
       122, 123, 126-128,
       152, 155, 163, 164,
       172, 176, 190, 192,
       205-207, 210, 218,
       221, 223, 225, 232,
       280, 310, 316, 320,
       323, 330-332, 337,
       339, 340, 368-370,
       388, 394, 396, 397
Schnatterly, S.E., 36, 39,
       42, 108, 109
Schonhammer, K., 385, 397
Schosser, C.L., 505, 518
Schrödinger, E., 182, 222
Schulz, H., 328, 331
Segal, G.A., 136, 156
Segall, B., 114, 120, 153,
       155, 205, 207, 225,
       324, 332
Segmuller, A., 260, 311
Seitz, F., 16, 36, 44, 46,
       56, 107, 110-112,
       176-178, 181, 187,
       199, 220, 221, 229,
       243, 246, 310, 311,
       476, 477, 479, 481,
       492, 501, 518, 547,
       562
Seiwatz, R., 374, 396
Selders, M., 215, 225
Selmke, M., 183, 184, 222
Selsmark, B., 328, 332
Shadwick, W.F., 124, 156
Sham, L.J., 12-16, 37, 47,
       53, 54, 58, 59, 67,
       107, 109, 111, 114-
       116, 118, 120-123,
       126-128, 131-133,
       135, 136, 145, 152-
       156, 157, 160, 161,
       164, 172, 176, 179,
       183, 186, 187, 220-
       222, 229, 284, 310,
       312, 313, 337, 338,
       388, 394, 397, 476,
       477, 492
Shapiro, S.D., 534, 561
Shaw, R.W., 54, 93, 112

Sheldon, B.J., 328, 332
Shvartz, 502, 518
Sikka, S.K., 209-211, 213,
     225
Silcox, J., 36, 43, 108
Silversmith, 367
Singh, B.D., 259, 267, 311
Singwi, K.S., 37, 43, 44,
     108, 110
Sinha, S.K., 183, 184, 222,
     416, 449, 560, 562
Sjölander, A., 37, 38,
     108, 109
Skillman, S., 306, 312
Skriver, H.L., 181, 191,
     192, 220, 222, 223
Slater, J.C., 116, 136,
     137, 140, 149, 155,
     156, 161, 172, 176,
     180, 182, 190, 222,
     223, 243, 311
Slattery, W.L., 541, 562
Smith, D.L., 493
Smith, H.G., 417, 449
Smith, J.R., 378, 397
Smith, N.V., 380, 382, 397
Snedecor, G.W., 534, 561
Solal, F., 368, 370, 396
Solbrig, A.W., 259, 267,
     311
Solin, S.A., 215, 225,
     353, 395
Song, J.J., 502, 518
Spain, I.L., 209-211, 225
Spanjaard, D., 403, 449
Spicer, W.E., 390, 391,
     398
Starkoff, T., 337, 394
Staun Olsen, J., 328, 333
Stedman, R., 85, 112
Steiner, P., 368, 370, 396
Sterne, P., 149, 156
Stewart, R.F., 348, 395
Stillinger, F.H., 560, 563
Stishov, S.M., 558, 562
Straub, G.K., 529, 535,
     540, 553, 554, 556-
     558, 561, 562

Strinati, G., 114, 115, 130-
     132, 134-137, 139-
     144, 146, 150, 153,
     155, 156, 388, 397
Stroud, D., 46, 111
Surratt, G.T., 151, 156
Swanson, R.E., 529, 553,
     557, 558, 561
Szasz, L., 46, 110
Szeftel, J.M., 402, 426,
     428, 449

Takeda, M., 484, 485, 492
Takeno, S., 38, 109
Talman, J.D., 124, 156
Tang, C.L., 511, 515, 517
Tang, Y.W., 477, 478, 492
Taylor, R., 37, 44, 108, 110
Temmerman, W., 242, 311
Terhune, R.W., 504, 518
Thiry, P., 380, 397
Thorpe, M.F., 215, 225
Thurston, R.N., 436, 449
Timmerhaus, K.D., 558, 562
Toennies, J.P., 402, 403,
     427, 428, 435, 448,
     449
Toigo, F., 42, 109
Toporov, 507, 517
Topiol, S., 46, 53, 110, 112
Topp, W., 163, 164, 169,
     172
Tosatti, E., 299, 302, 304,
     312
Tosi, M.P., 37, 108, 547,
     562
Trevol, R.L., 426, 449
Trickey, S.B., 193, 200,
     205, 224, 388, 397
Tripathy, D.N., 42, 43, 109
Trommer, R., 239, 311
Trucano, R., 348, 395
Tsvelick, A.M., 185, 222
Tubino, R., 267, 311, 353,
     395
Tucker, J.W., 496, 498,
     499, 518
Tuckevich, V.M., 502, 518

576

Turnbull, D., 36, 44, 46,
    56, 108, 110-112,
    176-178, 181, 187,
    221, 229, 243, 246,
    310, 311, 476, 477,
    479, 481, 492, 501,
    518, 547, 562

Ubaidullav, S.B., 507, 517
Uhrberg, R.I., 368, 370,
    396
Utsumi, K., 45, 110

Valbusa, U., 449
Van Camp, P.E., 35, 98,
    108, 112, 157, 159-
    161, 163, 168, 172,
    183, 184, 222, 230,
    238, 250, 252, 303,
    310, 312, 321, 333,
    561
Vanderbilt, D., 205, 217,
    225, 342, 350-354,
    359, 373-376, 395,
    396
Van der Veen, J.F., 368,
    370, 374, 376, 396
Van de Walle, C., 220
Van Doren, V.E., 35, 98,
    108, 112, 157, 159-
    161, 163, 168, 172,
    183, 184, 222, 230,
    238, 250, 252, 303,
    310, 312, 321, 333,
    561
Van Hove, L., 402, 449
Vanier, P.E., 451, 452,
    460, 466, 468, 483,
    488, 491
Vashishta, P., 37, 43, 108,
    109, 125, 126, 156,
    183, 184, 222, 560,
    562
Verges, J.A., 481, 492
Verlet, L., 527, 538, 561
Vineyard, G.H., 524, 561
Vinson, P.J., 451, 452,
    476, 491, 492
Vogl, P., 239, 311, 476,
    477, 492

Vogt, H., 514, 517
Von Barth, U., 114, 121,
    122, 155, 177, 178,
    187, 221, 315, 319,
    332, 337, 338, 394
Von der Linde, D., 502, 518
Vosko, S.H., 37, 108
Vredevoe, R., 499, 508, 518

Wainwright, T.E., 524, 561
Wallace, D.C., 10, 72, 85,
    108, 112, 157, 172,
    521, 529, 532, 534,
    535, 540, 545-547,
    549, 550, 553, 554,
    556-558, 561, 562
Wallis, R.F., 239, 241, 310
Wang, C.S., 115, 122, 153,
    155, 156, 176, 221,
    378, 388, 397
Wang, S.L., 397
Warren, J.L., 320, 333
Waugh, J.L.T., 239, 241,
    310
Weaire, D.L., 46, 111,
    215, 225, 229, 310,
    320, 332
Weber, T.A., 560, 563
Weber, W., 193, 205, 208,
    217, 224, 229, 242,
    252, 260, 310, 311,
    348, 350, 351, 395,
    481, 492
Webster, D., 364, 396
Weigmann, P.E., 185, 222
Weiler, H., 113
Weinert, M., 180, 193, 205,
    222, 224
Weiss, G.H., 244, 246, 260,
    261, 311
Wendel, H., 175, 181, 187,
    194, 205, 217, 220,
    229, 242, 310, 318,
    323, 326, 333
Wentorf, R.H., Jr., 214,
    215, 225
Wepfer, G.G., 151, 156
Wigner, E., 16, 43, 107,
    108, 176, 178, 189,
    197, 199, 210, 220,

Wigner, E., (continued)
        221, 223, 224, 337,
        338, 394
Wiley, J.D., 451, 491, 492
Wilhite, D.L., 151, 156
Wilk, M.B., 534, 561
Wilkens, J.W., 337, 338,
        395
Willardson, R.K., 451,
        491, 492
Williams, A.R., 114, 120-
        122, 155, 177, 178,
        187, 190-193, 200,
        201, 202, 219, 221,
        223, 224, 229, 310,
        315, 319, 332
Wilson, B.L., 451, 452,
        460, 488, 491
Wilson, K.G., 185, 222
Wimmer, E., 193, 205, 224
Wiser, N., 160, 172, 302,
        303, 312
Wolford, D., 490, 493
Woo, C.W., 197, 224
Wood, D.M., 197, 224, 321,
        333
Wood, W.W., 524, 527, 543,
        559, 561
Woods, A.D.B., 417, 449
Woodruff, T.O., 42, 109
Worlton, T.G., 320, 333
Wright, G.B., 239, 310
Wu, F.Y., 185, 222
Wynn, J.J., 516, 518

Xiong, J.J., 376, 377, 396

Yablonovith, E., 516, 518
Yafet, Y., 484, 492
Yang, W.S., 374, 396
Yashara, H., 38, 109
Yin, M.T., 114, 120, 155,
        191, 200, 205-207,
        209, 210, 213, 215,
        216-218, 223-225,
        229, 230, 234, 236,
        241-243, 246, 310,
        311, 324, 326, 328,
        332, 344-347, 350-
        353, 361-363, 376,
        388, 395-397
Yoshida, F., 38, 109
Yu, P.Y., 451, 452, 476,
        491, 492

Zacharias, P., 36, 108
Zaremba, E., 433, 449
Zierau, W., 183, 184, 222
Zerbi, G., 259, 267, 311
Ziman, J.M., 50, 111, 476,
        477, 492, 496, 518
Zinth, W., 514, 518
Zunger, A., 46, 53, 110,
        112, 121, 153-156,
        180, 187, 189, 191-
        193, 205-207, 222,
        223, 225, 243, 280,
        311, 312, 318, 319,
        321, 322, 331-333,
        337, 338, 343, 388,
        390, 394, 395, 397
Zwanzig, R., 559, 562

MATERIAL INDEX

Al, 42, 43, 54, 55, 61-63,
    65, 67, 71, 85, 96,
    97, 194, 200, 242,
    271, 347
Al on GaAs (110) surface, 376
AlAs, 490
AlSb, 490
α-Sn, 161

Be, 97, 347
BC8 Carbon, 217
BC8-structure, 214, 215, 363
β-Sn, 206, 209, 211, 212
β-Sn structure, 361

C, 153, 161, 215, 217

GaAlAs, 490
GaAs, 154, 207, 217, 218, 229,
    230, 233, 235, 237,
    238, 242, 254, 271,
    279, 281, 284, 289,
    294, 376, 476, 488,
    490
GaAsP, 490
GaP, 476, 488
Ge, 154, 161, 162, 169, 171,
    206, 207, 213-215,
    217, 218, 230, 231,
    241, 242, 247, 249,
    251, 254, 256, 260,
    263, 346, 480, 483,
    488, 490
2 X 1 Ge (111) surface, 373
Graphite, 348

H, 186, 197, 219, 382, 386

He, 38

InP, 488

K, 52, 65, 202

Li, 38, 65
LiF, 150, 154

Metals, 154, 364
Mg, 54, 55, 61-63, 200
Mo, 203, 204, 242, 347
Mo (001) 378, 380

Na, 42, 51, 54, 55, 61-63, 65, 67,
    68, 70, 96, 200, 347
NaCl, 38, 242
Nb, 3, 203, 204, 242
Nb (001), 378

Pd (111), 378, 380, 386

Rb, 202

Se, 217
Si, 35, 50, 52, 54, 55, 61-63, 67,
    69, 71, 130, 137, 138,
    153, 154, 159, 161, 162,
    166, 170, 181, 200, 206,
    209-211, 213-215, 217, 218,
    229, 245, 346, 351, 352,
    473, 480, 488, 490
ST12-structure, 214, 215

W, 347

Zr, 203, 204, 242

SUBJECT INDEX

Ab-initio
  calculations, 313
  potential, 163
  pseudopotentials, 53, 200,
     336, 337
Absorption, 133
  coefficient, 451, 452, 456
  data, 474
Acoustic, 480
Acoustic phonon, 82, 84
Acoustic sum rule, 159, 171
Adiabatic approximation, 29,
     228, 365
All-electron calculation,
     242, 306
Analysis of trends, 4
Angular forces, 406
Angular interactions, 267,
     293
Anharmonicity, 238, 248, 547
Anharmonic scattering, 502
Anharmonic terms, 217, 241,
     350, 353
Anharmonic thermodynamic
     functions, 550, 553
Antibonding, 135
Antiferromagnetic phase, 371
Armand effect, 432
ASA-LMTO, 207
Atomic potential, 478
Atomic scattering technique,
     402
Atomic sphere approximation
     (ASA), 191, 192
Atom surface potential, 401,
     430

Atom surface scattering cross
     section, 428
Auger process, 139
Augumented plane wave (APW) 190,
     202
Augumented spherical wave (ASW),
     190
Austin-Heine-Sham pseudopotential,
     49, 53, 58, 67

Bachelet-Hamann-Schlüter
     pseudopotential, 54, 55, 61,
     69, 70, 85, 97
Band
  energy, 27, 56, 76, 77
  gaps, 114, 120, 130
  width, 137
Bare exchange, 154
Bare four-phonon processes, 359
Bare HF-states, 113
Bare-ion pseudopotentials, 339
Basis sets, 341
Bethe Ansatz, 185
Bethe-Salpeter equation, 118, 127,
     131, 133
Beyond local density approximation,
     388
Binding energies, 131
Bloch function, 132
Bond-bending, 268
Π-bonded chain geometry, 371
Bonding, 135
Bonding orbitals, 142
Bond-stretching, 268
Born-Oppenheimer approximation, 350
Born-von Karman, 282

Boundary condition effects, 541
Bragg reflections, 133
Broyden's method, 195
Buckling model, 370, 374
Bulk modulus, 200, 201, 206,
        208, 233, 344
Bulk phonons, 402, 403
Bulk properties, 344

c/a ratio, 210, 211
Canonical ensemble, 524
Canonical ensemble average,
        525
Canonical partition function,
        524
Canonical statistical weight
        function, 524
CARS, 502
Causality, 26, 37
Ceperly-Alder exchange-
        correlation data,
        390
Chadi molecule model, 374
Charge density, 114, 166
    inhomogeneity, 132
Chemical potential, 122, 128
Chemisorption, 376
Closed diagrams, 125
Cluster expansion of free
        energy, 556
Coherent phonon packets, 502
Cohesive energy, 200, 201, 344
Collective coordinates, 33
Collective excitations, 36
Completeness relation, 152
Compressibility sum rule, 25
Computational physics, 1
Computational power, 176
Computer
    experiment, 4
    simulation, 521
Conduction bands, 151, 167,
        171
Conduction band minimum, 480
Confidence limits of MD data,
        532, 533
Configuration interaction,
        185
Continuity equation, 42, 44
Core charge, 339

Core-electrons, 45-47, 77
Core states, 58, 59
Correlation, 151, 153, 176, 186
Coulomb-hole, 138, 148
    contribution, 147
    plus screened-exchange (COHSEX),
        137, 144
Coupling constant integral, 127
Covalent crystals, 204, 206, 207,
        258, 264
Crystal structures, 209
Cubic anharmonicity, 243, 250
Current-conservation, 136
    criterion, 136

Debye Waller factor, 430
Decay rate, 117
Deformable-ion models, 477
Deformation potential, 218
Density functional, 152, 186, 304,
        305
    eigenvalues, 153
    formalism, 12, 336, 337
    method, 186, 295
    potential, 123
    theory, 113-115, 120, 125, 154
Density-of-states mass, 122
Density response
    function, 127
    matrix, 160, 161, 163
Dephasing time, 507
Depletion hole, 54
$\emptyset$-derivable approximations, 125
Diagrams, 360
Diagrammatic expansion, 44
Diamond structure, 346
Diamond (111) surface, 373
Diamond-type materials, 479
Dielectric constants, 177
Dielectric function, 36, 115, 122,
        133
Dielectric matrix, 19, 26, 34, 72,
        85, 91, 92, 97, 133, 160,
        184, 195, 227, 299, 304
Dielectric matrix tail, 168
Dielectric properties, 218
Dielectric response, 9, 80, 184,
        304
Dielectric screening method, 157,
        150, 160

582

Dimerization, 374
Dipole, 148
Dipole expansion, 149
Direct calculations, 217
Direct method, 184, 227
Dispersions relations, 26
Dynamic correlation, 45, 114,
        137, 150, 155
Dynamic melting, 548, 557,
        558
Dynamic structure factor,
        36, 37, 43
Dynamical exchange decoupling,
        38
Dynamical exchange effects,
        36, 40, 43, 44, 68,
        97
Dynamical matrix, 34, 83, 84,
        91, 92, 96, 101, 106,
        107, 161, 168, 244,
        352
Dynamical vibrations, 217
Dynamically screened Coulomb
        interaction, 117
Dynamically screened
        interaction, 112
Dyson equation, 114, 116, 125,
        128
Dyson integral equation, 132

Effect of stress along (001)
        and (111) on the
        indirect absorption
        edge, 460
Effective charges, 273, 277,
        278, 285, 304
Effective electronic
        perturbation, 456
Effective four-phonon
        interactions, 360
Effective ion-ion potential
        for Na, 547, 554
Effective mass approximation,
        131
Effective one particle
        potential, 154
Effective valence, 63
Eigenfrequencies, 217
Eigenvalue problem, 132
Eigenvectors, 177, 217

Elastic constants, 177, 208, 209,
        257, 406
Elastic potential energy density,
        498
Elastic properties, 207, 324
Electron, 141
Electron density, 61
Electron-hole
    attraction, 132, 133, 137
    interaction, 131
    pairs, 139
    polarization, 149
Electron-phonon coupling, 46
Electron-phonon matrix elements,
        473, 476, 488
Electron-test-charge dielectric
        function, 20, 72
Electron-test-charge dielectric
        matrix, 20, 22, 25, 84
Electronic energy, 27
Electronic states, 452, 478
Electrostatic energy, 98
Elementary excitations, 116
Empirical model potentials, 51
Empirical pseudopotential, 136
Empty-core model, 51, 52
Empty spheres, 207
Energy, 183, 184
    loss, 36
    wavenumber characteristic, 76,
        81, 91, 92, 101, 107
Ensemble
    average, 524, 527, 535
    corrections, 538, 540, 541
    corrections for fluctuations,
        538, 540
    weight function, 527, 543, 560
Enthalpy, 213, 215, 216, 362
Entropy, 126
Equations of motion, 115
Equilibrium, 529, 531, 532, 543
    lattice constant, 177
    melting, 555, 557, 558
    metastable, 531
Equivalent conduction band minima,
        461
Esbjerg-Norskov approximation, 430
Evolution of MD system, 529
Evolutionary process, 544

Ewald-Fuchs procedure, 35,
    78, 98, 101
Ewald term, 192
Exchange, 114, 176, 186, 302
  charge density, 152
  and correlation, 23, 44,
    72, 79, 80, 97
Exchange-correlation
  approximation, 171
  energy, 13, 116, 128, 154,
    161
  density, 338
  functional, 187
  hole, 120, 390
Exchange and correlation hole,
    24, 37, 62, 64
Exchange-correlation hole
    charge, 389
Exchange and correlation
    matrix, 23
Exchange and correlation
    polarizability, 21,
    24
Exchange-correlation potential,
    127
Exchange and correlation
    potential, 17, 20-22
Exchange
  energy, 15
  interaction, 40
  model, 151
  operator, 113
  potential, 11, 14, 15, 79
  self-energy, 151
Excitons, 131
Excitonic and free-pair
    energy regions, 474
Exclusion principle, 47
External potential, 298

Fermi energy, 73, 194
Fermi function, 126
Fermi sphere, 73
Fermi surface, 72-75, 89
Fermi wave vector, 73, 74
Finite-size effects, 541
First and second order angular
    force constants, 407
First order tangential force
    constant, 404

FLAPW, 205
Fluctuation, 526, 528, 536, 538,
    540, 560
Forces, 180, 183, 184, 186, 196,
    208, 209, 217, 218
Force
  constants, 183, 217, 230, 253
  theorem, 180, 182, 204, 314
Form factor, 481
  with an oscillating tail, 482
Fourier analyzed, 478
Four-phonon processes, 356
Free electron gas, 184
Free energy, 126
  of the fluid phase, 556
Frenkel, 131
Frequency moments, 42
Friedel oscillations, 134
Frozen-core approximation, 231, 339
Frozen phonon, 158, 228, 236, 255,
    304
  technique, 217, 350
Full potential linear (FLAPW), 193
Functional derivative, 126
Fundamental concepts, 2

Gap correction, 130, 152-153
Gaussian basis, 136, 205
Gaussian orbitals, 193
Generalized acoustic sum rule,
    159, 163, 171
Generalized canonical ensemble,
    535
Generalized canonical partition
    function, 535
Generalized elastic constants, 356
Generalized virial functions,
    526, 537
Geometrical structure factor, 30,
    58, 59, 66, 67, 74, 99, 101
Gibbs free energy, 365, 555
Gibbs line, 362
Gradient corrections, 388
Grand-potential, 125
Green's function, 44, 123, 130
  many-body approach, 115
Ground state
  density, 152
  energy, 181
  functional, 121

584

Ground state (continued)
  potential, 120
  properties of atoms, 120
  properties of molecules,
    120
  properties of solids, 120
GW-approximation, 122

Hamiltonian, 164, 460
Harmonic approximation, 102
Harmonic phonons, 29, 30,
    107, 158, 217
Harmonic phonon frequencies,
    184
Hartree approximation, 10,
    11, 118
Hartree calculation, 120
Hartree equation, 11
Hartree-Fock approximation,
    11, 14, 42, 117
Hartree-Fock band gap, 151
Hartree-Fock decoupling, 39
Hartree-Fock energy, 16
Hartree-Fock equation, 11, 14
Heine-Abarenkov pseudopotential,
    67
Heine-Animalu pseudopotential,
    54, 55, 59-61, 85
Hellmann-Feynman, 243
Hellmann-Feynman forces, 352
Hellmann-Feynman theorem, 180,
    218, 230, 284, 314
Helmholtz free energy, 366, 525
Hexagonal diamond structure,
    361
High resolution electron
    energy loss
    spectroscopy, 402
Hohenberg-Kohn-Sham, 305
Hohenberg-Kohn theorem, 12
Hole, 134
Hole-phonon scattering matrix
    element, 473
Homogeneous electron-gas,
    12, 14, 25, 91, 97,
    115, 130, 141, 149,
    152
Homopolar crystal, 273
Hydrogen on the Pd (111), 378

Improper self-energy, 128
Impurity screening, 135
Indirect gap, 452
  semiconductor, 456
Induced exchange and
    correlation potential, 20
Inelastic neutral atom scattering
    technique, 403
Infrared activity, 218
Interaction potential, 403
Interatomic force constants, 244,
    245
Interfaces, 177
Internal strain, 326
  parameter, 208
Interplanar force constants, 246,
    254
Intervalley electron-phonon and
    hole-phonon interactions,
    451
Insulator, 114, 154
Inverse dielectric matrix, 161,
    299, 301
Ionic potential, 169
Ionic pseudopotential, 58, 67, 82,
    91
Irreducible particle-hole
    interaction, 127

Jellium, 14, 36, 186

$\vec{k}$ point integration, 320
$\vec{k} \cdot \vec{p}$ calculations, 474
Kohn anomaly, 412
Kohn-Sham approximation, 16, 115
Kohn-Sham equation, 13, 116, 164,
    337
Kohn-Sham exchange potential, 17
Kohn-Sham gap, 388
Kohn-Sham local potential, 136,
    141, 152
Kohn-Sham theorem, 12
Korringa-Kohn-Rostocker (KKR),
    190, 200, 202

LA-phonon, 480
LA phonon assisted transition in
    Ge, 452, 490
LA and TA phonon-assisted
    transition in GaP, 452

LA and TA phonons in GaP,
  473, 488
Ladder diagrams, 132
Lateral cut-off, 432
Laterally averaged total
  potential, 434
Lattice constant, 167, 169,
  171, 200, 201, 208,
  232, 344
Lattice displacements, 452,
  477
Lattice dynamics of a slab,
  418
Lattice relaxation, 271
Lattice sum, 148
Lattice vibrations, 402
LCAO, 132, 242
LDA, 114, 120, 121
LDA functional, 130
LDA-gap, 153
LDA-Hartree, 151
LDF-theory, 205
Levine-Louie model, 390
Lifetime, 139
Lifetimes of optical phonons,
  496
Line-shape factor, 457
Line-shape of the derivative
  absorption spectrum
  of the indirect exciton
  in Si (and Ge), 470
Linear combination
  of atomiclike orbitals
    (LCAO), 342
  of Gaussian-type orbitals
    (LCGTO), 200
  of muffin-tin orbitals
    (LMTO), 190, 200, 202
Linear response theory, 160,
  227
Liquid, 178
LO-phonon, 480
Local-density approximation,
  13, 17, 23-25, 120,
  158, 160, 338
Local density functionals,
  186-189, 337
Local density of states, 381
Local fields, 271, 299
Local field correction, 36

Local-fields effect, 113, 132, 133
Local orbitals, 132
  representation, 115, 132
Local potentials, 154, 232
Local pseudopotential theory, 452,
  477, 479
Localized charges, 275
Long range attractive polarization,
  430
Longitudinal charge, 276
Löwdin perturbation theory, 232

Macroscopic dielectric function,
  161, 162
Macroscopic fields, 271, 280, 297
Madelung energy, 191, 192
Many-body perturbation theory,
  113, 116, 123, 132, 154
Mass operator, 122, 130
Materials science, 2
MD ensemble, 527, 538
  weight function, 527
Melting, 553, 557
Metallic screening, 389
Metastable, 177, 215
Minimum or extremal principle,
  123, 154
Minimum gap, 388
Mixed basis set, 205, 342
Mode-locked laser, 506
Model, 141, 150
Model for the self-energy, 141
Modification of the surface force
  constants, 426
Molecular-dynamic calculations for
  Na, 545
Molecular-dynamics at constant
  pressure, 559, 560
Molecular-dynamics at constant
  temperature, 559, 560
Molecular-dynamic simulations,
  521, 522
Moment expansion, 159
Momentum space, 343
  scheme, 337
Monte Carlo, 185
Muffin-tin, 132
Multipole series, 148
Multivalley indirect semiconductors,
  451, 452

Murnaghan equation of state,
206, 210, 215, 344

$n^{1/3}$-approximation, 13
Nonequilibrium molecular
dynamics, 559
Non-equilibrium thermal
Green's function,
124
Non-linear ionic
pseudopotentials,
341
Non-linear response of solids,
495
Non-local energy-dependent
self-energy, 122
Non-local pseudopotential,
229
Non-metals, 154
Non-uniform systems, 124
Norm conserving
pseudopotentials,
12, 13, 61, 190,
210, 232, 241
Normal mode
analysis, 31
coordinates, 355
Normalized WMA spectrum of
the TO indirect
exciton in Silicon,
466

One-electron approximation,
9, 10, 46, 97
One-particle Green's function,
114, 115, 123, 144, 145
One-particle propagator,
142, 143, 152
Optical absorption, 131, 133,
146, 154
Optical experiments, 137, 150
Optical mode, 480
Optical properties, 135
Optical response, 133
Orbitals, 135
Orbital strain, 460
Order parameter, 203, 204
Orthogonalization hole, 54, 64,
65, 67, 69, 72, 76, 77,
79, 80, 82, 85, 92

Orthogonalization potential, 65
Orthogonalized plane wave (OPW),
53, 191
Oscillator strengths, 131, 133
of the indirect transition, 456
Overlap, 151

Pair-correlation function, 25, 36,
45, 121
Pandey Π-bonded chain model, 374
Paramagnetic phase, 371
Particle-hole
attraction, 125
continuum, 44
(excitonic) effects, 113
interaction, 131
Partition function, 126, 527
Pauli exclusion principle, 10, 11
Penn-gap, 146
Penn model, 132
Penn-type model dielectric
functions, 138
Periodic boundary conditions,
541, 542
Perturbation theory, 27, 72, 74,
85, 101, 183, 360
ω-phase, 203, 204
Phase diagram, 377
Phase space trajectory, 543
Phase transition, 185
Phenomenological models, 259
Phillips-Kleinman approach, 53
Phonons, 102, 184, 217, 501
Phonon
assisted, 452
dispersions, 97, 350
dispersion curves, 184, 218, 351
frequencies, 83, 85, 96, 97, 177,
203, 204, 209, 217, 351
mode Grüneisen parameters, 351
-phonon interactions, 350, 353
polarization vectors, 479, 481
spectrum, 85
Photoemission, 116, 137, 150
spectra, 140
Piezoelectric constants, 177,
218, 304
Piezospectroscopy, 452
Piezospectroscopic experiments,
451, 457, 488, 490

Planar force constants,
217, 218, 352
Plane-wave basis, 164, 191,
200, 205, 210, 317,
320, 341
Plasma resonance, 139
Plasmon
dispersion, 97
energy, 42
pole, 137, 146
Point-charge, 134
Point-ion model potential, 51
Poisson ratio, 347
Polar crystal, 248, 271, 286
Polarizability, 25, 92–94,
96, 97
matrix, 11, 85–88, 159,
160, 304
Polarization, 44
Potential, 478
Pressure, 182, 196,
197, 209,
210, 213
Projected band structure,
378
Pseudo-plane-waves, 479
Pseudopotential, 9, 48, 50,
56–58, 66, 72, 77,
81, 92, 93, 139, 189,
190, 205, 206, 210,
475, 479
Pseudopotential approach,
336
Pseudopotential band
structure, 135
Pseudopotential concept, 45
Pseudopotential eigenstates,
113
Pseudopotential-LDA
calculation, 119
Pseudopotential method, 335
Pseudopotential model for
Na, 546
Pseudopotential
normconserving, 315,
320
Pseudopotential perturbation
theory, 56, 66, 92,
97

Pseudo Schrödinger equation, 67
Pseudo wave function, 48, 49,
56–58, 64, 86, 339, 481
Pump synchronously, 506

Quantum Monte Carlo, 185, 197, 198,
219
Quartic anharmonicities, 250, 252
Quartic anharmonic interactions,
354
Quasiergodic hypothesis, 528, 531
Quasi-holes, 139
Quasi-particle, 117, 134
bandstructure, 153
decay, 113, 130, 140
energies, 136, 137, 388
excitations, 122
model, 150
states, 113, 130, 132, 145, 154
wave functions, 136

Radiationless transition, 139
Random Phase Approximation, 36, 40,
42–44, 132, 133, 137, 139,
145, 304
Rare gasses, 202
Rayleigh peaks, 401
Reconstructions, 368
2 X 1 Reconstruction, 370
Relaxations, 368
Renormalization group method, 185,
377
Rigid-ion model, 452, 477
Rigid-pseudoion model, 452, 477, 483,
490
RPA screening, 67, 68, 85

Scaling, 182
Scatter
elastically, 501
inelastically, 501
Scattering matrix elements, 473
Schlüter-Chelikowsky-Louie-Cohen
potential, 163
Screened exchange, 147, 149
Screened interaction, 136, 138
Screening, 36, 133
Screening matrix, 143
Screening propagator, 144

Second nearest neighbor
    rigid-ion model,
    481
Second order radial force
    constant, 405
Seiwatz single-chain model,
    374
Selection rules for the
    optical transitions,
    457
Self-consistency interaction,
    321, 328
Self-consistent approximation,
    115, 123
Self-consistent band
    calculation, 165, 194
Self-consistent equations,
    188
Self-consistent potential,
    300
Self-consistent problem, 188
Self-consistent pseudopotential,
    56, 98, 164, 165
Self-energy, 44, 113, 117, 120,
    123, 131, 137, 141,
    147, 152, 154
  corrections, 132
  matrix elements, 141
  model, 151
  operator, 123, 154
Self-interaction corrections,
    122, 388
Semiconductor, 323
(III-V) Semiconductors, 204,
    219, 347, 364
Semiconductors Si, 114, 155
Semiconductor screening, 389
Shear horizontal, 435
Shear vertical, 426
Shift and splitting, 461
Short range repulsive, 430
Si (111) surface, 369
Simple cubic structure, 207
Simple hexagonal, 209, 211-213
Simple metals, 199, 202, 347
Single-particle
  excitations, 114, 154
  Green's function, 117
  HF equation, 119
  propagators, 132

Slater exchange, 162
Slater's exchange model, 149, 154
Slater's exchange potential, 15,
    162
Slater-Koster, 135
Slater potential, 16, 17
Small-core approximation, 47
Softening of the relevant force
    constant, 436
Solid-solid structural phase
    transformations, 346
Special points, 194
Spectral representation, 117
Static dielectric constant, 303
Static dielectric response, 295
Static equilibrium, 232
Static structural properties, 344
Statistical weight function, 527,
    535
Strain, 260
Strain parameter, 257
Stress, 183, 184, 186, 196, 205,
    208-210, 218, 234
  dependent eigenfunctions, 464
  dependent spin-orbit components,
    460
  effect, 460
  free crystal, 171
  theorem, 182, 183, 215, 218, 234,
    314
Structural phase transformations,
    361
Structure factor, 29
Substitutional point defect, 134
Sum rules, 42, 44, 45
Supercells, 194, 217, 245
Superconducting transition
    temperature, 197
Surface, 177
  Brillouin scattering, 402
  density of states, 385
Surfaces and chemisorption systems,
    367
(111) Surface of an FCC crystal,
    428
Surface phonons, 401, 402, 418
Surfaces of semiconductors and
    insulators, 368
Surface states, 368, 378
Surface state dispersions, 386

Surfaces of transition
        metals, 377
Systematic trends, 3

TA-mode, 480
Temperature dependent part
        of inverse lifetime,
        509
Temperature-induced structural
        transitions, 364
Test-charge-test-charge
        dielectric function,
        81, 89
Test-charge-test-charge
        dielectric matrix,
        29, 23, 25, 90
Tetrahedral bonding, 130, 134
Tetrahedrally bonded
        semiconductors, 153
Theory of atomic scattering,
        401
Thermodynamic functions, 525,
        527, 528, 543, 549
Third order nonlinear
        susceptibility, 511
Three-phonon decay, 508
Three-phonon processes, 356
Three-wave mixing signal,
        495
Tight-binding exchange, 151
Tight-binding model, 130, 141,
        146, 149, 150, 152,
        154
Time-dependent H approximation,
        132
Time-dependent screened
        Hartree-Fock (TDSHF),
        113, 131, 132, 139,
        141
TO-mode, 480
TO (Γ) mode, 237
Topp-Hopfield potential, 164
TO phonon assisted transition
        in Si, 452, 488, 490
Total energy, 76, 167, 171,
        203, 206, 313
  difference method, 158, 163
  expressions, 342
  techniques, 335
Trace, 125

Transferability, 53
Trajectory phase space, 528
Translation invariance, 26
Translation and rotation
        invariance, 179
Transition
  metals, 201, 202, 219, 347
  pressure, 200, 206, 210
Two-band model, 146
Two-particle
  (electron-hole) excitations, 114
  Green's function, 116, 117, 123,
        130, 154
  propagator, 131, 132, 142
  Wigner distribution function, 39
Two-phonon
  bound state theory, 353
  Raman anomaly in diamond, 353
Two-photon absorptive processes,
        514
Two-step second order nonlinear
        processes, 512

Ultra high pressure experiments,
        490
Uniaxial stress, 451, 452, 461
Uniform electron gas, 72, 81
Unscreened exchange, 151

Vacuum level, 153
Valence bands, 151
Valence charge densities, 348
Valence force field, 259, 267, 268
Variational principle, 179, 181,
        184, 313
Vector computer, 170
Vectorization of phonon programs,
        170
Vertex function, 117, 131
Vibrational properties, 348
Vibration spectrum, 83
Virial theorem, 182, 208, 314

Wannier exciton, 131, 132
Wannier function, 154
Ward identity, 131, 136
Wavelength modulated absorption
        spectrum, 466
Weighted density approximation,
        388

Weighted density functional
    scheme, 388
Weighted (non-local) density
    approximation (WDA),
    122
Widths, 121
Wigner distribution function,
    37-39, 44

Wigner interpolation formula,
    189
Wigner lattice, 16

XPS, 120
Xα-calculations, 136